Theory of Luminescence

Theory of Luminescence

B. I. Stepanov and V. P. Gribkovskii

Translated by Scripta Technica Ltd

Edited by S. Chomet,
King's College, London

LONDON ILIFFE BOOKS LTD

ILIFFE BOOKS LTD
42 RUSSELL SQUARE
LONDON, W.C.1

Originally published in Minsk, 1963, by
Izdatel'stvo akademii nauk BSSR
under the title
Vvedenie v teoriyu lyuminestsentsii

Translated and prepared for press by Scripta Technica Ltd

English edition first published in 1968 by
Iliffe Books Ltd

Printed in Great Britain by
J. W. Arrowsmith Ltd, Bristol

592 05046 7

CONTENTS

Preface

Although luminescence, as such, was known for a very considerable time, it was not until the middle of the nineteenth century that it was subjected to systematic study. Historically, the first law of luminescence was formulated by Stokes in 1852. It is known as Stokes' rule and states that the wavelength of luminescence excited by radiation is greater than the wavelength of the exciting radiation. At about the same time, Becquerel laid the foundations for the experimental investigation of the emission spectrum, the efficiency of excitation and the duration of luminescence afterglow (phosphorescence).

In 1889 Wiedemann introduced the term 'luminescence' and gave the first, although not entirely accurate, definition of this phenomenon as the excess emission over and above the thermal emission background.

Luminescence is a very varied phenomenon and all attempts at a theoretical explanation of its properties on the basis of classical theories were inevitably unsuccessful. It was only after the advent of quantum mechanics and quantum electrodynamics that a realistic basis became available for the understanding and quantitative description, from a unified point of view, of the many aspects of luminescence.

Preface

The Soviet school, led by Vavilov, played a prominent part in the study of luminescence. Soviet physicists were the first to solve many of the fundamental problems in the study of luminescence and in its applications.

There is now an enormous volume of published information on luminescence, including the three volumes of the collected works of Vavilov. Among the monographs the most fundamental are those due to Levshin [5] and Pringsheim [85]. However, there is still a need for review monographs of a theoretical nature. Special topics have been discussed by Feofilov [9] and Stepanov [83, 84].

In the present monograph, an attempt is made to give an account of the fundamentals of the theory of luminescence or, more precisely, the theory of photo-luminescence. The first three chapters are concerned with classical emission theory, and the quantum mechanics and quantum-electrodynamics which are necessary for the understanding of the physical processes leading to luminescence.

The next two chapters discuss the general principles of the theory of absorption and luminescence without reference to any specific models of matter. These chapters are devoted to a detailed study of the optical properties of the harmonic oscillator and of systems of particles with two, three or more energy levels.

Much of the material given in this book is based on the original work carried out at the Institute of Physics of the Academy of Sciences Byelorussian SSR. In particular, a detailed description is given of the effect of the thermal emission background, the properties of negative radiation fluxes and negative luminescence. Non-linear optical phenomena which arise in the interaction of matter with high, and occasionally with ordinary, fluxes of radiation are systematically investigated. They include departures from Bouguer's law, depolarisation, induced dichroism and amplification and generation of radiation in media with negative absorption coefficients.

1

Classical Theory of Absorption and Emission of Light

It is well known that luminescence cannot be accounted for within the framework of classical physics, and to some extent contradicts the classical theory. The classical theory of radiation has been very successful in the solution of a very broad range of optical problems, however, and correctly describes many of the characteristic features of luminescence. It is also distinguished by simplicity and susceptibility to mathematical treatment and physical interpretation of the final results. It is precisely for this reason that classical calculations are still used in conjunction with quantum-mechanical ideas in the analysis of many phenomena associated with luminescence.

1. ELEMENTARY SOURCES OF RADIATION IN CLASSICAL ELECTRODYNAMICS

In the classical description of absorption and emission of light, matter is usually replaced by a set of elementary sources of radiation. These sources can be electric or magnetic dipoles, quadrupoles, octupoles, rotators, and so on. Each of these model sources generates radiation of a definite intensity, polarisation and angular distribution.

Before we discuss each of these individual models, let us consider some general properties which do not depend on the choice of a particular model. We must first determine the electromagnetic field due to arbitrarily moving electric charges localised within a small volume. The causes responsible for the motion of the charges can for the moment be ignored. In classical electrodynamics, the determination of the laws of motion of a system, and of the emission of radiation by the system, can be resolved into two independent problems.

Let the origin of coordinates lie inside the system of charges and let $\mathbf{r}$ be the position vector of an element of charge $de = \rho\, dV$, where ρ is the charge density and dV a volume element. Moreover, let $\mathbf{R}_0$ be the position vector of the point of observation. The difference $\mathbf{R}_0 - \mathbf{r}$ will be denoted by $\mathbf{R}$. The retarded potentials at a point P due to the system of charges are then given by

$$\varphi = \int \frac{1}{R} \rho(t')\, dV$$

$$\mathbf{A} = \frac{1}{c} \int \frac{1}{R} \mathbf{j}(t')\, dV \tag{1.1}$$

where $t' = t - \frac{R}{c}$, c is the velocity of light and $\mathbf{j}$ is the current density. Since the velocity of propagation of electromagnetic signals is finite, the field at the point P at time t is determined by the state of the system of charges at time t'.

The electric and magnetic field strengths are related to the retarded potentials by the formulae

$$\mathbf{E} = -\frac{1}{c} \frac{\partial \mathbf{A}}{\partial t'} - \operatorname{grad} \varphi, \quad \mathbf{H} = \operatorname{curl} \mathbf{A} \tag{1.2}$$

If the point of observation is in a vacuum at a large distance $r \ll R_0$ from the radiating system (i.e. in the wave zone), then in a limited domain of space the emitted radiation may be regarded as consisting of plane electromagnetic waves. The field vectors $\mathbf{E}$ and $\mathbf{H}$ are then equal in magnitude, and both lie in the plane at right-angles to the direction of propagation, which may be represented by the unit vector $\mathbf{n} = \mathbf{R}/R$:

$$\mathbf{E} = [\mathbf{Hn}], \quad \mathbf{H} = \frac{1}{c}[\dot{\mathbf{A}}\mathbf{n}] = [\mathbf{nE}] \tag{1.3}$$

This shows that the field in the wave zone is completely defined by the vector potential $\mathbf{A}$, which is usually found approximately by recalling that when $r \ll R_0$, the distance R can be expanded into a rapidly converging series in powers of $\mathbf{r}$. If we confine our attention to the first-order approximation we have

$$R = |\mathbf{R}_0 - \mathbf{r}| = R_0 + \sum_i \frac{\partial}{\partial x_i} |\mathbf{R}_0 - \mathbf{r}|\, x_i = R_0 - \mathbf{rn} + \ldots \tag{1.4}$$

We can then neglect the difference between R and R_0 in the denominator of (1.1) and assume that these two distances are equal. However, this approximation cannot be made in the retarded time $t' = t - R/c$, since here the important quantity is the change in the velocity of the charges and not the absolute value of R/c and $\mathbf{rn}/c$. In general, ρ and $\mathbf{j}$ may undergo appreciable changes during the time $\mathbf{rn}/c$.

The approximate expression for $\mathbf{A}$ is

$$\mathbf{A} = \frac{1}{cR_0} \int \mathbf{j}(t - R_0/c + \mathbf{rn}/c)\, dV \tag{1.5}$$

If we expand $\mathbf{A}$ into a series in powers of $\mathbf{rn}/c$ and confine our attention to the first-order approximation, we obtain

$$\mathbf{A} = \frac{1}{cR_0} \int \mathbf{j}(t - R_0/c)\, dV + \frac{1}{c^2 R_0} \frac{\partial}{\partial t'} \int \mathbf{j}(t - R_0/c)(\mathbf{rn})\, dV \tag{1.6}$$

If the system consists of point charges, the integration in (1.6) must be replaced by summation over all the charges e_i, and the vector potential is given by

$$\mathbf{A} = \frac{1}{cR_0} \sum_i e_i \mathbf{v}_i' + \frac{1}{c^2 R} \frac{\partial}{\partial t'} \sum_i e_i \mathbf{v}_i (\mathbf{r}_i \mathbf{n}) \tag{1.7}$$

where $\mathbf{v}_i$ is the velocity of the i-th charge. The primes in these expressions indicate that when **A** is found at time t one must substitute on the right of (1.7) the values of the parameters at time $t' = t - R/c$. These primes will be omitted from now on.

Equation (1.7) may be rewritten in a more convenient form. According to (1.3), the electric and magnetic fields remain unaltered if an arbitrary vector proportional to **n** is added to **A**. We may therefore add to **A** the vector

$$-\frac{\mathbf{n}}{6c^2R}\frac{\partial^2}{\partial t^2}\sum_i e_i\mathbf{r}_i^2$$

and since

$$\mathbf{v}(\mathbf{rn}) = \frac{1}{2}\frac{\partial}{\partial t}\mathbf{r}(\mathbf{nr}) + \frac{1}{2}\mathbf{v}(\mathbf{nr}) - \frac{1}{2}\mathbf{r}(\mathbf{nv})$$

$$= \frac{1}{2}\frac{\partial}{\partial t}\mathbf{r}(\mathbf{nr}) + \frac{1}{2}[[\mathbf{rn}]\mathbf{n}]$$

we have

$$\mathbf{A}_1 - \mathbf{A}_1^{(0)} + \mathbf{A}_1^{(1)} = \frac{\dot{\mathbf{D}}}{cR_0} + \frac{1}{6c^2R_0}\ddot{\mathbf{Q}}_n + \frac{1}{cR_0}[\dot{\mathbf{M}}\mathbf{n}] \tag{1.8}$$

where $\mathbf{D} = \sum e_i\mathbf{r}_i$ is the dipole moment of the system, $\mathbf{M} = \frac{1}{2c}\sum_i e_i[\mathbf{r}_i\mathbf{v}_i]$ is the magnetic dipole moment and $\mathbf{Q}_n$ is the product of the vector **n** and the quadrupole moment tensor whose components are given by

$$Q_{\alpha\beta} = \sum_i e_i\,[3x_\alpha^{(i)}x_\beta^{(i)} - \delta_{\alpha\beta}r^2]$$

Radiation due to the first, second and third terms in (1.8) is respectively referred to as dipole, quadrupole and magnetic dipole radiation.

In the zero-order approximation, the vector potential **A** depends on the dipole moment only. It follows that, whenever it exists, dipole radiation is a quantity of the first-order. Quadrupole and magnetic dipole radiations are the first correction terms. Higher-order corrections can be obtained by expanding (1.5) into a series of powers of $(\mathbf{rn})/c$, each of which can be divided into an electric and magnetic part. In particular, the second correction will correspond to the electric octupole and the magnetic quadrupole radiation.

We shall confine our attention to those multipoles which correspond to the first approximation. The intensity associated with the emission by higher multipoles is quite small and is of minor importance. The relative contribution of each form of emission to the overall effect depends on the properties of the system, and in particular on the ratio v/c. If

$$v \ll c \tag{1.9}$$

then the quadrupole and the magnetic dipole radiation are usually several orders of magnitude lower in intensity than the dipole radiation.

The condition given by (1.9) may be written in a somewhat different form. Suppose, for example, that the system under consideration consists of a fixed nucleus and an electron executing a uniform motion on a circle of radius a. The velocity of the electron then equals $2\pi a\nu$ and the wavelength of the emitted radiation is $\lambda = c/\nu$. From (1.9) we then have

$$2\pi a \ll \lambda \tag{1.10}$$

It follows that the quadrupole and the magnetic dipole radiation may be neglected when the linear dimensions of the system are much smaller than the wavelengths of the emitted light. In the case of atoms and molecules for which $a \approx 10^{-8}$ cm, (1.10) is satisfied for all wavelengths in the visible range. There are, however, systems for which the dipole moment is zero or is time independent. Such systems do not emit dipole radiation and therefore most of the emitted energy is in the form of quadrupole or magnetic dipole radiation.

The classical theory of emission is also concerned with systems of moving charges which mainly or exclusively give rise to dipole radiation, or only quadrupole radiation, or generally n-pole radiation in pure form. In other words, the elementary source of radiation is such that its electromagnetic field is determined, for example, by one of the terms of (1.8). Let us now consider the various elementary sources of radiation in turn.

Electric dipole

The electric dipole is defined as a system of two charges which are equal in magnitude and opposite in sign. In optics, the distance between the charges is much smaller than the distance to the point of observation. The principal characteristic of a dipole is its dipole moment, which is a vector drawn from the negative to the positive charge, whose modulus is equal to the product of one of the charges and the distance between them. If the positive charge is at the origin and is practically fixed (consider, for example, the nucleus of an atom), then the position vector of the negative charge (electron) is opposite in direction to the dipole moment, which is given by $\mathbf{D}=-e\mathbf{r}$, where e is the positive charge.

Since the quadrupole and the magnetic dipole radiation due to an electric dipole are negligible, the vector potential due to a dipole may be taken to be

$$\mathbf{A}=\mathbf{A}_1^{(0)}=\frac{1}{cR_0}\dot{\mathbf{D}} \qquad (1.11)$$

Substituting (1.11) into (1.3), we have

$$\mathbf{H}=\frac{1}{c^2R_0}[\ddot{\mathbf{D}}\mathbf{n}]$$
$$\mathbf{E}=\frac{1}{c^2R_0}[[\ddot{\mathbf{D}}\mathbf{n}]\,\mathbf{n}] \qquad (1.12)$$

If the constituent charges of the dipole (or one of them) execute simple harmonic oscillations along its axis, the system is called a linear harmonic oscillator. The dipole moment is then $\mathbf{D}=\mathbf{D}_0\cos\omega t$ and therefore

$$\ddot{\mathbf{D}}=-\omega^2\mathbf{D}_0\cos\omega t=-\omega^2\mathbf{D} \qquad (1.13)$$

Substituting (1.13) into (1.12) we have

$$\mathbf{H}=-\frac{\omega^2}{c^2R_0}[\mathbf{D}\mathbf{n}]$$
$$\mathbf{E}=-\frac{\omega^2}{c^2R_0}[[\mathbf{D}\mathbf{n}]\mathbf{n}] \qquad (1.14)$$

According to (1.14) the harmonic oscillator emits linearly polarised radiation in all directions. The directions of $\mathbf{E}$, $\mathbf{H}$, dipole moment $\mathbf{D}$, and unit vector $\mathbf{n}$ are shown in Fig. 1.1.

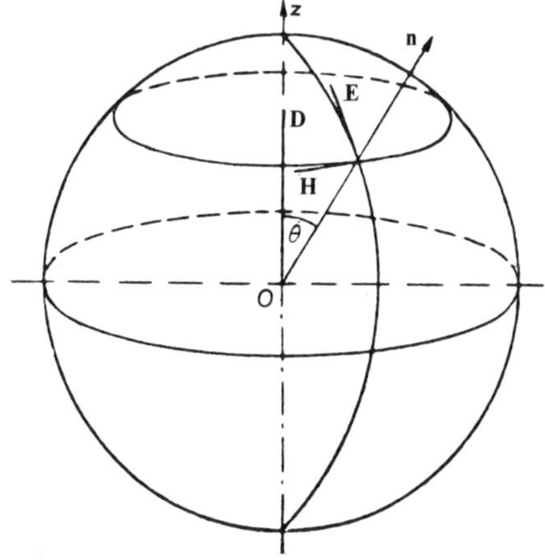

Fig. 1.1 Polarisation of radiation emitted by an electric dipole

Equation (1.12) may be used to find the flux of energy per unit area (Poynting's vector):

$$\mathbf{S} = \frac{c}{4\pi}[\mathbf{EH}] = \frac{\mathbf{n}}{4\pi c^3 R_0^2}\ddot{\mathbf{D}}^2 \sin^2\vartheta \qquad (1.15)$$

where ϑ is the angle between $\mathbf{n}$ and $\mathbf{D}$.

The angular distribution of radiation emitted by a harmonic oscillator in the plane containing $\mathbf{D}$ is shown in Fig. 1.2. The three-dimensional distribution may be obtained by rotating this diagram about the direction of the dipole. It can readily be seen that the intensity of the emitted radiation is a maximum at right-angles to $\mathbf{D}$ and is zero in the direction of the dipole moment. Integrating (1.15) over a sphere of unit radius, we obtain the following expression

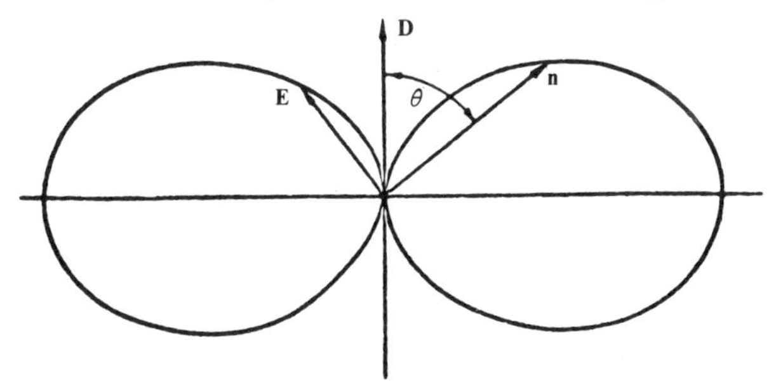

Fig. 1.2 Polar diagram of dipole radiation from a harmonic oscillator

for the total rate of emission of the dipole

$$W_{em} = \frac{2}{3c^3}\ddot{\mathbf{D}}^2 = \frac{2e^2}{3c^3}\ddot{\mathbf{r}}^2 \qquad (1.16)$$

This formula is valid not only for a harmonic oscillator giving a single sharp spectral line, but also for the dipole radiation from any system of charges having a total dipole moment **D**. For a harmonic oscillator of frequency ν, the average of (1.16) over one period of the oscillations is

$$W_{em} = \frac{\omega^4}{3c^3}\mathbf{D}_0^2 \qquad (1.16a)$$

To determine the dipole emission spectrum due to other systems, the dipole moment **D** must be expanded into a Fourier series. The intensities of the individual spectral lines can then be obtained by substituting the terms of this series into (1.16a). The spectral energy distribution of the emitted radiation will be discussed in greater detail in the next section.

The electric rotator

The simplest example consists of a plane rotator containing two charges, one fixed and the other executing uniform rotational motion in a plane circular orbit. The line passing through the fixed charge at right-angles to the plane of rotation is called the axis of the rotator. The polarisation of radiation emitted by a system of this kind depends on the direction of observation. Radiation emitted in the direction of the axis of the rotator is circularly polarised. If the radiation is observed at an angle to this axis, it is in general found to be elliptically polarised. It is linearly polarised when ϑ is $\pi/2$ (Fig. 1.3).

The rotational motion of a charge may be represented by the sum of two oscillatory motions along mutually perpendicular directions. The radiation from the rotator can therefore be regarded as being due to two harmonic oscillators whose frequencies and amplitudes are the same, but whose phases differ by $\pi/2$. The energy flux emitted in a direction $\mathbf{n}$ is then given by

$$\mathbf{S} = \frac{c}{4\pi}[\mathbf{E}_1 + \mathbf{E}_2,\ \mathbf{H}_1 + \mathbf{H}_2] = \frac{c}{4\pi}[\mathbf{E}_1\mathbf{H}_1] + \frac{c}{4\pi}[\mathbf{E}_2\mathbf{H}_2] + \frac{c}{4\pi}[\mathbf{E}_1\mathbf{H}_2] + \frac{c}{4\pi}[\mathbf{E}_2\mathbf{H}_1] \tag{1.17}$$

The subscripts 1 and 2 represent quantities which refer to the two oscillators with dipole moments $\mathbf{D}_1$ and $\mathbf{D}_2$ respectively. Since $\mathbf{D}_1$ and $\mathbf{D}_2$ are perpendicular, it follows that in

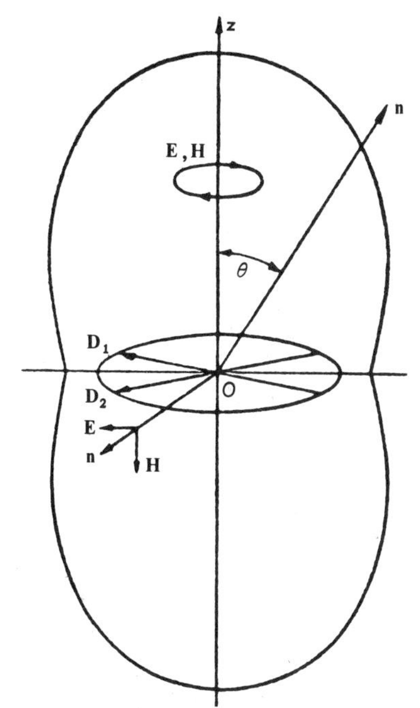

Fig. 1.3 Polarisation of radiation emitted by an electric rotator

view of (1.12), the last two components in (1.17) must be equal to zero. Since $\ddot{\mathbf{D}}_j = -\omega^2 \mathbf{D}_j$ we have from (1.17) and (1.12), after averaging over one period of oscillations,

$$\mathbf{S} = \frac{\omega^4 \mathbf{n}}{4\pi R_0 c^3}\left\{\overline{([\mathbf{D}_1\mathbf{n}])^2 + ([\mathbf{D}_2\mathbf{n}])^2}\right\} = \frac{\omega^4 \mathbf{n}}{8\pi c^3 R_0}\mathbf{D}_0^2(1+\cos^2\vartheta) \tag{1.18}$$

where $\mathbf{D}$ is the amplitude of the two oscillators.

In contrast to a simple oscillator, the rotator emits radiation in all directions, and the rate of emission is a maximum in the direction of the axis (Fig. 1.3). Integration of (1.18) over a sphere of unit radius yields the following expression for the total rate of emission of a rotator:

$$W_{em}^{rot} = \frac{2\,\omega^4}{3c^3} D_0^2 \tag{1.19}$$

This is higher by a factor of 2 than the mean rate of emission of the harmonic oscillator.

Electric quadrupole

To begin with, let us consider the quadrupole radiation of a harmonic oscillator, and then construct a system of charges for which the leading component of the emitted radiation is the quadrupole component. According to (1.7) and (1.8), the quadrupole part of the vector potential of a linear oscillator is given by

$$\mathbf{A}_q = \frac{e}{2c^2 R_0} \frac{\partial}{\partial t^2} ((\mathbf{nr})\mathbf{r}) \tag{1.20}$$

Using (1.3) and (1.20), we find that the energy flux of quadrupole radiation is

$$\begin{aligned} \mathbf{S}_q &= \frac{c}{4\pi} [\mathbf{E}_q \mathbf{H}_q] \\ &= \frac{e^2}{16\pi c^5 R_0^2} \mathbf{n} \cos^2\vartheta \sin^2\vartheta \left(\frac{\partial^3 r^2}{\partial t^3}\right)^2 \\ &= \frac{e^2}{64\pi c^5 R_0^2} \mathbf{n} (\dddot{r})^2 \sin^2 2\vartheta \end{aligned} \tag{1.21}$$

It follows from (1.21) that the axis of the oscillator is not the only direction in which no radiation is emitted; the intensity is also zero at right-angles to this axis. The angular distribution of quadrupole radiation emitted by a simple harmonic oscillator is shown in Fig. 1.4. Here again the electric vector lies in the plane containing the axis of the oscillator and the vector **n**.

From (1.15) and (1.21) we can show that the ratio of the

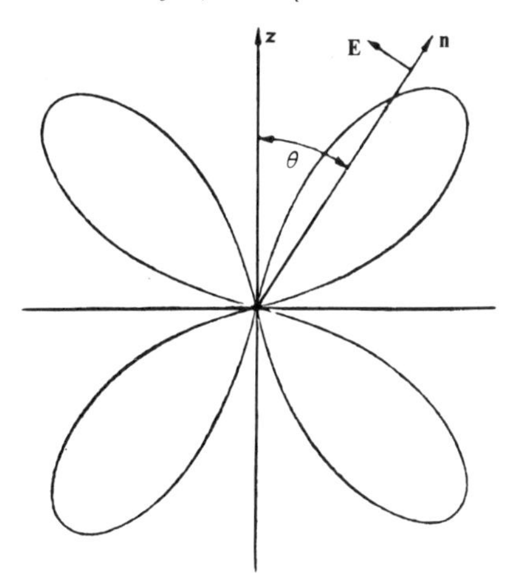

Fig. 1.4 Polar diagram of the quadrupole emission of an oscillator

quadrupole emission of a harmonic oscillator to its dipole emission is equal to $\left(\frac{a}{\lambda}\right)^2$, which for visible radiation ($\lambda = 5{,}000\,\text{Å}$) is equal to 10^{-8}. However, a system consisting of two collinear oscillators with the same frequencies but opposite phases (Fig. 1.5) emits no dipole radiation and is

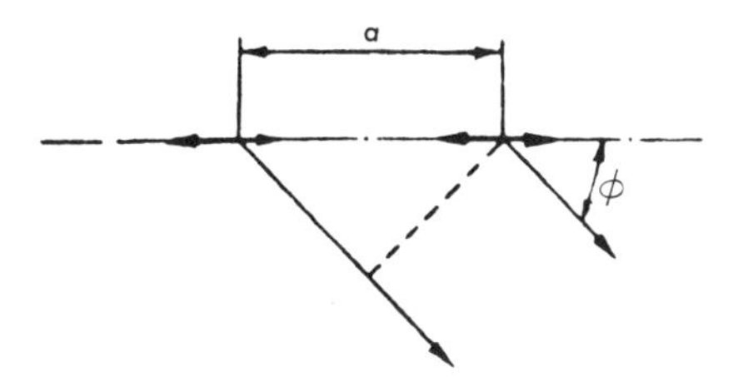

Fig. 1.5 A simple quadrupole

an example of a pure quadrupole source. Consider the radiation emitted by this system at an angle ϑ. The resultant electric field $\mathbf{E}$, is given by

$$\begin{aligned}\mathbf{E} = \mathbf{E}_1 + \mathbf{E}_2 &= \mathbf{E}_0 \sin\vartheta\,[\cos\omega t + \cos(\omega t + \pi + \varphi)] \\ &= 2\mathbf{E}_0 \sin\vartheta \sin(\omega t + \varphi/2)\sin\varphi/2\end{aligned} \tag{1.22}$$

where $\mathbf{E}_1$ and $\mathbf{E}_2$ are the electric vectors due to the first and second oscillators respectively and φ is the phase difference, which is determined by the mutual disposition of the oscillators and the position of the point of observation. As can be seen from Fig. 1.5, the phase difference is given by $\varphi = \frac{a}{\lambda} \cos \vartheta$. If the distance between the oscillators is much smaller than the wavelength $a \ll \lambda$, we can expand $\sin \varphi/2$ into a series and retain the first term only:

$$\sin \varphi/2 \sim \frac{a}{\lambda} \cos \vartheta \tag{1.23}$$

Substituting this expression into (1.22), we have finally

$$\mathbf{E} = \frac{1}{2} \mathbf{E}_0 \frac{a}{\lambda} \sin 2\vartheta \sin (\omega t + \varphi/2) \tag{1.24}$$

This shows that the emission of two collinear dipole oscillators separated by a small distance is identical in its properties with the emission of a single quadrupole oscillator. The model shown in Fig. 1.5 can therefore be regarded as an elementary quadrupole.

It should be noted that the quadrupole emission of an oscillator which is associated with a change in the quantity $(\mathbf{n}Q)$ does not exhaust the possible forms of quadrupole emission. It is possible to construct other elementary quadrupoles for which the angular distribution of the emitted radiation will be different from that shown in Fig. 1.4.

Magnetic dipole

The vector potential due to an elementary magnetic dipole is given by the last term in (1.8). Substituting $\mathbf{A}_m = \frac{1}{cR_0} [\dot{\mathbf{M}}, \mathbf{n}]$ in (1.3), we have

$$\mathbf{E} = \frac{1}{c^2 R_0} [\mathbf{n} \ddot{\mathbf{M}}], \quad \mathbf{H} = [\mathbf{nE}] = \frac{1}{c^2 R_0} [\mathbf{n}[\mathbf{n}\ddot{\mathbf{M}}]] \tag{1.25}$$

The magnetic dipole is similar to the electric dipole in that it emits radiation only when the second time derivative of the dipole moment is not zero. Comparison of (1.25) with (1.12) clearly shows that these formulae become identical if the electric dipole moment is replaced by the magnetic dipole moment and $\mathbf{E}$ is replaced by $\mathbf{H}$. It follows that the angular distribution of radiation emitted by a magnetic dipole is the same as that due to an electric dipole (Fig. 1.2), but the plane of polarisation is turned through $\pi/2$.

Magnetic rotator

Like the electric rotator, the plane magnetic rotator may be constructed from a set of two linear magnetic dipoles whose frequencies and amplitudes are the same but whose phases differ by $\pi/2$. The angular distributions emitted by the electric and the magnetic rotators are the same, but the planes of polarisation are at an angle of $\pi/2$.

The radiation emitted by a magnetic rotator in the direction of its axis is circularly polarised; in directions at right-angles to the axis it is plane polarised. In all other directions the polarisation is elliptical.

2. ABSORPTION AND EMISSION OF RADIATION BY A HARMONIC OSCILLATOR

Free oscillations of a dipole

Experiment shows that light emitted by atoms consists of individual, fairly sharp lines, so that the atomic charges must execute a motion resembling harmonic oscillations. This fact is used in the classical description of the interaction of light with matter as a basis for the replacement of the medium by a set of harmonic oscillators, i.e. electric dipoles which execute periodic oscillatory motion. In order to obtain such oscillations it is sufficient to assume that the potential energy of the dipole is of the form

$$V(x) = \frac{1}{2} kx^2 \tag{2.1}$$

where x is the deflection of the charge from the position of equilibrium and k is a constant. The force F_{el} acting on the charge when it is displaced from its position of equilibrium is then given by

$$F_{el} = -\frac{\partial V(x)}{\partial x} = -kx \tag{2.2}$$

If no other forces act on the oscillator, its equation of motion is

$$\frac{d^2x}{dt^2} + \omega_0^2 x = 0 \tag{2.3}$$

The solution of this differential equation is simply

$$x = x_0 \cos(\omega_0 t + \varphi) \tag{2.4}$$

where $\omega_0 = 2\pi\nu_0 = \sqrt{k/m}$ is the natural angular frequency of the oscillator, x_0 is the amplitude and φ is the initial phase of the oscillations. Both the latter are determined by the initial conditions. According to (2.4), a dipole having a potential energy of the form given by (2.1) executes undamped harmonic oscillations of frequency ν_0.

Radiation reaction. Damping

The assumption that all forces other than the elastic force given by (2.2) are absent is never strictly valid. The oscillatory motion of a charge is unavoidably accompanied by the emission of electromagnetic radiation which reacts on the charge and exerts a retarding force upon it. Therefore, even in the absence of external forces, the energy of an oscillator will continuously decrease, and this necessarily leads to the damping of the oscillations.

The reaction of the emitted radiation on the dipole producing the radiation is usually described by introducing the concept of radiative friction or resistance. The magnitude of this force can be calculated purely phenomenologically from the law of conservation of energy, since the work done by the frictional force at any given time must be equal to the energy emitted by the oscillator during that time. The force of radiative friction can be calculated more rigorously under certain special assumptions regarding the structure of the electron [1]. In either case it is given by

$$\mathbf{F}_{\mathrm{em}} = \frac{2e}{3c^3}\dddot{\mathbf{D}} \tag{2.5}$$

When this force is negligible in comparison with the elastic force, i.e. when

$$F_{\mathrm{el}} \gg F_{\mathrm{em}} \tag{2.6}$$

we may replace $\dddot{\mathbf{D}}$ by $-\omega^2 \dot{\mathbf{D}}$. In order to estimate the limits of applicability of (2.6), let us suppose that the oscillator executes simple harmonic oscillations of the form described by (2.4).

According to (2.5)

$$F_{\text{em}} = \frac{2e^2 \omega_0^3}{3c^3} x$$

and therefore

$$m \omega_0^2 \gg \frac{e^2 \omega_0^3}{3c^3}$$

which corresponds to the inequality

$$\lambda \gg \frac{e^2}{mc^2} = r_0 \tag{2.7}$$

where $\lambda = c/\nu_0$ is the wavelength of the emitted radiation, and r_0 is the classical radius of the electron (10^{-13}cm). In view of (2.7), Equation (2.6) is therefore satisfied for all frequencies in the optical region, and holds even for soft γ-rays. It loses its validity only for cosmic γ-rays.

Damping usually occurs not only as a result of the emission of radiation, but also as a result of collisions with other oscillators and of the decelerating effect of the surrounding medium. The effect of the medium on atoms and molecules may be very different in different systems. A rigorous theory of the interaction of radiating systems with the surrounding medium can only be constructed within the framework of quantum electrodynamics. However, in many cases the effect of the medium may be taken into account if it is replaced by a frictional force proportional to the velocity of the charges, i.e. $\mathbf{F}_{\text{fr}} = 2\gamma_{\text{fr}} m \dot{\mathbf{r}}$. This enables one to combine radiation reaction with the force acting on a moving charge and facilitates the solution of the equation of motion.

When frictional forces are taken into account, the equation of motion for an oscillator given by (2.3) must be replaced by

$$\ddot{x} + 2\gamma \dot{x} + \omega_0^2 x = 0 \tag{2.8}$$

where

$$\gamma = \gamma_{\text{em}} + \gamma_{\text{fr}} \quad 2\gamma_{\text{fr}} \dot{x} = \frac{1}{m} F_{\text{fr}}$$

and

$$\gamma_{em} = \frac{e^2 \omega_0^2}{3mc^3} \tag{2.9}$$

is the natural damping constant associated with the force of radiative friction. Similarly, γ_{fr} is referred to as the damping constant due to all other causes. The solution of (2.8) is known to be

$$x = x_0 e^{-\gamma t} \cos(\omega_1 t + \varphi) \tag{2.10}$$

where x_0 and φ denote, as before, the initial amplitude and the phase of the oscillations, and $\omega_1^2 = \omega_0^2 - \gamma^2$.

Equation (2.10) shows that frictional forces lead to a continuous reduction in the amplitude of the oscillations from the maximum value at $t = 0$. The oscillations practically vanish when $t \gg 1/\gamma$. In the absence of the medium ($\gamma_{fr}=0$), damping of dipole oscillations is due only to the emission of energy in the form of electromagnetic waves and occurs relatively slowly. The change in the amplitude during a single period T is quite negligible, as can be seen from the expansion

$$\frac{x_0 e^{-\gamma(n+1)T}}{x_0 e^{-\gamma nT}} = e^{-\gamma T} = 1 - \gamma T + \frac{1}{2}(\gamma T)^2 + \ldots$$

In the visible part of the spectrum (λ = 5,000Å) the damping constant is found from (2.9) to be $\gamma_{em}=4.42 \times 10^7$ sec^{-1}, and therefore

$$\gamma_{em} T = \frac{\gamma_{em}}{\nu} = 7.4 \cdot 10^{-8}$$

i.e. the amplitude is reduced in one period by less than one ten-millionth part of its initial value. Under real experimental conditions γ_{fr} is usually not equal to zero and may be much greater than γ_{em}. This leads to a more rapid damping of the oscillation.

In engineering calculations, the resonance properties of an oscillatory system are described by the so-called Q-factor. This factor is numerically equal to 2π multiplied by the ratio of the total energy of the oscillations E and the energy loss per period ΔE:

$$Q = 2\pi \frac{E}{\Delta E} \tag{2.11}$$

The energy associated with the free oscillations of a harmonic oscillator decreases in accordance with the expression $E = E_0 e^{-2\gamma t}$ (see Equation (2.10)), and therefore

$$|\Delta E| = 2\gamma E T = 2\gamma E \frac{1}{\nu_0} = 2\gamma E \frac{2\pi}{\omega_0}$$

where T is the period of the oscillations. Substituting E and ΔE into (2.11) we have

$$Q = \frac{\omega_0}{2\gamma} = \frac{\omega_0 t}{\Delta\omega} \qquad (2.11a)$$

which in the visible part of the spectrum yields $Q = 10^7 - 10^8$. It will be shown below that $\Delta\omega = 2\gamma$ is the width of a spectral line due to the oscillator. It follows that the Q-factor can also be defined as the ratio of the natural frequency of a system to the width of the emitted spectral line.

Spectral line profile

The damping of the oscillations of a dipole leads to a reduction in the amplitude of the electric field in the emitted radiation. According to (1.12) and (2.10) the function $E(t)$ is given by

$$E(t) = E_0 e^{-\gamma t} e^{i\omega_1 t} \qquad (2.12)$$

On entering a spectrograph a wave with decreasing amplitude such as that described by (2.12) will produce on the photographic plate, or any other recording device, a line of finite width. The dispersing element will analyse a composite oscillation into its harmonic components. This corresponds in mathematics to the integral Fourier expansion

$$E(t) = \frac{1}{2}\int_{-\infty}^{\infty} E(\omega)\, e^{i\omega t}\, d\omega \qquad (2.13)$$

where the monochromatic component $E(\omega)$ is given by

$$E(\omega) = \frac{1}{2\pi}\int_{0}^{\infty} E(t) e^{-i\omega t}\, dt \qquad (2.13a)$$

Substituting (2.12) into (2.13a) and completing the integration, we have

$$E(\omega)=\frac{1}{2\pi}\,\frac{E_0}{i(\omega-\omega_1)+\gamma} \tag{2.14}$$

The intensity $I(\omega)$ of a harmonic component of a spectral line, which is equal to the energy flux, is proportional to the square of the corresponding component of the Fourier integral:

$$I(\omega)=C\,|E(\omega)|^2=\frac{C}{4\pi^2}\,\frac{|E_0|^2}{(\omega-\omega_1)^2+\gamma^2} \tag{2.15}$$

The constant of proportionality C can be expressed in terms of the integral intensity I_0, of the line given by

$$I_0=\int_0^\infty I(\omega)d\omega=C\int_0^\infty |E(\omega)|^2 d\omega=C\,\frac{|E_0|^2}{4\pi}\,\frac{1}{\gamma} \tag{2.16}$$

If we determine C from (2.16) and substitute the result into (2.15) we obtain the following expression for the profile of a spectral line due to a harmonic oscillator:

$$I(\omega)=\frac{\gamma}{\pi}\,\frac{I_0}{(\omega-\omega_1)^2+\gamma^2} \tag{2.17}$$

In contrast to free oscillations which give rise to the emission of a strictly monochromatic wave of single frequency ν_0, a damped dipole will emit radiation of all frequencies. As can be seen from (2.17), the maximum intensity is emitted at the frequency $\nu_1=\nu_0-\frac{1}{8\pi^2}\frac{\gamma^2}{\nu_0}$. The intensity decreases rapidly on either side of this frequency (Fig. 1.6). Damping of oscillations leads not only to line broadening but also to a slight wavelength shift.

We note that (2.17) gives the spectral energy distribution obtained after the dipole has emitted all its energy. In order to find the line profile which is obtained at time t, the integral in (2.13) must be evaluated between 0 and t. Calculations show that the profile will vary continuously with time.

If damping of the oscillations occurs only as a result of radiation reaction, then the line profile is referred to as

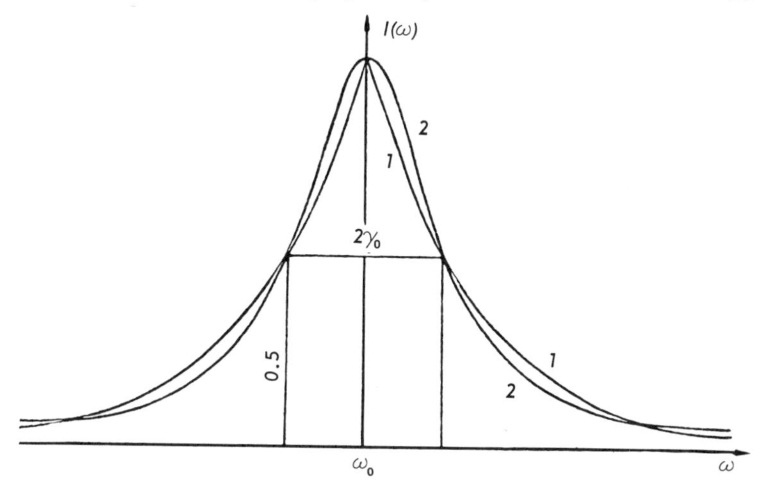

1 - Doppler; 2 - natural

Fig. 1.6 Spectral line profiles

the natural profile. It can be obtained from (2.17) by replacing $\gamma = \gamma_{em} + \gamma_{fr}$ by γ_{em}.

The width of a spectral (emission) line is characterised usually by the distance between the two points (on the frequency scale) at which the intensity is one half of the maximum intensity. It is evident from (2.17) that the natural width of a spectral line emitted by a harmonic oscillator is equal to twice the natural damping constant:

$$\Delta\omega = 2\gamma_{em} = \frac{2e^2\omega^2}{3mc^3}$$

On the wavelength scale the width of the line is

$$\Delta\lambda = \frac{2\pi c}{\omega^2}\Delta\omega = \frac{4\pi}{3}\left(\frac{e^2}{mc^2}\right)$$

and is independent of wavelength. For all dipoles in which the oscillating charge is an electron, the line width is equal to the universal constant

$$\Delta\lambda = \frac{4\pi}{3} r_0 = 1.17\cdot 10^{-4}\,\text{Å}$$

where, as before, r_0 is the classical radius of the electron.

On the ω scale we have the visible radiation ($\lambda = 5{,}000$ Å)

$$\Delta\omega = 8.87 \cdot 10^7 \text{ sec}^{-1}$$

If the oscillating charge is an ion whose mass is many thousands of times greater than the mass of the electron, the line width will be smaller by about 3 orders of magnitude, since γ_{em} is inversely proportional to the mass of the oscillating particle.

Under real experimental conditions the natural width of a line is of course never observed. The various additional factors responsible for line broadening can be reduced, but can never be entirely eliminated. Line broadening is mainly due to (a) collisions between radiating particles, (b) interaction between electric and magnetic fields, and (c) the Doppler effect.

The effect of external forces on the width of a spectral line has already been taken into account (at least on the first approximation) in (2.17), since the damping constant γ includes γ_{fr}. If the damping of the oscillations is due to collisions between the radiating particles in a gas, it is possible to show that the line profile is in fact given by (2.17), and to relate the line width to the temperature and pressure of the gas [2].

Calculations show that

$$\gamma_{fr} = \frac{1}{\bar{\tau}} \tag{2.18}$$

where $\bar{\tau}$ is the mean time interval between collisions. If we denote the mean velocity and the mean free path of the particles by $\bar{v}$ and l we obtain

$$\gamma_{fr} = \frac{1}{\bar{\tau}} = \frac{\bar{v}}{l} = 8\pi \frac{N_0\sigma^2}{\sqrt{R\mu}} \frac{p}{\sqrt{T}} \tag{2.19}$$

where N_0 is Loschmidt's number, p is the pressure of the gas, σ is the molecular diameter, R is the gas constant, μ is the molecular weight and T is the temperature.

If we express the pressure in atmospheres and substitute values appropriate to nitrogen ($\mu = 28$, $\sigma = 3.1 \times 10^{-8}$ cm) into (2.19), we obtain, for visible radiation

$$\gamma_{fr} = 8 \cdot 10^{10} \frac{p}{\sqrt{T}}$$

which at normal pressure and room temperature corresponds to

$$\Delta\omega_{fr} = 2\gamma_{fr} = 8.96 \cdot 10^9 \text{ sec}^{-1}$$

$$\Delta\lambda_{fr} = 1.19 \cdot 10^{-2} \text{ Å}$$

$\Delta\lambda_{fr}$ can be made much smaller than the natural line width by reducing the gas pressure. This effect is used in practice to determine the magnitude of radiation damping.

Equation (2.19) has been subjected to careful experimental tests. Experiments have shown that collision damping can be adequately described by (2.19), but the absolute line widths have been found to be much greater than one would expect from this formula. In order to achieve quantitative agreement it must be assumed that the collision frequency is greater by a factor of 2 to 3 than expected from the kinetic theory of gases.

Doppler line broadening

When we discussed free and damped oscillations it was assumed that the oscillator as a whole was at rest relative to the observer. In reality the atoms and molecules of a gas or liquid travel at high speeds which depend on the temperature. The motion of the particles gives rise to a Doppler effect, i.e. to a shift in the emitted frequencies. Since some of the atoms travel towards the observer while others travel in the opposite direction, and since there is a continuous distribution of particle velocities, it follows that even if the system were emitting strictly monochromatic waves, the recording instrument would still register a spectrum of frequencies.

Let us consider the line profile due to the Doppler effect on the assumption that the oscillator emits strictly monochromatic waves. If ν_0 is the frequency of the dipole at rest, while v is the component of its velocity in the direction of the observer, the recorded frequency is given by

$$\nu = \nu_0 (1 + v/c) \tag{2.20}$$

If the particles have a Maxwellian velocity distribution, the number of particles with velocities in the range between ν

and $\nu + d\nu$ is given by

$$dn = n(v)dv = n\sqrt{\frac{M}{2\pi RT}}\, e^{-\frac{mv^2}{2RT}}\, dv \tag{2.21}$$

where we have used the notation defined above. The coefficient of proportionality $\left(\frac{M}{2\pi RT}\right)^{1/2}$ in (2.21) is chosen so as to satisfy the normalisation conditions

$$\int_{-\infty}^{\infty} n(v)dv = n \tag{2.22}$$

It follows from (2.20) that

$$v = \frac{c}{\nu_0}(\nu - \nu_0); \quad dv = \frac{c}{\nu_0}\, d\nu \tag{2.23}$$

On substituting these expressions into (2.21) we obtain the number of oscillators emitting radiation of frequency in the range between v and dv,

$$dn = \sqrt{\frac{M}{2\pi RT}}\, ne^{-\frac{c^2(\nu-\nu_0)^2 M}{2\nu_0^2 RT}}\, d\nu \tag{2.24}$$

If it is assumed that all the oscillators have the same frequency and amplitude, i.e. the intensities emitted by all the dipoles are the same, the line profile is given by

$$I(\nu) = I_0 e^{-\left(\frac{\nu-\nu_0}{\delta}\right)^2} \tag{2.25}$$

where

$$\delta = \frac{\nu_0}{c}\sqrt{\frac{2RT}{M}} \tag{2.26}$$

is a parameter indicating the frequency difference from ν_0 at which the intensity is reduced by a factor e. The frequencies at which $I(\nu)$ is equal to one half of the maximum value I_0 are given by

$$\nu = \nu_0 \pm \sqrt{\ln 2}\,\delta$$

The Doppler half-width at room temperature and for

$\mu = 16$ and $\lambda = 5{,}000$ Å, is

$$\Delta\lambda_{Dop} = \frac{c\,\Delta\nu}{\nu^2} = 2\frac{c}{\nu^2}\sqrt{\ln 2}\,\delta \approx 1.56\cdot 10^{-2}\ \text{Å}$$

This is greater by a factor of 100 than the natural line width. The temperature at which Doppler line broadening would be equal to the natural line width is practically unattainable. Even for very heavy molecules it is equal to a few hundredths of a degree above absolute zero. The line profile given by (2.25) is symmetrical with respect to the frequency ν_0, but differs considerably from the natural line width. The intensity falls off exponentially on either side of ν_0, i.e. more rapidly than for the natural profile.

Forced oscillations

We shall now consider the forced oscillations of a dipole which are induced by external radiation. It is well known that the force acting on a charge e in an electromagnetic field is given by

$$\mathbf{F} = e\mathbf{E} + \frac{e}{c}[\mathbf{vH}] \tag{2.27}$$

where $\mathbf{v}$ is the velocity of the charge. Since emission of radiation in the optical region is associated with velocities which are lower by a few orders of magnitude than the velocity of light, the second term in (2.27) is much smaller than $e\mathbf{E}$ and is usually neglected. Suppose the axis of the oscillator is parallel to the x axis and that the position of equilibrium of the oscillating charge is at the origin. The equation of motion for a dipole interacting with linearly polarised light, whose electric vector is at an angle θ to the axis of the dipole, is

$$\ddot{x} + 2\gamma\dot{x} + \omega_0^2 x = \frac{e}{m}E_0 e^{i\omega t}\cos\theta \tag{2.28}$$

where ω is the angular frequency and E_0 the amplitude of the exciting radiation. As before,

$$\gamma = \gamma_{em} + \gamma_{fr}$$

The solution of (2.28) is

$$x = C_1 e^{i\,\omega_1 t - \gamma t} + C_2 e^{-i\,\omega_1 t - \gamma t} + \frac{\frac{e}{m} E_0 \cos\theta}{\omega_0^2 - \omega^2 + 2i\gamma\omega} e^{i\omega t} \qquad (2.29)$$

where $\omega_1^2 = \omega_0^2 - \gamma^2$, and C_1 and C_2 are constants of integration. If the oscillator was at rest at the initial instant of time $[x(0) = x(0) = 0]$, then

$$x = \frac{\frac{e}{m} E_0 \cos\theta}{\omega_0^2 - \omega^2 + 2i\,\gamma\omega} \left[-\frac{\omega + \omega_1 + i\,\gamma}{2\omega_1} e^{i\,\omega_1 t - \gamma t} + \frac{\omega - \omega_1 - i\,\gamma}{2\omega_1} e^{-i\,\omega_1 t - \gamma t} + e^{-i\,\omega t} \right] \qquad (2.30)$$

The solution given by (2.30) describes the steady-state behaviour of the oscillator after the source of radiation has been switched on. It contains two terms which include the factors $e^{-\gamma t}$, and which tend to zero for $\gamma t \gg 1$. When the latter condition is satisfied, the motion of the oscillator is determined by the last term and the dipole executes damped harmonic oscillations whose frequency is equal to the frequency of the incident radiation.

If we reject the damped terms in (2.30) the above solution may be written in the form

$$x = \frac{\frac{e}{m} E_0 \cos\theta}{\sqrt{(\omega_0^2 - \omega^2)^2 + 4\gamma^2\omega^2}} e^{i(\omega t - \varphi)} \qquad (2.31)$$

where $\tan\varphi = \frac{2\gamma\omega}{\omega_0^2 - \omega^2}$. This shows that in the steady state the phase of the oscillator lags behind the phase of the exciting electromagnetic wave by an angle φ, which reaches the maximum value of $\pi/2$ when $\omega = \omega_0$. The phase lag tends to zero with decreasing frictional forces ($\gamma \to 0$) or when $\gamma/(\omega - \omega_0)$ tends to zero. If one takes as the initial conditions the values of x and $\dot{x}$ at any given time during the steady-state oscillations and sets $E_0 = 0$ in (2.29). then it can readily be shown that the expression describing the damped oscillations is

$$x = \frac{\frac{e}{m} E_0 \cos\theta}{\sqrt{(\omega_0^2 - \omega^2)^2 + 4\gamma^2\omega^2}} e^{\gamma t} \left[\frac{\omega_1 + \omega - i\gamma}{2\omega_1} e^{i(\omega_1 t - \varphi)} + \frac{\omega_1 - \omega + i\gamma}{2\omega_1} e^{-i(\omega_1 t + \varphi)} \right] \tag{2.32}$$

This solution is equivalent to (2.10) where, apart from small corrections,

$$x_0 = \frac{\frac{e}{m} E_0 \cos\theta}{\sqrt{(\omega_0^2 - \omega^2)^2 + 4\gamma^2\omega^2}}$$

Absorption of external radiation

An electromagnetic wave interacting with a dipole does work which is used to maintain the undamped oscillations. The energy of the field is converted into mechanical energy of the moving charges which in its turn is partly returned in the form of radiation emitted by the oscillator. The rate of absorption of a dipole is equal to the work done by the light wave per unit of time. In complex notation it is given by

$$W_{\mathrm{abs}}(\omega, \theta) = \frac{1}{4}(\dot{x} + \dot{x}^*)\, e\, (E + E^*) \cos\theta \tag{2.33}$$

Differentiating (2.30) with respect to time and substituting for x and $\dot{x}^*$ in (2.33) we obtain

$$W_{\mathrm{abs}}(\omega, \theta) = \frac{\frac{e^2}{m} E_0^2 \gamma\omega^2 \cos^2\theta}{(\omega_0^2 - \omega^2)^2 + 4\gamma^2\omega^2} + \frac{\frac{e^2}{m} E_0^2 \cos^2\theta}{2\omega_1 [(\omega_0^2 - \omega^2)^2 + 4\gamma^2\omega^2]} \{ [4(\omega - \omega_1)\, \omega_0^2 - 8\gamma^2 \omega\omega_1] \times \sin(\omega - \omega_1)\, t - 8\gamma\omega\omega_1^2 \cos(\omega - \omega_1)\, t \} e^{-\gamma t} + R(\omega_1 + \omega) \tag{2.34}$$

where $R(\omega_1 + \omega)$ are terms which oscillate at the frequency $\omega_1 + \omega$. The average of these terms over one period is zero.

It follows from (2.34) that when a monochromatic wave is

absorbed by the oscillator, the rate of absorption initially increases, reaching the maximum value

$$W_{abs}(\omega, \theta) = \frac{\frac{e^2}{m} E_0^2 \gamma \omega^2 \cos^2 \theta}{(\omega_0^2 - \omega^2)^2 + 4\gamma^2\omega^2} \tag{2.35}$$

when $t \gg 1/\gamma$. The rate of absorption is very dependent on frequency difference between the incident radiation and the free oscillations of the oscillator. Maximum absorption is reached during resonance. The expression given by (2.35) determines the absorption line profile and its width decreases with decreasing γ (see Fig. 1.7).

If the incident wave is not monochromatic but has a continuous spectrum, the dipole will execute forced oscillations

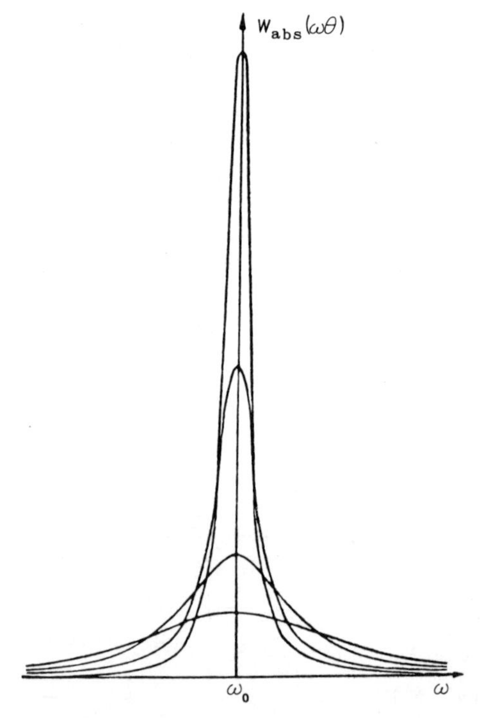

Fig. 1.7 Dependence of the absorption line profile on the damping constant γ. The areas under the curves are all equal

at all the frequencies and will absorb energy at all wavelengths in accordance with (2.35). In order to obtain the total absorbed power it is sufficient to integrate (2.34) with respect to the frequency ν. This operation is allowed since the motion of the dipole in the electromagnetic field is described by a linear equation. In view of the fact that the rates of absorption per unit frequency and per unit angular frequency are related by

$$W_{\text{abs}}(\omega)\, d\omega = W_{\text{abs}}(\nu)\, d\nu$$

the integral absorption is given by

$$\begin{aligned} W_{\text{abs}}(\theta) &= \int_0^\infty W_{\text{abs}}(\nu, \theta)\, d\nu \\ &= \int_0^\infty W_{\text{abs}}(\omega, \theta)\, d\omega \end{aligned} \tag{2.36}$$

The ingegrals encountered in (2.36) when (2.34) is substituted into it can easily be reduced to tabulated integrals if certain simplified assumptions are introduced. Thus, the function $W_{\text{abs}}(\nu, \theta)$ has a sharp maximum at $\nu = \nu_0$ and rapidly tends to zero on either side of ν_0. It may therefore be supposed that $E_0^2(\nu) = E_0^2(\nu_0)$, within the line profile and that the difference between ν and ν_0 in those expressions into which this difference enters is negligible. Substituting $\xi = 2\pi(\nu - \nu_0)/\gamma$ and $\eta = 2\pi \cdot (\nu - \nu_1)t$, and remembering that the extension of the limits of integration from ω/γ to infinity has practically no effect on the value of the integrals, we find that the integrals are reduced to the following standard forms:

$$\int_{-\infty}^{\infty} \frac{d\xi}{\xi^2 + 1} = \pi, \qquad \int_{-\infty}^{\infty} \frac{\eta \sin \eta\, d\eta}{\eta^2 + (\gamma t)^2} = \pi e^{-\gamma t}$$

$$\int_{-\infty}^{\infty} \frac{\gamma t \cos \eta}{\eta^2 + (\gamma t)^2}\, d\eta = \pi e^{-\gamma t} \tag{2.37}$$

These integrals may be used to show that, to within small corrections of the order of γ/ω_0 the total rate of absorption is given by

$$W_{abs} = \frac{\pi e^2}{m} u(\nu_0) \cos^2 \theta \quad (2.38)$$

where $u(\nu_0) = E_0^2(\nu_0)/8\pi$ is the energy density per unit frequency interval in the incident radiation.

It should be noted that when the oscillator is excited by radiation with a continuous spectrum, the integrated absorbed power does not depend on the initial conditions and is the same both in the steady state and during the transient period. This is fully confirmed by the quantum theory of radiation (see Section 27).

Equation (2.38) shows that the power absorbed by a dipole is independent of the constant γ, which characterises frictional forces. Whether the dipole gives up its energy to other bodies, or emits it in the form of light waves or simply retains it, the energy absorbed by a given dipole is always the same. In other words, the area under the absorption curve is constant and independent of the form of the curve. This important result of the classical theory is confirmed by experiment. Moreover, it follows from (2.38) that the rate of absorption is proportional to the density $u(\nu_0)$ of the incident energy. In most cases this result is also confirmed by experiment. However, in some conditions of excitation, which will be discussed in detail in the following chapters, experiment leads to quite different results.

The discrepancy between theory and experiment is reduced in the electron theory by introducing into (2.38) a number of empirical constants f_{abs} (the so-called oscillator strengths). These constants characterise the specific properties of individual absorption bands of given atoms or molecules. The number of such constants is equal to the total number of lines (bands) in the absorption spectrum.

Classical theory does not show how the oscillator strengths can be determined; they must be found directly from experiment and are therefore purely empirical. The best classical method of determining oscillator strengths is the investigation of anomalous dispersion by Rozhdestvenskii's 'hook' method.

Absorption by a given material is characterised by the absorption coefficient $k(\nu)$, which is related to the rate of absorption by the simple expression

$$k(\nu) = \frac{1}{cu(\nu)} W_{abs}(\nu) \quad (2.39)$$

One of the most important characteristics of a given medium is the integral of $k(\nu)$ with respect to the frequency. For a harmonic oscillator

$$k = \int_0^\infty k(\nu)\, d\nu = \frac{\pi e^2}{mc} \cos^2 \theta \qquad (2.40)$$

This integral was first evaluated by Kravets [3] in 1912, and has the important property that it is independent of the temperature of the medium and the density of the exciting radiation. It has played an important part in the development of spectroscopy and still retains its importance. In quantum theory, all the radiation, absorption and distribution formulae include oscillator strengths which are determined experimentally as quantities proportional to the Kravets absorption integral.

Dichroism

As can be seen from (2.38), the rate of absorption depends on the mutual orientation of the dipole axes and the electric vector in the incident light wave. The rate of absorption is equal to zero if the incident radiation is polarised along the z axis ($\theta = 90°$). Absorption reaches a maximum when the incident radiation is polarised along the x axis, i.e. along the axis of the dipole.

Dependence of the absorption coefficient on the polarisation of absorbed radiation is referred to as dichroism. Dichroism can be either natural or induced. In the former case the dependence of the absorption on the polarisation of incident radiation is due to internal properties of the absorbing material. Uniaxial and biaxial crystals whose absorption coefficients are tensors are characteristic examples of this. All the elementary sources discussed in Section 1 exhibit clearly-defined dichroism. Anisotropic absorption can also be observed in molecules if their absorbing centres have a preferred orientation. Induced dichroism is due to external anisotropic effects, for example electric and magnetic fields, mechanical deformation and excitation by beams of light. The latter effects will be discussed in detail in Chapters 6 to 8. The quantity

$$D_{\text{nat}} = \frac{k_{\text{max}} - k_{\text{min}}}{k_{\text{max}} + k_{\text{min}}} \qquad (2.41)$$

where k_{max} and k_{min} are respectively the maximum and minimum values of the absorption coefficients, can be taken as a measure of natural dichroism. Induced dichroism is more conveniently characterised by

$$D_{ind} = \frac{k_{\parallel} - k_{\perp}}{k_{\perp} + k_{\parallel}} \tag{2.42}$$

where $k_{\parallel}$ and $k_{\perp}$ are the absorption coefficients for two linearly polarised rays whose electric vectors are respectively parallel and perpendicular to the direction of maximum value of the driving agent. Whilst D_{nat} is a quantity which is essentially positive, D_{ind} may assume both positive and negative values between +1 and -1. For a linear harmonic oscillator $k_{min} = 0$ and therefore dichroism is equal to unity.

Consider the absorption coefficient of a set of n dipoles randomly distributed in space. If n is large enough, the number of oscillators whose dipole moments lie in a given solid angle is the same for all directions and is given by

$$n(\theta, \varphi) = \frac{n}{4\pi}$$

Suppose that two beams of linearly polarised radiation are incident on dipoles along the y axis. If the electric vector of one of the beams is parallel to the z axis, while the electric vector of the other beam is parallel to the x axis, the total rates of absorption by all the oscillators is given by the two integrals

$$W^{z}_{abs} = \int_0^{2\pi} d\varphi \int_0^{\pi} \frac{n}{4\pi} \frac{\pi e^2}{m} u_z \cos^2\theta \sin\theta \, d\theta = n \frac{\pi e^2}{3m} u_z \tag{2.43}$$

$$W^{x}_{abs} = \int_0^{2\pi} \cos^2\varphi \, d\varphi \int_0^{\pi} \frac{n}{4\pi} \frac{\pi e^2}{m} u_x \sin^3\theta \, d\theta = n \frac{\pi e^2}{3m} u_x \tag{2.44}$$

where u_z and u_x are the two respective incident energy fluxes. In obtaining these expressions we have used (2.38) and the fact that $\cos(\mathbf{D}, x) = \sin\theta \cos\varphi$.

We can now use the relation between k and W_{abs}, and with the aid of (2.43) and (2.44) show that the absorption coefficient per oscillator is

$$k_z = k_x = \frac{\pi e^2}{3mc} \tag{2.45}$$

which corresponds to the absence of dichroism. This result can be interpreted physically in two ways. It may be looked upon either as giving the absorption coefficient for linearly polarised light averaged over all the possible directions of E and D, or it may be regarded as the absorption coefficient for isotropic Planck radiation.

A system of randomly distributed oscillators does not therefore exhibit the property of dichroism. Dichroism will only appear when an anisotropy is introduced into the angular distribution of the dipole moments, e.g. by mechanical deformation of a solid.

Emission of radiation

The harmonic oscillator can execute two forms of oscillation, namely, natural damped oscillations in the absence of external forces, and forced oscillations in an external electromagnetic field. The emission of a dipole is therefore usually divided into spontaneous and induced parts. To begin with, let us consider induced emission in the steady state. If we differentiate (2.31) twice with respect to time and substitute it into (1.16) we have

$$W_{\text{em}}^{\text{st}}(\omega) = \frac{e^4 \omega^4 E_0^2}{3c^3 m^2} \frac{\cos^2\theta}{(\omega_0^2 - \omega^2)^2 + 4\gamma^2\omega^2} \tag{2.46}$$

This shows that the total amount of radiation emitted by a dipole in the steady state is proportional to the density of the incident radiation ($E_0^2 = 8\pi u$) and to the fourth power of the frequency. The resonance factor which gives the line profile for induced emission is of particular importance. Its form is the same as that of the resonance factor in (2.35) for absorption. Absorption of energy by a dipole and the induced emission of radiation by it must be considered in parallel as mutually accompanying processes. If we divide (2.46) by (2.35) we obtain the ratio of induced emission to induced absorption of energy during a given interval of time:

$$\frac{W_{\text{em}}^{\text{st}}(\omega)}{W_{\text{abs}}(\omega)} = \frac{1}{\gamma} \frac{e^2 \omega^2}{3mc^3} = \frac{\gamma_{\text{em}}}{\gamma_{\text{em}} + \gamma_{\text{fr}}} \tag{2.47}$$

It is characteristic that this ratio has no resonance properties. Its magnitude is determined above all by the nature of the damping. If the oscillator loses energy by radiation only, then $\gamma = \gamma_{em}$ and therefore the ratio given by (2.27) is equal to unity. In the steady state all the absorbed energy is emitted through induced emission. However, before the steady state is reached, the absorbed energy can also be expended in increasing the energy of the dipole. If in addition to emission the oscillator can lose energy in other ways, for example through collisions, then $W_{em}/W_{abs} < 1$.

On integrating (2.46) with respect to frequency we obtain the total power radiated by induced emission in the steady state:

$$W_{em}^{st} = \frac{\gamma_{em}}{\gamma} \frac{\pi e^2}{m} u(\nu_0) \cos^2 \theta \tag{2.48}$$

If the oscillator is initially at rest, the total power emitted by it during the excitation (exciting radiation on) and decay (exciting radiation off) processes is given by the following formulae:

$$W_{em}^{exc} = \frac{\gamma_{em}}{\gamma} \frac{\pi e^2}{m} u(\nu_0) (1 - e^{-2\gamma t}) \cos^2 \theta \tag{2.49}$$

$$W_{em}^{dec} = \frac{\gamma_{em}}{\gamma} \frac{\pi e^2}{m} u(\nu_0) e^{-2\gamma t} \cos^2 \theta \tag{2.50}$$

These expressions were obtained with the aid of (1.16), (2.30) and (2.32) by operations similar to those leading to (2.38).

As can be seen from (2.49), the rate of emission increases during excitation from zero at $t = 0$ to a maximum value which is reached in the steady state.

The energy E of the oscillator consists of the sum of the kinetic and potential energies:

$$\mathrm{E} = \frac{1}{2} m\dot{x}^2 + \frac{1}{2} kx^2 \tag{2.51}$$

In the steady state the energy of the dipole averaged over one period of the oscillation is a constant $\left(\frac{d}{dt}\mathrm{E} = 0\right)$. The rate of change of the energy of the dipole during the excitation and decay processes is given by

$$\frac{d}{dt}\mathrm{E}_{exc}=\frac{\pi e^2}{m}u(\nu_0)\,e^{-2\gamma t}\cos^2\theta \tag{2.52}$$

$$\frac{d}{dt}\mathrm{E}_{dec}=-\frac{\pi e^2}{m}u(\nu_0)\,e^{-2\gamma t}\cos^2\theta \tag{2.53}$$

If the oscillator loses its energy through the emission of electromagnetic waves only ($\gamma_{fr}=0$), then the law of conservation of energy

$$W_{abs}=W_{em}+\frac{d}{dt}\mathrm{E} \tag{2.54}$$

is satisfied both during excitation and decay.

It is evident that the energy balance must also be preserved when the frictional forces are introduced and some of the energy of the dipole is converted into other forms of energy, for example heat. The amount of heat liberated per unit time is given by

$$Q=W_{abs}-W_{em}-\frac{d}{dt}\mathrm{E} \tag{2.55}$$

All the quantities entering into the right-hand side of (2.55) have already been calculated for the various stages of emission by a dipole. It can readily be shown that

$$\begin{aligned} Q_{exc}&=\frac{\gamma_{em}}{\gamma}\,\frac{\pi e^2}{m}u(\nu_0)\,(1-e^{-2\gamma t})\cos^2\theta \\ Q_{dec}&=\frac{\gamma_{fr}}{\gamma}\,\frac{\pi e^2}{m}u(\nu_0)\,e^{-2\gamma t}\cos^2\theta \end{aligned} \tag{2.56}$$

The ratio of the energies encountered in the form of radiation and in the form of heat is always equal to γ_{em}/γ_{fr}. It should be noted that the law of conservation of energy given by (2.55) is a direct consequence of the equation of motion for the oscillator given by (2.28). In point of fact, on multiplying (2.28) by mx, we obtain

$$m\ddot{x}x+2\gamma m\dot{x}^2+\omega_0^2 m\dot{x}x=eE_0\dot{x}e^{i\omega t}\cos\theta \tag{2.57}$$

or

$$F_{\mathrm{fr}}\dot{x} = eE_0 x e^{i\omega t} \cos\theta - F_{\mathrm{em}}\dot{x} - \frac{d}{dt}\left(\frac{1}{2} m\dot{x}^2 + \frac{1}{2} kx^2\right) \quad (2.58)$$

This is equivalent to (2.55) and is valid for irradiation by either a monochromatic wave or a continuous spectrum.

3. ANISOTROPIC ABSORPTION AND EMISSION OF RADIATION BY A SET OF OSCILLATORS

Coherent and incoherent emission

The emission and absorption of radiation by a single harmonic oscillator was calculated in the preceding section. Under real experimental conditions the incident radiation interacts with a large number of atoms and molecules. The intensity, polarisation and angular distribution of the radiation emitted by a single oscillator and by a set of oscillators are quite different. We have already seen in Section 1 that even in the case of only two dipoles the total emitted radiation differs substantially from the emission by a single oscillator, and resembles the emission of an elementary quadrupole or a plane rotator with a certain mutual orientation and phase difference of the two oscillators.

The emission of radiation by a set of non-interacting oscillators may be either coherent or incoherent. For example, under normal thermal excitation, the phases of individual oscillators vary independently of each other and thus the emitted waves are incoherent. Conversely, light waves will not only excite the oscillations of individual dipoles, but they will also synchronise the oscillations. As can be seen from (2.31), the phase of forced oscillations is completely determined by the incident radiation. Dipoles located at different points in space will have different phases at a given instant of time, but the phase difference will remain constant across the wave front.

An important property of coherent waves is their ability to interfere, i.e. reinforce or cancel each other. Partly coherent waves can also give rise to interference, but the corresponding interference pattern will vary with time. It follows that in calculating the total emission by a set of coherent sources we must add the amplitudes of the individual waves with allowance for their phases and then use the result to calculate the energy flux. Moreover, the total emission by a set of incoherent sources is equal to the sum of the individual emissions and therefore one can simply add the intensities or fluxes.

Interference of secondary radiation provides a simple explanation of phenomena such as the propagation of light through matter, reflection and refraction.

Serious difficulties arise in the classical theory when an attempt is made to use it to explain photoluminescence, which is known from experiment to be an incoherent process; emission induced by incident radiation should be completely coherent according to the classical theory of the harmonic oscillator. A satisfactory explanation of all the properties of photoluminescence can only be obtained in quantum electrodynamics. However, it will be shown that the oscillator model of a molecule suffices for the interpretation of a large number of experimental effects if it is postulated that the emission by the individual oscillators is not coherent. The reasons for this lack of coherence are not usually discussed in classical theory.

Propagation of light through matter

To begin with, let us consider only the absorption of primary incident radiation and the emission of secondary light waves by a set of harmonic oscillators. We shall assume for simplicity that the dipoles are distributed randomly in a given layer, and that they do not interact with each other. The thickness of the layer is such that the attenuation of the exciting radiation and secondary absorption may be neglected.

Suppose that the electric vector in the incident electromagnetic wave varies in space and time in accordance with the formula

$$\mathbf{E} = \mathbf{E}_0 \cos(\omega t - \boldsymbol{\varkappa} \mathbf{r}) \qquad (3.1)$$

where $\mathbf{r}$ is the radius vector of the point under consideration and $\boldsymbol{\varkappa}$ is the wave vector.

In our discussion of the interaction of radiation with a single oscillator, the change in the field strength in space was not taken into account because the dimensions of the dipole were assumed to be much smaller than the wavelength. The constant quantity $\varkappa r$ could simply be included in the initial phase of the oscillations. In the present case, on the other hand, the transverse dimensions of the layer are very much greater than the wavelength and therefore the phases of the oscillations at different points within the layer may differ substantially from each other.

The wave given by (3.1) is called a plane wave, since the locus of points of equal phase (the wave front) takes the form of a plane. In fact, the requirement that the phase should be constant at a time t_1, i.e.

$$\omega t_1 - \boldsymbol{\varkappa} \mathbf{r} = \text{const}$$

yields

$$\boldsymbol{\varkappa}\, \mathbf{r} = a \tag{3.2}$$

which is the equation of a plane M perpendicular to the wave vector $\boldsymbol{\varkappa}$.

If we increase a by 2π, we obtain the equation of another plane N, parallel to M and separated from it by one wavelength λ:

$$\boldsymbol{\varkappa}\, \mathbf{r}' = a + 2\pi \tag{3.3}$$

On subtracting (3.2) from (3.3) and recalling that the projection of ($\mathbf{r}$—$\mathbf{r}'$) on to the direction of $\boldsymbol{\varkappa}$ is equal to λ, we have

$$\boldsymbol{\varkappa}\, \lambda = 2\pi \tag{3.4}$$

or

$$\varkappa = \frac{2\pi}{\lambda} = \frac{2\pi}{\lambda_0} n = n \frac{\omega}{c},$$

where λ_0 is the wavelength in vacuum and n is the refractive index of the medium. It follows that the wave vector is given by

$$\boldsymbol{\varkappa} = \frac{2\pi}{\lambda} [\mathbf{i} \cos(\boldsymbol{\varkappa}\, x) + \mathbf{j} \cos(\boldsymbol{\varkappa}\, y) + \mathbf{k} \cos(\boldsymbol{\varkappa}, z)]. \tag{3.5}$$

Under the above assumptions the total absorption in the layer is equal simply to the sum of the absorptions by the individual dipoles. When the incident radiation is plane polarised, the absorption is given by (2.43).

Consider the emission by the dipoles in the direction at an

angle θ to the direction of propagation of the incident wave (Fig. 1.8). It is evident that interference between secondary waves emitted by all the oscillators must be taken into account. To do this, consider the dipoles lying on a particular wave front at a time t. They will all have the same phases. Let us divide the wave front bc into sections of length

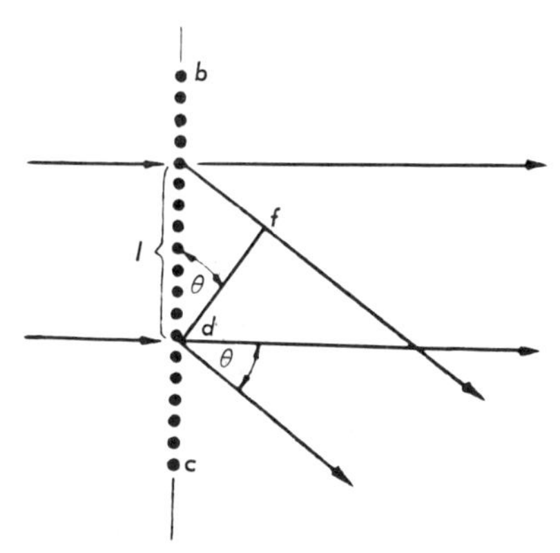

Fig. 1.8 Propagation of radiation through an isotropic medium

$t = \lambda/\sin\theta$, which is possible for all $\theta \neq 0$, since the linear dimensions of the layer occupied by the oscillators are much greater than the wavelength λ. Along the line df the phase of the rays travelling in the direction θ, will change by 2π, and if the dipoles lie at equal distances from each other, the phase of the k-th oscillator is

$$\varphi_k = \omega t + \frac{2\pi}{m} k \tag{3.6}$$

where m is the number of oscillators in the length l, and the particles are numbered in the direction of cb. The electric field in the wave propagating in the direction θ and emitted by n dipoles is given by

$$E = \sum_k E_k \cos\left(\omega t + \frac{2\pi}{m} k\right) \tag{3.7}$$

If the dipoles are identical we have $\overline{E_k = E_0}$, and it can readily be shown that

$$E = \frac{1}{2} E_0 \sum_{k=1}^{n} \left[e^{i\left(\omega t + \frac{2\pi}{m} k\right)} + e^{-i\left(\omega t + \frac{2\pi}{m} k\right)} \right]$$

$$= \frac{1}{2} E_0 e^{i\omega t} \sum_{k=1}^{n} e^{i\frac{2\pi}{m} k} + \frac{1}{2} E_0 e^{-i\omega t} \sum_{k=1}^{n} e^{-i\frac{2\pi}{m} k}$$

$$= \frac{1}{2} E_0 e^{i\left(\omega t + \frac{2\pi}{m}\right)} \frac{1 - e^{2\pi i}}{1 - e^{2\pi i/m}} \qquad (3.8)$$

$$+ \frac{1}{2} E_0 e^{-i(\omega t + 2\pi/m)} \frac{1 - e^{-2\pi i}}{1 - e^{-2\pi i/m}} = 0.$$

It follows that there is no induced emission in directions different from the direction of propagation of the incident radiation for a set of oscillators which form an isotropic medium. The field E has non-zero values only for $\theta = 0$. Experimentally this is observed as transmission of light through a medium without scattering.

Polarisability

Let us consider briefly the polarisability of molecules, which is a very important characteristic of the interaction of radiation with matter. We have already seen (Section 1.2) that an oscillator will execute forced oscillations proportional to the component of the external field vector along the axis of the dipole. If we denote the induced dipole moment of the oscillator by $\mathbf{p}$, and substitute $\mathbf{E}_p$, for $\mathbf{E} \cos \theta$, we obtain from (2.30) in the steady state

$$\mathbf{p} = \frac{e^2/m}{\omega_0^2 - \omega^2 + 2i\gamma\omega} \mathbf{E}_p = \alpha_1 \mathbf{E}_p \qquad (3.9)$$

where

$$\alpha_1 = \frac{e^2/m}{\omega_0^2 - \omega^2 + 2i\gamma\omega}$$

is the polarisability of the oscillator. If the oscillator absorbs external radiation ($\gamma \neq 0$), the polarisability is complex. This means that the phase of the oscillations of the dipole is not equal to the phase of the incident radiation (see (2.31)).

In contrast to a linear oscillator, the electron cloud of a molecule will execute three-dimensional oscillations under the action of incident radiation. An electric field parallel to, say the z axis, will cause a displacement of the charges not only along this axis but also along the other two axes, since the charges in the molecule interact with each other. It follows that the components of the dipole moment which are induced by the external field are related to its components by the equations

$$\begin{aligned} p_x &= \alpha_{xx} E_x + \alpha_{xy} E_y \quad \alpha_{xz} E_z \\ p_y &= \alpha_{yx} E_x + \alpha_{yy} E_y + \alpha_{yz} E_z \\ p_z &= \alpha_{zx} E_x + \alpha_{zy} E_y + \alpha_{zz} E_z \end{aligned} \tag{3.10}$$

If we introduce the polarisability tensor

$$\alpha = \begin{pmatrix} \alpha_{xx} & \alpha_{xy} & \alpha_{xz} \\ \alpha_{yx} & \alpha_{yy} & \alpha_{yz} \\ \alpha_{zx} & \alpha_{zy} & \alpha_{zz} \end{pmatrix} \tag{3.11}$$

we can rewrite (3.10) in the form of the single equation

$$\mathbf{p} = \alpha \mathbf{E}. \tag{3.12}$$

The polarisability α represents the ability of the electrons in the molecule to become displaced by the field $\mathbf{E}$ and has the dimensions of a volume. Usually, α is of the order of the cube of the molecular radius, i.e. about 10^{-24} cm^3. In the case of a linear oscillator parallel to the x-axis, all the components of the tensor (3.11) are zero except $\alpha_{xx} = \alpha_1$.

The total induced dipole moment $\mathbf{P}$ of all the particles in a unit volume is known as the polarisability vector and is given by

$$\mathbf{P} = N_1 \alpha \mathbf{E} \tag{3.13}$$

where N_1 is the number of particles per unit volume.

By definition, the induction $\mathbf{D}$ is

$$\mathbf{D} = \varepsilon \mathbf{E} = \mathbf{E} + 4\pi \mathbf{P} \tag{3.14}$$

and therefore

$$\mathbf{P} = \frac{\varepsilon - 1}{4\pi} \mathbf{E} \tag{3.15}$$

From (3.13) and (3.14) we have the following relation between the dielectric constant and polarisability

$$\varepsilon = 1 + 4\pi N_1 \alpha \tag{3.16}$$

Refractive index

The relation between the refractive index and the dielectric constant given by (3.16) may be rewritten in the form

$$\tilde{n} = \sqrt{\varepsilon} = \sqrt{1 + 4\pi N_1 \alpha} \tag{3.16a}$$

where $\tilde{n}$ is in general complex. If we substitute the value of α from (3.9) into this expression and extract the square root approximately, we obtain the expression for the refractive index of a gas consisting of harmonic oscillators:

$$\tilde{n} = n + ik_1 \tag{3.17}$$

where

$$n = 1 + 2\pi N_1 \frac{e^2}{m} \frac{\omega_0^2 - \omega^2}{(\omega_0^2 - \omega^2)^2 + 4\gamma^2 \omega^2}$$

$$k_1 = 4\pi N_1 \frac{e^2}{m} \frac{\gamma\omega}{(\omega_0^2 - \omega^2)^2 + 4\gamma^2 \omega^2}$$

A plane electromagnetic wave propagating in a medium with a complex refractive index may be represented by

$$E = E_0 e^{-\frac{k_1 \omega}{c} r} e^{i\omega\left(t - \frac{nr}{c}\right)} \tag{3.18}$$

It follows that the imaginary part of the refractive index characterises the damping of the wave. The coefficient k_1 is often referred to as the extinction coefficient. The intensity of the wave falls off exponentially in accordance with the expression

$$I = I_0 e^{-kr}$$

where the absorption coefficient k is given by

$$k = k_1 \frac{\omega}{c}$$

Figure 1.9 illustrates the variation of n and k_1 with frequency. As can be seen, the form of the function $k_1(\omega)$ is determined by the absorption line profile. The graph of $n(\omega)$ has a more complicated shape. If the oscillators have different natural frequencies, the number $\tilde{n}$ will be replaced by the sum $\Sigma \tilde{n}_i$.

Rayleigh scattering by density fluctuations

In our discussion of the propagation of radiation through matter, we assumed that the medium was optically isotropic. Strictly speaking, this assumption is never valid. Even if the medium as a whole is uniform, the thermal motion of the atoms and molecules gives rise to density fluctuations which lead to optical irregularities. In other words, if we subdivide the total volume V occupied by the

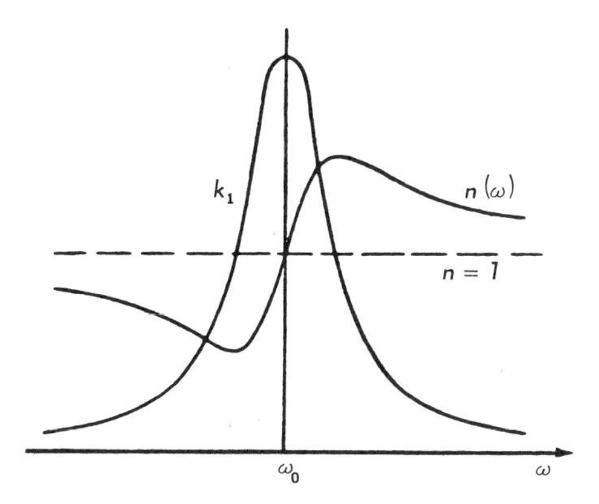

Fig. 1.9 Frequency dependence of the refractive index and the absorption coefficient near the natural frequency of a dipole

molecules into elementary volumes v_i, the number of particles N_i, the density ρ_i and the dielectric constant ε_i, will fluctuate about their mean values $\overline{N}_i$, $\overline{\rho_i}$ and $\overline{\varepsilon_i}$ respectively. The linear dimensions of the elementary volumes are chosen to be much smaller than the wavelength of the incident radiation λ. Owing to the non-uniformity of the medium, secondary waves emitted by the dipoles cannot extinguish each other, not only in the direction of the incident radiation, but also in all other directions. The result is that the incident radiation is scattered.

Density fluctuations are not the only cause of scattering. Scattering can also arise as a result of angular anisotropy in the distribution of dipoles within an elementary volume. However, here we shall confine our attention to Rayleigh scattering by density fluctuations.

We can calculate the scattered radiation for a more general model than the linear harmonic oscillator without undue complication of the mathematical derivation. If we write the dielectric constant in the form

$$\varepsilon = \bar{\varepsilon} + \Delta\varepsilon \tag{3.19}$$

where $\bar{\varepsilon}$ is the mean of ε, and $\Delta\varepsilon$ is the departure from the mean, we can rewrite (3.15) in the form

$$\mathbf{P} = \frac{\bar{\varepsilon} - 1}{4\pi}\mathbf{E} + \frac{\Delta\varepsilon}{4\pi}\mathbf{E} \tag{3.20}$$

The first term in (3.20) gives the average of the dipole moment over the entire volume. As was pointed out above, if all the dipoles are identically excited they do not give rise to scattering. In the present case the part of the individual oscillators is played by the elementary volumes. Consequently, the scattering process is determined by the second term in (3.20). The dipole moment of the i-th elementary volume is

$$\mathbf{p}_i = \frac{\Delta\varepsilon_i}{4\pi}\mathbf{E}\, v_i \tag{3.21}$$

According to (1.14), the electric field in the scattered wave propagating in the direction $\mathbf{n}$ at large distances from the dipole moment (3.21) is given by

$$\mathbf{E}_i^{\text{scat}} = -\frac{\omega^2}{c^2 R}[\mathbf{n}[\mathbf{n}\mathbf{p}]] = -\frac{\omega^2 \Delta\varepsilon_i}{4\pi c^2 R}[\mathbf{n}[\mathbf{n}\mathbf{E}]]\, v_i \tag{3.22}$$

Since the dielectric constant ε is a function of the density ρ and temperature T, it follows that

$$\Delta\varepsilon_i = \frac{\partial\varepsilon}{\partial\rho}\Delta\rho_i + \frac{\partial\varepsilon}{\partial T}\Delta T_i \tag{3.23}$$

Calculations show that the temperature fluctuations are of minor significance, and so we can set approximately [4]

$$\Delta\varepsilon_i = \frac{d\varepsilon}{d\rho}\Delta\rho_i \tag{3.23a}$$

If we now substitute this expression into (3.22) and sum over all the elementary volumes, we obtain the electric field strength in the wave produced by all the oscillators, which is propagating in the direction $\mathbf{n}$:

$$\mathbf{E}_{\text{scat}} = \sum_i E_i^{\text{scat}} = -\frac{\omega^2}{4\pi c^2 R}\frac{d\varepsilon}{d\rho}[\mathbf{n}[\mathbf{n}\mathbf{E}]]\sum_i \Delta\rho_i v_i \tag{3.24}$$

This expression leads to the following mean flux of radiation in this direction

$$S_0 = \frac{c}{4\pi}[\mathbf{E}_{\text{scat}}\mathbf{H}_{\text{scat}}] = \frac{c}{4\pi}\mathbf{n}\overline{(\mathbf{E}_{\text{scat}})^2}$$

$$= \frac{\omega^4}{64\pi^3 c^3 R^2}\mathbf{n}\left(\frac{d\varepsilon}{d\rho}\right)^2[\mathbf{n}[\mathbf{n}\mathbf{E}]]^2\overline{\left(\sum_i \Delta\rho_i v_i\right)^2} \tag{3.25}$$

$$= \frac{\omega^4}{64\pi^3 c^3 R^2}\mathbf{n}\left(\frac{d\varepsilon}{d\rho}\right)^2 E^2\sin^2\theta\,\overline{\left(\sum_i \Delta\rho_i v_i\right)^2}$$

where θ is the angle between $\mathbf{E}$ and $\mathbf{n}$. The mean square in this expression can be shown [4] to be given by

$$\overline{\left(\sum_i \Delta\rho_i v_i\right)^2} = \sum_i \overline{\rho^2}\frac{v_i}{\overline{N_i}} \tag{3.26}$$

and therefore

$$S(\theta) = \frac{\omega^4}{64\pi^3 c^3 R^2}\left(\frac{d\varepsilon}{d\rho}\bar{\rho}\right)^2 E^2\sin^2\theta\sum_i\frac{v_i^2}{\overline{N_1}} \tag{3.27}$$

Since $\overline{N_i} = N_1 v_i$, (3.27) may be written in the form

$$S(\theta) = \frac{\omega^4}{16\pi^2 c^3 R^2}\left(\frac{d\varepsilon}{d\rho}\bar{\rho}\right)^2 u\sin^2\theta\,\frac{V}{N_1} \tag{3.28}$$

where $u = E^2/4\pi$ is the density of the incident radiation.

If $\bar{n}$ is the refractive index, we have the following relation for gases

$$\frac{d\varepsilon}{d\rho}\rho = n^2 - 1 \approx 2(n-1) \qquad (3.29)$$

and the flux scattered radiation is given by

$$S(\theta) = \frac{4\pi^2 c}{\lambda^4 R^2}(n-1)\sin^2\theta \, \frac{V}{N_1} u \qquad (3.30)$$

which is known as Rayleigh's formula.

Comparison of (3.30) with (1.15) shows that when the medium is illuminated by linearly polarised radiation, the angular distribution of Rayleigh scattered radiation is the same as the angular distribution in the case of dipole emission (see Fig. 1.2). According to Rayleigh's formula, the intensity of the scattered radiation is inversely proportional to the fourth power of the wavelength of the incident radiation. Therefore, if the incident radiation is white, the scattered radiation should be mainly blue, whilst the transmitted radiation is mainly red. If we integrate (3.30) over a sphere we obtain the flux of scattered radiation propagating in all directions:

$$S = \int_0^{2\pi} d\varphi \int_0^{\pi} S(\theta) R^2 \sin\theta \, d\theta = \frac{16\pi^3 c}{3\lambda^4} \frac{(n-1)^2}{N_1} Vu \qquad (3.31)$$

It should be noted that this dependence of S on λ is valid only if the frequency of the incident radiation is very different from the natural frequency of the oscillators. If the frequency approaches the resonance frequency, it is necessary to take into account the strong dependence of the refractive index on λ.

Polarisation

In order to define a particular radiation completely, one must specify five independent quantities, namely, the direction of propagation, the frequency (or the spectrum), the intensity, the phase of the oscillations and the polarisation. The last is an important characteristic of radiation. It indicates the anisotropic effect of light in the plane perpendicular to the direction of propagation, and contains a very considerable amount of information about the properties of the emitting atoms and molecules. The nature of the polarisation is determined by the locus of the end point of the electric vector of the wave. For example, if the vector executes oscillations in a given plane, the light is said to be plane polarised.

Radiation of any polarisation may be represented as the sum of two waves which are plane polarised in two mutually perpendicular directions. For a monochromatic wave, the components of the electric vector along these directions are given by

$$E_1 = a_1 \cos(\omega t + \varphi_1); \; E_2 = a_2 \cos(\omega t + \varphi_2) \tag{3.32}$$

If we eliminate the time between these two expressions we obtain the equation for the locus of the end point of the electric vector:

$$\frac{E_1^2}{a_1^2} + \frac{E_2^2}{a_2^2} - 2\frac{E_1 E_2}{a_1 a_2}\cos^2(\varphi_2 - \varphi_1) = \sin^2(\varphi_2 - \varphi_1) \tag{3.33}$$

In general, this is the equation of an ellipse. It assumes the simpler form

$$\frac{E_1^2}{a_1^2} + \frac{E_2^2}{a_2^2} = 1 \tag{3.33a}$$

when the phase difference $\varphi_2 - \varphi_1 = \pi/2$. The coefficients a_1 and a_2 are the semi-axes of the ellipse, and when they are equal the light is circularly polarised. If, on the other hand, $\varphi_2 - \varphi_1 = k\,2\pi$, where k is an integer including zero, then we have from (3.33)

$$\frac{E_1}{a_1} = \pm\frac{E_2}{a_2} \tag{3.33b}$$

which are the equations of straight lines.

The radiation emitted by hot bodies is due to the emission by a large number of atoms and molecules and consists of waves of various frequences. The phase difference $\varphi_2 - \varphi_1$ can then assume any value with equal probability. Natural radiation is therefore completely isotropic in the plane perpendicular to the direction of propagation.

It should be noted that circularly polarised light and natural radiation are externally indistinguishable and have an analogous effect on polarisation instruments. This is also true for elliptically polarised light and partly polarised natural light. In order to establish the true nature of polarisation, it is necessary to perform special experiments involving phase changes.

It is not always convenient to describe the polarisation with the aid of equations of the type given by (3.33). If the radiation consists of many waves this is practically impossible. Polarisation is therefore sometimes characterised by the degree of polarisation, whilst in other cases it is described by the degree of depolarisation.

If we pass a beam of light through a Nicol prism we can easily establish the two mutually perpendicular directions in which the oscillations of the electric vector have the maximum and minimum values. The degree of depolarisation is then defined by

$$\Delta_1 = \frac{I_{min}}{I_{max}} \tag{3.34}$$

If the radiation is plane polarised we have $I_{min} = 0$ and therefore $\Delta_1 = 0$. For natural radiation, or for circularly polarised radiation $I_{min} = I_{max}$ and $\Delta_1 = 1$. In addition to the degree of depolarisation, radiation is frequently characterised by the degree of polarisation, which is defined by

$$P_1 = \frac{I_{max} - I_{min}}{I_{max} + I_{min}} \tag{3.35}$$

The degree of polarisation defined in this way is always positive and may assume any value between 0 and 1. It is easy to see that it is related to the degree of depolarisation by the expressions

$$P_1 = \frac{1 - \Delta_1}{1 + \Delta_1}; \quad \Delta_1 = \frac{1 - P_1}{1 + P_1} \tag{3.36}$$

When one of the two quantities (P_1 or Δ_1) is increased from 0 to 1 the other decreases from 1 to 0.

Polarisation of secondary emission

Polarisation studies provide important information about the properties of atoms and molecules. Both the degree of polarisation defined by (3.35) and the direction of the maximum oscillations of the electric field in the secondary wave are significant in this connection. In polarisation measurements the exciting radiation is usually plane polarised, and the direction of observation lies in the plane perpendicular to the electric field in the incident wave. The degree of depolarisation and the degree of polarisation of the secondary waves are then defined somewhat differently:

$$\Delta = \frac{I_x}{I_z} \tag{3.37}$$

$$P = \frac{I_z - I_x}{I_z + I_x} \tag{3.38}$$

As before, I_z and I_x are the intensities of two components which are polarised in mutually perpendicular planes. However, in distinction to the preceding case, the orientations of the planes of polarisation is determined by the directions of maximum and minimum oscillations of the electric field not in the secondary waves but in the exciting radiation. The direction of the oscillations of the electric field in the component I_z is parallel to the direction of maximum oscillations of the electric field in the incident radiation, while the direction of the oscillations in I_x is parallel to the direction of the minimum oscillations.

The parameters Δ and P are related by formulae analogous to (3.36). However, they differ from Δ_1 and P_1 both in their range of values and their physical significance. In point of fact, while Δ_1 and P_1 vary between 0 and 1, the new parameters Δ and P can in principle assume any values within the limits

$$0 \leqslant \Delta \leqslant \infty, -1 \leqslant P \leqslant 1$$

Without reference to the exciting radiation, only Δ_1 and P_1 have a physical meaning, whilst Δ and P are undefined. For example, Δ can assume any value between 0 and infinity for plane-polarised radiation depending on the angle between the plane of polarisation and the z-axis, i.e. depending on the resolution into the components I_z and I_x, which is quite arbitrary without reference to the exciting radiation.

Light emitted by elementary sources is always polarised (see Section 1). However, in practice one measures the radiation emitted by an enormous number of atoms and molecules. In order that this radiation should be polarised it is not sufficient to have anisotropic sources. Radiation is polarised when at least one of the following conditions is satisfied: (1) the elementary sources have preferred orientations, (2) the excitation is anisotropic, i.e. sources of definite orientation are excited, or the degree of excitation depends on the orientation.

The first condition is satisfied in solids, for example crystals, where the internal regular fields force the molecules into certain definite orientations. Anisotropic films of cellophane and certain organic fibres are also found to exhibit preferred orientations. The anisotropy of a macroscopic system may be produced by various mechanical methods, as well as by electrical and magnetic fields.

We shall confine our attention to the polarised emission due to macroscopically isotropic systems. Their emission is polarised because the effect of external radiation on the molecules depends on their orientation.

Polarisation in Rayleigh scattering

We were concerned above with Rayleigh scattering by density fluctuation. Suppose now that the incident radiation propagates along the x axis, while the scattered radiation is received along the y axis. The electric field of the incident wave will then lie in the yz-plane, while the field in the scattered radiation, will, in view of (3.22), be parallel to the z axis. Consequently, both for plane-polarised and for natural incident radiation, the degree of depolarisation of the scattered radiation is equal to zero, whilst the degree of polarisation is equal to unity. Experiment shows that for gases the degree of depolarisation does not, in fact, exceed a few hundredths. However, the x-component is never completely zero, and can assume an appreciable value in scattering by liquids. The reason for the appearance of the x-component is that if the molecules are anisotropic there is a fluctuation in the anisotropy in elementary volumes, and this is analogous to the density fluctuations considered above.

In the absence of absorption, the polarisability tensor of many molecules (the exceptions are mainly the optically active molecules) can be reduced to the diagonal form [4]

$$\alpha = \begin{pmatrix} \alpha_{xx} & 0 & 0 \\ 0 & \alpha_{yy} & 0 \\ 0 & 0 & \alpha_{zz} \end{pmatrix} \tag{3.39}$$

by rotating the coordinate axes. If the fluctuations in the anisotropy are taken into account in this case, the degree of depolarisation of light scattered in the transverse direction is given by

$$\Delta = \frac{3g^2}{5b^2 + 4g^2} \tag{3.40}$$

$$\Delta_e = \frac{6g^2}{5b^2 + 7g^2} \tag{3.41}$$

where the former refers to illumination by plane-polarised radiation, while the latter corresponds to natural incident radiation. The quantities b and g are related to the components of the polarisability tensor by

$$\begin{aligned} b &= \alpha_{xx} + \alpha_{yy} + \alpha_{zz} \\ g^2 &= \frac{1}{2}\left[(\alpha_{xx} - \alpha_{yy})^2 + (\alpha_{yy} - \alpha_{zz})^2 + (\alpha_{zz} - \alpha_{xx})^2\right] \end{aligned} \tag{3.42}$$

The parameter Δ_e may be expressed in terms of Δ with the aid of (3.40) and (3.41):

$$\Delta_e = \frac{2\Delta}{1 + \Delta} \tag{3.43}$$

For completely isotropic molecules $\alpha_{xx} = \alpha_{yy} = \alpha_{zz}$, $g = 0$ and therefore $\Delta = \Delta_e = 0$. If, on the other hand, the molecules are completely anisotropic, as for the linear oscillator, i.e. if for example $\alpha_{xx} \neq 0$, $\alpha_{yy} = \alpha_{zz} = 0$, then the degree of depolarisation will reach the maximum values $\Delta = 1/3$, $\Delta_e = 1/2$. Intermediate values are obtained in all the remaining cases.

The degree of depolarisation of scattered radiation will therefore lie within the range

$$\begin{aligned} 0 &\leqslant \Delta \leqslant \frac{1}{3} \\ 0 &\leqslant \Delta_e \leqslant \frac{1}{2} \end{aligned} \tag{3.44}$$

depending on the properties of the scattering molecules.

The degree of polarisation, on the other hand, will, in accordance with (3.36), lie within the ranges

$$\frac{1}{2} \leqslant P \leqslant 1$$
$$\frac{1}{3} \leqslant P_e \leqslant 1 \tag{3.45}$$

This is valid only for the scattered radiation. The polarisation of luminescence is always much lower and, as will be shown in the following section, is never greater than 1/2.

Polarisation of luminescence excited by plane-polarised radiation

Let us consider the polarisation of the luminescence emitted under the two most important methods of excitation, namely, excitation by natural and by plane-polarised radiation. These calculations were first carried out by Levshin [5] and Vavilov. Consider, to begin with, a set of linear harmonic oscillators. As has already been pointed out, this model is capable of explaining a large number of experimental facts on polarised luminescence. Some generalisations of this model of a luminescing molecule (two-dimensional and three-dimensional oscillations) will be considered below.

Since luminescence is not a coherent process, its total intensity is equal to the simple sum of the intensities due to the individual oscillators, and can be readily calculated.

Suppose that the harmonic oscillators under consideration are randomly distributed in space. If all the electric field vectors are drawn from the origin, their end points will lie with uniform density on the surface of a sphere (parallel displacement of the dipole moment vector within the specimen having no effect on the polarisation of the emitted luminescence). Suppose further that the oscillators are excited by plane-polarised radiation incident along the x axis. The electric field in the incident radiation is parallel to the z axis and the luminescence is measured in the direction of the y axis.

In calculating the polarisation of luminescence we shall suppose that the molecules do not execute rotational motion, so that the direction of the dipole moment of a given oscillator will remain the same. The effect of Brownian rotations on the polarisation of luminescence will be considered at the end of this section.

Consider an oscillator in a set of dipoles, and let its direction in space be defined by the angles θ and φ. According to (2.31), external radiation will excite oscillations in the dipole of amplitude proportional to $\cos\theta$. The amplitude of the electromagnetic wave is also proportional to $\cos\theta$, and the components of the electric field in the emitted radiation along the z and x axes are

$$\begin{aligned} E_z &= E\cos^2\theta \\ E_x &= E\cos\theta\sin\theta\cos\varphi \end{aligned} \tag{3.46}$$

where E is the field strength in radiation emitted by an oscillator lying along the z axis.

In order to find the intensity of luminescence emitted by all the dipoles in the direction of the y axis and polarised along the z or x axes, we must square (3.46) and integrate with respect to θ and φ. The result is

$$\begin{aligned} I_z = C\int\limits_{\Omega}|E_z|^2\,d\Omega &= C|E|^2\int\limits_0^{2\pi}d\varphi\int\limits_0^{\pi}\cos^4\theta\sin\theta\,d\theta \\ &= C|E|^2\,2\pi\frac{1}{5} \end{aligned} \tag{3.47}$$

$$\begin{aligned} I_x = C\int\limits_{\Omega}|E_x|^2\,d\Omega &= C|E|^2\int\limits_0^{2\pi}\cos^2\varphi\,d\varphi\int\limits_0^{\pi}\cos^2\theta\sin^3\theta\,d\theta \\ &= C|E|^2\,2\pi\frac{1}{15} \end{aligned} \tag{3.48}$$

where C is a proportionality constant. The degree of polarisation is given by

$$P = \frac{I_z - I_x}{I_z + I_x} = \frac{1}{2} \tag{3.49}$$

The magnitude of the polarisation depends on the direction of observation. If we determine the polarisation of luminescence emitted in the direction of the z axis then, in the case under consideration, the radiation will be found to be completely depolarised ($P=0$). In all other directions it will be found to lie between 0 and 0.5. The maximum degree of polarisation which is possible for the chosen set of elementary sources in the absence of depolarisation factors will be called the limiting degree of polarisation. It is equal to 0.5 for linear harmonic oscillators.

Polarisation of luminescence excited by natural radiation

Suppose that the oscillators are excited by natural radiation incident along the x axis. The radiation may be represented by a set of linearly polarised rays, the electric field vectors of which lie in the plane perpendicular to the direction of propagation (Fig. 1.10).

The amplitude of oscillations excited by one of the plane-polarised rays in the dipole with angles θ and φ is proportional to

$$\cos(\mathbf{ED}) = (\cos\eta\cos\theta + \sin\eta\sin\theta\sin\varphi) \tag{3.50}$$

In view of (3.46), the intensities may be written in the form

$$I_z^e = C\,|\,E\,|^2 \int_0^\pi d\eta \int_\Omega \cos^2(\mathbf{ED})\cos^2\theta\, d\Omega \tag{3.51}$$

$$I_x^e = C\,|\,E\,|^2 \int_0^\pi d\eta \int_\Omega \cos(\mathbf{ED})\sin^2\theta\cos^2\varphi\, d\Omega \tag{3.52}$$

where the integration is carried out over all the orientations of the electric field in the exciting radiation and over all the dipoles in the system. The superscript e represents the fact that the system is excited by natural radiation. These integrals can easily be rewritten with the aid of (3.50) in the form

$$I_z^e = \int_0^\pi [\cos^2\eta\, I_z + \sin^2\eta\, I_x]\, d\eta = \frac{\pi}{2}(I_z + I_x) \tag{3.53}$$

$$I_x^e = \int_0^\pi I_x\, d\eta = \pi I_x \tag{3.54}$$

Accordingly, the degrees of polarisation of luminescence

excited by natural and linearly polarised radiation are related by

$$P_e = \frac{P}{2-P} \qquad (3.55)$$

On substituting $P = 1/2$ in (3.55) we obtain $P_e = 1/3$.

The expression given by (3.55) is valid not only for luminescence, but also for scattered radiation. This can be easily verified with the aid of (3.36) and (3.43).

Dependence of polarisation on the angle between absorption and emission dipoles

The degree of polarisation of luminescence determined experimentally depends not only on the nature of the elementary sources and the method of excitation, but also on various depolarising factors. Therefore, if the luminescence is characterised by the degree of polarisation $P < 1/2$ or $P_e < 1/3$, this can be explained as a result of depolarisation. However, none of the depolarisation factors, including Brownian rotations, can lead to negative values of P. The appearance

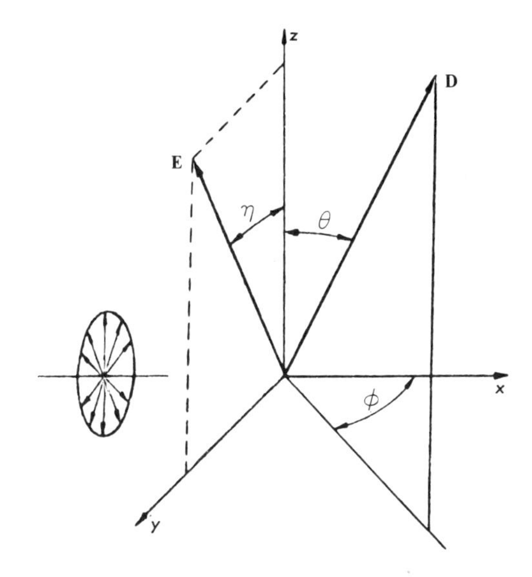

Fig. 1.10 Excitation of oscillators by unpolarised radiation

of negative values of the degree of polarisation can easily be explained if it is assumed that the absorption and emission of radiation is associated with different non-collinear dipoles in the molecule.

Consider a set of particles, each of which consists of an absorbing and an emitting linear oscillator, at an angle ξ to each other. The amplitudes of the oscillators in each particle will be assumed to be proportional to each other. Moreover, we shall suppose that the amplitudes are completely determined by external excitation and the spatial orientation of the absorbing dipole only. The emitting dipole will not interact with the incident radiation. If we now consider a narrow bundle of absorbing dipoles lying in the direction θ, φ the associated emitting oscillators will lie on a cone as shown in Fig. 1.11.

The total rate of absorption by all the oscillators whose dipole moments lie on the cone is proportional to the power absorbed by the dipoles lying along the axis of the cone and is therefore proportional to $\cos^2\theta$ if the incident light is plane polarised. The problem is to find the components of

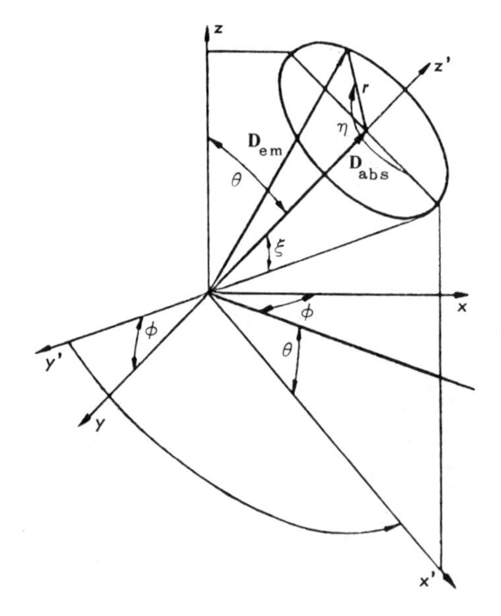

Fig. 1.11 Orientation of emission oscillators relative to a given direction of absorption oscillators

$\mathbf{D}_{em}$ along the z the x axes, average them with respect to η and then integrate with respect to θ and φ. In order to find the components D^z_{em} and D^x_{em} it is convenient to use an auxiliary system of coordinates whose z' axis is parallel to $\mathbf{D}_{abs}$, whose y' axis lies in the yx-plane and whose x' axis is in the plane $z\,O\,\mathbf{D}_{abs}$. We then have

$$\begin{aligned} D^z_{em} &= D_{x'}\cos(x'z) + D_{y'}\cos(y'z) + D_{z'}\cos(z'z) \\ &= D_{em}(-\sin\xi\cos\eta\sin\theta + \cos\xi\cos\theta) \\ D^x_{em} &= D_{x'}\cos(x'x) + D_{y'}\cos(y'x) + D_{z'}\cos(z'x) \\ &= D_{em}(\sin\xi\cos\eta\cos\theta\cos\varphi + \sin\xi\sin\eta\sin\varphi + \cos\xi\sin\theta\cos\varphi) \end{aligned} \tag{3.56}$$

from which it is easy to show that the luminescence of all the dipoles, which is polarised along the z- and x-axes and propagates along the y axis is given by

$$\begin{aligned} I_z(\Omega) &= C|E|^2\cos^2\theta\,\overline{\cos^2(\mathbf{D}_{em}z)}^{\eta} \\ &= C|E|^2\cos^2\theta\left(\cos^2\theta\cos^2\xi + \frac{1}{2}\sin^2\theta\sin^2\xi\right) \\ I_x(\Omega) &= C|E|^2\cos^2\theta\,\overline{\cos^2(\mathbf{D}_{em}x)}^{\eta} \\ &= C|E|^2\cos^2\theta\left(\frac{1}{2}\cos^2\theta\cos^2\varphi\sin^2\xi\right. \\ &\quad \left. + \frac{1}{2}\sin^2\varphi\sin^2\xi + \sin^2\theta\cos^2\varphi\cos^2\xi\right) \end{aligned} \tag{3.57}$$

where the bar over the cosines represents averages with respect to η. If we now integrate with respect to θ and φ we obtain

$$\begin{aligned} I_z &= \frac{4\pi}{5}C|E|^2\left(\cos^2\xi + \frac{1}{3}\sin^2\xi\right) \\ I_x &= \frac{4\pi}{5}C|E|^2\left(\frac{1}{3}\cos^2\xi + \frac{2}{3}\sin^2\xi\right) \end{aligned} \tag{3.58}$$

which corresponds to

$$P = \frac{2 - 3\sin^2\xi}{4 - \sin^2\xi} \tag{3.59}$$

Levshin's formula (3.59) gives the polarisation of luminescence as a function of the angle between the absorption and emission dipoles, and is one of the fundamental formulae in the classical theory of polarised luminescence. According to this formula, the degree of polarisation should vary between 1/2 and -1/3 when the angle ξ is increased from 0 to 90°. It will be shown in Chapter 8 that the formula given by (3.59) is the limiting case of a more general quantum-mechanical expression.

Polarisation of luminescence due to plane and three-dimensional oscillators [6]

So far we have discussed luminescence in terms of the linear harmonic model of a molecule. This model is based on two assumptions: (1) that the dipole moment of the molecule induced by the external field is linearly related to the field strength, and (2) that the incident radiation excites one-dimensional oscillations along a certain axis in the molecule.

The first assumption is satisfied within very wide limits. The determination of these limits is a problem in the quantum theory of radiation and will be considered later. The second hypothesis is much more specialised. In general, the effect of the incident radiation will be to excite three-dimensional motion of the molecule, which can be represented as a superposition of three mutually perpendicular oscillations. In point of fact, scattering experiments have shown that the polarisability of molecules is in general a tensor quantity, and therefore the electric vector of a plane-polarised wave excites oscillations along all three axes (see (3.10)). This is due to the interaction between all the charges in the molecule.

The tensorial nature of polarisability can be seen not only in scattering, but also in luminescence. The polarisation of luminescence must occasionally be interpreted by assuming that two or three mutually perpendicular oscillators in each molecule are responsible for the emission of luminescence [7]. The set of such dipoles in a particular molecule is usually referred to as a plane or a three-dimensional oscillator.

The theory of polarised luminescence of three-dimensional oscillators can be constructed by analogy with the theory of scattering. It is only necessary that we introduce the single assumption (though somewhat artificial in classical theory) that absorption and emission are separated by intermediate processes which disrupt the relation between the phases of the incident and emitted waves.

In general, a luminescing molecule must be represented by a model consisting of three perpendicular oscillators, i.e. two spatial dipoles, one of which is responsible for absorption and the other for emission. In other words, for a given density of incident radiation, the rate of absorption by a molecule is determined by the properties and the orientation of the absorption dipole only. This dipole must therefore transmit its energy in some way to the oscillator responsible for emission. The polarisation of luminescence is fully determined by the properties and orientation of the oscillator responsible for emission and does not directly depend on the external radiation. Let the absorption oscillators be represented by $\mathbf{A}_1$, $\mathbf{A}_2$, $\mathbf{A}_3$ and the emission oscillators by $\mathbf{B}_1$, $\mathbf{B}_2$, $\mathbf{B}_3$. The first set of vectors is chosen so that the ratios $\mathbf{A}_1 : \mathbf{A}_2 : \mathbf{A}_3$ are equal to the ratios of the principle polarisabilities of the molecule $\alpha_1 : \alpha_2 : \alpha_3$. The proportion of luminescence with a given direction of the electric field is determined by the quantities B_1, B_2 and B_3. It will be convenient to use the notation

$$\begin{aligned} &\cos(x, \mathbf{A}_i) = a_{ix}, \quad \cos(z, \mathbf{A}_i) = a_{iz} \\ &\cos(y, \mathbf{A}_i) = a_{iy}, \quad \cos(\mathbf{B}_j\ \mathbf{A}_i) = b_{ji} \end{aligned} \tag{3.60}$$

Suppose that isotropically oriented molecules are illuminated by external radiation in which the electric field is parallel to the z axis ($E = E_z$), whilst the luminescence is observed along the y axis. The rate of absorption by molecules with given orientation of the absorbing dipoles is given by

$$\begin{aligned} W_{\text{abs}} &= C\,|E|^2\,[A_1^2 a_{1z}^2 + A_2^2 a_{2z}^2 + A_3^2 a_{3z}^2] \\ &= C\,|E|^2 \sum_i A_i^2\, a_{iz}^2 \end{aligned} \tag{3.61}$$

By analogy with the above discussion, the z- and x-components of the intensity of luminescence may be written in the form

$$\begin{aligned} I_z(\Omega) &= CW_{\text{abs}}\,[B_1^2\cos^2(B_1 z) + B_2^2\cos^2(B_z z) \\ &+ B_3^2\cos^2(B_3 z)] = C'\,|E|^2 \sum_i A_i^2\, a_{iz}^2 \sum_j B_j^2 \cos^2(B_j\, z) \end{aligned} \tag{3.62}$$

$$I_x(\Omega) = CW_{\text{abs}}\,[B_1^2 \cos^2(\mathbf{B}_1 x) + B_2^2 \cos^2(\mathbf{B}_2 x) + B_3^2 \cos^2(\mathbf{B}_3 x)]$$

$$= C' \,|E|^2 \sum_i A_i^2\, a_{iz}^2 \sum_j B_j^2 \cos^2(\mathbf{B}_j\, x) \tag{3.63}$$

where C' is a coefficient of proportionality. Since

$$\cos^2(\mathbf{B}_j\, z) = \left(\sum_k b_{jk}\, a_{kz}\right)^2 = \sum_k \sum_l b_{jk}\, a_{kz}\, b_{jl}\, a_{lz}$$

$$\cos^2(\mathbf{B}_j\, x) = \left(\sum_k b_{jk}\, a_{kx}\right)^2 = \sum_k \sum_l b_{jk}\, a_{kx}\, b_{jl}\, a_{lx}$$

it follows from (3.62) and (3.63) that

$$I_z(\Omega) = C' \,|E|^2 \sum_i \sum_j \sum_k \sum_l A_i^2 \times a_{iz}^2\, B_j^2 b_{jk}\, a_{kz}\, b_{jl}\, a_{lz} \tag{3.64}$$

$$I_x(\Omega) = C' \,|E|^2 \sum_i \sum_j \sum_k \sum_l A_i^2 \times a_{iz}^2\, B_j^2\, b_{jk}\, a_{kx}\, b_{jl}\, a_{lx} \tag{3.65}$$

To obtain the total intensities, the expressions given by (3.64) and (3.65) must be integrated over all angles. This can easily be done if the orientation of a molecule is specified by the Euler angles (Fig. 1.12). It follows directly from the geometrical construction that

$$\begin{aligned}
a_{1z} &= \sin\psi \sin\theta, & a_{1x} &= \cos\varphi \cos\psi - \sin\varphi \sin\psi \cos\theta \\
a_{2z} &= \cos\psi \sin\theta, & a_{2x} &= \cos\varphi \sin\psi + \sin\psi \cos\psi \cos\theta \\
a_{3z} &= \cos\theta, & a_{3x} &= \sin\varphi \sin\theta
\end{aligned} \tag{3.66}$$

and hence

$$I_z = \int_0^{2\pi} d\varphi \int_0^{2\pi} d\psi \int_0^{\pi} I_z(\Omega) \sin\theta\, d\theta$$

$$= 8\pi^2 C' \,|E|^2 \left[\frac{1}{5} \sum_i \sum_j A_i^2\, B_j^2\, b_{ji}^2 + \frac{1}{15} \sum_i \sum_j \sum_{k \neq i} A_i^2\, B_j^2\, b_{jk}^2 \right] \tag{3.67}$$

$$I_x = \int_0^{2\pi} d\varphi \int_0^{2\pi} d\psi \int_0^{\pi} I_x(\Omega) \sin\theta \, d\theta$$
$$= 8\pi^2 C' |E|^2 \left[\frac{1}{15} \sum_i \sum_j A_i^2 B_j^2 b_{ji}^2 + \frac{2}{15} \sum_i \sum_j \sum_{k \neq i} A_i^2 B_j^2 b_{jk}^2 \right] \quad (3.68)$$

which corresponds to the following degree of polarisation:

$$P = \frac{2 \sum_i \sum_j A_i^2 B_j^2 b_{ji}^2 - \sum_i \sum_j \sum_{k \neq i} A_i^2 B_j^2 b_{jk}^2}{4 \sum_i \sum_j A_i^2 B_j^2 b_{ji}^2 + 3 \sum_i \sum_j \sum_{k \neq i} A_i^2 B_j^2 b_{ik}^2} \quad (3.69)$$

This expression may be used to find the degree of polarisation P_e, of luminescence excited by natural radiation, since P and P_e are related by the simple formula given by (3.55). The expression given by (3.69) is the most general classical formula for the degree of polarisation. It is valid as long as there is a linear relationship between the induced

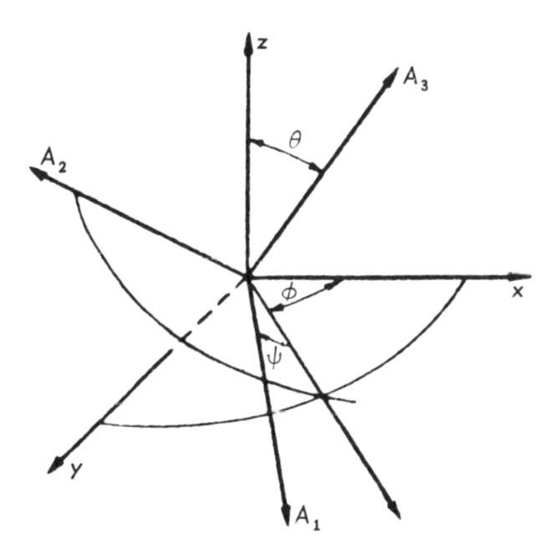

Fig. 1.12 Euler angles

dipole moment and the electric field in the incident wave, i.e. within very wide limits. Let us consider (3.69) in a number of special cases.

1. Firstly, let us suppose that the absorbing oscillator or the emitting oscillator are completely isotropic ($A_1 = A_2 = A_3$ or $B_1 = B_2 = B_3$). We can then take A_i or B_j outside the summation sign. The numerator in (3.69) will be zero, which

represents complete depolarisation of the luminescence. This is a natural conclusion, since the assumptions from which it is derived ensure that either the molecules are excited in the same way whatever their orientation, or they emit isotropic luminescence independently of the degree of excitation.

2. Secondly, suppose that absorption is due to a linear dipole ($A_1 \neq 0$, $A_2 = A_3 = 0$), whilst the emission is due to a plane oscillator ($B_1 = B_2 \neq 0, B_3 = 0$). We can now easily transform (3.69) into the simpler form

$$\frac{3\sin^2\beta - 2}{\sin^2\beta + 6} \tag{3.70}$$

where β is the angle between the absorbing dipole and the perpendicular to the plane of the emitting oscillators. As β increases from 0° to 90°, the degree of the polarisation varies within the limits $-1/3 \leqslant P \leqslant 1/7$. An analogous result is obtained if a plane oscillator is responsible for absorption ($A_1 = A_2 \neq 0$, $A_3 = 0$), whilst the linear dipole is responsible for emission ($B_1 \neq 0$, $B_2 = B_3 = 0$).

3. If the dipoles responsible for absorption and for emission are completely anisotropic i.e. are linear oscillators, ($A_1 \neq 0, B_1 \neq 0, A_2 = A_3 = B_2 = B_3 = 0$), we have from (3.69)

$$\begin{aligned} P &= \frac{2b_{11}^2 - (b_{12}^2 + b_{13}^2)}{4b_{11}^2 + 3(b_{12}^2 + b_{13}^2)} = \frac{3b_{11}^2 - 1}{b_{11}^2 + 3} \\ &= \frac{3\cos^2\xi - 1}{\cos^2\xi + 3} = \frac{2 - 3\sin^2\xi}{4 - \sin^2\xi} \end{aligned} \tag{3.71}$$

where we have taken into account $b_{11}^2 + b_{12}^2 + b_{13}^2 = 1$. As was to be expected, (3.71) is identical with equation (3.59) above.

When the absorption and emission oscillators are completely anisotropic, the variation in the polarisation is the maximum possible, namely $-1/3 \leqslant P \leqslant 1/2$. Any degree of polarisation observed experimentally may therefore be attributed to the luminescence of linear harmonic oscillators. However, it does not follow from this that the optical properties of molecules will always be best represented by linear dipoles.

It should be noted that the dependence of the degree of polarisation of luminescence and of scattered light on the anisotropy of the molecules is quite different. The degree of polarisation of scattered radiation decreases with increasing anisotropy, whilst the degree of polarisation of luminescence increases. Moreover, the two ranges never overlap, and it is only in the limiting case of completely anisotropic molecules that the maximum degree of polarisation of luminescence is equal to the minimum value of P for scattered light, namely 1/2.

Luminescence method of determining the nature of an elementary source

The nature of elementary sources may be determined by studying the emission spectra under normal conditions or in magnetic fields, by establishing the duration of luminescence afterglow, by observing interference patterns and so on. However, these methods are often inadequate for complicated systems.

It was shown in Section 1.1 that the most clearly defined characteristic of an elementary source is the angular distribution of the radiation emitted by it. This distribution is characteristic not only of the emission of individual particles, but also of large sets of particles. Even when the elementary sources are distributed isotropically in space, the anisotropy of the exciting radiation leads to a selective excitation which depends on the particle orientation. The anisotropy in the angular distribution of excited particles depends on the nature of the elementary sources, and is clearly reflected in the polarisation of the luminescence. This is the basis of the polarisation method for the determination of the nature of elementary sources which was put forward by Vavilov [8]. Vavilov also derived the polarisation diagrams for practically all the most important examples when the absorbing and emitting elementary sources are dipoles or quadrupoles. In essence his method involves the measurement of the polarisation of luminescence as a function of η and φ, where η is the angle between the z axis, and the electric field in the exciting radiation, whilst φ defines the direction of observation in the xy-plane (Fig. 1.13). By comparing the experimental curve with the calculated polarisation diagrams it is possible to establish the nature of the elementary source.

Let us consider a simple example of the construction of polarisation diagrams. Above, we calculated the polarisation

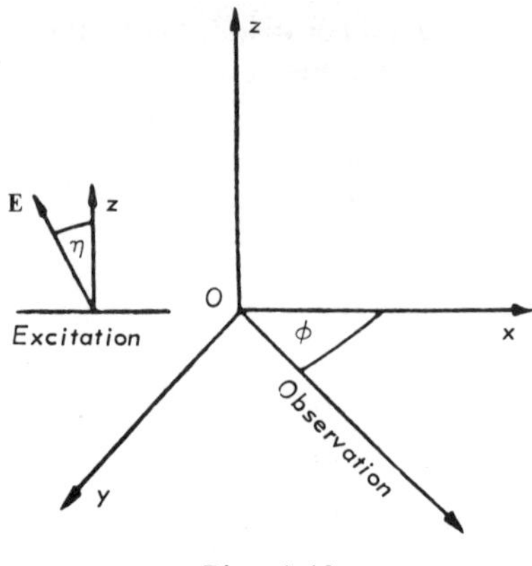

Fig. 1.13

of the luminescence emitted by a set of harmonic oscillators excited by plane-polarised light. It was assumed that $\eta=0$ and $\varphi=90°$. In general the degree of polarisation is given by

$$P=\frac{\cos^2\eta-\sin^2\eta\cos^2\varphi}{2-\sin^2\eta\sin^2\varphi} \tag{3.72}$$

If the electric field in the exciting radiation is parallel to the z axis ($\eta=0$), then (3.72) yields $P={}^1/_2$ for all φ. This was to be expected, since in the present example the z axis is the symmetry axis of the angular distribution of the excited particles. If the system is observed along the y axis ($\varphi=90°$) and the angle η is increased from 0 to 90°, the polarisation changes from ${}^1/_2$ to 0 in accordance with the formula

$$P=\frac{\cos^2\eta}{2-\sin^2\eta} \tag{3.72a}$$

If, instead of a set of dipoles, we take a set of quadrupoles, the polarisation of the luminescence will be given by a completely different formula, namely,

$$P=\frac{2\cos^2\eta\cos^2\varphi-\sin^2\eta(1+\cos 4\varphi)}{2+2\cos^2\varphi\cos^2\eta+(1+\cos 4\varphi)\sin^2\eta} \tag{3.73}$$

In distinction to (3.72), the latter expression shows that P is a function of φ even when $\eta=0$. The polarisation varies from ${}^1/_2$ to 0 as φ varies from 0 to $\pi/2$. Other examples may be found in the literature [9]. A number of polarisation diagrams are shown in Fig. 1.14. Polarisation diagrams for non-coincident absorbing and emitting dipoles have been given by Gurinovich and Sevchenko [10].

Effect of Brownian rotations on the polarisation of luminescence

It was assumed above in the calculation of the polarisation of luminescence that the molecules had no rotational degrees of freedom. The assumption is not valid especially for vapours and non-viscous solutions. However, Brownian rotations are possible even in the solid phase and under certain conditions may have an important effect on the anisotropy of the emitted radiation.

To begin with, let us consider rotational Brownian motion of a spherical particle with an attached linear harmonic oscillator. If at the initial instant of time the direction of the dipole moment of the oscillator **D** is specified by the angles $\theta_0 = \varphi_0 = 0$, then at some subsequent time there is a definite probability that **D** will have a different direction. Since Brownian motion is isotropic, this probability is symmetrical with respect to the initial position of the vector **D**, i.e. its value is independent of φ. Let us denote this function by $f(\theta, t)$. It is evident that the angular distribution function for the vector **D** should satisfy the normalisation condition

$$\int_0^{2\pi} d\varphi \int_0^{\pi} f(\theta, t)\sin\theta\, d\theta = 2\pi \int_0^{2\pi} f(\theta, t)\sin\theta\, d\theta = 1 \qquad (3.74)$$

The rotational Brownian motion is, by analogy with the translational motion, described by the diffusion equation

$$\frac{\partial f}{\partial t} = D_{rot}\Delta(\theta, \varphi) f \qquad (3.75)$$

where D_{rot} is the diffusion coefficient and $\Delta(\theta, \varphi)$ is the Laplace operator

$$\Delta(\theta; \varphi) = \frac{1}{\sin\theta}\frac{\partial}{\partial\theta}\left(\sin\theta\frac{\partial}{\partial\theta}\right) + \frac{1}{\sin^2\theta}\frac{\partial^2}{\partial\varphi^2} \qquad (3.76)$$

For spherical particles, the diffusion coefficient D_{rot} is

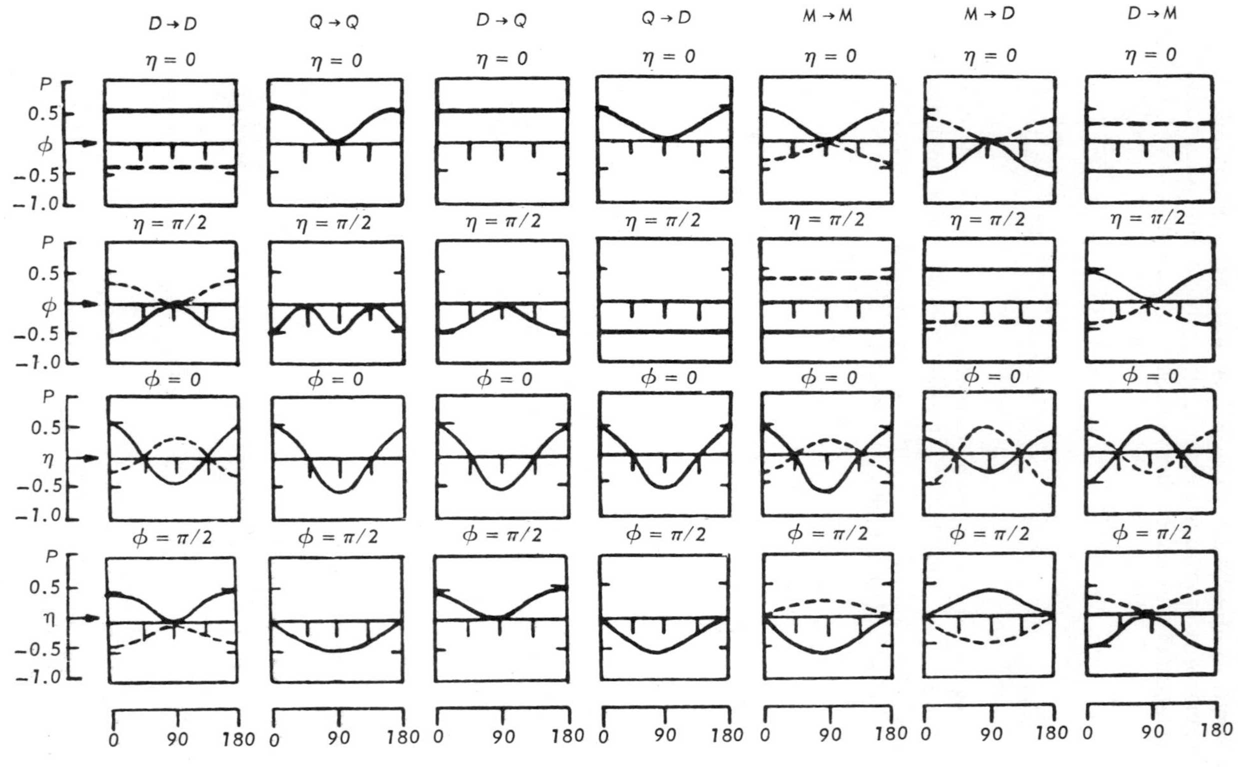

Fig. 1.14 Polarisation diagrams

related to Boltzmann's constant, the temperature, the volume V of a particle and the viscosity η of the solvent by the equation [11]

$$D_{rot} = \frac{1}{6} \frac{kT}{V\eta} \tag{3.77}$$

Since f is independent of φ it follows from (3.75) that

$$\frac{\partial f}{\partial t} = D_{rot} \left[\frac{1}{\sin\theta} \frac{\partial}{\partial\theta} \left(\sin\theta \frac{\partial f}{\partial\theta} \right) \right] \tag{3.78}$$

In the ensuing analysis we shall need the mean

$$v(t) = \overline{\cos^2\theta} = 2\pi \int_0^\pi \cos^2\varphi\, f(\theta, t) \sin\theta\, d\theta \tag{3.79}$$

which may be evaluated without a knowledge of the explicit form of the function f. To find the mean, let us differentiate (3.79) under the integral sign with respect to time:

$$\frac{d}{dt} v(t) = 2\pi \int_0^\pi \cos^2\theta \frac{\partial f}{\partial t} \sin\theta\, d\theta \tag{3.80}$$

If we now substitute for $\frac{\partial f}{\partial t}$ from (3.75) into this expression and integrate twice by parts, we have

$$\frac{d}{dt} v(t) = 2 D_{rot} - 6 D_{rot} v(t) \tag{3.81}$$

The solution of this equation, subject to the initial condition $v(0) = 1$, is

$$v(t) = \overline{\cos^2\theta} = \frac{1}{3} \left(1 + 2e^{-6D_{rot}t}\right) \tag{3.82}$$

When $D_{rot}t \ll 1$, it follows from (3.82) that

$$\overline{\theta^2} = 2D_{rot}t$$

which is analogous to Einstein's law for translational one-dimensional motion:

$$\overline{\Delta x^2} = 2D\Delta t \tag{3.83}$$

where $\overline{\Delta x^2}$ is the mean value of the square of the displacement in time Δt, and D is the diffusion coefficient for translational Brownian motion.

The orientation of a Brownian particle may be specified by nine direction cosines b_{il}, which determine at the time t the position of the three rectangular coordinate axes of the sphere relative to the initial position of these coordinates. It is evident that $\overline{b^2_{ii}} = \overline{b^2_{jj}} = \overline{b^2_{kk}} = \overline{\cos^2 \theta}$. Since in isotropic Brownian motion any two of the axes are equivalent with respect to the third, it follows that $\overline{b^2_{ij}} = \overline{b^2_{ik}}$.

Since

$$b^2_{ii} + b^2_{ij} + b^2_{ik} = 1$$

we have, using (3.82)

$$\overline{b^2_{ii}} = \frac{1}{3} \left(1 + 2 e^{-6D_{rot} t}\right)$$
$$\overline{b^2_{ik}} = \frac{1}{3} \left(1 - e^{-6D_{rot} t}\right) \qquad (3.84)$$

The last two expressions reflect the chief properties of rotational Brownian motion for a spherical particle and may be used to determine the polarisation of luminescence.

Let us suppose that up to the time $t_0 = 0$ the particles were illuminated by plane-polarised light ($E = E_z$) propagating along the x axis, and that luminiscence is observed along the y axis. The excitation terminates at time $t = 0$ and at the same time the particles begin to undergo rotations. The components of luminescence emitted at time t by particles which at the initial instant of time had a definite orientation in space are related to the components of luminescence at $t=0$ by the expressions

$$I_z(\Omega, t) = \alpha(t) \left[I_{x0}(\Omega) \overline{b^2_{xz}} + I_{y0}(\Omega) \overline{b^2_{yz}} + I_{z0}(\Omega) \overline{b^2_{zz}}\right]$$
$$I_x(\Omega, t) = \alpha(t) \left[I_{x0}(\Omega) \overline{b^2_{xx}} + I_{y0}(\Omega) \overline{b^2_{yx}} + I_{z0}(\Omega) \overline{b^2_{zx}}\right] \qquad (3.85)$$

where $\alpha(t)$ is the proportionality factor which determines the decay of the afterglow. In order to obtain the z- and x-components of the luminescence of all the particles, we must integrate (3.85) over all the initial particle orientations.

The result is

$$I_z(t) = \alpha(t)\,[I_{x0}\,\overline{b^2_{xz}} + I_{y0}\,\overline{b^2_{yz}} + I_{z0}\,\overline{b^2_{zz}}]$$
$$I_x(t) = \alpha(t)\,[I_{x0}\,\overline{b^2_{xx}} + I_{y0}\,\overline{b^2_{yx}} + I_{z0}\,\overline{b^2_{zx}}] \tag{3.86}$$

where

$$I_{j0} = \int_{\Omega} I_{j0}(\Omega)\,d\Omega$$

Since, on excitation by plane-polarised light, $\overline{I_{yo} = I_{xo}}$, it follows from (3.84) and (3.86) that

$$I_z(t) = \alpha(t)\frac{1}{3}\,[2(1 - e^{-6D_{rot}t})\,I_{x0} + (1 + 2e^{-6D_{rot}t})\,I_{z0}] \tag{3.87}$$

$$I_x(t) = \alpha(t)\frac{1}{3}\,[(2 + e^{-6D_{rot}t})\,I_{x0} + (1 - e^{-6D_{rot}t})\,I_{z0}] \tag{3.88}$$

If P_0 is the polarisation at the initial instant of time, then according to (3.87) and (3.88), it is given at time t by the expression

$$P(t) = \frac{6}{6 + 2(e^{-6D_{rot}t} - 1)\,P_0}\,P_0 e^{-6D_{rot}t} \tag{3.89}$$

The factor

$$\frac{6}{6 + 2(e^{-6D_{rot}t} - 1)\,P_0}$$

varies only slightly and is always very nearly equal to unity. For $P_0 = 1/2$ it increases from 1 to 1.2 as t increases from 0 to infinity. In very approximate calculations it may therefore be assumed that the polarisation of luminescence afterglow falls off in accordance with the exponential law

$$P(t) \approx P_0\,e^{-6D_{rot}t} \tag{3.90}$$

although strictly speaking the decrease should occur more slowly.

In order to determine the polarisation of luminescence arising as a result of a complete de-excitation of the dipoles, we must find the explicit form of the function $\alpha(t)$. From

(2.10), the energy of the oscillator decreases in accordance with the expressions

$$E = E_0 e^{-2\gamma t} = E_0 e^{-t/\tau} \qquad (3.91)$$

Hence the intensity of the emitted radiation is proportional to

$$I \sim \left(\frac{dE}{dt}\right) \sim E_0 \frac{1}{\tau} e^{-t/\tau}.$$

Therefore, for harmonic oscillators

$$\alpha(t) = C \frac{1}{\tau} e^{-t/\tau} \qquad (3.92)$$

where $C = \text{const.}$

Substituting (3.92) into (3.87) and (3.88) and integrating with respect to time between 0 and infinity, we have

$$I_z = \int_0^\infty I_z(t)dt = C \frac{1}{1 + 6 D_{rot}\tau} \times [4 D_{rot}\tau I_{x0} + (1 + 2D_{rot}\tau) I_{z0}] \qquad (3.93)$$

$$I_x = \int_0^\infty I_x(t)dt = C \frac{1}{1 + 6 D_{rot}\tau} \times [(1 + 4 D_{rot}\tau) I_{x0} + 2 D_{rot}\tau I_{z0}] \qquad (3.94)$$

If we denote by P_0 the polarisation of luminescence in the absence of Brownian rotations, then in view of (3.77) we have from the last two expressions

$$\left(\frac{1}{P} - \frac{1}{3}\right) = \left(\frac{1}{P_0} - \frac{1}{3}\right)\left(1 + \frac{kT}{V\eta}\tau\right) \qquad (3.95)$$

or

$$\frac{1}{P} = \frac{1}{P_0} + \left(\frac{1}{P_0} - \frac{1}{3}\right)\frac{kT}{V\eta}\tau \qquad (3.96)$$

This is the well-known formula of Levshin and Perrin, which gives the dependence of polarisation on the temperature and viscosity of the solution, the volume of the particles and the mean duration of luminescence afterglow. This formula has been verified experimentally for many systems. It is frequently used for the experimental determination of the mean duration of afterglow. In order to determine τ one measures $1/P$ as a function of T/η. The slope of the straight line then yields the magnitude of τ. It is also possible to find V in a similar way if the other quantities entering into (3.96) are known.

In conclusion, we note that the Levshin-Perrin formula was obtained for the de-excitation process, and therefore its use under steady-state conditions of illumination has not yet been justified.

4. THE HAMILTONIAN FORM OF THE FIELD EQUATIONS

Before we proceed to the fundamentals of the quantum theory of emission of radiation, we must consider the Hamiltonian form of the field equations. In quantum electrodynamics the medium and the radiation interacting with it are regarded as a single system. Since the quantum mechanical operators are constructed on the basis of the classical Hamiltonian function, we must commence by expressing the equations for the electromagnetic field as equations for the Hamiltonian function.

In the classical theory of emission there are two basic forms of the field equations. The first of these are the well-known Maxwell-Lorentz equations

$$\begin{aligned} &\operatorname{curl} \mathbf{E} + \frac{1}{c}\dot{\mathbf{H}} = 0, && \operatorname{div} \mathbf{H} = 0 \\ &\operatorname{curl} \mathbf{H} - \frac{1}{c}\dot{\mathbf{E}} = \frac{4\pi}{c}\rho\mathbf{v} && \operatorname{div} \mathbf{E} = 4\pi\rho \end{aligned} \tag{4.1}$$

where ρ and $\mathbf{v}$ are the density and velocity of the charges respectively. The vector and scalar potentials $\mathbf{A}$ and φ which are related to the field intensities $\mathbf{E}$ and $\mathbf{H}$ by the expressions given by (1.2) can be used to rewrite (4.1) in the form

$$\begin{aligned} &\frac{1}{c^2}\ddot{\mathbf{A}} - \nabla^2\mathbf{A} + \operatorname{grad}\left(\operatorname{div}\mathbf{A} + \frac{1}{c}\dot{\varphi}\right) = \frac{4\pi}{c}\rho\mathbf{v} \\ &-\nabla^2\varphi - \frac{1}{c}\operatorname{div}\dot{\mathbf{A}} = 4\pi\rho \end{aligned} \tag{4.2}$$

The potentials $\mathbf{A}$ and φ are not unambiguously related to the field. There are various ways of choosing $\mathbf{A}$ and φ without changing $\mathbf{E}$ and $\mathbf{H}$. The invariance of $\mathbf{E}$ and $\mathbf{H}$ with respect to transformations of $\mathbf{A}$ and φ is referred to as guage invariance.

If in the space under consideration there are no charges, we may set $\varphi = 0$ and

$$\operatorname{div} \mathbf{A} = 0 \tag{4.3}$$

The set of field equations can then be reduced to the single differential equation

$$\nabla^2 \mathbf{A} - \frac{1}{c^2} \ddot{\mathbf{A}} = 0 \tag{4.4}$$

This equation is called the wave equation, or the d'Alembert equation.

In order to define the vector potential $\mathbf{A}$ in terms of canonical variables for all points in space and time, we would require an infinite number of such variables. Usually, $\mathbf{A}$ is defined within a restricted region, for example a cube of side L. The linear dimensions of the cube must be large in comparison with the radiating system so that the physical behaviour of the system is independent of L.

If the boundary conditions require that the potential $\mathbf{A}$ and its derivatives should be the same on opposite faces of the cube, the general solution of (4.4) may be written as the superposition of orthogonal waves

$$\mathbf{A} = \sum_{\lambda} q_{\lambda}(t)\, \mathbf{A}_{\lambda}(\mathbf{r}) \tag{4.5}$$

where $\mathbf{A}_{\lambda}(\mathbf{r})$ are functions which depend on the coordinates only, and $q_{\lambda}(t)$ are functions of time only. It follows from (4.4) that $q_{\lambda}(t)$ and $A_{\lambda}(\mathbf{r})$ should satisfy the equations

$$\nabla \mathbf{A}_{\lambda} + \frac{(2\pi\nu_{\lambda})^2}{c^2} \mathbf{A}_{\lambda} = 0 \tag{4.6}$$

$$\ddot{q}_{\lambda} + (2\pi\nu_{\lambda})^2 q_{\lambda} = 0 \tag{4.7}$$

where $\mathbf{A}_{\lambda}$ must be periodic within the cube and

$$\operatorname{div} \mathbf{A}_{\lambda} = 0 \tag{4.3a}$$

The solutions of (4.6) subject to (4.3a) may be sought in

the form

$$\mathbf{A}_\lambda = \sqrt{\frac{8\pi c^2}{V}}\,\mathbf{e}_\lambda \cos(\boldsymbol{\varkappa}\,\mathbf{r});\quad \mathbf{A}_\lambda = \sqrt{\frac{8\pi c^2}{V}}\,\mathbf{e}_\lambda \sin(\boldsymbol{\varkappa}_\lambda\,\mathbf{r}) \quad (4.8)$$

where $\boldsymbol{\varkappa}_\lambda$ and $\mathbf{e}_\lambda$ are vectors respectively specifying the direction of propagation of the wave and its polarisation, $V = L^3$ is the volume of the cube, and $|\boldsymbol{\varkappa}_\lambda| = 2\pi\nu_\lambda/c$. In view of (4.3a), the vectors $\boldsymbol{\varkappa}_\lambda$ and $\mathbf{e}_\lambda$ are always perpendicular. The factors in front of the functions in (4.8) are chosen so as to satisfy the normalisation condition

$$\int_V (\mathbf{A}_\lambda \mathbf{A}_\mu) dV = 4\pi c^2 \delta_{\lambda\mu} \quad (4.9)$$

The requirement that $\mathbf{A}_\lambda$ should be periodic leads to the fact that the vectors $\boldsymbol{\varkappa}$ can assume only the discrete values

$$\varkappa_{\lambda x} = \frac{2\pi}{L} n_{\lambda x},\quad \varkappa_{\lambda y} = \frac{2\pi}{L} n_{\lambda y},\quad \varkappa_{\lambda z} = \frac{2\pi}{L} n_{\lambda z} \quad (4.10)$$

where $n_{\lambda x}$, $n_{\lambda y}$ and $n_{\lambda z}$ are positive or negative integers. To each wave with a given direction of $\boldsymbol{\varkappa}$ there corresponds two perpendicular polarisations $\mathbf{e}_{\lambda_1}$ and $\mathbf{e}_{\lambda_2}$, whose directions can be arbitrary.

Thus, by confining our attention to the field within a cube we can select an enumerable set of plane waves which can be represented by a complete closed system of orthonormal functions. Therefore, any electromagnetic wave $\mathbf{A}$ may be represented within a given volume by a linear combination of plane waves of the form (4.5). Since the behaviour of $\mathbf{A}_\lambda$ is specified and remains unaltered once the cube has been chosen, the field under consideration will be characterised by the amplitudes $q_\lambda(t)$. The field equations can in fact be reduced to equations of the form given by (4.7), which can easily be expressed in the canonical form. In point of fact, (4.7) is equivalent to Hamilton's equations

$$\frac{\partial H_\lambda}{\partial q_\lambda} = -\dot{p}_\lambda,\quad \frac{\partial H_\lambda}{\partial p_\lambda} = \dot{q}_\lambda = p_\lambda \quad (4.11)$$

where H_λ is the Hamiltonian for the harmonic oscillator and is given by

$$H_\lambda = \frac{1}{2}\left[p_\lambda^2 + (2\pi\nu_\lambda)^2 q_\lambda^2\right] \quad (4.12)$$

It follows that any field enclosed in the volume under consideration will now be described by an infinite number of canonical variables q_λ and p_λ and the total Hamiltonian

$$H = \sum_\lambda H_\lambda = \frac{1}{2}\sum_\lambda [p_\lambda^2 + (2\pi\nu_\lambda)^2 q_\lambda^2] \quad (4.13)$$

i.e. the electromagnetic field is now represented by a set of independent oscillators (plane waves).

In classical mechanics H_λ is equal to the oscillator energy. We shall now show that the total energy of the field is also equal to the sum of all the H_λ. Thus, the total energy is given by

$$U = \frac{1}{8\pi}\int_V (\mathbf{E}^2 + \mathbf{H}^2)\, dV \quad (4.14)$$

If we substitute into this expression $\mathbf{E}$ and $\mathbf{H}$ expressed in terms of $\mathbf{A}$ (see (1.2) with $\varphi = 0$), we have in view of (4.5)

$$U = \frac{1}{8\pi}\int\Big(\sum_\lambda p_\lambda \mathbf{A}_\lambda\Big)^2 dV + \frac{1}{8\pi}\int\Big(\sum_\lambda q_\lambda \operatorname{curl} \mathbf{A}_\lambda\Big)^2 dV \quad (4.15)$$

Since $\mathbf{A}_\lambda$ are orthonormal functions, it follows from (4.9) that

$$\int\Big(\sum_\lambda p_\lambda \mathbf{A}_\lambda\Big)^2 dV = \sum_\lambda p_\lambda^2 \int \mathbf{A}_\lambda^2\, dV = 4\pi c^2 \sum p_\lambda^2 \quad (4.16)$$

The integrals in the second term in (4.14) may be split into two parts as follows:

$$\int \operatorname{curl} \mathbf{A}_\lambda \operatorname{curl} \mathbf{A}_\mu\, dV = \oint [\mathbf{A}_\lambda \operatorname{curl} \mathbf{A}_\mu]_n\, ds + \int_V \mathbf{A}_\lambda \operatorname{curl}\operatorname{curl} \mathbf{A}_\mu dV \quad (4.17)$$

Since $\mathbf{A}_\lambda$ and $\mathbf{A}_\mu$ are periodic on the surface of the cube, the first term in (4.17) is zero, and since curl curl = grad div $- \nabla^2$, we have, in view of (4.6) and (4.3a)

$$\int_V \operatorname{curl} \mathbf{A}_\lambda \operatorname{curl} \mathbf{A}_\mu dV = \frac{(2\pi\nu_\lambda)^2}{c^2}\int (\mathbf{A}_\lambda \mathbf{A}_\mu) dV = 16\pi^3\nu_\lambda^2\delta_{\lambda\mu} \quad (4.18)$$

Evaluation of (4.15) then yields

$$U = \frac{1}{2}\sum_\lambda [p_\lambda^2 + (2\pi\nu_\lambda)^2 q_\lambda^2] = \sum_\lambda H_\lambda \tag{4.19}$$

In quantum theory the field is usually described not by trigonometric functions but by complex exponential functions. In terms of complex functions the real vector potential **A** can be written in the form

$$\mathbf{A} = \sum_\lambda [q_\lambda(t)\mathbf{A}_\lambda + q_\lambda^*(t)\mathbf{A}_\lambda^*] \tag{4.20}$$

where

$$\mathbf{A}_\lambda = \sqrt{\frac{4\pi c^2}{V}}\, \mathbf{e}_\lambda e^{i(\boldsymbol{\varkappa}, \mathbf{r})}$$
$$q_\lambda(t) = q_\lambda e^{-2\pi i \nu_\lambda t} \tag{4.21}$$

As before, the quantities $\mathbf{A}_\lambda$ are orthonormal functions satisfying the conditions given by (4.9). However, on this definition of **A**, the quantities q_λ are no longer canonical variables. One must therefore introduce the new canonical variables

$$Q_\lambda = q_\lambda + q_\lambda^*, \quad P_\lambda = -2\pi i \nu_\lambda (q_\lambda - q_\lambda^*) = \dot{Q}_\lambda \tag{4.22}$$

The field equation given by (4.7), which is valid both for q_λ and for their complex conjugates, can be derived from the Hamiltonian

$$H_\lambda = 8\pi^2\nu_\lambda^2 q_\lambda q_\lambda^* = \frac{1}{2}[P_\lambda^2 + (2\pi\nu_\lambda)^2 Q_\lambda^2] \tag{4.23}$$

It is quite easy to show that as in the case discussed above the field energy is equal to $\sum_\lambda H_\lambda$.

Let us determine now the number of field oscillators with given planes of polarisation, direction of propagation (within a solid angle $d\Omega$), and frequency (in the range between ν and $\nu + d\nu$) per unit volume of the cube. This quantity is called the density of states, and will be denoted by $\rho(\nu, \Omega)$. As can be seen from (4.21), the frequency of oscillations and the direction of propagation of a plane wave oscillator is fully specified by the wave vector $\boldsymbol{\varkappa}_\lambda$ which is defined by the three integers n_x, n_y, n_z. In the space of these numbers the size of a particular volume is equal to the number of possible values of $\varkappa_\lambda$, since each unit cell of the volume can be associated with a particular combination of n_x, n_y and n_z. The required number of oscillators is therefore equal to the volume element in this space, which is given by

$$\rho_\lambda(\nu, \Omega)\, d\Omega\, V = n^2 dn \sin\theta\, d\theta\, d\varphi = n^2 dn\, d\Omega \tag{4.24}$$

where $n^2 = n_x^2 + n_y^2 + n_z^2$ is the square of the radius vector. According to (4.10)

$$n^2 dn = \left(\frac{L}{2\pi}\right)^3 \varkappa^2 d\varkappa = \frac{V}{c^3} \nu^2 d\nu$$

and therefore

$$\rho(\nu, \Omega) = \frac{\nu^2}{c^3} \tag{4.25}$$

If the radiation enclosed in the cube is isotropic (Planck radiation), then integration over all directions of propagation and summation over the two polarisations yields

$$\rho(\nu) = 8\pi\rho(\nu, \Omega) = \frac{8\pi\nu^2}{c^3} \tag{4.26}$$

Multiplying this expression by the mean energy kT, associated, on the classical theory, with each degree of freedom, we obtain the Rayleigh-Jeans formula

$$u(\nu) = \frac{8\pi\nu^2}{c^3} kT \tag{4.27}$$

which is a special case of the quantum formula of Planck.

Once the radiation has been represented by a set of oscillators, its quantisation can easily be effected by analogy with the quantisation of ordinary mechanical oscillators.

2

Quantum Theory of Absorption and Emission of Light

5. ENERGY LEVELS

Energy levels and spectra

According to quantum theory, the possible values of the energy of a system are determined by its internal properties, i.e. by its structure.

Energy spectra can be divided into two main groups, namely, continuous and discrete spectra. Complicated condensed systems and even complicated molecules have a continuous spectrum of energy levels. Such systems can have any energy, and this brings them close to systems obeying purely classical laws. As a rule, isolated atoms and simple molecules have discrete energy-level spectra, and this determines their specific properties. However, there are no systems in nature for which the spectrum of possible energy values is completely discrete. Even in the simple case of the hydrogen atom, the energy levels are discrete only within a certain range of values. Beyond this range the energy spectrum is continuous. This corresponds to a real physical process, namely the removal of the electron from the atom and its motion relative to the nucleus in the absence of a stable coupling between them.

A similar classification of energy levels can be used for molecules. If an energy level lies below the dissociation limit of the molecule, it is discrete. If, on the other hand, the energy level lies above the dissociation energy, it belongs to the continuum.

Strictly discrete or strictly continuous spectra are only limiting cases. There are no absolutely discrete energy levels in nature. They would be encountered only in isolated systems, and even then only if the natural level widths were ignored. Energy levels usually have a finite width. A detailed discussion of the width of energy levels will be given in Section 10. The level width is sometimes very large because of a strong interaction between the individual degrees of freedom. The energy spectrum then exhibits a band structure, and the system can assume only those energies which lie within the bands.

The specific character of the energy spectrum is reflected in the absorption and emission spectra. The basic relation in spectroscopy is the Bohr formula

$$\nu_{ij} = \frac{E_i - E_j}{h} \tag{5.1}$$

which gives the frequencies of electromagnetic waves emitted or absorbed by a given system.

Electromagnetic waves are not emitted and absorbed continuously (as predicted by classical electrodynamics), but in discontinuous steps when the system undergoes transitions between the available energy levels. If at least one of the two levels E_i or E_j in (5.1) lie in the continuum, the emission and absorption spectra are also continuous, i.e. the system can absorb or emit electromagnetic waves of all frequencies. If both E_i and E_j lie within the discrete spectrum, the corresponding absorption and emission spectra are also discrete, and the system can absorb or emit electromagnetic waves at certain frequencies only, i.e. in more or less narrow spectral lines. Such spectra are called line spectra. As a rule, line spectra are emitted by atoms and simple molecules. The individual lines in molecular spectra appear in groups which are referred to as bands.

If all the energy levels were simply discrete, the spectral lines would be infinitely narrow. In reality, all the energy levels have a finite width, and this is reflected in the broadening of spectral lines. If the line width exceeds the distance between individual lines, the two lines blend together. This phenomenon is encountered in heavy diatomic molecules; it is common in complicated molecules and in solids. For all such systems, the spectrum exhibits more or less broadly spread formations.

Schroedinger equations for stationary states. Eigenvalues and eigenfunctions

The possible values of the energy of a system may be determined by solving the Schroedinger equation

$$\mathbf{H}\psi = (\mathbf{T} + \mathbf{U})\psi = \mathrm{E}\psi \tag{5.2}$$

where $\mathbf{H}$ is the energy operator representing all the forms of motion and interaction inside the system and its interaction with the surrounding medium; $\mathbf{T}$ and $\mathbf{U}$ are the kinetic energy and potential energy operators, ψ is the wave function for the system and E is the energy.

According to the basic postulates of quantum mechanics, the form of the operator $\mathbf{H}$ can easily be found by writing down the classical Hamiltonian for the system $\mathbf{H} = \mathbf{T} + \mathbf{U}$, i.e. the expression for the energy as a function of generalised momenta and coordinates for all the degrees of freedom, and then replacing the momenta p_{x_i} by the momentum operators $\frac{ih}{2\pi}\frac{\partial}{\partial x_i}$. Thus, the Schroedinger equation for the one-dimensional harmonic oscillator is

$$-\frac{h^2}{8\pi^2\mu}\frac{\partial^2\psi}{\partial x^2} + \frac{1}{2}kx^2\psi = \mathrm{E}\psi \tag{5.3}$$

The first term on the left is the kinetic energy operator acting on the required eigenfunction ψ, whilst the second term is the potential energy multiplied by the function ψ.

Equation (5.2) does not have solutions for all values of E but only for certain discrete values of this parameter, say, E_1, E_2, E_3.... These eigenvalues of the Schroedinger equation determine the energies which the system can have.

Unfortunately, the solution of (5.2) for specific systems is exceedingly complicated. The solutions are relatively simple only for a free particle, the hydrogen atom, hydrogen-like ions, the harmonic oscillator, and the rigid rotator. Approximate methods must be employed for all other systems.

Most of our information about energy levels of atoms, molecules and other systems was obtained experimentally by studying the absorption, emission and Raman spectra. The composite semi-empirical method has also been widely used. The principle of this method is as follows. By solving the Schroedinger equation for a number of special cases, it is sometimes possible to find not the energies themselves but certain regularities in their distribution. The expressions obtained in this way contain constants which can be determined experimentally. For example, the solution of the Schroedinger equation for the harmonic oscillator can be expressed relatively simply in terms of well-known analytical functions. All the eigenfunctions for this problem are given by

$$E_v = a(v + {}^1/_2) \qquad v = 0,\ 1,\ 2, \ldots \tag{5.4}$$

where a is a constant which is determined by the reduced mass and the quasi-elastic constant. Since the properties of a diatomic molecule resemble those of a harmonic oscillator, it follows that the energy levels associated with small oscillations of the nuclei can also be represented by (5.4). If the values of E_v are known from experiment for two or three levels, it is possible to determine a and use Equation (5.4) to compute a large number of other levels.

According to quantum mechanics, any state of a system with a definite energy E_i is a stationary state, i.e. a state which does not vary with time. It is described by an eigenfunction $\psi_i(x)$, which may be obtained from the solution of the Schroedinger equation (5.2). In the one-dimensional problem the quantity $|\psi_i(x)|^2 dx$ gives the probability that when the coordinates of the system are measured, they will be found to lie between x and $x + dx$. For a multidimensional system, the quantity $|\psi_i(x_1, x_2, x_3, \ldots)|^2 dx_1dx_2dx_3 \ldots$ equals the probability of finding the system in the coordinate intervals between $x_1, x_2, x_3, \ldots$ and $x_1 + dx_1, x_2 + dx_2, x_3 + dx_3, \ldots$.

The eigenfunctions $\psi_i(x)$ for different states of a given system are linearly independent, expressed mathematically as being orthogonal, i.e. they satisfy the relation

$$\int \psi_i^*(x)\,\psi_j(x)\,dx = \delta_{ij}$$

The square of the modulus of the eigenfunction is normalised to unity in accordance with its physical significance as a probability density.

The eigenfunction gives a complete specification of the stationary state of a system. It can be used to calculate the mean value of any physical quantity measured experimentally, or to determine the probability of obtaining experimentally any specific value of this quantity.

Not all the physical quantities in a stationary state of given energy have definite values. In particular, the coordinates of heavy nuclei and of the electrons in a molecule do not have definite values. Moreover, in addition to energy, there are also certain other physical quantities which do have definite values in a given stationary state. Among them are physical quantities whose operators $\mathbf{L}$ commute with the energy operator, i.e. satisfy the relation

$$\mathbf{HL} - \mathbf{LH} = 0$$

Examples of such operators for isolated systems are the square of the angular momentum operator and the operator representing the component of the angular momentum along a special direction in space. Consequently, the energy E_i, the square of the angular momentum M_i^2 and its component $(M_z)_i$ have definite values in a stationary state $\psi_i(x)$. For some systems there are also other operators which commute with $\mathbf{H}$. This facilitates the systematisation and interpretation of experimental data. The introduction of quantum numbers is based on these properties. Specification of the quantum numbers is unambiguously related to the specification of the values of physical quantities which are discrete in the given state.

Separation of variables and classification of energy levels

By solving the Schroedinger equation for any given system, and thus deriving its eigenvalues, we obtain the possible energies of the system as a whole. It is sometimes possible, by introducing special coordinates, to separate the variables in the Schroedinger equation and use this as a basis for the classification of the energy levels.

Suppose, for example, that the operator $\mathbf{H}(x)$ can be written as the sum of a number of other operators which are functions of different variables:

$$\mathbf{H}(x_1, x_2, x_3, \ldots) = \mathbf{H}_1(x_1) + \mathbf{H}_2(x_2) + \mathbf{H}_3(x_3) + \ldots \tag{5.5}$$

If we seek the solution of (5.2) in the form $\psi = \psi_1(x_1)\psi_2(x_2)\psi_3(x_3)\ldots$, the equation splits into a number of independent equations as follows:

$$\mathbf{H}_1(x_1)\psi_1(x_1) = E_1\psi_1(x_1); \quad \mathbf{H}_2(x_2)\psi_2(x_2) = E_2\psi_2(x_2) \quad \mathbf{H}_3(x_3)\psi_3(x_3) = E_3\psi_3(x_3)\ldots \tag{5.6}$$

where

$$E = E_1 + E_2 + E_3 + \ldots \tag{5.6'}$$

The total energy of the system is then the sum of the individual terms, each of which has a definite physical meaning. This separation of variables is possible for different non-interacting particles or, for example, for the individual normal vibrations of a molecule. For a single molecule or atom it is always possible to separate out the translational motion of the system as a whole. Usually, the operator $\mathbf{H}(x)$ for the system under investigation cannot be transformed into an expression of the form of (5.5). However, the Schroedinger equation can frequently be reduced to the form

$$[\mathbf{H}_1(x_1) + \mathbf{H}_2(x_2) + \mathbf{H}_3(x_3) + \ldots + \mathbf{H}_{int}(x_1, x_2, x_3, \ldots)]\,\psi(x_1 x_2 \ldots) = E\psi(x_1, x_2 \ldots) \tag{5.7}$$

If the interaction operator $\mathbf{H}_{int}(x_1, x_2, x_3, \ldots)$ between the different degrees of freedom is relatively small, the variables can be separated approximately, and the total energy of the molecule can conventionally be regarded as the sum of the individual parts. For example, although the motion of electrons in a molecule is always connected with the rotation of the nuclei, this interaction can sometimes be neglected and the rotational and electronic energies may be regarded as independent. Similarly, the rotation of the nuclei in a molecule is connected with their oscillations, since the moments of inertia depend on the distances between the nuclei. However, when the oscillations of the nuclei are small, this interaction is also small, and therefore one can speak approximately of the vibrational and rotational energies of the molecule.

On the other hand, if the interaction operator is appreciable, the relation given by (5.6') is no longer valid and the total energy of the system cannot, in principle, be written as the sum of the energies associated with the individual degrees of freedom, i.e. with the various types of motion.

The approximate separation of variables in the Schroedinger equation is an important procedure because it serves as the basis for the systematisation and analysis of experimental data. It enables one to classify the various types of energy levels into electronic, vibrational, rotational, translational, intermolecular groups and so on. All these types of energy level reflect the different processes occurring in atomic systems exhibiting specific properties and require special study.

When the energy operator is expanded in powers of the small parameter m/M, where M is the mean mass of the nuclei and m is the mass of the electron, it can be shown that the energy of the molecule is given approximately by [12]

$$E \approx E_0 + \sqrt{\frac{m}{M}}\, E_1 + \frac{m}{M} E_2 + \ldots \tag{5.8}$$

The zero-order term E_0 gives the electronic energy of the molecule when the nuclei are fixed, whereas the first-order term

$$\sqrt{\frac{m}{M}}\, E_1$$

gives the vibrational energy. The second-order term includes the rotational energy of the molecule and a proportion of the vibrational energy. The ratios E_{vib}/E_{el} and E_{rot}/E_{vib} are of the order of magnitude of the square root of the ratio of the mass of the electron to the mean mass of the nuclei.

Separation of variables in the Schroedinger equation may be of practical importance even for systems for which the interaction operator is relatively large. It enables one to obtain the physically useful zero-order approximation which may be used with the aid of the perturbation theory to obtain a more rigorous solution of the problem.

Separation of electronic and vibrational coordinates of molecules and crystals on the adiabatic approximation [13]

The separation of variables describing the motion of electrons from those characterising the vibrations of nuclei relative to each other is of particular importance in connection with the optical properties of molecules and crystals. If we ignore the rotation of the molecule, the Schroedinger equation may be written in the form

$$\left[\mathbf{T}_{el}(x) + \mathbf{T}_{vib}(q) + \mathrm{U}(x, q)\right] \psi(x,q) = \mathrm{E}\, \psi(x, q) \tag{5.9}$$

where $\mathbf{T}_{el}$ and $\mathbf{T}_{vib}$ are the kinetic energy operators for the motion of the electrons and the vibrations of the nuclei, and $\mathrm{U}(x, q)$ is the potential energy associated with the Coulomb interaction between the electrons and the nuclei. The symbols x and q represent the coordinates of the electrons and nuclei respectively. In molecules and crystals the interaction between the motion of the electrons and the vibrations of the nuclei is frequently very large, and the operators $\mathbf{H}_{el}(x)$ and $\mathbf{H}_{vib}(q)$ cannot be written down.

A good zero-order solution of (5.9) may be obtained through the so-called adiabatic approximation, which is based on the expansion given by (5.8), and assumes that the motion of electrons relative to nuclei is faster by one or two orders of magnitude than the vibrations of the nuclei. The approximate solution of (5.9) is sought in the form

$$\psi(x, q) = \psi_{el}(x, q)\, \psi_{vib}(q) \tag{5.10}$$

where $\psi_{el}(x, q)$ is a rapidly varying function of x. The nuclear coordinates q enter into $\psi_{el}(x, q)$ as parameters (to each possible distance between the nuclei there corresponds a particular function ψ_{el}). Substituting (5.10) into (5.9) we have

$$\psi_{vib}(q)\, \mathbf{T}_{el}(x, q) + \mathrm{U}(x, q)\, \psi_{el}(x, q)\, \psi_{vib}(q) + \mathbf{T}_{vib}\, \psi_{el}(x, q) \psi_{vib}(q) = \mathrm{E}\, \psi_{el}(x, q)\, \psi_{vib}(q) \tag{5.11}$$

This equation is still exact. In order to separate the variables, we must assume that the operator $\mathbf{T}_{vib}$ has only a slight effect on $\psi_{el}(x, q)$, and that in the zero-order approximation this effect may be neglected. If we now divide (5.11) by $\psi_{el}(x, q)\psi_{vib}(q)$, we obtain the following approximate equation

$$\frac{1}{\psi_{el}(x, q)}\mathbf{T}_{el}\,\psi_{el}(x, q) + \mathbf{U}(x, q) + \frac{1}{\psi_{vib}(q)}\mathbf{T}_{vib}\,\psi_{vib}(q) = \mathrm{E} \quad (5.12)$$

The first two terms in this equation depend on the coordinates of the electrons and of the nuclei, whereas the last term is a function of q only. Their sum can only be equal to the constant E if

$$[\mathbf{T}_{el} + \mathbf{U}(x, q)]\,\psi_{el}(x, q) = \mathrm{E}_{el}(q)\,\psi_{el}(x, q) \quad (5.13)$$

$$[\mathbf{T}_{vib} + \mathrm{E}_{el}(q)]\,\psi_{vib}(q) = \mathrm{E}\,\psi_{vib}(q) \quad (5.14)$$

Equation (5.13) is the wave equation describing the motion of the electrons for a fixed distance between the nuclei. To each configuration of the nuclei there corresponds a particular electron energy $\mathrm{E}_{el}(q)$, which is equal to the sum of the kinetic and potential energies of the electrons at the constant internuclear distance and the potential energy of interaction between the nuclei. The wave equation given by (5.14) determines the oscillations of the nuclei relative to each other. It shows directly that the eigenvalue of (5.13) averaged over all the electron coordinates plays the role of a potential energy:

$$\mathrm{E}_{el}(q) = U_{vib}(q) \quad (5.15)$$

Thus, the forces acting between the nuclei are determined by the properties of the electron shell of the molecule.

It follows from Equation (5.14) that before we can solve the problem of the oscillations of the nuclei we must find the eigenvalues of (5.13). This can only be done for a small number of specific systems (for example, H_2 and H_2^+ molecules). However, the potential function given by (5.15) can frequently be approximated by some simple analytical function. For small oscillations of a diatomic molecule, the potential function can be approximately replaced by the harmonic function $U_{vib} = \frac{1}{2}k(q-q_0)^2$. A more general form is the Morse function

$$U_{vib}(q) = D\left[1 - e^{-\beta(q-q_0)}\right]^2 - D \qquad (5.16)$$

where D, β and q_0 are constants which are determined experimentally and characterise the properties of the electron shell (D is the dissociation energy, q_0 is the coordinate at the minimum of the potential energy curve and β is a constant which determines the form of the curve near the minimum).

It should be emphasised that the simultaneous solution of (5.13) and (5.14) gives only the total energy of the molecules; it cannot be separated into the electronic and vibrational terms. If the oscillations of the nuclei are regarded from the purely classical point of view, there is a continuous transformation of the kinetic energy into potential energy and vice versa. According to (5.15), any change in potential energy associated with the vibrations must be equal to the change in the electron energy, and therefore vibrations of the nuclei and changes in the state of the electron cloud are inseparable and constitute a single process. At the same time, separation of Equation (5.9) into (5.13) and (5.14) simplifies the solution of the problem considerably and provides a correct interpretation of experimental data. This is connected with the isolation of the purely vibrational eigenfunction $\psi_{vib}(q)$ and the resulting possibility of calculating the relative intensities of spectral lines. Moreover, separation of variables on the adiabatic approximation enables one to isolate the electronic part of the energy in the absence of vibrations, i.e. the energy of the molecule for those values of the vibrational coordinates which correspond to the minimum of the potential function. It is precisely in this sense that one often distinguishes between the electronic and vibrational energies and constructs the corresponding energy-level diagrams for the excited electron state (Fig. 2.1). The total energy E of the molecule can conventionally be divided into E_{el} and E_{vib}, where $E_{el} = E_{el}(q_i^{eq})$.

The electron equation (5.13) has, in general, various sets of solutions corresponding to different electron states. Each state corresponds to definite functions $E_n^{el}(q) = U(q)$, where n is the number of the electron energy level. The vibrational equation (5.14) must be solved for each electron state characterised by the particular function $U_n(q)$.

The expression given by (5.15) is the basic formula for all spectroscopic studies of molecules and crystals. It must however, be noted that its range of applicability is limited, particularly in complex systems. In point of fact, the operator $\mathbf{T}_{vib}$ in (5.11) acts on $\psi_{\text{эл}}(x, q)$ and therefore the third term in (5.11) must be written in the form

$$\mathbf{T}_{vib}\,\psi_{el}(x, q)\psi_{vib}(q) = \psi_{el}(x, q)\,\mathbf{T}_{vib}\,\psi_{vib}(q) + \mathbf{V}\,\psi_{el}(x, q)\,\psi_{vib}(q) \quad (5.17)$$

where $\mathbf{V}$ is the operator representing the interaction between electrons and nuclei. In many cases the operator $\mathbf{V}$ cannot be neglected. If it is particularly large, the variables x and q cannot be separated and the concept of potential energy loses its meaning altogether. For most luminescent molecules the adiabatic approximation leads to good results if the departure from the adiabatic approximation is subsequently compensated. Departure from this approximation is one of the causes of non-optical transitions (see Section 9) [14]. We have described the separation of variables for the

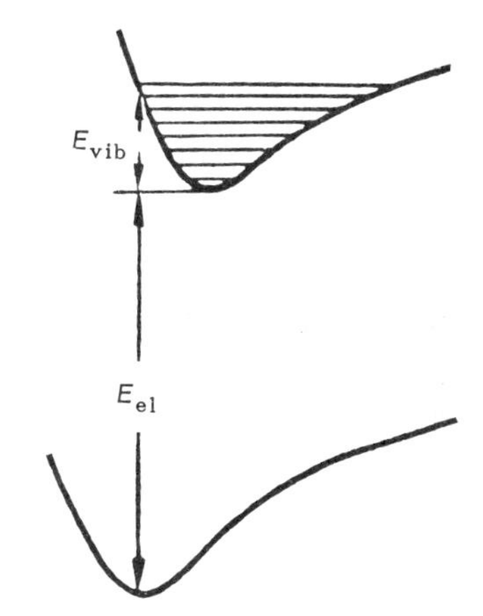

Fig. 2.1 Separation of the total energy of a molecule into the energy of electrons and the vibrational energy of nuclei

electronic and vibrational degrees of freedom of a molecule on the adiabatic approximation. An analogous separation of variables for other systems is possible provided they can be split into slow and fast sub-systems. For example, the fast sub-system may be taken to be the molecules in a solution, whilst the slow sub-system may represent the solvent (relative motion of the molecules). It is often convenient to define a number of sub-systems in increasing order of rapidity (electron and vibrational coordinates of a molecule, coordinates of intermolecular vibrations and so on).

Statistical weights

It turns out that for the majority of actual systems there are a number of independent solutions of the Schroedinger equation (5.2) for given E_i. Each of these solutions ψ_i^λ ($\lambda = 1, 2, \ldots, g_i$) is orthogonal to the other solutions. The total number of solutions, g_i is called the degree of degeneracy of the level i, or its statistical weight. In certain problems, for example, those which are encountered in connection with thermodynamic equilibrium, it is possible to ignore the specific properties of the individual components of a degenerate level and consider only the degree of degeneracy. The statistical weight is usually quite small for simple systems. In atoms it is determined by the total angular momentum ($g = 2J + 1$). The vibrational levels of diatomic molecules are not degenerate ($g = 1$) in general. For the rotational levels of diatomic molecules, $g = 2K + 1$ where K is the rotational angular momentum. The statistical weight may be quite large in complicated systems with a large number of degrees of freedom.

Representation of real systems by harmonic oscillators and by systems of particles with two or three energy levels

Any real system has an infinite number of energy levels and an infinite variety of optical properties. One important way of analysing experimental data is to divide the energy levels into levels and sub-levels corresponding to different types of motion, and then to investigate the electronic vibrational, rotational and other energy levels with subsequent corrections for the interactions between them.

Another widely used procedure is to introduce simplified models with a finite number of energy levels. Some of the characteristic features of the interaction of light with matter can be established by analysing the properties of a system of particles with only two energy levels. Such systems do not exist in nature, but there are many concrete situations in which only two energy levels participate in the absorption and emission of light, the remaining levels being of secondary importance. This occurs, for example, in resonance luminescence of atoms, or in transitions between the first and second vibrational levels of a molecule. Although results obtained in this way are only approximate, they nevertheless correctly reflect the fundamental properties of the physical processes which are involved.

The use of a simple system of particles with two or three levels as a model for real systems is sometimes based on the separation of variables and on the isolation of the level structure. For example, each electronic level of a molecule has a set of vibrational sub-levels, and an optical transition between two electronic levels can be associated with an infinite number of transitions between different vibrational sub-levels in the upper and lower electron states. Nevertheless, on the first approximation, the vibrational structure of the bands may be ignored and only the properties associated with the electronic levels considered.

At first sight it is not surprising that the above model gives better results for complicated molecules with a very large number of atoms than for simple molecules, although some information about the latter can, of course, also be obtained. However, the different vibrational sub-levels of simple systems have clearly defined specific properties which cannot be accounted for on the basis of simple models. In complex molecules, on the other hand, absorption and emission processes, involving transitions between different electronic states, are separated by a rapid redistribution of vibrational energy between the degrees of freedom and the individual vibrational levels lose their identity. As a result, electronic-vibrational transitions can be satisfactorily described by transitions between two levels which may be referred to as generalised or mean (over the vibrations). A number of important details connected with the structure of an electronic-vibrational level, for example, the shape of absorption and emission bands, cannot be accounted for in this way.

The two-level model is widely used in other problems of a similar nature. The real energy levels usually have a structure associated with various kinds of motion and with the interactions between them. This structure can be replaced by a single-level model in most cases, by performing the appropriate averaging over the sub-levels.

The three-level model is the natural extension of the two-level scheme. The three-level model correctly reflects the properties of a number of concrete systems, especially molecules with a metastable energy level which can exhibit the phenomena of fluorescence and phosphorescence. The three-level model is of basic importance for masers, the interpretation of anti-Stokes fluorescence, problems in non-linear optics and a number of other spectroscopic phenomena.

The harmonic oscillator is one of the most important models which have been successfully used to describe the properties of real systems. It has been widely used in classical theory (Chapter 1), but it is equally important in quantum theory. Although a pure harmonic oscillator does not exist in nature, there are many systems which approximate to it, in particular diatomic molecules with a small store of vibrational energy. Small oscillations of polyatomic molecules can be described by the motion of a set of harmonic oscillators. Even the properties of electronic spectra can be described with the aid of the harmonic oscillator model, and this explains the success of some of the results obtained in classical theory. The success of this model in the interpretation of experimental data has served as a basis for the so-called principle of correspondence between quantum and classical theories and it has been successfully used in the quantum theory of the electromagnetic field.

The great importance of the above models in the interpretation and systematisation of a wide range of experimental data has led to extensive studies of the properties of these models. Despite the relative simplicity of the calculations, this work has only been undertaken in recent years. Previous applications to the analysis of spectra were somewhat one-sided and inadequately justified. A detailed discussion of systems of particles with two or three energy levels, and of the harmonic oscillator model, is in Chapters 6-8.

6. PARTITION FUNCTIONS

Population of energy levels

The energy level population is an important characteristic of an atomic or molecular system and has an important influence on its optical properties.

Suppose that the total number of particles per unit volume is n, and that the total energy of the system is zero when all the particles occupy the lowest unexcited energy state. If, on the other hand, the system as a whole has a store of energy, the various particles will occupy different energy states, for example, the i-th energy level will be occupied by, say, n_i, particles with energies E_i. The total number of particles, n, is equal to the sum of these populations

$$n_1 + n_2 + n_3 + \ldots = \sum_i n_i = n \tag{6.1}$$

If we divide both sides of (6.1) by n, we obtain

$$\sum_i \rho_i = 1 \tag{6.2}$$

where

$$\rho_i = \frac{n_i}{\sum_i n_i} = \frac{n_i}{n} \tag{6.3}$$

Equations (6.2) and (6.3) are valid as given for a discrete energy spectrum, and must be modified somewhat for a continuous spectrum. Instead of (6.3) we then have

$$\rho(E)\, dE = \frac{dn(E)}{n} \tag{6.4}$$

where $dn(E)$ is the number of particles with energies between E and $E+dE$, and $\rho(E)$ is the partition function for the continuous spectrum. The function $\rho(E)$ gives the number of particles per unit energy interval, its dimensions being those of reciprocal energy. The normalisation condition is

$$\int_E \rho(E)\, dE = 1 \tag{6.5}$$

If both discrete and continuous levels are present in the energy spectrum, equations (6.4) and (6.2) must be combined.

When we introduced the concept of level population, we did not take into account the degeneracy of the levels, and used n_i to represent the number of particles occupying a given level i. However, if this level is degenerate, n_i gives the number of particles occupying all the sub-levels of the degenerate level. This is often sufficient to characterise the properties of the system of particles. In some cases, however, a more detailed description is necessary, and the population of each of the sub-levels must be specified. If the number of particles occupying the sub-level λ of the degenerate level of energy E_i is represented by n_i^λ, we have instead of (6.1)

$$\sum_i \sum_{\lambda=1}^{g_i} n_i^\lambda = n \tag{6.6}$$

$$\sum_{\lambda=1}^{g_i} n_i^\lambda = n_i \tag{6.7}$$

When the population of each sub-level is taken into account, the partition function for the i-th level will be determined by a set of quantities of the form

$$\rho_i^\lambda = \frac{n_i^\lambda}{n} \tag{6.8}$$

where

$$\sum_i \sum_{\lambda=1}^{g_i} \rho_i^\lambda = 1 \tag{6.9}$$

We have based our definition of the partition function on the normalisation condition given by equation (6.1). Since the levels may be classified by a number of variables, it is occasionally convenient to use a different normalisation procedure. For example, for the vibrational structure of an electronic level the partition function may be normalised not to the total number of particles occupying all the levels, but only to those particles which occupy the vibrational levels of the given electronic state. The form of formulae such as (6.3) is then practically the same as before:

$$\rho_{\alpha i}^{\text{vib}} = \frac{n_{\alpha i}^{\text{vib}}}{\sum_i n_{\alpha i}^{\text{vib}}} \tag{6.10}$$

where α represents the particular electronic state.

Fundamental partition functions

The various partition functions can be divided into two groups, namely, equilibrium and non-equilibrium functions. The equilibrium distribution over the energy levels occurs in thermodynamic equilibrium with the surrounding medium. It is quite universal in character and is well known.

Studies of the equilibrium distribution are very important in connection with thermal emission and the absorption and scattering of light.

The properties of the equilibrium distribution are also important in other respects, since the system may have been in thermodynamic equilibrium prior to a disturbance and the initial conditions are reflected in the form of all the subsequent distributions.

Non-equilibrium distributions arise as a result of departures from thermodynamic equilibrium, for example, as a result of illumination, the incidence of electrons, ions or neutral particles, the onset of chemical reactions and so on. It follows that the form of the non-equilibrium partition functions depends both on the nature and on the intensity of the external disturbances. Moreover, non-equilibrium partition functions depend on the properties of the system itself and on the ability of the system to react to a given external disturbance. For example, if the incident radiation is not absorbed by the system, there is no departure from equilibrium. The form of non-equilibrium partition function depends also on the temperature of the system, or to be more precise, on the temperature of the medium with which it was in equilibrium prior to the introduction of the external disturbance. (The temperature of the system cannot be defined for a large departure from equilibrium.) We shall discuss these general properties of partition functions after we have investigated the properties of the systems themselves and their interactions with radiation.

Non-equilibrium partition functions may be divided into two main sub-groups, namely, non-stationary and stationary. The first characterises distributions which arise when the intensity of the external disturbance is time-dependent and when a constant external disturbance is either switched on or switched off. In all such systems the partition function is non-stationary and depends on time.

Stationary partition functions are time-independent. They characterise the steady-state conditions some time after a constant external disturbance has been switched on. Here, there is dynamic equilibrium in the system which is different from the thermal equilibrium. The system receives energy continually from the external source and gives up some of it through various channels.

Partition function for thermodynamic equilibrium

According to the basic postulates of statistical physics the population of any particular level E_i is given by

$$\rho_i^\lambda = C(T)\, e^{-E_i/kT} \qquad (6.11)$$

The population of all the degenerate sub-levels is the same. When degeneracy is taken into account this formula may be rewritten in the form

$$\rho_i = \sum_{\lambda=1}^{g_i} \rho_i^\lambda = C(T)\, g_i e^{-E_i/kT} \qquad (6.12)$$

where g_i is the statistical weight of the i-th level. The population ratio for the two levels is

$$\frac{\rho_i}{\rho_j} = \frac{g_i}{g_j}\, e^{-(E_i-E_j)/kT} \qquad (6.13)$$

If the distribution of all the levels and their statistical weights are known, the constant $C(T)$ in (6.12) can be determined from the normalisation condition given by (6.9).

For a continuous energy spectrum the equilibrium partition function is of the form

$$\rho(E)\, dE = C(T)\, g(E)\, e^{-E/kT}\, dE \qquad (6.14)$$

where $g(E)$ is the number of independent states per unit energy.

It follows from the form of the partition functions given by (6.12) and (6.14) that at absolute zero all the particles are in the lowest state, i.e. they are not excited. Any increase in temperature leads to an increase in the population of higher-lying levels. When $T \to \infty$, the level populations are proportional to their statistical weights, and when $g_i = \text{const}$ they are all the same.

Let us consider some numerical examples. Suppose that the separation between the energy levels is 20,000 cm^{-1} which corresponds to electronic levels. The value of kT at room temperature is about 200 cm^{-1} so that according to (6.13)

$$\frac{n_2}{n_1} \approx \frac{g_2}{g_1} \cdot 10^{-42}$$

Most of the particles occupy the lowest possible electronic state. If it is assumed that the total number of particles per unit volume is of the order of 10^{19}, the number of particles occupying the level E_2 is only 10^{-23}. An appreciable increase in the number of excited particles is achieved at high temperatures.

The distribution of particles over the vibrational levels is quite different in diatomic molecules. Suppose that the separation between neighbouring levels is 3,000 cm^{-1}. The ratio of the populations for $kT \sim$ 200 cm^{-1} is then

$$\frac{n_2}{n_1} \approx 10^{-6}$$

and therefore the number of excited particles is still very small. For heavy molecules, the separation between vibrational levels is much smaller. If $E_2 - E_1 \approx E_3 - E_2 = \ldots \approx 200$ cm^{-1}, we have

$$\frac{n_2}{n_1} \approx 0.37, \quad \frac{n_3}{n_2} \approx 0.37, \ \ldots$$

The ratio of the number of molecules in successive vibrational levels to the number n_1 on the harmonic oscillator approximation is $1 : 0.37 : (0.37)^2 : (0.37)^3 \ldots$ The constant C is equal to 0.63. In the ground state there are 63% of all molecules, while the next state is occupied by 22%, the third by 8% and so on.

The distribution of the molecules over the vibrational energy levels is very dependent on temperature. For example, when $E_2 - E_1 = 3{,}000$ cm^{-1} and $T = 6{,}000$ °K, we have $n_2/n_1 = 0.49$. The lower energy level of the molecule is then occupied by 51% of all molecules while the second is occupied by 0.25%. If $E_2 - E_1 = 200$ cm^{-1} and $T = 6{,}000$ °K, n_2/n_1 = 95%, the first level is occupied by 5% of all molecules and the second by almost the same number.

The separation of neighbouring rotational levels is very small as a rule. This means that even at room temperature, the Boltzmann factor $e^{-E_i/kT}$ is very nearly equal to unity. Since the statistical weights of rotational levels increase with the level number, the level population at first increases and then falls off as E increases.

A characteristic feature of thermodynamic equilibrium is the so-called detailed balancing, i.e. the number of collisions of the first kind is equal to the number of collisions of the second kind, chemical reactions proceeding in one direction are compensated by reactions proceeding in the opposite direction and so on. This is also valid for optical processes since matter is in equilibrium with the thermal radiation described by Planck's function. According to the principle of detailed balancing, the number of elementary acts of absorption of Planck radiation is equal to the number of acts of emission at each frequency and in each direction of propagation. It should be noted that the principle of detailed balancing is not obeyed under stationary conditions. The level population remains constant only because there is a mutual compensation of different processes, for example, compensation of the excitation of molecules by electron impact through the emission of photons.

7. RADIATIVE TRANSITION PROBABILITIES

Differential Einstein coefficients

If a particular system is illuminated by radiation of frequency ν_{ij}, then, as a result of the interaction between the field and the medium, the molecules undergo transitions to higher-lying levels, for example, from level j to level i (Fig. 2.2). These transitions are accompanied by an increase in the energy of the particle system and, in accordance with the law of conservation of energy, there is a reduction in the energy of the incident radiation, i.e. the incident radiation is absorbed. The number of $j \to i$, transitions depends on the number of particles in the level j, the density of the incident radiation $u(\nu_{ij})$, the time t and the properties of the molecules. Different molecules interact differently with radiation. Their ability to absorb radiation of different frequency is characterised by transition probabilities. The ability of atoms and molecules to emit radiation, and thereby lose a certain amount of their energy, is characterised by analogous transition probabilities.

The concept of transition probability for transitions between energy levels was introduced by Einstein in a purely phenomenological fashion well before it was justified in quantum mechanics and quantum electrodynamics. Transition probabilities play a fundamental part in spectroscopy and characterise the basic optical properties of matter. If the transition probabilities between the levels of a molecule are

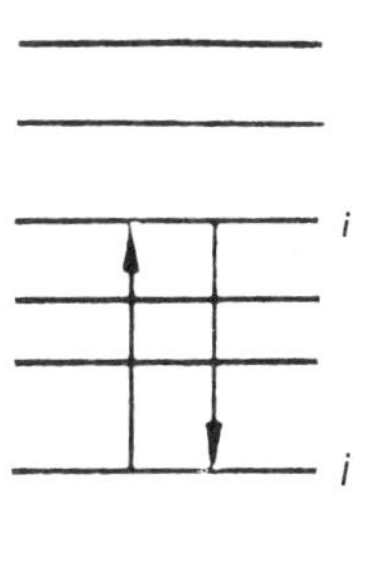

Fig. 2.2 Transitions leading to absorption and emission

known, it is quite easy to calculate most of the experimentally observed quantities.

To begin with, let us consider the general case of arbitrarily oriented molecules and arbitrary anisotropy of external radiation. We shall suppose for the sake of simplicity that the molecule has one special axis, whose position can be specified by two angles in a spherical system of coordinates. The orientation of the molecules in space can be conveniently represented by the partition functions $n_j(\Omega_1)$. The quantity $n_j(\Omega_1)d\Omega_1$ is equal to the number of molecules occupying the level E_j with the special axis lying within the solid angle between Ω_1 and $\Omega_1 + d\Omega_1$, where

$$\int_{\Omega_1} n_j(\Omega_1)\,d\Omega_1 = n_j \tag{7.1}$$

The anistropy of the radiation interacting with the molecules under investigation may be specified by a distribution function $u^\alpha(\nu_{ij}, \Omega_2)$, where $u^\alpha(\nu_{ij}, \Omega_2)d\Omega_2$ is the density of radiation of frequency ν_{ij} propagating in the solid angle between Ω_2 and $\Omega_2 + d\Omega_2$ with polarisation α. It is related to the total density of radiation $u(\nu_{ij})$ at a particular point in space by the expression

$$\sum_{\alpha=1}^{2} \int_{\Omega_2} u^\alpha(\nu_{ij}, \Omega_2)\,d\Omega_2 = u(\nu_{ij}) \tag{7.2}$$

Now we suppose that the number of $j \to i$ transitions which lead to the absorption of radiation in a time dt is proportional to the number of molecules in the j-th state, to the time and to the density of the incident radiation. These assumptions form the basis of the so-called probabilistic method and will be justified later when we come to discuss the quantum theory of the interaction of light with matter. They have been verified in most applications although they do have certain (still not fully understood) limits of applicability. Thus,

$$dn_{ji\alpha}(\Omega_1,\ \Omega_2) = b_{ji}^{\alpha}(\Omega_1, \Omega_2)\, u^{\alpha}(\nu_{ij},\ \Omega_2)\, d\,\Omega_2\, n_j(\Omega_1)\, dV dt\, d\,\Omega_1 \qquad (7.3)$$

where $dn_{ji\alpha}(\Omega_1,\ \Omega_2)$ is the number of $j \to i$ transitions in a time dt which are due to the effect of radiation of density $u^{\alpha}(\nu_{ij}, \Omega_2)\, d\,\Omega_2$ experienced by molecules occupying the volume dV and lying within the solid angle $d\,\Omega_1$.

Each transition is accompanied by the absorption of a photon of energy $h\nu_{ij}$ and therefore the absorbed energy is equal to $h\nu_{ij}\, dn_{ji\alpha}(\Omega_1, \Omega_2)$.

Let us rewrite (7.3) in the form

$$b_{ji}^{\alpha}(\Omega_1,\ \Omega_2)\, u^{\alpha}(\nu_{ij},\ \Omega_2)\, d\,\Omega_2\, dt = \frac{dn_{ji\alpha}(\Omega_1,\ \Omega_2)}{n_j(\Omega_1)\, d\,\Omega_1\, dV} \qquad (7.4)$$

The right-hand side of this expression gives the ratio of the number of particles which undergo $j \to i$ transitions in a time dt to the total number of particles in the state j. It follows that $b_{ji}^{\alpha}(\Omega_1,\ \Omega_2) \times u^{\alpha}(\nu_{ij},\ \Omega_2)\, d\,\Omega_2$ is the number of $j \to i$ transitions per second per molecule. It is called the transition probability. A transition probability has the dimensions of the reciprocal of time and is characterised by a large number (in $\sec^{-1}$).

From our original assumption we see that the transition probability $b_{ji}^{\alpha}(\Omega_1,\ \Omega_2)\ u^{\alpha}(\nu_{ij},\ \Omega_2)\, d\,\Omega_2$ is proportional to the density $u^{\alpha}(\nu_{ij},\ \Omega_2)\, d\,\Omega_2$ of the incident radiation. The coefficient of proportionality $b_{ji}^{\alpha}(\Omega_1, \Omega_2)$ is called the differential Einstein coefficient for the absorption of light during the transition $j \to i$. It depends not only on the properties of the absorbing particle but also on its orientation relative to the direction of the electric field in the incident wave. Quantum-mechanical calculations of the interaction of light with matter show (Section 8) that incident radiation of frequency ν_{ij} can induce optical transitions not only in the upward direction ($j \to i$) but also in the downward direction ($i \to j$). Since this reduces the energy of the system, it follows that in accordance with the law of conservation of energy, the field energy should increase. An induced $i \to j$ transition, should give rise to the emission of a photon of energy $h\nu_{ij}$. This process is referred to as stimulated emission. The possibility of such transitions can also be established from general considerations based on the principle of detailed balancing (see Chapter 3). The absorption of radiation during $j \to i$ transitions and the stimulated emission of radiation during $i \to j$ transitions will always take place, since the two phenomena are different aspects of the same process of interaction between light and matter.

Stimulated emission has one important special property whose essence can only be established in quantum electrodynamics (Section 14). Thus, the quanta of light emitted by a molecule under the influence of external radiation do not propagate in all directions but only in the direction of the primary beam, and cannot have arbitrary polarisation but only the polarisation of the incident radiation. It is for this reason that stimulated emission is frequently referred to as negative absorption; a fraction of the energy taken from the incident beam as a result of $j \to i$ transitions is compensated for by stimulated emission through $i \to j$ transitions.

In accordance with our original assumptions, the number of stimulated $i \to j$ transitions in time dt in the molecules $n_i(\Omega_1) \times d\Omega_1 \, dV$ under the action of radiation of density $u^\alpha(\nu_{ij}, \Omega_2) \times d\Omega_2$ is

$$\begin{aligned} dn_{ij\alpha}^{\text{stim}}(\Omega_1, \Omega_2) &= b_{ij}^{\alpha}(\Omega_1, \Omega_2) \\ &\times u^\alpha(\nu_{ij}, \Omega_2) \, d\Omega_2 \, n_i(\Omega_1) \, d\Omega_1 \, dV dt \end{aligned} \tag{7.5}$$

where $b_{ij}^{\alpha}(\Omega_1, \Omega_2) u^\alpha(\nu_{ij}, \Omega_2) \, d\Omega_2$ is the probability of stimulated emission of radiation and $b_{ji}^{\alpha}(\Omega_1, \Omega_2)$ is the corresponding differential Einstein coefficient.

The processes of absorption and stimulated emission occur as a result of the effect of external radiation. However, an excited molecule will also undergo spontaneous transitions to lower-lying states. These occur both in the presence and in the absence of external radiation. Spontaneous transitions are accompanied by a reduction in the energy of the molecule and the simultaneous emission of light in different directions with different polarisations.

Evidence for the existence of spontaneous transitions is supplied in the first instance by experiment (emission of light by bodies after preliminary illumination, emission of radiation under the action of non-radiative agents and so on). The existence of spontaneous transitions is explained in quantum electrodynamics: they arise as a result of interactions with the so-called zero-point fields.

The number of spontaneous $i \rightarrow j$ transitions in a time dt which are accompanied by the emission of radiation with polarisation α between the solid angle Ω_3 and $\Omega_3 + d\Omega_3$ is

$$dn_{ij\alpha}^{\text{spon}}(\Omega_1, \Omega_2) = a_{ij}^{\alpha}(\Omega_1, \Omega_3)\, d\Omega_3\, n_i(\Omega_1)\, d\Omega_1\, dV dt \qquad (7.6)$$

The proportionality factor $a_{ij}^{\alpha}(\Omega_1, \Omega_3)$ is called the differential Einstein coefficient of spontaneous emission and determines the probability of emission of a photon by a single excited particle per unit solid angle.

The expressions given by (7.3) to (7.6) have been written out for discrete levels. In some cases it is convenient to have a more rigorous formulation which takes into account the profiles of the resulting spectral lines. In such cases the Einstein coefficients are functions of frequency. For example, the probability of transition from a level j to a level i under the action of radiation in a frequency interval $d\nu$ is equal to

$$b_{ji}^{\alpha}(\Omega_1, \Omega_2, \nu)\, u^{\alpha}(\nu, \Omega_2)\, d\nu\, d\Omega_2$$

The various frequencies are absorbed in different ways within the limits of a given line. If the system is illuminated by non-monochromatic radiation, then by integrating this expression with respect to frequency, we arrive at (7.4) where

$$\int_{\nu} b_{ji}^{\alpha}(\Omega_1, \Omega_2, \nu)\, u^{\alpha}(\nu, \Omega_2)\, d\nu = b_{ji}^{\alpha}(\Omega_1, \Omega_2)\, u^{\alpha}(\nu_{ij}, \Omega_2) \qquad (7.7)$$

If the radiation density per unit frequency interval ν_{ij} within the confines of the line is constant, we have

$$\int_{\nu} b_{ji}^{\alpha}(\Omega_1, \Omega_2, \nu)\, d\nu = b_{ji}^{\alpha}(\Omega_1, \Omega_2) \qquad (7.7a)$$

The probability of spontaneous emission is also a function of frequency and

$$a_{ij}^{\alpha}(\Omega_1, \Omega_3) = \int a_{ij}^{\alpha}(\Omega_1, \Omega_3, \nu)\, d\nu \qquad (7.8)$$

Universal relationships between the differential Einstein coefficients

It is quite easy to show that, when a set of molecules is in thermodynamic equilibrium with electromagnetic radiation, the Einstein coefficients $b^{\alpha}_{ji}(\Omega_1,\Omega_2)$, $b^{\alpha}_{ij}(\Omega_1,\Omega_2)$ and $a^{\alpha}_{ij}(\Omega_1,\Omega_2)$ are connected by universal relationships which are valid for all systems.

Consider the optical transitions between any two levels j and i of an atom or molecule. The principle of detailed balancing requires that, under thermodynamic equilibrium, the number of acts of absorption should be equal to the number of acts of emission in an arbitrary volume ΔV and an arbitrary time interval Δt. This is valid not only for the total number of $i \to j$ and $j \to i$ transitions but also for each set of angles Ω_1 and Ω_2 and for each polarisation individually. Therefore, in thermodynamic equilibrium

$$dn_{ji\alpha}(\Omega_1,\ \Omega_2) = dn^{\text{stim}}_{ij\alpha}(\Omega_1,\ \Omega_2) + dn^{\text{spon}}_{ij\alpha}(\Omega_1,\ \Omega_2) \qquad (7.9)$$

Substituting the expression $dn_{ji\alpha}(\Omega_1,\ \Omega_2)$ and $dn^{\text{stim}}_{ij\alpha}(\Omega_1,\ \Omega_2)$ from (7.3), (7.5) and (7.6), and the expression $dn^{\text{spon}}_{ij}(\Omega_1,\ \Omega_2)$ (with $\Omega_3 = \Omega_2$) and abbreviating with dt, dV, $d\Omega_1$ and $d\Omega_2$, we have

$$\begin{aligned} & b^{\alpha}_{ji}(\Omega_1,\ \Omega_2)\, u^{\alpha}(\nu_{ij},\ \Omega_2)\, n_j(\Omega_1) \\ &= b^{\alpha}_{ij}(\Omega_1,\ \Omega_2)\, u^{\alpha}(\nu_{ij},\ \Omega_2)\, n_i(\Omega_1) + a^{\alpha}_{ij}(\Omega_1,\ \Omega_2)\, n_i(\Omega_1) \end{aligned} \qquad (7.10)$$

where $n_j(\Omega_1)$ and $n_i(\Omega_1)$ are the level populations which are related by the Boltzmann equation (6.13) and $u^{\alpha}(\nu_{ij},\ \Omega_2)$ is the density of equilibrium radiation, which is related to Planck's function by (7.2). Since equilibrium radiation is completely isotropic, we have

$$u^{\alpha}(\nu_{ij},\ \Omega_2) = \frac{1}{8\pi}\, u(\nu_{ij}) = \frac{h\nu^3_{ij}}{c^3}\ \frac{1}{e^{h\nu_{ij}/kT} - 1} \qquad (7.11)$$

From (7.10) it follows that

$$u^{\alpha}(\nu_{ij},\ \Omega_2) = \frac{a^{\alpha}_{ij}(\Omega_1,\ \Omega_2)\, n_i(\Omega_1)}{b^{\alpha}_{ji}(\Omega_1,\ \Omega_2)\, n_j(\Omega_1) - b^{\alpha}_{ij}(\Omega_1,\ \Omega_2)\, n_i(\Omega_1)} \qquad (7.12)$$

If we divide both the numerator and the denominator on the

right-hand side of (7.12) by $b^{\alpha}_{ij}(\Omega_1, \Omega_2)\, n_i(\Omega_1)$ and substitute for $n_j(\Omega_1)/n_i(\Omega_1)$ from (6.13), then, on equating the right-hand sides of (7.11) and (7.12), we have

$$\frac{a^{\alpha}_{ij}(\Omega_1, \Omega_2)}{b^{\alpha}_{ij}(\Omega_1, \Omega_2)} = \frac{h\nu^3_{ij}}{c^3} \tag{7.13}$$

$$\frac{b^{\alpha}_{ji}(\Omega_1, \Omega_2)}{b_{ij}(\Omega_1, \Omega_2)} = \frac{g_i}{g_j} \tag{7.14}$$

The two Einstein relations given by (7.13) and (7.14) were derived on the basis of (7.10), which is valid in thermodynamic equilibrium. However, the final formulae do not contain parameters which depend on the partition functions or parameters characterising the properties of the external field. The ratios of the Einstein coefficients depend only on the properties of the system itself, i.e. on the separation between the levels ν_{ij} and their statistical weights g_i, g_j. Hence, it follows that (7.13) and (7.14) are useful relationships which are valid for any partition functions and any external fields. If (7.13) and (7.14) were not valid, it would be impossible to satisfy (7.10), which would mean that there could not be thermodynamic equilibrium between matter and radiation.

The relationships given by (7.13) and (7.14) show that for theoretical calculations or experimental determinations of the Einstein coefficients it is sufficient to limit one's attention to the determination of only one out of the three coefficients characterising a given pair of levels.

In writing down (7.5), we assumed that the interaction of light with matter leads not only to absorption but also to stimulated emission of radiation. Equation (7.14) gives direct confirmation of this assumption. According to (7.9) and (7.10), the absorption of light through $j \to i$ transitions under thermodynamic equilibrium cannot, in principle, be compensated by spontaneous emission only. Moreover, it follows from (7.14) that whatever the density $u(\nu_{ij})$ of the exciting radiation, the absorption probabilities $b^{\alpha}_{ji}(\Omega_1, \Omega_2)\, u^{\alpha}(\nu_{ij}, \Omega_2)\, d\Omega_2$ are equal within the statistical weights to the spontaneous emission probabilities $b^{\alpha}_{ij}(\Omega_1, \Omega_2)\, u^{\alpha}(\nu_{ij}, \Omega_2)\, d\Omega_2$. These probabilities are equal in general for non-degenerate levels.

To estimate the ratio of the probabilities of spontaneous and stimulated emission in $i \to j$ transitions, consider the expression

$$\frac{a_{ij}^{\alpha}(\Omega_1, \Omega_2)}{b_{ij}^{\alpha}(\Omega_1, \Omega_2)\, u^{\alpha}(\nu_{ij}, \Omega_2)\, d\Omega_2} = \frac{1}{u^{\alpha}(\nu_{ij}, \Omega_2)\, d\Omega_2} \frac{h\nu_{ij}^3}{c^3} \tag{7.15}$$

where $u^{\alpha}(\nu_{ij}, \Omega_2)$ is given by the Planck formula. This yields

$$\frac{a_{ij}}{b_{ij}u(\nu_{ij})} = e^{h\nu_{ij}/\kappa T} - 1 \tag{7.15a}$$

When $h\nu/kT \gg 1$, i.e. in the visible part of the spectrum at normal temperatures, stimulated emission is negligible. If, on the other hand, $h\nu/kT \ll 1$, which occurs, for example, at radio frequencies, the right-hand side of (7.15a) is very small and therefore the probabilities of spontaneous emission may be neglected.

The relations given by (7.13) and (7.14) were derived here in a purely phenomenological fashion. They will be derived later by the methods of quantum mechanics and quantum electrodynamics. Formulae will also be established for the absolute values of the Einstein coefficients.

Integral Einstein coefficients

In the solution of specific problems it is frequently necessary to know the total number of optical transitions $i \to j$ and $j \to i$ without introducing detailed considerations concerned with the anisotropy of the radiation and the angular distribution of the molecules. These numbers can easily be obtained by integrating (7.3) and (7.5) with respect to the two angles Ω_1 and Ω_2, and (7.6) with respect to Ω_1 and Ω_3 and summing over the two resolved polarisations. The result is

$$dn_{ji} = dt dV \int n_j(\Omega_1)\, d\Omega_1 \sum_{\alpha} \int b_{ji}^{\alpha}(\Omega_1, \Omega_2)\, u^{\alpha}(\nu_{ij}, \Omega_2)\, d\Omega_2 \tag{7.16}$$

$$dn_{ij}^{\text{stim}} = dt dV \int n_i(\Omega_1)\, d\Omega_1 \sum_{\alpha} b_{ij}^{\alpha}(\Omega_1, \Omega_2)\, u^{\alpha}(\nu_{ij}, \Omega_2)\, d\Omega_2 \tag{7.17}$$

$$dn_{ij}^{\text{spon}} = dt dV \int n_i(\Omega_1)\, d\Omega_1 \sum_{\alpha} \int a_{ij}^{\alpha}(\Omega_1, \Omega_2)\, d\Omega_3 \tag{7.18}$$

The result of the integration is independent of the order of integration with respect to Ω_1, Ω_2 (Ω_3).

Let us suppose to begin with that the orientation of the particles is arbitrary and that the external radiation acting upon them is isotropic. This is encountered for completely diffuse radiation (and in particular for equilibrium radiation). If the radiation is isotropic we have

$$u^{\alpha}(\nu_{ij}, \Omega_2) = \frac{1}{8\pi} u(\nu_{ij}) \tag{7.19}$$

and therefore the integrals which enter into (7.16) and (7.17) become much simpler. Let us substitute

$$B_{ji} = \frac{1}{8\pi} \sum_{\alpha} \int_{\Omega_2} b^{\alpha}_{ji}(\Omega_1, \Omega_2)\, d\Omega_2 \tag{7.20}$$

$$B_{ij} = \frac{1}{8\pi} \sum_{\alpha} \int_{\Omega_2} b^{\alpha}_{ij}(\Omega_1, \Omega_2)\, d\Omega_2 \tag{7.21}$$

The quantities B_{ij} and B_{ji} are independent of Ω_1 because the state of a molecule is independent of its orientation with respect to the external isotropic radiation. The significance of the two parameters B_{ji} and B_{ij} can be elucidated by substituting (7.20) and (7.21) into (7.16) and (7.17) and bearing in mind (7.1):

$$dn_{ji} = B_{ji} u(\nu_{ij})\, n_j\, dt dV \tag{7.22}$$

$$dn_{ij}^{\text{spon}} = B_{ij} u(\nu_{ij})\, n_i dt dV \tag{7.23}$$

It follows from (7.22) that $B_{ji} u(\nu_{ij})$ determines the total probability of the $j-i$ transition. The quantity $B_{ij} u(\nu_{ij})$ has an analogous meaning. The two coefficients B_{ij} and B_{ji} are called the integral Einstein coefficients for absorption and stimulated emission of radiation.

In the above special case of arbitrarily oriented molecules, the spontaneous emission by the entire set of molecules is anisotropic, i.e. it exhibits a specific distribution of the emitted radiation with respect to the directions of propagation and polarisation. However, the total emission of each individual molecule is the same and therefore the integral

$$A_{ij} = \sum_{\alpha} \int a^{\alpha}_{ij}(\Omega_1, \Omega_3)\, d\Omega_3 \tag{7.24}$$

in (7.18) is independent of Ω_1. Substituting (7.24) into (7.18) we have, in view of (7.1),

$$dn_{ij}^{\text{spon}} = dtdV \int n_i(\Omega_1)\, d\Omega_1 A_{ij} = A_{ij} n_i dV dt \tag{7.25}$$

The coefficient A_{ij} has the meaning of a total probability of all spontaneous $i \rightarrow j$ transitions and is called the integral Einstein coefficient for spontaneous emission.

Consider now the second special case which is frequently encountered in practice and involves a considerable simplification of the calculations. Suppose that the particles in a volume dV are oriented randomly both in the upper and lower state and therefore in accordance with (7.1)

$$n_j(\Omega_1) = \frac{1}{4\pi} n_j, \quad n_i(\Omega_1) = \frac{1}{4\pi} n_i \tag{7.26}$$

The radiation acting on the molecules may be anisotropic. If this is so, by integrating (7.16)-(7.18) with respect to Ω_1 and bearing in mind (7.2), it is quite easy to obtain the relations which are similar to (7.22), (7.23) and (7.25) except that the three coefficients B_{ji}, B_{ij} and A_{ij} have a different meaning:

$$B_{ji} = \frac{1}{4\pi} \int\limits_{\Omega_1} b_{ji}^{\alpha}(\Omega_1, \Omega_2)\, d\Omega_1 \tag{7.27}$$

$$B_{ij} = \frac{1}{4\pi} \int\limits_{\Omega_1} b_{ij}^{\alpha}(\Omega_1, \Omega_2)\, d\Omega_1 \tag{7.28}$$

$$A_{ij} = 2 \int\limits_{\Omega_1} a_{ij}^{\alpha}(\Omega_1, \Omega_2)\, d\Omega_1 \tag{7.29}$$

In contrast to the preceding case, the transition probabilities $B_{ji}u(\nu_{ij})$, $B_{ij}u(\nu_{ij})$ and A_{ij} refer not to a single molecule but to a set of molecules, and they determine the mean value of the transition probability in the various molecules. Since the orientations of the molecules are random, these average probabilities depend only on the total intensity of the external radiation and are independent of Ω_2 and α.

As the differential Einstein coefficients are symmetrical with respect to Ω_1 and Ω_2 (though this is valid only for those molecules whose orientations in space can be defined by a single axis), integration with respect to Ω_1 and Ω_2 leads to the same result. If the polarisation is also taken into account, it is easy to see that the coefficients B_{ji}, B_{ij} and A_{ij} in (7.27)-(7.29) equal numerically the integral Einstein coefficients calculated from (7.20), (7.21) and (7.24) respectively.

The integral Einstein coefficients B_{ji}, B_{ij} and A_{ij} are related by universal expressions analogous to (7.13) and (7.14). Substituting $b^{\alpha}_{ji}(\Omega_1, \Omega_2)$ from (7.20) into (7.14), and comparing with (7.21), we have

$$\frac{B_{ji}}{B_{ij}} = \frac{g_i}{g_j} \tag{7.30}$$

As in the comparison of (7.24) with (7.21) it follows, in view of (7.13), that

$$\frac{A_{ij}}{B_{ij}} = \frac{8\pi h \nu^3_{ij}}{c^3} \tag{7.31}$$

Equations (7.16), (7.17) and (7.18) may be written in the form of (7.22), (7.23) and (7.25) even when the radiation field and the orientation of the molecules are simultaneously anisotropic. It is quite easy to show from these formulae that

$$B'_{ji} = \frac{\sum\limits_{\alpha}\int\limits_{\Omega_1}\int\limits_{\Omega_2} u^{\alpha}(\nu_{ij}, \Omega_2)\, b^{\alpha}_{ji}(\Omega_1, \Omega_2)\, n_j(\Omega_1)\, d\Omega_1\, d\Omega_2}{\left[\sum\limits_{\alpha}\int u^{\alpha}(\nu_{ij}, \Omega_2)\, d\Omega_2\right]\left[\int n_j(\Omega_1)\, d\Omega_1\right]} \tag{7.32}$$

$$B'_{ij} = \frac{\sum\limits_{\alpha}\int\limits_{\Omega_1}\int\limits_{\Omega_2} u^{\alpha}(\nu_{ij}, \Omega_2)\, b^{\alpha}_{ij}(\Omega_1, \Omega_2)\, n_i(\Omega_1)\, d\Omega_1\, d\Omega_2}{\left[\sum\limits_{\alpha}\int u^{\alpha}(\nu_{ij}, \Omega_2)\, d\Omega_2\right]\left[\int n_i(\Omega_1)\, d\Omega_1\right]} \tag{7.33}$$

$$A'_{ij} = \frac{\sum\limits_{\alpha}\int\limits_{\Omega_1}\int\limits_{\Omega_2} a^{\alpha}_{ij}(\Omega_1, \Omega_2)\, n_i(\Omega_1)\, d\Omega_1\, d\Omega_2}{\int n_i(\Omega_1)\, d\Omega_1} \tag{7.34}$$

In the present case, the coefficients B'_{ji}, B'_{ij} and A'_{ij} cannot be represented by numbers characterising only a given type of molecule. They explicitly depend on the form of the distributions $n_j(\Omega_1)$, $n_i(\Omega_1)$ and $u^{\alpha}(\Omega_2)$. Nevertheless calculation of

dn_{ij}^{stim}, dn_{ij}^{spon} and dn_{ji} from (7.22), (7.23) and (7.25) together with (7.32)-(7.34), has definite advantages. The probabilities $B'_{ji}u(\nu_{ij})$ and $B'_{ij}(\nu_{ji})$ refer as before to a single molecule, and represent averages of all actual transition probabilities in individual molecules over all the orientations of the molecules, all directions of propagation and polarisation. In contrast, the spontaneous transition probability A'_{ij} was obtained by averaging over all the orientations of the molecules and summing over all directions of emission. The coefficients B'_{ji}, B'_{ij} and A'_{ij} are not related by universal expressions of the form of (7.13) and (7.14). Such relationships can be obtained only in special cases. For example, if the distribution of the molecules with respect to the angles Ω_1 is the same in states i and j we have

$$B'_{ji}/B'_{ij} = g_i/g_j$$

Equations (7.32) and (7.33) become much simpler when polarised radiation travelling in a particular direction Ω_0 falls on a system of arbitrarily oriented particles. The density of such radiation may be represented by a δ-function:

$$u^{\alpha}(\nu_{ij}, \Omega_2) = u_0^{\alpha}\,\delta(\Omega_2 - \Omega_0) \tag{7.35}$$

Substituting (7.35) into (7.32) and (7.33), and using the well-known properties of δ-functions, we have

$$B'_{ji} = \frac{\int b_{ji}(\Omega_1, \Omega_0)\, n_j(\Omega_1)\, d\Omega_1}{\int n_j(\Omega_1)\, d\Omega_1} \tag{7.36}$$

$$B'_{ij} = \frac{\int B_{ij}(\Omega_1. \Omega_0)\, n_i(\Omega_1)\, d\Omega_1}{\int n_i(\Omega_1)\, d\Omega_1} \tag{7.37}$$

Degenerate levels

Let us suppose that the energy levels i and j are degenerate (Fig. 2.3). The degrees of degeneracy are respectively equal to g_i and g_j. The populations of the sub-levels E_i^{μ} and E_j^{λ} will be denoted by n_i^{μ} and n_j^{λ} respectively, where

$$\sum_{\mu=1}^{g_i} n_i^{\mu} = n_i, \quad \sum_{\lambda=1}^{g_j} n_j^{\lambda} = n_j \tag{7.38}$$

The expressions given by (7.22), (7.23) and (7.25) determine the total number of $i \to j$ or $j \to i$ transitions from all sub-levels of a particular state to all sub-levels of another state.

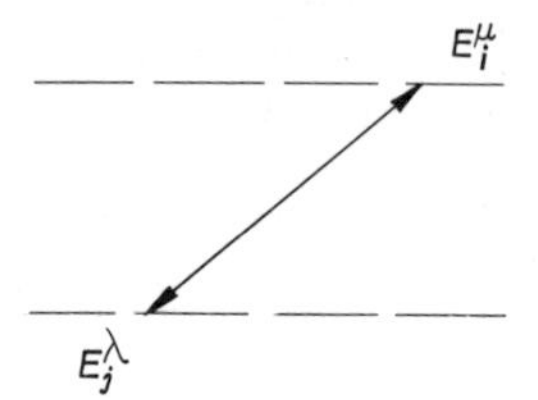

Fig. 2.3 Radiative transitions between sub-levels of degenerate levels

If we consider transitions between the sub-level E_i^{μ} and the sub-level E_j^{λ}, these relationships assume the form

$$dn_{(j\lambda)(i\mu)} = B_{ji}^{\lambda\mu} u_{ij} n_j^{\lambda} dV dt \tag{7.39}$$

$$dn_{(i\mu)(j\lambda)}^{\text{stim}} = B_{ij}^{\mu\lambda} u_{ij} n_i^{\mu} dV dt \tag{7.40}$$

$$dn_{(i\mu)(j\lambda)}^{\text{spon}} = A_{ij}^{\mu\lambda} n_i^{\mu} dV dt \tag{7.41}$$

The coefficients $B_{ji}^{\lambda\mu}$, $B_{ij}^{\mu\lambda}$ and $A_{ij}^{\mu\lambda}$ may be different for different combinations of μ and λ.

To obtain the total number of $i \leftrightarrows j$ transitions we must sum (7.39) to (7.41) over μ and λ. This leads to the original formulae (7.22), (7.23) and (7.25), where

$$B_{ji} = \frac{\sum_{\lambda} n_j^{\lambda} B_{ji}^{\lambda}}{\sum_{\lambda} n_j^{\lambda}} = \frac{\sum_{\lambda} n_j^{\lambda} \sum_{\mu} B_{ji}^{\lambda\mu}}{n_j} \tag{7.42}$$

$$B_{ij} = \frac{\sum_{\mu} n_i^{\mu} B_{ij}^{\mu}}{\sum_{\mu} n_i^{\mu}} = \frac{\sum_{\mu} n_i^{\mu} \sum_{\lambda} B_{ij}^{\mu\lambda}}{n_i} \tag{7.43}$$

$$A_{ij} = \frac{\sum_{\mu} n_i^{\mu} A_{ij}^{\mu}}{\sum_{\mu} n_i^{\mu}} = \frac{\sum_{\mu} n_i^{\mu} \sum_{\lambda} A_{ij}^{\mu\lambda}}{n_i} \tag{7.44}$$

It follows from these expressions that summation over initial sub-levels yields a mean rather than a sum. For example, in (7.44), the probability of spontaneous transition $i \to j$ is equal to the sum of the probabilities of transitions to all sub-levels of the set j from one sub-level of the set i averaged over the population.

8. QUANTUM-MECHANICAL THEORY OF ABSORPTION AND EMISSION OF RADIATION

Formulation of the problem

The quantum-mechanical theory of the interaction of light and matter is semi-classical in character. The properties of matter are described from the quantum-mechanical point of view but the properties of the electromagnetic field are described purely classically. It is precisely for this reason that quantum mechanics cannot account completely for the process. Nevertheless, it is very effective in many respects and provides, among other things, a very good description of the absorption of light and of the main features of stimulated emission.

According to the basic assumptions of quantum mechanics, any isolated system in a stationary state of energy E_i described by an eigenfunction $\psi_i(x)$ will remain in that state indefinitely until it is disturbed by an external agent. There can be no spontaneous transitions from a level i to other levels j which are associated with a change in the energy of the system. It is precisely for this reason that quantum mechanics cannot, even in principle, describe spontaneous emission of radiation.

The limitation is not accidental. In this particular respect, quantum mechanics is analogous to classical mechanics: in an isolated classical system the total energy is also conserved. Classical mechanics provides a good description of some features of the interaction of charged particles with electromagnetic fields when these fields are regarded as the cause of changes in the particle motion. However, it does not take into account the emission of radiation by charged particles executing accelerated or decelerated motion. Such processes are described only by classical electrodynamics. Similarly, a rigorous description of the interaction of light with matter, including spontaneous emission, can only be given within the framework of quantum electrodynamics.

The difficulties which arise in the quantum-mechanical investigation of emission may be greatly reduced by using Einstein's phenomenological theory in conjunction with quantum mechanics (see Section 7). The phenomenological theory is based on the generalisation of experimental data, and can be used together with the results of quantum mechanics to obtain a quantitative description of the process of emission.

Schroedinger's equation

Let us suppose that the particle experiences a variable electromagnetic field. The Hamiltonian for the system can then be written in the form $\mathbf{H}(x, t) = \mathbf{H}_0(x) + \mathbf{V}(x, t)$ where $\mathbf{H}_0(x)$ is the Hamiltonian for the system in the absence of the field and $\mathbf{V}$ is the operator representing the interaction of the particle with the field. Since the Hamiltonian depends on the time, it follows that possible non-stationary states of the system are characterised by a time-dependent wave function $\psi(x, t)$. The form of this function may be found by solving Schroedinger's equation

$$i \frac{h}{2\pi} \frac{\partial \psi(x,t)}{\partial t} = \mathbf{H} \psi(x, t) = [\mathbf{H}_0(x) + \mathbf{V}(x, t)] \psi(x, t) \quad (8.1)$$

When the interaction operator is absent, e.g. before the light source is switched on, the solution of the equation

$$i \frac{h}{2\pi} \frac{\partial \psi(x,t)}{\partial t} = \mathbf{H}_0(x) \psi(x, t) \quad (8.2)$$

reduces to the solution of the time-independent Schroedinger equation (5.2). Knowing the solution of (5.2) it is quite easy to obtain the solution of (8.2) simply by multiplying $\psi_i(x)$ by the time factor $e^{-2\pi i (E_i/h)t}$. Thus, when $V = 0$, we have

$$\psi_i(x, t) = \psi_i(x)\, e^{-2\pi i \frac{E_i}{h} t} \tag{8.3}$$

The quantity $|\psi(x, t)|^2 dx$ is time-independent and represents the probability of finding the system with energy E_i in the interval of coordinates between x and $x + dx$.

In accordance with the general postulates of quantum mechanics, the eigenfunctions are orthogonal and normalised, that is,

$$\int \psi_i^*(x, t)\, \psi_j(x, t)\, dx = \delta_{ij} \tag{8.4}$$

The set of eigenfunctions for the stationary states of the system is complete and closed and therefore any function $\psi(x, t)$ may be expanded in terms of the functions given by (8.3):

$$\psi(x, t) = \sum_i C_i(t)\, \psi_i(x)\, e^{-2\pi i \frac{E_i}{h} t} \tag{8.5}$$

Since the Schroedinger equation (8.2) is linear, the state function given by (8.5) is also a solution. However, in this case the state of the system is not stationary and its energy does not have a definite value. By measuring the energy we shall change the state of the system, taking it to a stationary state with one of the possible definite values of the energy, and therefore experiment yields one of the possible values of E_i which are admitted by the Schroedinger equation (5.2). The probability of obtaining experimentally a particular value of E_i, or of finding the system in the stationary state

$$\psi_i(x) e^{-2\pi i \frac{E_i}{h} t}$$

is determined by

$$W_i(t) = |C_i(t)|^2 \tag{8.6}$$

which is time-dependent. The mean energy of the system which is obtained as a result of a large number of measurements is given by

$$\int \psi^*(x, t)\, \mathbf{H}_0\, \psi(x, t)\, dx = \sum_i E_i\, |C_i(t)|^2 \tag{8.7}$$

This is also a function of time.

Solution of Schroedinger's equation by perturbation theory methods

Suppose that the external disturbance has been switched on and that we require the corresponding solution of (8.1). It will be a function of coordinates and of time and may therefore be sought in the form of (8.5). Substituting (8.5) into (8.1) we have

$$\frac{ih}{2\pi}\left[\sum_i \frac{\partial C_i}{\partial t}\psi_i(x,t) + C_i(t)\frac{\partial\psi(x,t)}{\partial t}\right] = \sum_i C_i(t)\,\mathbf{H}_0(x)\,\psi_i(x,t) + \sum_i C_i(t)\,\mathbf{V}(x,t)\,\psi_i(x,t) \tag{8.8}$$

On replacing the derivative $\frac{\partial\psi(x,t)}{\partial t}$ by the expression given by (8.2), we have

$$\frac{ih}{2\pi}\sum_i \frac{\partial C_i}{\partial t}\psi_i(x,t) = \sum_i C_i(t)\,\mathbf{V}(x,t)\,\psi_i(x,t) \tag{8.9}$$

If we now multiply both sides of (8.9), by $\psi_j(x,t)$, integrate with respect to x over all space and use the orthonormal property of the functions expressed by (8.4), we can easily obtain the following set of simple equations for the coefficient $C_j(t)$

$$\frac{ih}{2\pi}\frac{\partial C_j}{\partial t} = \sum_i C_i(t)\,e^{2\pi i\frac{E_j-E_i}{h}t}\,V_{ji}(t) \tag{8.10}$$

where

$$V_{ji}(t) = \int \psi_j^*(x)\,\mathbf{V}(x,t)\psi_i(x)\,dx \tag{8.11}$$

are the matrix elements of the interaction operator for the states j and i.

The system of equations given by (8.10) replaces the original Schroedinger equation (8.1). By determining the values of $C_j(t)$ for all t, we completely determine the state of the system.

When the interaction operator $\mathbf{V}(x, t)$ is relatively small, the solutions of (8.10) are more simply obtained from perturbation theory. Let us suppose that at time $t = 0$ the system is in a stationary state k and therefore, in (8.5),

$$C_i^0 = 1 \quad \text{for} \quad i = k; \quad C_i^0 = 0 \quad \text{for} \quad i \neq k \tag{8.12}$$

Substituting (8.12) into the right-hand side of (8.10) we obtain a system of equations for the first approximation $C_j^{\mathrm{I}}(t)$

$$\frac{ih}{2\pi} \frac{dC_j^{\mathrm{I}}(t)}{dt} = V_{jk}(t)\, e^{2\pi i \frac{E_j - E_k}{h} t} \tag{8.13}$$

Hence, in view of the initial conditions we have

$$C_j^{\mathrm{I}}(t) = \frac{2\pi}{ih} \int_0^t V_{jk}(t)\, e^{2\pi i \frac{E_j - E_k}{h} t} dt + \delta_{jk} \tag{8.14}$$

If we substitute this first approximation into the right-hand side of (8.10), we can easily find a system of equations for the second approximation:

$$\frac{ih}{2\pi} \frac{dC_j^{\mathrm{II}}(t)}{dt} = \sum_k V_{jk}(t)\, e^{2\pi i \frac{E_j - E_k}{h} t} C_k^{\mathrm{I}}(t) \tag{8.15}$$

By integrating (8.15) with respect to time, we obtain $C_j^{\mathrm{II}}(t)$. This procedure may be continued and will yield successively more accurate solutions for $C_j(t)$. For the interaction of radiation with matter, the operator $\mathbf{V}(x, t)$ is relatively small and it is frequently possible to stop after the first or second approximation.

The above calculation is, in principle, suitable for any perturbation. For interaction with an electromagnetic field $E = E_0 \cos 2\pi\nu t$ of given frequency ν, the operator $\mathbf{V}(x, t)$ is a periodic function of time and a detailed solution is possible. The operator $\mathbf{V}$ can then be written in the form

$$\mathbf{V}(x, t) = \mathbf{V}(x)\, e^{-2\pi i\nu t} + \mathbf{V}^*(x)\, e^{2\pi i\nu t} \tag{8.16}$$

and is determined by the properties of the system, the density of the electromagnetic field and the properties of the interaction between them. Substituting (8.16) and (8.11) we obtain

$$V_{ji}(x, t) = V_{ji} e^{-2\pi i\nu t} + V_{ji}^* e^{2\pi i\nu t} \tag{8.17}$$

where

$$V_{ji} = \int \psi_j^*(x) \mathbf{V}(x) \psi_i(x)\, dx, \quad V_{ji}^* = \int \psi_j^*(x) \mathbf{V}^*(x) \psi_i(x)\, dx \tag{8.18}$$

are the matrix elements of $\mathbf{V}(x)$ and $\mathbf{V}^*(x)$.

Substituting (8.17) into (8.14) and performing the integration with respect to time subject to the given initial conditions, we obtain the first-approximation coefficients $C_j^{\mathrm{I}}(t)$. To indicate the initial level k we shall write $C_{kj}^{\mathrm{I}}(t)$ instead of $C_{kj}^{\mathrm{I}}(t)$. We then have for $j \neq k$

$$C_{kj}^{\mathrm{I}}(t) = -\frac{V_{jk}\, e^{2\pi i\left(\frac{E_j - E_k}{h} - \nu\right)t} - 1}{h\left(\frac{E_j - E_k}{h} - \nu\right)} - \frac{V_{jk}^{*} e^{2\pi i\left(\frac{E_j - E_k}{h} + \nu\right)t} - 1}{h\left(\frac{E_j - E_k}{h} + \nu\right)} \qquad (8.19)$$

Some special cases

Let us suppose that the frequency ν of the incident radiation is equal to the $j \rightarrow k$ transition frequency, i.e.

$$\nu = \frac{E_j - E_k}{h} = \nu_{jk} \qquad (8.20)$$

Moreover, let $E_j > E_k$. The denominator in the first term in (8.19) will then be zero. Since

$$\lim_{\nu \rightarrow \nu_{jk}} \frac{e^{2\pi i(\nu_{jk} - \nu)t} - 1}{\nu_{jk} - \nu} = 2\pi i t \qquad (8.21)$$

the first term in (8.19) will continuously increase with time and may become very large. The second term in (8.19) is negligible. Similarly, small values of $C_i^{\mathrm{I}}(t)$ for all i do not specify (8.20).

According to (8.6) the quantity $|C_{kj}^{\mathrm{I}}(t)|^2$ is the probability of finding the system at time t in the state j if at $t=0$ it was in the initial state k with energy $E_k < E_j$. It therefore follows from (8.19) and (8.21) that the energy of the system increases under the action of the incident radiation. This energy could only be taken at the expense of the incident flux, and therefore the above result describes the process of absorption of radiation which is accompanied by a change in the state of the system. If the energy of the initial state

$\psi_k(x)$ is greater than the energy of the j-th level, the first term in (8.19) will be negligible. When $\nu = \nu_{kj}$, the second term in (8.19) becomes important. After the light source has been switched on, the system may then be found in the state j of lower energy. The energy difference is emitted in the form of radiation, and therefore equation (8.19) describes the process of stimulated emission.

For monochromatic radiation and a discrete energy spectrum it follows from (8.21) and (8.19) that

$$|C^{\mathrm{I}}_{kj}(t)|^2 = \frac{4\pi^2 |V_{kj}|^2}{h^2} t^2 \tag{8.22}$$

The probability of finding the system in the state j is a quadratic function of t. This holds for small t when the first approximation of the perturbation theory is adequate $(|C^{\mathrm{I}}_{kj}(t)|^2 \ll 1)$.

The quadratic dependence on time is rarely found in practice, although it must be taken into account in the solution of some problems. The other special cases are encountered much more frequently. In the first, the system is illuminated by non-monochromatic light with a fairly narrow range of frequencies. In the second case, the incident flux is monochromatic, but the energy levels form a continuum. In order to establish the effect of the incident radiation in these two cases, we must investigate the frequency dependence of $|C^{I}_{kj}(t)|^2$. If we neglect the second term in (8.19), i.e. if we consider processes accompanied by an increase in the energy of the system, we obtain

$$|C^{\mathrm{I}}_{kj}(t, \nu)|^2 = \frac{4\,|V_{ji}|^2}{h^2} \times \frac{\sin^2 \pi (\nu_{jk} - \nu)\, t}{(\nu_{jk} - \nu)^2} \tag{8.23}$$

Figure 2.4 shows the probability $|C^{\mathrm{I}}_{kj}(t)|^2$ as a function of the frequency of incident radiation. The principal maximum lies at $\nu = \nu_{jk}$. The height of the principal peak is proportional to t^2 and the area under the curve is proportional to t. The first minima of (8.23) lie at the points

$$\nu_{\min} = \nu_{jk} \pm \frac{1}{t}$$

where t is the time measured from the instant at which the light source was switched on. For large values of t, the width of the principal maximum is very small. The minimum exposure time which it is possible to use is of the order of a

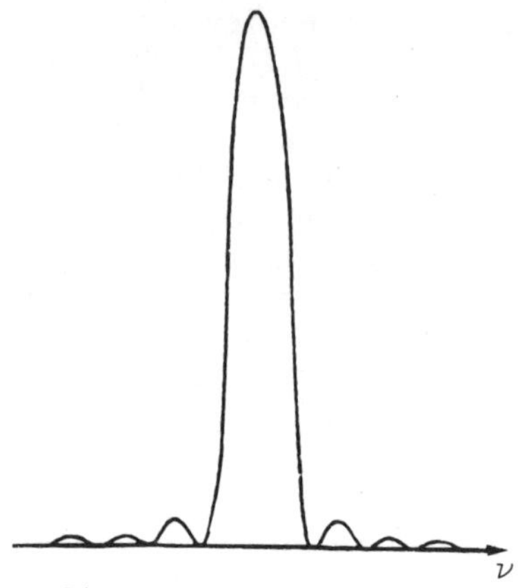

Fig. 2.4 Plot of the function (8.23)

microsecond and therefore the width of the principal maximum is of the order of 10^6 sec^{-1}.

All the usual sources of light produce non-monochromatic radiation and even for the best of them there is a spread of $\sim 10^7$ sec^{-1} in the visible region. Therefore, in studying the absorption at discrete levels we must take into account the fact that the incident light has a continuous spectrum, but the density of the incident radiation is independent of frequency under the principal peak of (8.23).

The basic equation (8.19) is also valid when matter is illuminated by radiation with a continuous frequency spectrum. It need only be remembered that interaction operators such as (8.16) will include a set of frequencies. Since these frequencies are very close together, the operators $\mathbf{V}(x)$ and $\mathbf{V}^*(x)$ in (8.16) and matrix element (8.18) may be regarded as frequency-independent. The resultant effect of the incident radiation is obtained by integrating (8.23) with respect to frequency at constant $|V_{ji}|^2$. Since

$$\int_{-\infty}^{\infty} \frac{\sin^2 ax}{x^2} dx = a\pi \tag{8.24}$$

it follows that

$$|C^{\text{I}}_{kj}(t)|^2 = \int |C_{kj}(t, \nu)|^2 d\nu = \frac{4\pi^2 |V_{jk}|^2}{h^2} t \tag{8.25}$$

The probability of finding the particle in the state j at time t (if at $t = 0$ it was in the state k) under the action of radiation with a continuous frequency spectrum is therefore proportional to the time t.

An analogous result can easily be obtained for the second special case, i.e. the interaction of monochromatic radiation with particles having a continuous energy spectrum. Equations (8.19) and (8.23) are also valid in this case. Under the action of radiation of frequency ν, a particle which at $t = 0$ was in the state k, may occupy one of the continuum of levels $E_j = E_k + h\nu$. The probability of finding the particle in states with energy $E_j \neq E_k + h\nu$ is given by (8.23), and the central peak of the probability curve is very narrow. Integrating (8.23) with respect to E_j on the assumption that $|V_{ji}|^2$ remains constant, we have

$$|C^{\mathrm{I}}_{kE_j}(t)|^2 = \int |C_{kE_j}(t)|^2 \, dE_j = \frac{4\pi^2}{h} |V_{kE_j}|^2 t \qquad (8.26)$$

In the case under consideration, the probability $|C_{kE_j}(t)|^2$ is also proportional to t. The transition occurs in a very narrow energy interval near

$$E = E_k + h\nu \qquad (8.27)$$

So far, we have been concerned with the solution of the system of equations given by (8.10) subject to the initial conditions (8.12), i.e. $C_k(t = 0) = 1$; $C_j(t = 0) = 0$. If at the initial instant of time $C_j(t = 0) = 1$, $C_k(t = 0) = 0$ and $E_j > E_k$, an interchange of the subscripts j and k in (8.10) should have no effect on the solution of (8.10). When the system is illuminated by light with a continuous frequency distribution concentrated near $\nu = \nu_{jk}$ and the process of stimulated emission takes place, the system may be found in an energetically lower state with a probability

$$|C^{\mathrm{I}}_{jk}(t)|^2 = \frac{4\pi^2 |V_{kj}|^2}{h^2} t \qquad (8.28)$$

Comparison of (8.25) with (8.28) will show that the qualitative characteristics of the processes of absorption and of stimulated emission are identical. Other conditions being equal, the two processes proceed at the same rate.

The concept of transition probability

According to (8.25) and (8.26), first-order quantum-mechanical theory of the interaction of radiation with matter predicts that the probability of finding the system in a higher energy state is proportional to the time. Since at $t = 0$ the particle was in the state k while at time t it may be found in the state j, it is possible to speak of a transition from the state k to the state j.

When $|C^{\mathrm{I}}_{kj}(t)|^2$ is proportional to t, we can introduce the concept of the $k \to j$ transition probability which is defined by

$$p_{kj} = \frac{d}{dt} |C^{\mathrm{I}}_{kj}(t)|^2 = \frac{4\pi^2}{h^2} |V_{jk}|^2 \tag{8.29}$$

The transition probability p_{kj} is independent of time and this justifies its name. It must be emphasised that the word 'probability' is used here in two different ways. The quantity $|C_{kj}(t)|^2$ is the probability in the usual sense of the word given to it in mathematics. It is equal to the ratio of the number of cases in which measurements show the particle to be in the state j to the total number of measurements. The transition probability p_{kj} is a dimensional quantity (sec^{-1}) and may assume large values. It is, in fact, the proportionality factor determining the rate of change of $|C^{\mathrm{I}}_{kj}(t)|^2$.

The transition probability (8.29) characterises the process of absorption. The probability of stimulated emission associated with $j \to k$ transitions can be introduced in a similar way. According to (8.28), the probability

$$p_{jk} = \frac{d}{dt} |C^{\mathrm{I}}_{jk}(t)|^2 = \frac{4\pi^2}{h^2} |V_{jk}|^2 . \tag{8.30}$$

is equal to p_{kj}. This is the quantum-mechanical justification of the relation given by (7.30) for the Einstein coefficients characterising the absorption and stimulated emission of radiation by a particle.

It follows from the discussion in the preceding section that the concept of transition probability is not universally valid, but is frequently applicable nevertheless. It is invalid when the frequency of the incident radiation ν is not equal to the frequency ν_{ik} corresponding to any pair of levels, i.e. for the phenomenon of scattering of light. Again, it is invalid for resonance transitions under the action of monochromatic radiation of frequency ν_{kj} and in many other cases. At the same time, the transition probability can conveniently be used for the description of the absorption and stimulated emission of radiation, since the experimental conditions under which these processes occur approach those which were postulated in the derivation of the solution which is set out above of the Schroedinger equation in the presence of an external perturbation.

As indicated by the symbol $|C^{I}_{kj}(t)|^2$, all the results obtained above are valid in first-order perturbation theory. Whether a particular approximation is valid or not can be easily checked with the aid of the inequality

$$|C^{I}_{kj}(t)|^2 = p_{kj}t \ll 1 \tag{8.30a}$$

which should hold either for small values of p_{kj} or for short times. In the first approximation it is assumed that the probability of finding the system in the initial state k is equal to unity, although there is already a finite probability of finding the system at a higher energy level j. This internal contradiction can be resolved only in higher-order approximations.

For large t the actual process is quite complicated. As $|C_j(t)|^2$ initially increases, there is a reduction in $|C_k(t)|^2$ which leads to a reduction in the rate of increase of $|C_j(t)|^2$. As a result of the stimulated emission of radiation, there is also a tendency for $|C_k(t)|^2$ to increase, and this may become appreciable for large $|C_j(t)|^2$. As t increases further and the system departs from its initial state, the importance of both levels, j and k, becomes comparable. Moreover, the quantities $|C_k(t)|^2$ and $|C_j(t)|^2$ become equal after a sufficient length of time, and the instantaneous values of $|C_k(t)|^2$ and $|C_j(t)|^2$ fluctuate about a mean value. When $|C_k(t)|^2$ and $|C_j(t)|^2$ remain constant, the state of the system can naturally be referred to as a stationary state, whereas the preceding period during which $|C_j(t)|^2$ increases may be called a transient period. The time necessary to reach the stationary state is determined by the density of the exciting radiation and the degree of interaction between the field and the medium, i.e. by the matrix elements (8.18).

In a real experiment the particles may be illuminated with a very short pulse of radiation or with a pulsating radiation. If the period of illumination is short, the first approximation may give satisfactory results and may provide a complete description of the process. However, the illumination is usually continuous, and the interaction of incident radiation with the medium continues indefinitely. The stationary state is then definitely reached and the first approximation to the Schroedinger equation (8.1) is, of course, invalid. At the same time, higher-order approximations are exceedingly difficult to obtain. It is only for particles with two energy levels that a precise solution of (8.1) can be obtained for illumination with monochromatic radiation.

A semi-phenomenological probabilistic method has been developed in the literature for the quantitative calculation of the rates of absorption and emission of radiation, and of its other properties. The method originated in Einstein's work, in which the concept of transition probability was first formulated (see Section 7). The probabilistic method has been found useful in practice, and whenever it has led to a quantitative result it has been verified experimentally. The accuracy of the method and its limits of applicability are still not clear however.

The probabilistic method, which is essentially based on (8.29), can be justified by quantum mechanics. It is assumed that the method is valid not only for the initial instant of time (which is immediately apparent from its derivation), but also for all times. Let us suppose that at time $t = t_0$, the probability of finding the system in state k which is described by $\psi_k(x)$ is equal to $|C_k(t_0)|^2$, while in the state j it is equal to $|C_j(t_0)|^2$. The complete eigenfunction for the system at this time is

$$\psi(x, t_0) = \sum_i C_i(t_0)\,\psi_i(x)e^{-2\pi i \frac{E_i}{h} t_0} \tag{8.31}$$

In order to find the change in the state of the system in the time interval between t_0 and $t_0 + dt$, we must solve the system of equations (8.10) subject to initial conditions which differ from (8.12) and take into account not only the real but also the complex parts of $C_i(t_0)$. By analogy with (8.29) we have

$$d\,|C_j(t)|^2 = p'_j dt \tag{8.32}$$

where p'_j is different from p_{kj} in (8.29). It depends not only on the interaction operator $\mathbf{V}(x, t)$, but also on the initial conditions $C_i(t_0)$ and $C_k(t_0)$. In the probabilistic method it is assumed that p'_j may be expressed directly in terms of the transition probability (8.29) subject to the conditions $|C_k(0)|^2 = 1$, and $|C_j(0)|^2 = 0$, and in terms of the probability of finding the system in states k and j at time $t = t_0$:

$$p'_j = |C_k(t_0)|^2 p_{kj} - |C_j(t_0)|^2 p_{jk}. \tag{8.33}$$

Similarly, the change in the probability $|C_k(t)|^2$ is given by

$$d|C_k(t)|^2 = p'_k\,dt \tag{8.34}$$

where

$$p'_k = |C_j(t_0)|^2 p_{jk} - |C_k(t_0)|^2 p_{kj} \tag{8.35}$$

The time dependence of the probabilities p'_j and p'_k is therefore transferred to the quantities $|C_j(t)|^2$ and $|C_k(t)|^2$, while the transition probability itself is supposed to be constant. The values of p'_j in (8.33) and of p'_k in (8.35) are written as differences because absorption and stimulated emission proceed in parallel and are merely two aspects of the interaction of radiation with matter. A justification of the probabilistic method will be given in Section 16.

Operator for the interaction of radiation with matter

Let us consider in greater detail the operator for the interaction of a charged particle with the electromagnetic field. An electromagnetic wave may be specified in a vacuum by a single vector potential $\mathbf{A}$, which may be chosen so that $\operatorname{div}\mathbf{A} = 0$. The interaction operator is then of the form

$$\mathbf{V}_{\text{int}} = -\frac{e}{mc}(\mathbf{pA}) + \frac{e}{2mc^2}\mathbf{A}^2 \tag{8.36}$$

This expression is usually derived from the classical Hamiltonian function. If in addition to the electromagnetic potentials $\mathbf{A}$, φ the particle is in the field of some other forces specified by the force function $\mathbf{U}$, the correct equation of motion is obtained from the Hamiltonian

$$\mathbf{H} = \frac{1}{2m}\left(\mathbf{p} - \frac{e}{c}\mathbf{A}\right)^2 + e\varphi + \mathbf{U} \tag{8.37}$$

where e and m are the charge and mass of the particle and $\mathbf{p}$ is the generalised momentum. In quantum theory, $\mathbf{p}$ is replaced by the momentum operator

$$\mathbf{p} = -i\frac{h}{2\pi}\left(\mathbf{i}\frac{\partial}{\partial x} + \mathbf{j}\frac{\partial}{\partial y} + \mathbf{k}\frac{\partial}{\partial z}\right) = -i\frac{h}{2\pi}\nabla$$

where ∇ is the so-called nabla operator. Since the momentum operators and the function $\mathbf{A}$ obey the commutation relations

$$p_i\mathbf{A}_i - \mathbf{A}_i p_i = -i\frac{h}{2\pi}\frac{\partial \mathbf{A}_i}{\partial x} \tag{8.38}$$

it follows from (8.37) that

$$\dot{\mathbf{H}} = \frac{1}{2m}\mathbf{p}^2 + \mathbf{U} - \frac{e}{m}(\mathbf{pA}) + \frac{e}{2mc^2}\mathbf{A}^2 - i\frac{h}{4\pi m} \times \operatorname{div}\mathbf{A} + e\varphi \tag{8.39}$$

The first two terms in this expression do not depend on the presence of the light wave. They form an operator which may be regarded as a Hamiltonian for a free quantum system. The interaction operator is given by the remaining terms

$$\mathbf{V}_{\text{int}} = -\frac{e}{mc}(\mathbf{pA}) + \frac{e}{2mc^2}\mathbf{A}^2 - i\frac{h}{4\pi m}\operatorname{div}\mathbf{A} + e\varphi \tag{8.40}$$

When $\varphi = 0$ and $\operatorname{div}\mathbf{A} = 0$ (gauge invariance), equation (8.40) becomes identical with (8.36) and $\mathbf{p}$ commutes with $\mathbf{A}$.

The second term in (8.36) is smaller than the first by a few orders of magnitude, and can be neglected for single-photon processes such as absorption or emission of radiation. The term containing $\mathbf{A}^2$ need only be taken into account in calculations of multi-photon processes such as, for example, scattering of light. It follows that in first-order perturbation theory we may write

$$\mathbf{V}_{\text{int}} = \frac{e}{mc}(\mathbf{pA}) = \frac{ieh}{2\pi mc}\mathbf{A}\nabla \tag{8.41}$$

Suppose that a charged particle (the optical electron of an atom) interacts with a plane monochromatic wave

$$E = E_0 \sin 2\pi \left[\nu t - \frac{(\mathbf{nr})}{\lambda} \right] \tag{8.42}$$

where $\mathbf{n}$ is a unit vector specifying the direction of propagation of the wave and $\nu = \nu_{ij}$. It can readily be verified using (1.2) that the vector potential for this wave is

$$\mathbf{A} = \frac{c}{2\pi\nu} \mathbf{E}_0 \cos 2\pi \left[\nu t - \frac{(\mathbf{nr})}{\lambda} \right] \tag{8.43}$$

while the interaction operator is

$$\mathbf{V}_{\text{int}}(\mathbf{r}, t) = -\frac{e}{2\pi m\nu} (\mathbf{pE}_0) \cos 2\pi \left[\nu t - \frac{(\mathbf{nr})}{\lambda} \right]$$

$$= -\frac{e}{4\pi m\nu} (\mathbf{pE}_0)\, e^{2\pi i \frac{(\mathbf{nr})}{\lambda}} e^{-2\pi i \nu t} \tag{8.44}$$

$$-\frac{e}{4\pi m\nu} (\mathbf{p}^*\mathbf{E}_0^*)\, e^{-2\pi i \frac{(\mathbf{nr})}{\lambda}} e^{2\pi i \nu t}$$

Comparing (8.16) with (8.44) we obtain

$$\mathbf{V}(x) = -\frac{e}{4\pi m\nu} (\mathbf{pE}_0)\, e^{2\pi i \left(\frac{\mathbf{nr}}{\lambda}\right)} = \frac{ieh}{8\pi^2 m\nu} \mathbf{E}_0 e^{2\pi i \left(\frac{\mathbf{nr}}{\lambda}\right)} \nabla \tag{8.45}$$

Multipolarity

Some of the general properties of the matrix elements of the interaction operator, and therefore of the transition probabilities, can easily be established with the aid of (8.45). Substituting $\mathbf{V}(x)$ into (8.45) and (8.18) we have

$$V_{ji} = \frac{ieh}{8\pi^2 m\nu} \mathbf{E}_0 \int \psi_j^*(x)\, e^{2\pi i \left(\frac{\mathbf{nr}}{\lambda}\right)} \nabla \psi_i(x)\, dx \tag{8.46}$$

where the integration should be performed over all x. Since the wave functions have appreciable values only within the limits of the atom (molecule) it may be assumed that r does not exceed the radius a of the atom. This means that for wavelengths in the optical region $\mathbf{nr} \approx 2a$ is much smaller than unity. The factor $\exp\left(2\pi i \frac{\mathbf{nr}}{\lambda}\right)$, which gives the change in the phase of the incident wave within the atom, may therefore be expanded in powers of $2\pi\left(\frac{\mathbf{nr}}{\lambda}\right)$ ...

$$e^{2\pi i\left(\frac{\mathbf{nr}}{\lambda}\right)} = 1 + 2\pi i\left(\frac{\mathbf{nr}}{\lambda}\right)..$$

In the first approximation, equation (8.46) will then read

$$V_{ji} = \frac{ieh}{8\pi^2 m\nu}\mathbf{E}_0 \int \psi_j^* \nabla \psi_i \, dx - \frac{he}{4\pi \, mc}\mathbf{E}_0 \int \psi_j^* (\mathbf{nr}) \nabla\psi_i dx + ... \quad (8.47)$$

The first term in this expression is the matrix element of the momentum operator

$$\mathbf{p}_{ji} = -i\frac{h}{2\pi}\int \psi_j^* \nabla\psi_i \, dx \quad (8.48)$$

It may be expressed in terms of the matrix elements of the position vector and therefore of the dipole moment [15]:

$$\mathbf{p}_{ji} = 2\pi \, im \nu_{ji} \, \mathbf{r}_{ji} = 2\pi \, im \nu_{ji} \frac{1}{e}\mathbf{D}_{ji} \quad (8.49)$$

The second term in (8.47) contains the matrix element of the operator

$$(\mathbf{nr})\frac{\mathbf{p}}{m} = (\mathbf{nr})\frac{dr}{dt}$$

which can easily be written in the form

$$(\mathbf{nr})\frac{d\mathbf{r}}{dt} = \frac{1}{2}\frac{d}{dt}((\mathbf{nr})\,\mathbf{r}) - \frac{1}{2}\left[\mathbf{n}\left[\mathbf{r}\frac{d\mathbf{r}}{dt}\right]\right] \quad (8.50)$$

The quantity $3e(\mathbf{nr})\mathbf{r}$ is the product of $\mathbf{n}$ and the second-rank tensor

$$Q = \begin{vmatrix} 3ex^2 & 3exy & 3exz \\ 3eyx & 3ey^2 & 3eyz \\ 3ezx & 3ezy & 3ez^2 \end{vmatrix} \quad (8.51)$$

which is analogous to the quadrupole-moment tensor (see Section 1), while $\left[\mathbf{r}\frac{d\mathbf{r}}{dt}\right]$ is the angular momentum operator

$$\left[\mathbf{r}\frac{d\mathbf{r}}{dt}\right] = \frac{1}{m}[\mathbf{r}\mathbf{p}] = \frac{1}{m}\mathbf{M}' \tag{8.52}$$

It is related to the magnetic moment $\mathbf{M}$ of the atom by the simple expression

$$\mathbf{M} = -\frac{e}{2mc}\mathbf{M}' \tag{8.53}$$

Since, moreover, the time-independent matrix element of the derivative of $\mathbf{L}$ is given by

$$\left(\frac{d\mathbf{L}}{dt}\right)_{ji} = i\,2\pi\nu_{ji}L_{ji} \tag{8.54}$$

we can rewrite (8.47) in the form

$$\begin{aligned} V_{ji}(x) &= -i\frac{1}{2}(\mathbf{E}_0\mathbf{D}_{ji}) \\ &+ \frac{\pi}{6\lambda}(\mathbf{E}_0(\mathbf{n}\mathbf{Q}_{ji})) - \frac{i}{2}(\mathbf{E}_0[\mathbf{n}\mathbf{M}_{ji}]) + \ldots \end{aligned} \tag{8.55}$$

If we neglect the change in phase within the atom, the matrix elements of the interaction operator will depend only on the matrix elements of the dipole moment of the atom so that

$$V_{ji}(x) = -\frac{i}{2}(\mathbf{E}_0\mathbf{D}_{ji}) \tag{8.56}$$

The absorption and emission of radiation, associated with changes in the dipole moment, are usually referred to as dipole absorption and emission.

The quadrupole and magnetic dipole emission, associated with changes in $\mathbf{Q}$ and $\mathbf{M}$ respectively, arise when the phase change is taken into account on the first approximation. The next terms in the expansion in (8.47) give the octupole, magnetic quadrupole and higher-order emission. If the dipole emission at a particular frequency is allowed by the selection rules, it is usually much stronger than the quadrupole and magnetic dipole emissions. The ratio of the intensities associated with quadrupole and dipole emissions is of the order of $\left(\frac{2\pi a}{\lambda}\right)^2$. Classical electrodynamics leads to a similar conclusion (Section 1).

If transitions with dipole emission cannot take place from a given energy level, then owing to the low probability of quadrupole transitions, the corresponding lifetime τ is relatively long. In the case of visible radiation and $a = 1$ Å, it is approximately equal to 0.01 sec. States with lifetimes of this order are called metastable states. The quadrupole and magnetic dipole emission must be allowed for only when dipole emission is forbidden.

Selection rules

It was noted above [see (8.29) and (8.30)] that the probability of stimulated transitions depends on the properties of the quantum system, the density of the incident radiation and the nature of the interaction. Equation (8.55) may be used to investigate these effects individually. It is easy to see that V_{ji} depends on the properties of the system only through the matrix elements of the dipole moment, the quadrupole moment and so on. If, for example, a particular matrix element $\mathbf{D}_{ml}$ is equal to zero, a transition between levels m and l with dipole absorption or stimulated emission of radiation is not allowed. If $\mathbf{Q}_{ml}$ or $\mathbf{M}_{ml}$ are equal to zero, transitions with quadrupole or magnetic dipole absorption or emission are forbidden.

Each quantum system is characterised by certain selection rules which indicate which transitions are allowed. All other transitions are forbidden. There are no selection rules which would apply to all quantum systems. The selection rules are not connected with the structural details of a particular system, however, but are determined by certain fundamental properties which are common to certain classes of systems.

Consider the selection rules for a particle moving in a centrally symmetrical field (optical electron of the hydrogen atom or alkali-metal atom; or, approximately, certain more complicated atoms). The stationary states of a system of this kind are described by the wave functions [15]

$$\psi_{nlm}(r, \theta, \varphi) = R_{nl}(r) P_l^m(\cos\theta) e^{im\varphi} \qquad (8.57)$$

where $R_{nl}(r)$ are the radial functions, $P_l^m(\cos\theta)$ are the spherical harmonics and n, l, m are the principal, orbital and magnetic quantum numbers respectively.

To begin with, let us consider the selection rules for dipole emission. Since the matrix elements of the dipole moment and of the position-coordinate operator differ by the constant factor e, it is sufficient to calculate only the position-coordinate matrix elements. In order to reduce the algebra, it is convenient to calculate the matrix elements of the following combinations of coordinates:

$$\begin{aligned} \xi &= x + iy = r\sin\theta\, e^{i\varphi} \\ \eta &= x - iy = r\sin\theta\, e^{-i\varphi} \\ z &= r\cos\theta \end{aligned} \tag{8.58}$$

Hence, we have

$$\xi_{nlm,\, n'l'm'} = \int_0^\infty R_{nl}R_{n'l'}r^3\,dr \int_0^\pi P_l^m P_{l'}^{m'} \sin^2\theta\,d\theta \int_0^{2\pi} e^{i(m-m')\varphi + i\varphi}\,d\varphi$$

$$\begin{aligned} \eta_{nlm,\, n'l'm'} &= \int_0^\infty R_{nl}R_{n'l'}r^3dr \int_0^\pi P_l^m P_{l'}^{m'} \sin^2\theta\,d\theta \\ &\times \int_0^{2\pi} e^{i(m-m')\varphi - i\varphi}\,d\varphi \end{aligned} \tag{8.59}$$

$$\begin{aligned} z_{nlm,\, n'l'm'} &= \int_0^\infty R_{nl}R_{n'l'}\,r^3dr \int_0^\pi P_l^m P_{l'}^{m'} \cos\theta\sin\theta\,d\theta \\ &\times \int_0^{2\pi} e^{i(m-m')\varphi}\,d\varphi \end{aligned}$$

The integrals with respect to φ are clearly given by

$$\begin{aligned} \int_0^{2\pi} e^{i(m-m')\pm i\varphi}\,d\varphi &= 2\pi\delta_{m'\mp 1,\, m} \\ \int_0^{2\pi} e^{i(m-m')\varphi}\,d\varphi &= 2\pi\delta_{m',m} \end{aligned} \tag{8.60}$$

Since the fact that these integrals are not equal to zero is a necessary condition that the integrals in (8.59) should have non-zero values, it follows immediately from (8.25) that the selection rules for the magnetic quantum number are

$$m' - m = 0 \text{ or } \pm 1 \tag{8.61}$$

This means that only those dipole transitions are possible for which the magnetic moment of the system remains either unchanged or changes by $\pm$ 1 (in the appropriate units). When $\Delta m = 0$ the emitted photon is linearly polarised along the z-axis. When $\Delta m = \pm 1$ the polarisation may be circular or elliptical, depending on the direction of observation, as in the case of a rotator (see Section 1).

In order to find the selection rules for the orbital quantum number l we must establish the conditions under which the integrals with respect to ϑ do not vanish. When $m = m'$ they are of the form

$$C_{ll'}^{mm} = \int_0^\pi P_l^m P_{l'}^m \cos\vartheta \sin\vartheta \, d\vartheta \tag{8.62}$$

If we substitute $x = \cos\theta$ into this expression we have

$$C_{ll'}^{mm} = \int_{-1}^{1} P_l^m(x) P_{l'}^m(x) \, x dx \tag{8.62a}$$

Using the following property of spherical harmonics

$$xP_l^m(x) = a_{lm} P_{l+1}^m(x) + b_{lm} P_{l-1}^m(x) \tag{8.63}$$

where a_{lm} and b_{lm} are certain coefficients which depend on m and l. Recalling that P_l^m are orthogonal, we have from (8.62)

$$C_{ll'}^{mm} = a_{lm} \delta_{l', l+1} + b_{lm} \delta_{l', l-1} \tag{8.64}$$

and hence the following selection rules for $m = m'$:

$$l' - l = \pm 1 \tag{8.65}$$

When $m' = m \pm 1$ we need only to evaluate the integral

$$S_{ll'}^{m, m\pm 1} = \int_{-1}^{+1} P_l^{m\pm 1}(x) \sqrt{1 - x^2} \, P_l(x) \, dx \tag{8.66}$$

Since

$$\sqrt{1-x^2}\; P_l^m(x) = a'_{lm} P_{l-1}^{m+1}(x) + b'_{lm} P_{l+1}^{m+1}(x) \qquad (8.67)$$

we have from (8.66)

$$\begin{aligned} S_{ll'}^{m,\,m+1} &= a'_{lm}\,\delta_{l-1,\,l'} + b'_{lm}\,\delta_{l+1,\,l'} \\ S_{ll'}^{m,\,m-1} &= a'_{l,\,m-1}\,\delta_{l,\,l'-1} + b'_{l,\,m-1}\,\delta_{l,\,l'+1} \end{aligned} \qquad (8.68)$$

These two expressions again lead to the selection rules (8.65). We thus see that only those transitions are allowed in dipole emission or absorption for which the orbital quantum number changes by ± 1.

We recall that in spectroscopy the state with orbital quantum number $l = 0$ is called the s-term, the state with $l = 1$ is called the p-term, the state with $l = 2$ is called the d-term and so on. The selection rules state that the only transitions possible are those between terms s and p, p and d, d and f and so on.

There are no selection rules for the principal quantum number. The difference $n-n'$ can assume any value.

In order to find the selection rules for quadrupole emission, we must evaluate the matrix elements of the components of the quadrupole moment tensor given by (8.51). Since the matrix elements for x, y and z have already been calculated, this can easily be done using the rule for the multiplication of matrices. For example,

$$(x^2)_{l'l} = \sum_{l''} x_{l'l''} x_{l''l}.$$

Since $l' = l'' \pm 1$ and $l'' = l \pm 1$, it follows that $l - l' = 0$ or ± 2. Similar results are obtained for the other components of the tensor Q. Calculations lead to the following selection rules for quadrupole emission:

$$\begin{aligned} m - m' &= 0,\ \pm 1,\ \pm 2 \\ l - l' &= 0,\ \pm 2 \end{aligned} \qquad (8.69)$$

These rules have one exception: transitions between states for which $l = l' = 0$ are forbidden, since the corresponding matrix elements are zero. There is no magnetic dipole emission in atomic systems with single optically active electrons.

Classical theory leads to the same conclusion. Selection rules for more complicated systems lie outside the scope of the present book [16].

Dipole emission. Correspondence principle

If we substitute (8.56) into (8.29) and confine our attention to systems with non-degenerate levels, we obtain the transition probabilities for dipole absorption and stimulated emission:

$$p_{kj} = p_{jk} = \frac{4\pi^2}{h^2} |V_{jk}|^2 = \frac{\pi^2}{h^2} |(\mathbf{E}_0 \mathbf{D}_{kj})|^2 \tag{8.70}$$

If we denote the angle between $\mathbf{D}_{kj}$ and $\mathbf{E}_0$ by θ_{kj}, this expression may be written in the form

$$p_{kj} = \frac{\pi^2}{h^2} |\mathbf{D}_{kj}|^2 \mathbf{E}_0^2 \cos^2 \theta_{kj} = \frac{4\pi^3}{h^2} |\mathbf{D}_{kj}|^2 \cos^2 \theta_{kj} u(\nu) \tag{8.71}$$

where

$$u(\nu) = \frac{1}{8\pi} \mathbf{E}_0^2$$

is the density of the incident radiation. Comparison of (8.71) with (7.4) leads to the following expression for the differential Einstein coefficients for the absorption and stimulated emission in the direction of the incident radiation:

$$b_{kj}(\nu) = b_{jk}(\nu) = \frac{8\pi^3}{h^2} |\mathbf{D}_{kj}|^2 \cos^2 \theta_{kj} \tag{8.72}$$

It should be noted that the coefficients b_{kj} depend on the frequency scale (ν or ω) of u. Since unit intervals on the ν and ω scales differ by a factor of 2π, it follows that $b_{kj}(\nu)$ and $b_{kj}(\omega)$ are related by

$$b_{kj}(\omega) = 2\pi\, b_{kj}(\nu)$$

In the ensuing analysis we usually employ the ν scale.

Using (7.13) and (8.72), we obtain the differential Einstein coefficient for spontaneous emission ($j > k$):

$$a_{jk} = \frac{h\nu_{jk}^3}{c^3} b_{jk} = \frac{8\pi^3 \nu_{jk}^3}{hc^3} |\mathbf{D}_{kj}|^2 \cos^2 \theta_{kj} \tag{8.73}$$

The coefficients b_{jk} and a_{jk} given by (8.72) and (8.73) are functions of the angle between $\mathbf{E}_0$ and $\mathbf{D}_{kj}$. However, before we can determine the integral Einstein coefficients, the differential coefficients must be referred to two independent polarisations, and must be written as functions of angles ϑ and φ which define the direction of propagation. Let $\mathbf{e}_1$ be the first of the two polarisation vectors. It is perpendicular to the ray direction and lies in the plane containing the ray and the vector $\mathbf{D}_{kj}$ (Fig. 2.5). The second polarisation vector $\mathbf{e}_2$ will then be perpendicular to both $\mathbf{e}_1$ and the wave normal $\mathbf{n}$, and therefore also to the vector $\mathbf{D}_{kj}$. As can be seen from

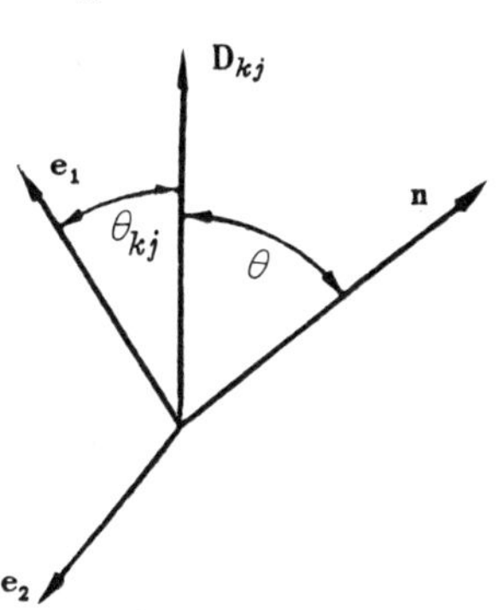

Fig. 2.5 Direction of polarisation vectors

Fig. 2.5, $\vartheta = \dfrac{\pi}{2} - \theta_{kj}$, and hence, using (8.72) and (8.73) we have

$$b_{jk}^{(1)} = b_{kj}^{(1)} = \frac{8\pi^3}{h^2}|\mathbf{D}_{kj}|^2 \sin^2\vartheta, \quad b_{jk}^{(2)} = b_{kj}^{(2)} = 0 \tag{8.74}$$

$$a_{jk}^{(1)} = \frac{8\pi^3\nu_{jk}^3}{hc^3}|\mathbf{D}_{kj}|^2 \sin^2\vartheta, \quad a_{jk}^{(2)} = 0 \tag{8.75}$$

Substituting these formulae into (7.20) and (7.24) and integrating over all directions of propagation, we obtain the integral Einstein coefficients

$$B_{kj} = B_{jk} = \frac{8\pi^3}{3h^2}|\mathbf{D}_{kj}| \tag{8.76}$$

$$A_{jk} = \frac{64\pi^4\nu_{jk}^3}{3hc^3}|\mathbf{D}_{kj}|^2 = \frac{8\pi h\,\nu_{jk}^3}{c^3}B_{kj} \tag{8.77}$$

Suppose now that the system under consideration was in the excited state j at time $t = 0$. It follows that soon after this instant ($t \approx 0$) the rate of emission of radiation per unit solid angle in the direction ϑ is

$$W_{\overline{em}}(\vartheta) = a_{jk} h \nu_{jk} = \frac{8\pi^3 \nu_{jk}^4}{c^3} |\mathbf{D}_{kj}|^2 \sin^2 \theta \tag{8.78}$$

which corresponds to a total of emission

$$W_{\overline{em}} = \int_\Omega W_{\overline{em}}(\vartheta)\, d\Omega = \frac{32\pi^2 \nu_{jk}^4}{3c^3} |\mathbf{D}_{kj}|^2$$
$$= \frac{4\omega_{jk}^4}{3c^3} |\mathbf{D}_{kj}|^2 \tag{8.79}$$

Substituting the quantum mechanical formulae (8.78) and (8.79) into the classical formulae (1.15) and (1.16a) we see that both the angular distribution of the emitted energy and the total rate of emission by the quantum-mechanical system during $j \to k$ transitions is the same as for a harmonic oscillator of frequency ν_{jk} and dipole-moment amplitude $(D_0)^2 = 4 |D_{jk}|^2$. This is the basis for the correspondence principle, according to which a quantum-mechanical system absorbs and emits radiation as a set of classical harmonic oscillators. This principle has been very fruitful in the solution of a wide range of optical problems. Its limits of applicability will be discussed in detail later.

Oscillator strengths. The sum rule

Well before the emergence of quantum mechanics and the formulation of the correspondence principle, the classical theory of the harmonic oscillator was widely used for the investigation of the interaction of radiation with matter. By associating a harmonic oscillator with each spectral line of an atom, classical theory led to a series of correct predictions, including the dependence of the absorption coefficient and the refractive index on the frequency of external radiation. There was, however, a quantitative discrepancy between theory and experiment. As has already been mentioned (Section 2), the discrepancy was partly overcome by introducing dimensionless correction factors which are otherwise known as oscillator strengths. Quantum theory leads to explicit expressions for the oscillator strengths and also to a clearer interpretation of their physical meaning. This is achieved by comparing classical and quantum-mechanical formulae for the induced dipole moment, polarisability tensor, absorption coefficient and so on.

Consider, for example, the absorption of plane-polarised radiation. According to (2.38) the rate of absorption by a linear oscillator of frequency $\nu = \nu_{jk}$ is given by

$$W_{\rm abs}^{\rm el}(\nu_{jk}) = f'_{kj}\frac{\pi e^2}{m}\cos^2\theta_{kj}\,u(\nu_{jk}) \tag{8.80}$$

where the oscillator strength f'_{kj} is introduced in order to achieve agreement with experiment. According to quantum theory the rate of absorption by a system which at the initial instant of time was in the lower level k is given by

$$W_{\rm abs}^{\rm qm}(\nu_{jk}) = b_{kj}u(\nu_{jk})\,h\nu_{jk} = \frac{8\pi^3}{h^2}|\mathbf{D}_{kj}|^2\cos^2\theta_{kj}\,h\nu_{jk}u(\nu_{jk}) \tag{8.81}$$

When $k>j$ this expression gives the stimulated emission. A more precise definition of the concept of the rate of absorption with allowance for stimulated emission will be given in Section 17.

Comparison of (8.80) with (8.81) shows that

$$f'_{kj} = \frac{8\pi^2 m}{he^2}\nu_{jk}|\mathbf{D}_{kj}|^2 = \frac{8\pi^2 m}{h}\nu_{jk}|x_{kj}|^2 \tag{8.82}$$

Using (8.76) and (8.77), we can express oscillator strengths in terms of the integral Einstein coefficients:

$$f'_{kj} = \frac{3mc^3}{8\pi^2 e^2\nu_{jk}^2}A_{jk} = \frac{3mh\nu_{jk}}{\pi e^2}B_{kj} \tag{8.83}$$

These expressions may be written in the following more convenient form

$$f'_{kj} = \frac{A_{jk}}{A_{10}^{\rm osc}} = \frac{B_{kj}}{B_{01}^{\rm osc}} \tag{8.84}$$

where we have used the integral Einstein coefficients for $1 \rightleftarrows 0$ transitions in the harmonic oscillator (see Chapter 6).

This illustrates the physical meaning of the oscillator strength; it gives the ratio of $j \to k$ and $1 \rightleftarrows 0$ transition probabilities for a harmonic oscillator of natural frequency ν_{jk}. In other words, if we choose the harmonic-oscillator transition probability as the unit of measurement, the transition probabilities for a given quantum system will be numerically equal to the oscillator strengths. It will be shown later that A_{10} is equal to $2\gamma_{em}$ which is the classical damping constant.

The second equality in (8.82) is valid for molecules and other systems in which the direction of the dipole moment is independent of the external field and may be parallel to a chosen axis. In atoms disturbed by an external field $\mathbf{E}$, the direction of $\mathbf{D}_{jk}$ will, to a greater or lesser extent, depend on the direction of $\mathbf{E}$, and therefore the vectors $\mathbf{D}_{jk}$ always have non-zero components in a fixed system of coordinates. The isotropic harmonic oscillator is then more convenient than the linear oscillator as a standard for comparison. The direction of its induced dipole moment is always parallel to the direction of the external field, and the rate of absorption does not contain $\cos^2\theta$. Instead of (8.80) we then have

$$W^{\text{cl}}_{\text{abs}} = f_{kj}\frac{\pi e^2}{m}u(\nu_{jk}) \tag{8.85}$$

Comparison of this expression with the rate of absorption averaged over θ_{kj}, which is given by

$$W^{\text{qm}}_{\text{abs}} = \frac{8\pi^3}{3h}\nu_{jk}|\mathbf{D}_{kj}|^2 u(\nu_{jk}) \tag{8.86}$$

yields the following new expression for the oscillator strength

$$f_{kj} = \frac{8\pi^2 m\nu_{jk}}{3h}|\mathbf{D}_{kj}|^2 = \frac{8\pi^2 m\nu_{jk}}{3h}\left(|x_{kj}|^2 + |y_{kj}|^2 + |z_{kj}|^2\right) \tag{8.87}$$

The quantities f_{kj} satisfy (8.84) if A_{10} and B_{01} represent the Einstein coefficients for an isotropic (three-dimensional) harmonic oscillator. They are greater by factors of 3 than the Einstein coefficients for the linear oscillator.

The oscillator strengths f_{kj} satisfy the so-called sum rule which states that the sum of all the f_{kj} for transitions from the level k is equal to the number N of optical electrons:

$$\sum_j f_{kj} = N \tag{8.88}$$

This rule, like other rules of its kind, is based on the fact that the functions ψ_j in the expansion for the perturbed function form a complete set. We shall confine our attention to the one-electron system $(N = 1)$ and will show that

$$\sum_j \frac{8\pi^2 m}{h} \nu_{jk} |x_{jk}|^2 = 1 \tag{8.89}$$

The sum in this expression includes both positive terms corresponding to transitions to higher levels $j > k$ and negative terms associated with transitions to lower-lying levels $j < k$. The sign of a particular term is determined by the sign of $\nu_{jk} = (E_j - E_k)/h$.

In view of the first equation in (8.49), the left-hand side of (8.89) may be written in the form

$$\sum_j \frac{8\pi^2 m}{h} \nu_{jk} |x_{jk}|^2 = \sum_j (x_{jk}x^*_{jk} + x^*_{jk}x_{jk}) \frac{4\pi^2 m}{h} \nu_{jk}$$
$$= \sum_j \frac{2\pi}{ih} (p_{jk}x^*_{jk} - p^*_{jk}x_{jk}) \tag{8.90}$$

The matrix elements x_{jk} and p_{jk} are the coefficients of the expansion in powers of the eigenfunctions which arise as a result of the application of the position and momentum operators to ψ_k:

$$\mathbf{x}\,\psi_k = \sum_j x_{jk}\,\psi_j \tag{8.91}$$

$$\mathbf{p}_\lambda \psi_k = -i \frac{h}{2\pi} \frac{\partial}{\partial x} \psi_k = \sum_l p_{lk}\,\psi_l \tag{8.92}$$

Similar equations hold for the complex conjugate quantities:

$$\mathbf{x}\,\psi^*_k = \sum_j x^*_{jk}\,\psi^*_j \tag{8.93}$$

$$i \frac{h}{2\pi} \frac{\partial}{\partial x} \psi^*_k = \sum_l p^*_{lk}\,\psi^*_l \tag{8.94}$$

If we multiply together the right- and left-hand sides of (8.91) and (8.94) and also (8.92) and (8.93), and integrate over all space, we have

$$-i\frac{h}{2\pi}\int \mathbf{x}\,\psi_k \frac{\partial}{\partial x}\psi_k^* \,dx = \sum_j p_{jk}x_{jk}^* \tag{8.95}$$

$$i\frac{h}{2\pi}\int \mathbf{x}\,\psi_k^* \frac{\partial}{\partial x}\psi_k \,dx = \sum_j p_{jk}^* x_{jk} \tag{8.96}$$

from which

$$\sum_j \frac{2\pi}{ih}(p_{jk}x_{jk}^* - p_{jk}^* x_{jk}) = -\int \mathbf{x}\frac{\partial}{\partial x}|\psi_k|^2\,dx = 1 \tag{8.97}$$

The validity of the second equation in (8.97) can easily be verified by integrating by parts and remembering that ψ_k vanishes at infinity. Since (8.89) is valid for all the three position coordinates of the electron, it follows from (8.87) that for the special case $N = 1$, we have the sum rule

$$\sum_j f_{kj} = 1 \tag{8.88a}$$

Franck-Condon principle

Consider the optical transitions between vibrational sub-levels of two electronic levels of a molecule. The discussion will be valid for any system consisting of a fast and a slow sub-system. It was shown in Section 5 that on the adiabatic approximation the eigenfunction for the electron-vibrational state may be written in the form

$$\psi_{n,j}(x, q) = \psi_n^{\text{el}}(x, q)\,\psi_{nj}^{\text{vib}}(q) \tag{8.98}$$

where n and j represent the electronic and vibrational levels, x are the position coordinates of the electrons and q are the position coordinates of the nuclei. The probability of transition from the j-th sub-level of the first electronic state to the j-th level of the second state is given by

$$p_{ji} = C\left|\int \psi_{2i}^*(x, q)\,\mathbf{V}(x, q)\,\psi_{1j}(x, q)\,dx\,dq\right|^2 \tag{8.99}$$

where $\mathbf{V}(x, q)$ is the operator which represents the interaction of the medium with the field and is responsible for optical transitions. Substituting (8.98) into (8.99) we obtain

$$p_{ji} = C \left| \int dq\, \psi_{2i}^{\text{vib}}(q)^* \psi_{1j}^{\text{vib}}(q) \times \int \psi_{2i}^{\text{el}}(x, q)^* \mathbf{V}(x,q)\, \psi_{1j}^{\text{el}}(x, q)\, dx \right|^2 \tag{8.100}$$

If we denote the result of integration with respect to x (matrix element for the electronic transition) by $V(q)$ we have

$$p_{ji} = C \left| \int \psi_{2i}^{\text{vib}}(q)^* V(q)\, \psi_{1j}^{\text{vib}}(q) dq \right|^2 \tag{8.101}$$

This formula determines the distribution of the probability of transitions from a given original state $(1j)$ to all sublevels $(2i)$ of the final electron state. The dependence of p on i and j is called the Franck-Condon factor.

On the first, and frequently adequate, approximation it may be assumed that the optical transition operator depends only on the coordinates of the electrons. On this approximation (the Franck-Condon approximation),

$$p_{ji} = |V_{12}|^2 \left| \int \psi_{2i}^{\text{vib}}(q)^* \psi_{1j}^{\text{vib}}(q)\, dq \right|^2 \tag{8.102}$$

where

$$|V_{12}|^2 = C \left| \int \psi_2^{\text{el}*} \mathbf{V}(x)\, \psi_1^{\text{el}}\, dx \right|^2 \tag{8.103}$$

is a constant which characterises the probability of transition between electronic levels, and is given by

$$|V_{12}|^2 = \sum_i p_{ji} \tag{8.104}$$

This result may be established by expanding the eigenfunction $\psi_{1j}^{\text{vib}}(q)$ for the initial state into a series in powers of the vibrational eigenfunctions ψ_{2i}^{vib} for the upper electronic state:

$$\psi_{1j}^{\text{vib}}(q) = \sum_k C_k \psi_{2k}^{\text{vib}}(q) \tag{8.105}$$

On multiplying (8.105) by $\psi_{2i}^{\text{vib}}(q)^*$, integrating with respect to q and recalling that the functions are orthonormal, we obtain

$$\int \psi_{2i}^{\text{vib}}(q)^* \psi_{1j}(q)\, dq = C_i \tag{8.106}$$

Since $\Sigma |C_i|^2 = 1$ we find that (8.104) is in fact correct. It follows that on the Franck-Condon approximation, the sum of transition probabilities from a vibrational sub-level j of the original state to all the vibrational sub-levels of the final state is independent of the number j and is the same for all vibrational levels. Moreover, the intensity distribution of the individual vibrational bands depends only on the properties of $\psi_{1j}^{vib}(q)$ and $\psi_{2i}^{vib}(q)$, i.e. on the mutual disposition of the potential surfaces of the upper and lower electronic states.

The expression given by (8.102) has a simple classical interpretation. We shall see in Chapter 6 that for large E_{vib} the vibrational eigenfunctions are appreciable only in one or two narrow intervals of q. The integral in (8.102) is large when the eigenfunctions $\psi_{2i}^{vib}(q)$ and $\psi_{1j}^{vib}(q)$ have appreciable values in the same intervals of q. In other words, the system will undergo transitions with a large probability only to that vibrational level of the final electronic state for which the most probable values of the coordinates are the same as for the initial vibrational level. This corresponds to a situation in which the electronic-vibrational transitions are such that the distance between the nuclei remains practically constant during the transition.

The Franck-Condon approximation is widely used to explain the intensity distributions in the spectra of diatomic molecules, and establish on the basis of experimental data potential curves for the upper and lower electronic states.

9. NON-RADIATIVE TRANSITIONS

Formulation of the problem

Non-radiative transitions between energy levels are frequently encountered in practice. They are absent only in very simple isolated systems, but are important whenever one is concerned with real systems which interact with the surrounding medium. The probabilities of non-radiative transitions may have different values depending on the nature of the interaction. In complicated systems the $i \rightarrow j$ transitions ($E_i > E_j$) will also occur as the result of internal interactions and the released energy is transformed into forms of energy other than electromagnetic radiation.

The simplest type of transition in which there is no absorption or emission radiation occurs during collisions with electrons under conditions where the kinetic energy of the electrons can be transformed into the internal energy of atoms or molecules, and the excited particles then give up part of the energy in the form of radiation. Similar phenomena occur in many light sources. Transfers of kinetic energy from incident electrons to atoms or molecules which result in transitions to electronic excited states are referred to as collisions of the first kind.

When an excited particle collides with an electron the reverse process may occur, i.e. the excitation energy may be transferred to the electron. The kinetic energy of the electron will increase, while the excited particle will undergo a transition to a lower state. This is not accompanied by the emission of radiation and is thus a non-radiative transition. Such collisions are known as collisions of the second kind.

The existence of collisions of the second kind can easily be established experimentally for electrons. They are a necessary consequence of general physical principles, e.g. as a means of establishing the thermodynamic equilibrium in a mixture of atoms and electrons. Moreover, it will be shown below that the probability of collisions of the second kind is much greater than that of collisions of the first kind.

Non-radiative transitions are observed not only in collisions with electrons but also in collisions with other particles. Consider for example the case of the so-called sensitised fluorescence of a mixture of mercury and thallium vapour illuminated by light from a quartz mercury lamp. The incident radiation can be absorbed only by the mercury atoms which undergo transitions to an excited state. In the absence of the thallium vapour, transitions to lower energy levels would result in the emission of mercury lines. In the presence of thallium, however, the emission spectra exhibit thallium lines as well as mercury lines. This has a simple interpretation. Some of the excited mercury atoms collide with thallium atoms and communicate their excitation energy to them. As a result, the mercury atoms undergo transitions to lower states without the emission of radiation, and the thallium atoms become excited without absorption of radiation. The excess energy is transformed into the kinetic energy of the atoms. The thallium atoms subsequently undergo the usual radiative transitions to lower energy levels, and this gives rise to the emission of the thallium line spectrum.

Sensitised fluorescence is frequently encountered in nature and is being systematically studied at the present time.

Non-radiative transitions induced by collisions between particles are also known to occur during the so-called luminescence quenching. Let us consider as an example the luminescence of optically excited mercury vapour. For a given source intensity there is a definite intensity of luminescence, but the intensity is sharply reduced when a small amount of a foreign gas is added to the mercury vapour. Figure 2.6 shows the ratio of intensities in the presence and absence of the impurity as a function of the partial pressure of the foreign gas. It is evident from the figure that the

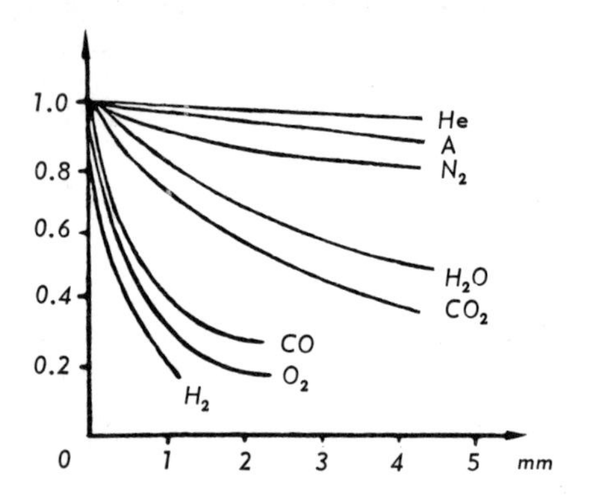

Fig. 2.6 Quenching by impurities of the luminescence of mercury vapour

excited atoms transfer their excitation energy to the foreign atoms during the collisions. This process is significant for collisions with hydrogen and oxygen, but is practically absent in collisions with inert gases. The effectiveness of the collisions is different for different gases. This is connected with differences in their chemical composition, electron shell structure and energy level schemes.

Non-radiative transitions are produced not only during collisions, but generally in all kinds of interaction between an excited particle and other systems, for example in interactions with a solvent, with other molecules in a crystal and in the intermolecular interaction in an amorphous solid. Any transfer of energy from one system to another which is not accompanied by the emission or absorption of light constitutes a non-radiative transition. Absorption and emission of light are influenced by accompanying non-radiative transitions. This fact must be taken into account in spectroscopy. Conversely, the study of the optical properties of matter is sometimes an excellent method of studying other processes involving energy transfer.

An important example of non-radiative transitions which occur in solvents, crystals and biological objects, is the resonant transfer of excitation energy from one molecule to another. Having absorbed a photon, the initially unexcited molecule transfers its excitation energy to another similar molecule, even when the interaction is small. Since the energy levels of the interacting molecules are the same, transfer of energy has a resonant characteristic. In some systems this kind of energy transfer is repeated hundreds and thousands of times, until a photon is emitted or a non-radiative transition takes place. In crystals, the process is referred to as the motion of an exciton, while in solutions of molecules it is called migration of excitation energy.

Non-radiative transitions occurring within a given system, for example a molecule and occasionally even an atom, have an important influence on the optical properties of matter. It was shown in Section 5 that the total energy of a system may be approximately divided into the energy of electronic motion, the vibrational energy of the nuclei, the rotational energy of the molecules, and the energy of translational motion. In simple systems the accuracy of this approximation is high, but in complicated systems it is low. Suppose that at the initial instant of time the total energy of a system is E and that it can be divided into a number of components, e.g. electronic, vibrational and rotational. If there were no interaction between these types of motion, radiative transitions between energy levels could be discussed separately for each of these groups of degrees of freedom. Transitions between electronic levels would occur independently of transitions between vibrational and rotational levels. In reality the interaction between them is not zero and the electronic energy may become converted into vibrational and rotational energies, and vice versa. These processes unavoidably affect the optical properties of matter. If we confine our attention to transitions between electronic levels, for example transitions from an excited state to the ground state, then the transformation of electronic energy into vibrational energy is represented as a transition from one level to another without the emission of radiation.

The probability of such processes depends on the degree of complexity of the system. These processes are rare in atoms, but they are encountered in the form of the so-called pre-ionisation, in which the total energy of two electrons is transformed into the energy of one electron. In diatomic and simple polyatomic molecules they are encountered very much more frequently, e.g. in pre-dissociation, but are still not typical (the electronic energy is released in the form of vibrational energy, or is used to disrupt a valence bond).

Non-radiative processes play a leading part in the properties of complicated polyatomic molecules and crystals, and frequently dominate the emission of radiation. The electron energy is transformed either into vibrational energy of the molecule or into the vibrational energy of the crystal lattice, i.e. into heat. This phenomenon can be referred to as internal de-activation of electronic energy.

Processes accompanied by chemical transformations are a particular type of non-radiative transition. Suppose, for example, that a molecule becomes excited as a result of absorption of light. It is sometimes found that the excited electronic state is chemically unstable and the excited molecule decays into parts, or there is a change in the structure of the molecule. The original molecule then ceases to exist, and therefore the radiative transition from an excited state to a lower state cannot occur. This causes a reduction in the emitted intensity, and can again be taken as a non-radiative transition from the excited state. However, the molecule can no longer participate in the subsequent absorption and emission of radiation, and there is therefore no true non-radiative transition. Nevertheless, from the point of view of emission of radiation, the existence of the photochemical process formally leads to the same results as the usual non-radiative transition, and the two may be discussed in parallel.

Probability of non-radiative transitions. Ratio of probabilities for direct and reverse transitions

By analogy with the transition probabilities for the absorption or emission of radiation, it is convenient to introduce the concept of a non-radiative transition probability. Any non-radiative transition within a complicated system can in principle be reduced to the transfer of energy between two (or several) sub-systems. We shall denote the energy levels

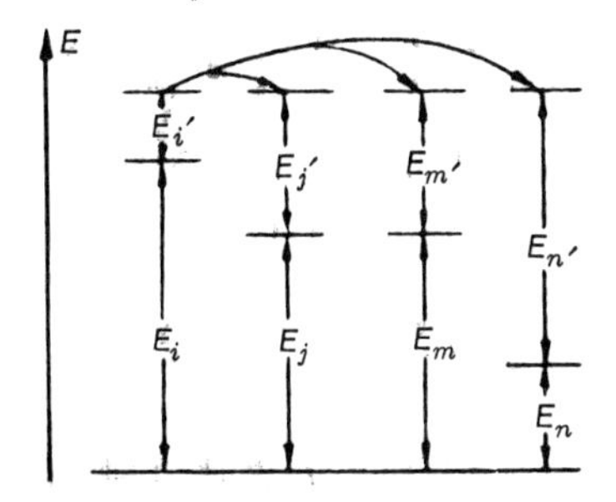

Fig. 2.7 Non-radiative transitions

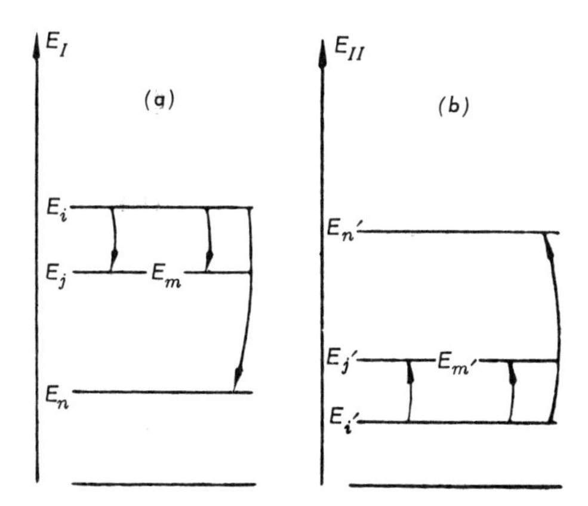

Fig. 2.8 Non-radiative transitions in the first (a) and second (b) sub-systems

of the first system by $i, j, m, \ldots$, and those of the second system by $i', j', m', \ldots$. The energy of the system as a whole will be represented by $E_i + E_{i'}$. Possible non-radiative transitions from state ii' to the states $jj', mm', nn', \ldots$ are shown in Fig. 2.7. In the absence of radiative transitions

$$E_i + E_{i'} = E_j + E_{j'} = E_m + E_{m'} = \ldots \tag{9.1}$$

The system as a whole undergoes transitions only between different sub-levels of a given degenerate level. In accordance with the law of conservation of energy, non-radiative transitions are forbidden if the levels of the complete system are non-degenerate. Non-radiative transitions may be represented graphically in another way. According to (9.1)

$$E_i - E_j = -(E_{i'} - E_{j'}); \; E_i - E_m = -(E_{i'} - E_{m'}) \ldots \qquad (9.2)$$

A reduction (increase) in the energy of the first sub-system is compensated by an increase (decrease) in the energy of the second sub-system. It is precisely for this reason that we can concentrate our attention on the main processes in which we are interested, and ignore the complicated and frequently uncontrollable transformations of energy of the second sub-system. In Fig. 2.8a the possible energies of the first sub-system are plotted along the ordinate axis. The non-radiative transitions $E_i \to E_j$, $E_i \to E_m$, $E_i \to E_n$ are represented by the curved arrows. They correspond to the transitions indicated in Fig. 2.7. Figure 2.8b shows the accompanying transitions in the second sub-system.

Let us suppose that the number of non-radiative transitions from state ii' to state jj' in time dt is proportional to the number of the first sub-systems which occupy the level E_i, to the probability $\rho_{i'}$ that the second sub-system has the energy $E_{i'}$, to the transition probability $p_{ii',jj'}$ and to the time interval dt:

$$dn_{ii',jj'} = p_{ii',jj'} n_i \rho_{i'} \, dt \qquad (9.3)$$

Similarly, the number of reverse non-radiative transitions is given by

$$dn_{jj',ii'} = p_{jj',ii'} n_j \rho_{j'} \, dt \qquad (9.4)$$

Let us consider the processes occurring in the first sub-system. A transition from level i to a lower level j may occur simultaneously with various transitions in the second sub-system, depending on its initial state. In order to obtain the total number of $i \to j$ transitions it is necessary to sum (9.3) over all i':

$$dn_{ij} = \sum_{i'} p_{ii',jj'} n_i \rho_{i'} \, dt \, \delta \left[(E_{i'} - E_{j'}) - (E_i - E_j) \right] \qquad (9.5)$$

We can now use the δ-function to separate the final state j' which is unambiguously related to i'. If we let

$$\sum_{i'} p_{ii',jj'} \rho_{i'} \delta \left[(E_{i'} - E_{j'}) - (E_i - E_j) \right] = d_{ij} \qquad (9.6)$$

we obtain

$$dn_{ij} = d_{ij} n_i dt \tag{9.7}$$

The quantity d_{ij} can conveniently be referred to as the probability of a non-radiative transition in the first sub-system. We shall now omit the words 'in the first sub-system'.

For the reverse transitions $j \to i$, which are associated with an increase in the energy of the first sub-system, we can write

$$dn_{ji} = d_{ji} n_j dt \tag{9.8}$$

where

$$d_{ji} = \sum_{j'} p_{jj',ii'} \rho_{j'} \delta [(E_{j'} - E_{i'}) - (E_j - E_i)] \tag{9.9}$$

As it is possible to establish thermodynamic equilibrium, the probabilities $p_{ii',jj'}$, $p_{jj',ii'}$, d_{ij} and d_{ji} are not all independent. In fact, in view of the principle of detailed balancing we have at thermodynamic equilibrium $dn_{ii',jj'} = dn_{jj',ii'}$. Equating (9.3) and (9.4), we have, using (9.2) and the Boltzmann relation (6.13),

$$p_{ii',jj'} = p_{jj',ii'} \tag{9.10}$$

The probabilities of direct and reverse transitions are thus found to be equal.

The relation between the probabilities of non-radiative transitions in the first sub-system with allowance for statistical weights is of the form

$$\frac{d_{ji}}{d_{ij}} = \frac{g_i}{g_j} e^{-(E_i - E_j)/kT} \tag{9.11}$$

The statistical weights of the second sub-system are automatically allowed for in the summation of (9.6) and (9.9) over i' and j'.

The relation given by (9.11) is equivalent to the relations (7.13) and (7.14) between the Einstein coefficients, although it is not of an equally universal character. The Einstein coefficients are fully determined by the properties of the system itself. At the same time according to (9.6) and (9.9), the probabilities of non-radiative transitions, d_{ji} and d_{ij}, depend on external factors, including in particular the distribution of the second sub-systems over the energy levels.

The relation given by (9.11) is strictly valid only when the first and second sub-systems (molecules and medium) are in thermodynamic equilibrium. It may be used in the solution of many problems if it is assumed that the external disturbance which upsets the thermodynamic equilibrium in the first sub-system does not affect the equilibrium in the surrounding medium (second sub-system), and thus does not change the interaction between the molecules and the medium. Since according to (9.6) and (9.9) the probabilities d_{ij} and d_{ji} depend only on $n_{i'}$ and $n_{j'}$, it follows that in such a departure from equilibrium they remain constant and explicitly depend on temperature.

A similar situation frequently arises in radiative excitation (photoluminescence). Let us suppose that before the external excitation was introduced the system was in thermodynamic equilibrium and the number of radiative $j \rightarrow i$ transitions associated with the absorption of Planck radiation was equal to the number of radiative $i \rightarrow j$ transitions; direct and reverse non-radiative transitions were also compensated by each other. Under the action of the external electromagnetic field the number of particles occupying the upper level will increase, and if the external radiation has no effect on the state of the medium, and the non-radiative transitions are as before determined by the thermal motion of the particles, the values of d_{ji} and d_{ij} will remain unaltered, and will depend on the density of the incident radiation.

It follows from (9.11) that when $E_i - E_j \gg kT$ the probability of a non-radiative $i \rightarrow j$ transition associated with a loss of excitation energy is much smaller than the probability of the reverse transition.

Calculation of non-radiative transition probabilities

The theoretical calculation of non-radiative transition probabilities is made difficult by the variety and complexity of the problem. In most cases the theory is semi-empirical and is limited to the formulation of general regularities. The resulting formulae contain certain constants which must be determined experimentally.

The classical theory (see Chapter 1) treats this problem phenomenologically in a very simple way. It is assumed that the damping constant γ consists of two parts

$$\gamma = \gamma_{em} + \gamma_{fr} \tag{9.11a}$$

where γ_{em} represents the energy loss by dipole emission and γ_{fr} represents the energy loss through interaction of the dipole with other objects and must be known in advance. The latter constant characterises all non-radiative processes and represents a statistical average of the interaction of the dipole with the medium.

Real interactions with the medium are usually functions of time. However, the interaction time for a dipole is much shorter than the interval between the interactions and in order to simplify the calculations it is assumed that the effect of the medium is continuous and the constant γ_{fr} is independent of time. Despite these approximations, the results of the classical theory are in qualitative agreement with experiment. Non-radiative transitions are treated quantitatively in quantum mechanics. In precise calculations it is necessary that we know, however, not only the eigenfunctions for the system under consideration but also the eigenfunctions for all the other systems interacting with it (electron flux, solvent, etc.) as well as the form of the interaction operator. It is precisely for this reason that the calculation of non-radiative transition probabilities is of particular difficulty.

Let us consider the simplest problem, i.e. the interaction of two groups of degrees of freedom of a single system (electronic and vibrational motion in a molecule, molecules and solvent and so on). The results obtained can easily be generalised. Suppose that the Schroedinger equation is of the form

$$\frac{ih}{2\pi}\frac{\partial}{\partial t}\psi(x, q) = [\mathbf{H}_1(x) + \mathbf{H}_2(q) + \mathbf{H}_{int}(x, q)]\,\psi(x, q) \tag{9.12}$$

where x and q are the coordinates of the first and second sub-systems and $\mathbf{H}_{int}$ is the operator representing the interaction between them. If we are concerned with transitions associated with the transformation of electronic energy into the vibrational energy of the nuclei, then the operator $\mathbf{H}_{int}(x, q)$ is the non-adiabatic operator in (5.17).

The solution of (9.12) is analogous to the solution of (8.1) and we shall therefore confine our attention to the final results only. On the zero-order approximation, when $\mathbf{H}_{int}(x, q)$ is not taken into account, the solution of (9.12) is given by

$$\psi_{ii'}(x, q) = \psi_i(x) e^{-2\pi i \frac{E_i}{h} t} \psi_{i'}(q) e^{-2\pi i \frac{E_{i'}}{h} t} \quad (9.13)$$

where E_i and $E_{i'}$ are the possible values of the energy of each sub-system. When $\mathbf{H}_{int}(x, q) = 0$, radiative transitions in both systems proceed independently of each other.

The precise solution of (9.12) will be written in the usual form

$$\psi(x, q, t) = \sum_{ii'} C_{ii'}(t) \psi_i(x) \psi_{i'}(q) e^{-2\pi i (E_i + E_{i'}) t/h} \quad (9.14)$$

By analogy with (8.23) we have

$$|C^{I}_{ii', jj'}(t)|^2 = 4 |V_{jj', ii'}|^2 \frac{\sin^2 \frac{\pi (E_j + E_{j'} - E_i - E_{i'}) t}{h}}{(E_j + E_{j'} - E_i - E_{i'})^2} \quad (9.15)$$

where

$$V_{jj', ii'} = \int \psi_j^*(x) \psi_{j'}^*(q) \mathbf{H}_{int}(x, q) \psi_i(x) \psi_{i'}(q) \, d\tau \quad (9.16)$$

The expression given by (9.15) has the characteristic properties of a δ-function: it has non-zero values only at $E_j + E_{j'} = E_i + E_{i'}$. It follows that non-radiative transitions occur only between different sub-levels of a given degenerate level. This result has already been used above in our discussion of equations (9.1) and (9.2).

If the energy levels of the system under consideration form a discrete set of degenerate levels, then for small t we have from (9.15), by analogy with (8.22)

$$|C_{ii', jj'}(t)|^2 = \frac{4\pi^2 |V_{ii', jj'}|^2}{h^2} t^2 \quad (9.17)$$

The probability of finding the system in the state jj' is proportional to the square of the time. The concept of time-independent transition probability cannot be introduced.

If the set of energy levels $E_j + E_{j'}$ forms a continuum (one of the substances should have a continuous spectrum), the probability of finding the system in one of the levels of the set $E_j + E_{j'}$ can be found by integrating (9.15) over $E_j + E_{j'}$. This yields

$$|C^{I}_{ii', jj'}(t)|^2 = \frac{4\pi^2}{h} |V_{jj', ii'}|^2 g(E_{j'}) \delta (E_j + E_{j'} - E_i - E_i) t \quad (9.18)$$

where $g(E_{j'})$ is the level density of the second sub-system.

The $ii' \to jj'$ transition probability is given by

$$p_{ii',\,jj'} = \frac{d}{dt}\,|C^{I}_{ii',\,jj'}(t)|^2$$

$$= \frac{4\pi^2}{h}\,|V_{jj',\,ii'}|^2\, g(E_{j'})\,\delta(E_j + E_{j'} - E_i - E_{i'}) \qquad (9.19)$$

The probability of the reverse transition $jj' \to ii'$ is equal to the probability of the direct transition. This result was obtained above (see (9.12)) when we discussed the possibility of thermodynamic equilibrium.

Determination of the non-radiative transition probability $p_{ii',\,jj'}$ is thus reduced to the determination of the interaction matrix element (9.16). In some cases the calculation can be successfully completed. Occasionally it is possible to elucidate the main characteristic properties of the matrix elements and find under what conditions they have non-zero values. It is then possible to formulate selection rules for non-radiative transitions which are analogous to the selection rules for emission.

Equation (9.19) gives the probability of transition from level ii' to only one of the sub-levels jj'. The number of degenerate sub-levels is very large in complicated systems, and hence in order to determine the transition probabilities for the first sub-system, the expressions given by (9.6) and (9.9) must be summed over all possible states of the second sub-system, allowance being made for the corresponding distribution function. Should the second sub-system be in a state of equilibrium, the probabilities d_{ij} and d_{ji} are temperature-dependent.

Subsequent calculations are only possible under some definite assumptions about the properties of the interacting sub-systems [17].

Propagation of energy in linear chains

If at least one of the sub-systems under consideration has a continuous energy spectrum, then in accordance with (9.18) it is convenient to introduce the concept of non-radiative transition probability, enabling us to investigate the time dependence of the state of the system by the probabilistic method which will in fact be used below in the solution of various special problems.

If the energy levels of the interacting sub-systems are discrete, then the probabilistic method may be used to investigate the time dependence of the process. Many real processes do in fact involve resonant transfers of energy from one system to another, and it is then necessary to know the exact solution of (9.12) for any value of the time.

In this section we shall give a brief outline of the exact solution of Schroedinger's equation for a simple model. Let us suppose that the system consists of N identical sub-systems, and that the interaction occurs only between neighbouring sub-systems whose indices differ by one. This is equivalent to considering a linear chain. For the sake of simplicity we shall also assume that the interaction between any pair of adjacent sub-systems is the same.

The solution of the Schroedinger equation (9.12) for the problem in hand will be sought in the form

$$\psi(x_1, x_2, x_3, \ldots, t) = \sum_j C_j(t)\,\psi_j \tag{9.20}$$

where ψ_j is the zero-order eigenfunction which is equal to the product of the eigenfunctions for all sub-systems, only one of which (the j-th) is in an excited state, all others being in the ground state. The system of differential equations equivalent to (8.13) may be written in the form

$$\frac{ih}{2\pi}\,\frac{d}{dt}\,C_j(t) = V\,[C_{j-1}(t) + C_{j+1}(t)] \tag{9.21}$$

where V is the matrix element of the interaction between the state ψ_j and the states ψ_{j-1} and ψ_{j+1}. In view of our original assumption this interaction is independent of j, and the total number of levels is determined by the total number N of elements in the chain.

The exact solution of (9.21) which is valid for any value of the time subject to the initial conditions $C_1(t=0)=1 \cdot C_2(t=0) = 0, \ldots C_N(t=0) = 0$ (excitation at the end of the chain) is of the form

$$C_j = \frac{2}{N+1}\sum_{k=1}^{N} \sin\frac{k}{N+1}\pi \sin\frac{k_j}{N+1}\pi\, e^{D_k t} \tag{9.22}$$

where

$$D_k = \frac{4\pi V}{ih} \cos \frac{k}{N+1} \pi \tag{9.23}$$

If the number of elements in the chain is infinite, equation (9.22) becomes very simple and can be expressed in terms of special functions

$$C_j = I_{j-1}\left(\frac{4\pi Vt}{h}\right) i^{(j-1)} - I_{j+1}\left(\frac{4\pi Vt}{h}\right) i^{j+1} \tag{9.24}$$

The probability of localisation of the excitation energy in the j-th sub-system is

$$\begin{aligned} |C_j|^2 &= \left[I_{j-1}\left(\frac{4\pi Vt}{h}\right) + I_{j+1}\left(\frac{4\pi Vt}{h}\right)\right]^2 \\ &= \frac{jh}{2\pi Vt} I_j\left(\frac{4\pi Vt}{h}\right) \end{aligned} \tag{9.25}$$

The formulae are also very simple in the other limiting case ($N = 2$) which corresponds to the interaction between the components of a two-fold degenerate level

$$C_1 = \cos \frac{2\pi V}{h} t, \quad C_2 = -i \sin \frac{2\pi V}{h} t \tag{9.26}$$

Let us consider these solutions in greater detail. It follows from (9.26) that the probability $|C_2|^2$ is at first proportional to t^2 (and not t). The rate of increase falls off rapidly, and at time $t = \frac{h}{4V}$ the system goes over completely from the state ψ_1 to the state ψ_2. The original state is re-established after a comparable interval of time, and the excitation energy continuously pulsates between the two sub-systems. The larger the matrix element of the interaction, the smaller the pulsation time. For $V \sim 1$ cm^{-1}, the period of the oscillatory process is $\sim 10^{-11}$ sec. The mean values of $|C_1|^2$ and $|C_2|^2$ are equal over the pulsation period and therefore the two sub-systems are equivalent.

The expressions given by (9.22) are thus easy to interpret for a three-element chain. Figure 2.9 shows plots of $|C_1|^2$, $|C_2|^2$ and $|C_3|^2$ as functions of time. All of them are harmonic in nature and after the short interval of time

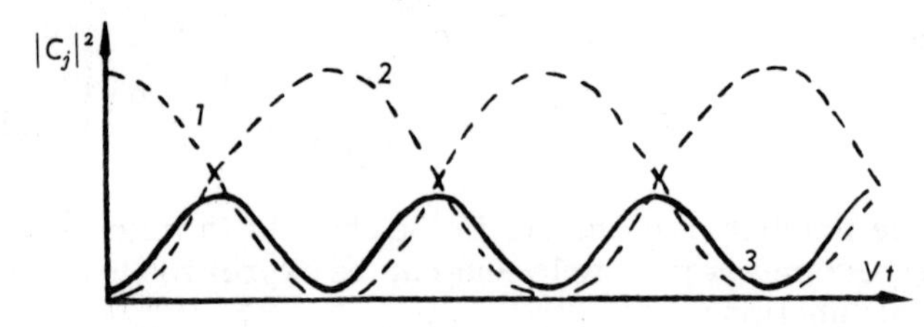

1, 2 - end elements, 3 - middle element

Fig. 2.9 Linear chain consisting of three elements

$h/\sqrt{2}V$ the initial distribution is re-established. The mean probabilities of finding the excitation at the extremities of the chain $\overline{|C_1|^2}$ and $\overline{|C_3|^2}$, are equal to 3/8, while the mean probability of finding it in the second element is $\overline{|C_2|^2} = 1/4$.

Further increase in the degree of degeneracy of an energy level (number of elements in the chain) leads to qualitative

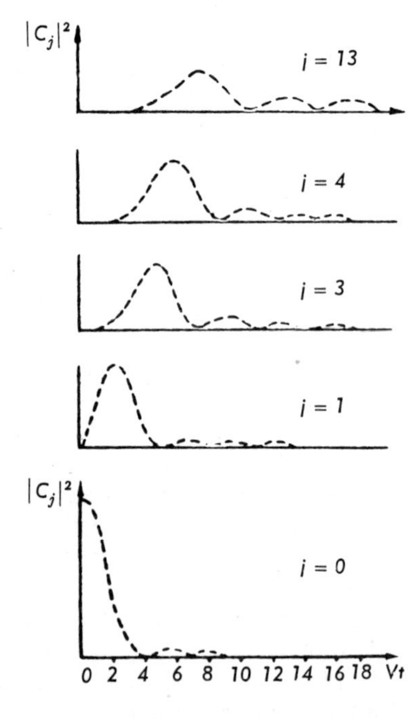

Fig. 2.10 Infinite linear chain excited at one end (j as parameter)

changes. The periodicity of the process is lost and the initial distribution of energy over the sub-systems is never re-established. When the number of elements is large the quantities $|C_j|^2$ fluctuate about their mean values without any characteristic periodicity.

The distribution of excitation energy along an infinite linear chain described by (9.25) is of particular interest. Figure 2.10 shows the dependence of $|C_j|^2$ on time for a number of elements of an infinite chain, while Fig. 2.11 shows plots of $|C_j|^2$ as functions of j for different times. These graphs represent the instantaneous states of the system.

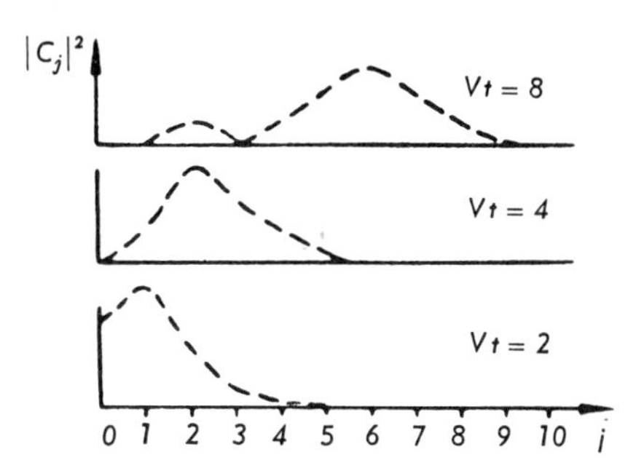

Fig. 2.11 Infinite linear chain excited at one end (Vt *as parameter*)

It follows from Fig. 2.9 that the wave packet travelling along a chain is complicated. As t increases the first maximum propagates into the region of large values of j, and its intensity decreases while its width increases. The remaining maxima exhibit similar changes, but differ from each other in their intensities which decrease with distance from the first maximum. For large values of $4\pi Vt/h$ the first maximum travels along the chain with the constant velocity

$$v = \frac{d}{dt} j_{\max} = \frac{4\pi V}{h} \tag{9.27}$$

The initial velocity of propagation of the excitation energy is somewhat smaller than that given by (9.27).

As can be seen from (9.22) the probabilities under investigation are functions of the product Vt. This means that the same distribution of probabilities will be established along the chain for different V though occurring at different times.

In finite chains the motion of the wave packet is limited; it is reflected from the ends and the resultant effect is a superposition of waves travelling in opposite directions.

10. PROFILES OF SPECTRAL LINES AND BANDS

Time-independent perturbation theory

The quantum-mechanical theory of spectral line profiles may be developed by solving the Schroedinger equation for stationary states. Let us suppose that we know the solution of the equation

$$\mathbf{H}_0(x)\psi_i(x) = E_i\psi_i(x) \tag{10.1}$$

which is valid in the zero-order approximation. The solution of the exact equation

$$[\mathbf{H}_0(x) + \mathbf{H}_{\text{int}}(x)]\,\psi(x) = E\,\psi(x) \tag{10.2}$$

which takes into account internal interactions in the system can be sought in the form

$$\psi(x) = \sum_i C_i\psi_i(x) \tag{10.3}$$

Substituting (10.3) into (10.2) and using (10.1), we have

$$\sum_i C_iE_i\psi_i(x) + C_i\mathbf{H}_{\text{int}}\psi_i(x) = \sum_i E\psi_i(x) \tag{10.4}$$

In view of the orthogonality of the eigenfunctions, if we multiply both sides of (10.4) by $\psi_j^*(x)$ and integrate with respect to x we obtain the following system of linear equations for the coefficients C_i:

$$(E_j - E)C_j + \sum_i C_iV_{ji} = 0,\ i = 1, 2, 3, \ldots, \tag{10.5}$$

where

$$V_{ji} = \int\psi_j^*(x)H_{\text{int}}(x)\,\psi_i(x)\,dx \tag{10.6}$$

$$V_{ji} = V_{ij}^* \tag{10.7}$$

The system given by (10.5) has a non-zero solution providing

$$\begin{vmatrix} E_1 - E & V_{12} & V_{13} \ldots & V_{1i} \ldots \\ V_{21} & E_2 - E & V_{23} \ldots & V_{2i} \ldots \\ V_{31} & V_{32} & E_3 - E \ldots & V_{3i} \ldots \\ \ldots & \ldots & \ldots & \ldots \\ V_{i1} & V_{i2} & V_{i3} \ldots & E_i - E \ldots \\ \ldots & \ldots & \ldots & \ldots \end{vmatrix} = 0 \qquad (10.8)$$

This equation is the so-called secular equation and determines the possible values of the energy of the system when the interaction operator is taken into account. To each value of E there corresponds a system of the form of (10.5) and therefore the coefficients C_i are functions of E.

The solution of the secular equation (10.8) and the system of equations (10.5) can be obtained in a straightforward way only in simple cases. Let us suppose that all the $V_{ii'}$ are zero except for V_{12} and V_{21}. This means that we confine our attention to the 'interaction' between two levels and ignore the effect of all other levels (for example, in the study of the splitting of degenerate levels). It then follows from the secular equation

$$\begin{vmatrix} E_1 - E & V_{12} \\ V_{21} & E_2 - E \end{vmatrix} = 0 \qquad (10.9)$$

that

$$E_{\pm} = \frac{E_1 + E_2}{2} \pm \sqrt{\left(\frac{E_2 - E_1}{2}\right)^2 + |V_{21}|^2} \qquad (10.10)$$

When $E_2 - E_1 \gg |V_{21}|$, we have $E_+ = E_2$, $E_- = E_1$, there is no 'interaction' between the levels and the zero-order approximation is satisfactory. Maximum departure from zero-order approximation occurs at resonance ($E_2 = E_1$) when $E_+ = E + |V_{21}|$, $E_1 = E - |V_{21}|$ and $E_+ - E_- = 2|V_{21}|$. The greater the interaction matrix element, the greater the shift of the levels.

The system of equations for C_1 and C_2 corresponding to (10.9) is

$$\begin{aligned} (E_1 - E_{\pm}) C_1 + V_{12} C_2 &= 0 \\ V_{21} C_1 + (E_2 - E_{\pm}) C_2 &= 0 \end{aligned} \qquad (10.11)$$

and has the simple solutions

$$\psi_{\pm} = \frac{|V_{21}|}{\sqrt{(E_1 - E_{\pm})^2 + |V_{21}|^2}} \psi_1 + \frac{E_{\pm} - E_1}{\sqrt{(E_1 - E_{\pm})^2 + |V_{21}|^2}} \psi_2 \qquad (10.12)$$

When $E_2 - E_1 \gg |V_{12}|$, we have $\psi_+ \to \psi_2$ and $\psi_- \to \psi_1$. At resonance $(E_2 = E_1)$

$$\psi_+ = \frac{1}{\sqrt{2}}(\psi_1 + \psi_2)$$
$$\psi_- = \frac{1}{\sqrt{2}}(\psi_1 - \psi_2) \tag{10.13}$$

It follows that when the square of the interaction matrix element $|V_{21}|^2$ is small in comparison with the distance between the levels, the inclusion of the operator $\mathbf{H}_{int}$ gives only a small correction to the zero-order solution. On the other hand, if all energy levels coincide, for example in the presence of degeneracy, the zero-order approximation is invalid. This is so even for small $[V_{21}]$ because the two states in (10.13) have equal weights.

Level widths

Absorption and emission lines always exhibit a finite width, so that the emitted radiation is never exactly monochromatic. Classical theory (Section 2) relates line broadening with the damping of dipole oscillations during the emission process, or with the transfer of dipole energy to other bodies. Quantum theory provides an analogous interpretation.

The basic Bohr relation (5.1) predicts that if the energy levels involved in a transition are strictly discrete the absorption and emission lines must be infinitely narrow. This is never observed in practice, and therefore at least one of the two levels must have a finite width. The profiles of spectral lines or bands depend on the profiles of the energy levels. The latter are quite complicated, and cannot always be calculated.

As has already been emphasised, the broadening of energy levels is associated with the interaction of the emitting particles with the surrounding medium. If the concept of a medium includes zero-point electromagnetic fields, the effect of the medium can never be fully eliminated in any practical experiment.

In order to determine the level profiles by the methods of quantum mechanics, we must use the perturbation theory and consider the interaction between two sub-systems.

We shall suppose that one of them (atom or molecule) has discrete levels, while the other (medium) has continuous levels. In order to simplify the calculations we shall use continuous-spectrum eigenfunctions, and replace the continuous distribution of levels of the second sub-system by a discrete set of closely spaced levels.

The Schroedinger equation for stationary states is of the form

$$[\mathbf{H}(x) + \mathbf{H}(q) + \mathbf{H}_{\text{int}}(x, q)]\,\psi(x, q) = \mathrm{E}\,\psi(x, q) \qquad (10.14)$$

In the zero-order approximation

$$\psi_{ii'}(x, q) = \psi_i(x)\,\psi_{i'}(q), \quad \mathrm{E} = \mathrm{E}_i + \mathrm{E}_{i'} \qquad (10.15)$$

A representation of the energy levels on the zero-order approximation is shown in Fig. 2.12. To each level of the molecule there corresponds an infinite set of sub-levels of the medium.

The solution of the exact equation (10.14) is

$$\psi_{\mathrm{E}}(x, q) = \sum_i \sum_{i'} C_{ii'}(\mathrm{E})\,\psi_i(x)\,\psi_{i'}(q) \qquad (10.16)$$

where E is the energy of the system as a whole and contains all the states of the zero-order approximation. If on the zero-order approximation the energy of each sub-system

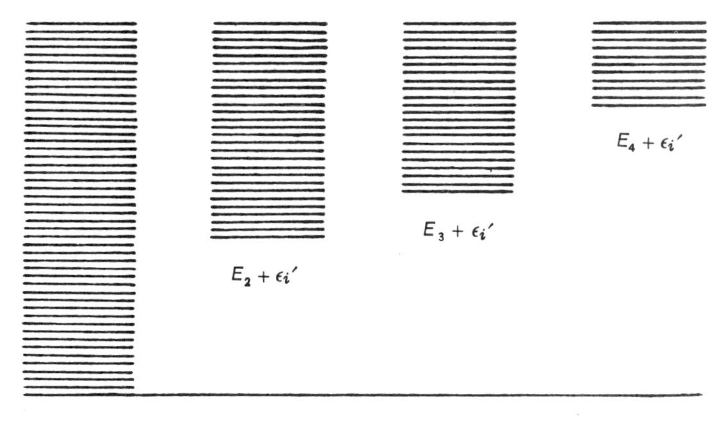

Fig. 2.12 Energy level diagram for a system consisting of two sub-systems (zero-order approximation)

were strictly determined, then when the interaction between the sub-systems is taken into account the energy becomes undetermined. The quantity $|C_{ii'}(E)|^2$ is the probability of finding the first sub-system in the level E_i and the second in the level $\varepsilon_{i'}$ where the sum $E_i + \varepsilon_{i'}$ may not equal the energy E of the complete system. Each zero-order state $\psi_i(x)\,\psi_{i'}(q)$ can be detected with a finite probability in all states $\psi_E(x, q)$ with any E.

The values of $C_{jj'}$ can be determined by solving a system of equations equivalent to (10.5):

$$(E_j + \varepsilon_{j'} - E)\,C_{jj'}(E) + \sum_i \sum_{i'} C_{ii'}(E)\,V_{jj',\,ii'} = 0 \qquad (10.17)$$

The secular equation for the eigenvalues E is obtained by setting the determinant of (10.17) equal to zero.

It follows from (10.17) that the $C_{jj'}(E)$ depend on the interaction matrix elements between all the jj' and ii' levels shown in Fig. 2.12. The solution of (10.17) for $C_{jj'}(E)$ is exceedingly difficult, and can only be obtained under simplifying assumptions. Here we shall discuss only the first-order approximation, which is equivalent to the first-order theory of non-radiative transitions.

Let us suppose that we are interested in the effect of the operator $\mathbf{H}_{\text{int}}$ on a particular level in the zero-order of approximation which in Fig. 2.13 is indicated by the symbol E^0. On the first approximation it is convenient to ignore the interaction between the levels and discuss the effect of all remaining levels on this particular level.

For the level scheme shown in Fig. 2.13 the system of equations (10.17) assumes the simpler form

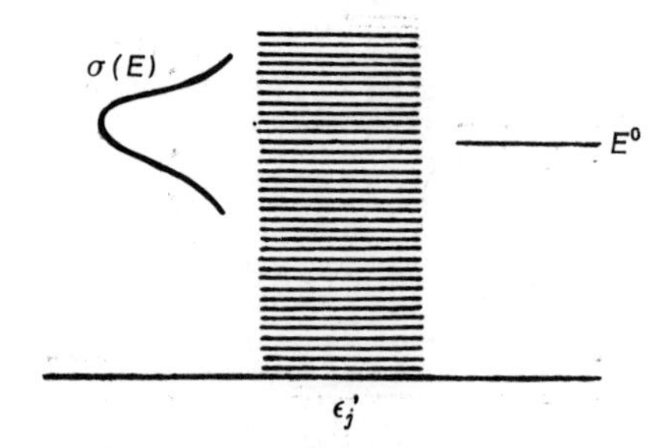

Fig. 2.13 Profile of an energy level

$$(E^0 - E')\, C^0(E) + \sum_{j'} C_{j'}(E)\, V_{j'} = 0,$$
$$V^*_{j'} C^0(E) + (\varepsilon_{j'} - E)\, C_{j'}(E) = 0,\ j' = 1,\ 2,\ 3, \ldots \tag{10.18}$$

The eigenfunction which is suitable for the first-order approximation is of the form

$$\psi_E(x, q) = C^0(E)\, \psi_{E^0}(x, q) + \sum_{j'} C_{j'}(E)\, \psi_{j'}(x, q). \tag{10.19}$$

This approximation gives correct estimates of $C^0(E)$, but introduces considerable errors into $C_{j'}(E)$. This is due to the interaction of levels $\varepsilon_{j'}$ not only with the level E^0 but also with all other levels, which is not allowed for in first-order theory.

The possible energies of the system may be obtained from the secular equation for (10.18):

$$\begin{vmatrix} E^0 - E & V_1 & V_2 & V_3 \ldots \\ V_1^* & \varepsilon_1 - E & 0 & 0 \ldots \\ V_2^* & 0 & \varepsilon_2 - E & 0 \ldots \\ V_3^* & 0 & 0 & \varepsilon_3 - E \ldots \\ \ldots & \ldots & \ldots & \ldots \end{vmatrix} = 0 \tag{10.20}$$

An exact solution of (10.20) and (10.18), subject to certain further assumptions, will be obtained in the next section. The main conclusions associated with the introduction of the level width may be obtained without detailed calculations. Since the second sub-system has a continuous energy spectrum on the zero-order of approximation, it follows that the energies $\varepsilon_1, \varepsilon_2, \ldots$ differ by a very small (in the limit, infinitely small) quantity. Interaction between the first and the second sub-systems does not affect the nature of the energy spectrum; it remains continuous. The number of energy levels is also unaffected. It follows that the coefficients $C^0(E)$ and $C_{j'}(E)$ are continuous functions of the energy of the system.

Let us consider in greater detail the physical interpretation of $|C^0(E)|^2$. We have frequently emphasised that this quantity determines the probability of finding in $\psi_E(x, q)$

the true state of the system, properties which are characteristic of the zero-order state ψ_{E^0}. As a result of the interaction between the sub-systems, properties of the level E^0 can be detected not only for $E = E^0$ but for all other values of E with the probability $\sigma(E) = |C^0(E)|^2$. It follows that the discrete level spreads into a band. The emission and absorption of radiation, which in the zero-order approximation is characteristic only of the single discrete level E^0, is thus extended to all other energy levels.

The probability $|C^0(E)|^2$ may be represented by the $\sigma(E)$ curve, which can be conveniently referred to as the profile of the level E^0. An example of this is illustrated in Fig. 2.13. The precise form of the curve can be obtained by solving (10.18) for given values of the interaction matrix elements. $\sigma(E)$ usually reaches a maximum either at $E = E^0$ or near this point. The latter effect is associated with the reduction in the 'interaction' between levels $\varepsilon_{j'}$ and E^0 as they separate.

Calculation of level widths

In order to determine $\sigma(E)$ we shall introduce two simplifying assumptions:

1. The level separation in the second sub-system is constant, i.e. the quantity $\Delta = \varepsilon_{j'+1} - \varepsilon_{j'}$ is independent of j'. This is not always valid. Variations in the level density at large values of $\varepsilon - E^0$ will not, however, affect results of the calculation, since the contributions from distant levels are small.
2. The interaction matrix elements V_j of the level E^0 and the levels $\varepsilon_{j'}$ are independent of j'. This assumption is also valid for $\varepsilon_{j'}$ approaching E^0.

These assumptions are equivalent to the assumption that the matrix elements are independent of E. This assumption is widely used in the theory of radiative and non-radiative transitions, for example in the integration of (8.23).

Since $V_{j'} = V$ it follows from (10.18) that

$$C_{j'} = \frac{V^* C^0}{\varepsilon_{j'} - E} \tag{10.21}$$

$$E^0 - E = -|V|^2 \sum_{j'} \frac{1}{\varepsilon_{j'} - E} \tag{10.22}$$

Equation (10.22) is identical with the secular equation (10.20). Let E be one of the solutions of (10.22). The system of equations (10.21) has an exact solution (for $\Delta \to 0$) when the normalisation of the eigenfunctions is taken into account [18]. Here we shall only quote the final result:

$$\sigma(\mathrm{E}) = |C^0(\mathrm{E})|^2 = \frac{|V|^2}{\pi^2 |V|^4 + (\mathrm{E}^0 - \mathrm{E})^2} \tag{10.23}$$

In contrast to (10.21) and (10.22) the values of $|V|^2$ and $|C^0(\mathrm{E})|^2$ are calculated per unit frequency interval in the entire range of j. This removes the arbitrary assumption associated with the selection of the distance Δ between the levels of the continuous spectrum.

The expression (10.23) determines the profile of an energy level on the first approximation of the perturbation theory. The function $\sigma(\mathrm{E})$ has the usual dispersion form. The maximum of $\sigma(\mathrm{E})$ occurs at $\mathrm{E} = \mathrm{E}^0$ and the height of the maximum is

$$\sigma_{\max} = \frac{1}{|V|^2 \pi^2}$$

In the absence of interaction with the medium ($V = 0$), the half-width $2\pi |V|^2$ is equal to zero. As the interaction increases there is a continuous increase in the width of the level at E^0 and a reduction in $\sigma_{\max}$. The level E^0 disappears altogether as $|V|^2 \to \infty$.

The integral over the level profile for any V is

$$\int \sigma(\mathrm{E})\, d\mathrm{E} = \int\limits_{-\infty}^{\infty} \frac{\sigma_{\max}}{1 + (\mathrm{E}^0 - \mathrm{E})^2 \pi^2 \sigma_{\max}^2}\, d\mathrm{E} = 1 \tag{10.24}$$

This means that the sum of probabilities of finding the properties characteristic of the level E^0 among the levels E of the system is equal to unity. A discrete level becomes broadened but does not disappear altogether - its properties are conserved throughout. In order that (10.23) should determine the true level profile, it is necessary that the conditions formulated at the beginning of this section be satisfied, at least near the profile maximum. These conditions are usually satisfied for small V, but for strong interactions the maximum of the $\sigma(\mathrm{E})$ curve may be shifted relative to E^0 and the profile may become asymmetric.

Connection between level profiles and the profiles of spectral bands

Consider the absorption of radiation by an atom or molecule. If there is no interaction with a medium, the probability of radiative transition between discrete levels j and i is given by (8.29). On the dipole approximation the matrix element V_{ij} is proportional to the matrix element of the dipole-moment operator

$$\mathbf{D}_{ij} = \int \psi_i^*(x)\,\mathbf{D}\,\psi_j(x)\,dx \tag{10.25}$$

Let us suppose that the medium does not absorb in the spectral region under investigation, so that the matrix elements for the transition

$$\int \psi_{i'}(q)\,\mathbf{D}_{\mathrm{m}}\,\psi_{j'}(q)dq \tag{10.26}$$

are all zero. If we regard the molecule and the medium as two parts of a single system, (10.25) and (10.26) may be combined into the single expression

$$\mathbf{D}_{ii',\,jj'} = \int \psi_i^*(x)\,\psi_{i'}^*(q)\,\mathbf{D}\,\psi_j(x)\,\psi_{j'}(q)dxdq \tag{10.27}$$

where $\mathbf{D}_{ii',\,jj'}$ is not zero except when $i' = j'$, so that when the molecule absorbs radiation there is no change in the state of the medium.

The expression given by (10.27) is valid only in the absence of interaction between the molecule and the medium. This can only occur in reality if the absorption spectrum contains strictly discrete lines. Figure 2.14a illustrates a transition of this kind. The transition frequency is independent of the original energy of the medium (ε^0).

Let us now consider the transition between a discrete level E_1 and an energy level E_2 with a profile $\sigma(E)$. Only the unexcited energy levels of the molecule can be regarded as strictly discrete. The eigenfunctions for the first and second levels are (see (10.15) and (10.19))

$$\psi_{\varepsilon^0}(x, q) = \psi_1(x)\,\psi_{\varepsilon^0}(q) \tag{10.28}$$

$$\psi_{E}(x, q) = C^0(E)\,\psi_2(x)\,\psi_{\varepsilon^0}(q) + \sum_{\varepsilon'} C_{\varepsilon'}(E)\,\psi_1(x)\,\psi_{\varepsilon'}(q). \tag{10.29}$$

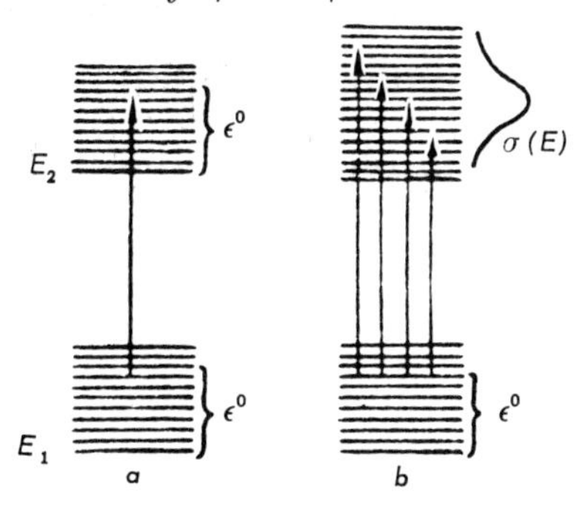

Fig. 2.14 Transitions between the energy levels of a system on the zero-order approximation when the interaction between the subsystems is allowed for

In the state $\psi_{\varepsilon^0}(x, q)$, the energy of the molecule is zero and the energy of the medium is arbitrary (ε^0). In the state $\psi_E(x, q)$ only the total energy of the system as a whole is precisely determined and cannot be separated into the energy of the molecule and the energy of the medium.

The matrix element for a typical transition between the level $E=\varepsilon^0$ of the system and any other level E corresponding to the excitation of the molecule is

$$\begin{aligned} \mathbf{D}_{\varepsilon^0 E} &= \int \psi^*_{\varepsilon^0}(x, q)\, \mathbf{D}\, \psi_E(x, q) dx dq \\ &= C^0(E) \int \psi^*_1(x)\, \mathbf{D}\, \psi_2(x) dx \int \psi^*_{\varepsilon^0}(q)\, \psi_{\varepsilon^0}(q) dq \\ &+ \sum_{\varepsilon'} C_{\varepsilon'}(E) \int \psi^*_1(x)\, \psi^*_{\varepsilon}(q)\, \mathbf{D}\, \psi_1(x)\, \psi_{\varepsilon'}(q) dx dq \\ &= C^0(E)\, \mathbf{D}_{12} \end{aligned} \tag{10.30}$$

The integrals under the summation sign are zero since the medium does not absorb. The integral

$$\mathbf{D}_{12} = \int \psi^*_1(x)\, \mathbf{D}\, \psi_2(x) dx \tag{10.31}$$

is determined by the properties of the molecule on the zero-order approximation.

The quantity

$$|\mathbf{D}_{\varepsilon^0 E}|^2 = |C^0(E)|^2 |\mathbf{D}_{12}|^2 \tag{10.32}$$

is proportional to the probability of the radiative transition $\varepsilon^0 \rightarrow E$. Instead of a discrete line we now have a band with a definite profile.

A number of possible transitions are shown in Fig. 2.14b. As $|\mathbf{D}_{12}|^2 = \text{const}$, it follows that the distribution of the transition probabilities over the frequencies

$$\nu_{E\,\varepsilon^0} = (E - \varepsilon^0)/h$$

is the same in form as the profile $\sigma(E) = |C^0(E)|^2$ of an excited level. Substituting for $|C_0(E)|^2$ from (10.23), we obtain the following expression for the frequency dependence of the transition probability

$$p(\nu) = \text{const}\,|\mathbf{D}_{12}|^2 \frac{|V|^2/h^2}{(\nu_{21} - \nu)^2 + \pi^2 |V|^4/h^2} \tag{10.33}$$

where V is the matrix element of the interaction between the molecule and the medium per unit energy interval in the continuous spectrum, and ν_{21} is the frequency of a discrete line in the absence of the interaction.

The form of the function given by (10.33) has been discussed earlier. Maximum absorption occurs at the frequency ν_{21}, the band is symmetrical, the probability of absorption at the maximum is equal to const $|\mathbf{D}_{12}|^2 \dfrac{1}{\pi^2 |V|^2}$ and the half-width is $\dfrac{2\pi |V|^2}{h}$. The spectral line broadens as the interaction with the medium increases. It is important, however, that the integrated absorption remains constant, i.e.

$$\int p(\nu)\, d\nu = \text{const}\,|\mathbf{D}_{12}|^2$$

and is equal to the absorption of radiation before the interaction is 'switched on'.

So far, we have been concerned with the spectral line profile for a transition between a discrete and a broadened level. If the width of one level is much greater than the width of another, the line profile will be determined almost entirely by the profile of the broader level. If the widths are comparable, the line profile depends on the profiles of both lines $\sigma'(E')$ and $\sigma''(E'')$. The probability of radiative transitions from an unexcited level in the range between E' and

$E'+dE'$ to an excited level between E'' and $E''+dE''$ is proportional to

$$|\mathbf{D}_{12}|^2 \sigma'(E')\,dE'\sigma''(E'')dE'' \tag{10.34}$$

The corresponding frequency is $\nu = \frac{E''-E}{h}$. The same frequency can be emitted as a result of transitions from other levels E'. The total probability for the frequency interval $d\nu$ may be obtained by substituting $E''=E'+h\nu$, $dE''=hd\nu$ and integrating (10.34) with respect to E'

$$p(\nu)\,d\nu = \text{const}\,|\mathbf{D}_{12}|^2 hd\nu \int_{E'} \sigma'(E')\,\sigma''(E'+h\nu)\,dE' \tag{10.35}$$

The integral can be evaluated by substituting expressions similar to (10.23) for $\sigma'(E')$ and $\sigma''(E'')$ into (10.35). It is found that the width of the resulting line is approximately equal to the sum of the two level widths. This result is valid only in first-order perturbation theory. In exact calculations it is necessary to take into account the interaction between the two levels which occurs through the medium and the mutual dependence of the two level profiles.

Relation between lifetime and half-width

As we have seen, there are two ways of studying the interaction between a molecule and a medium (the first and second sub-systems). In the first of these (Section 9) the interaction with the medium (or any other interaction) which is ignored on the zero-order approximation, is looked upon as the cause of transitions between levels. In the second approach (Section 10), the effect of the interactions on the stationary states is characterised by a shift or broadening of the levels.

It can readily be shown that the two methods are closely related and constitute two different ways of describing the same physical situation and the same result of the interaction. Studies of the shift and broadening of levels can serve as a reliable means of determining transition probabilities, while experimental studies of the temporal characteristics of such processes enable us to predict the structure of energy levels and hence the emission and absorption spectra.

Let us verify this statement for some simple examples. Suppose we have two levels of equal energy and the matrix element for the interaction between them is V. We then see from (10.10) that the exact energy is given by $E = E_0 \pm |V|$ and the level separation is $-2|V|$. On the other hand, according to (9.26), the time necessary for the transition to take place from one state to the other on the zero-order approximation is $\tau = \frac{h}{4|V|}$. It follows that the 'perturbed' level separation ΔE and the time taken by the transition are related by the simple expression

$$\Delta \bar{E} \tau = \frac{h}{2} \tag{10.36}$$

The larger the level shift, the smaller is the time interval. Moreover, the product $\Delta E \tau$ is independent of V. An analogous result is obtained by considering the 'interaction' between three components of a degenerate level, or the 'interaction' between two non-degenerate levels.

Consider another limiting case, namely, the 'interaction' of a single discrete level with levels in a continuous spectrum. According to (10.23), the discrete level becomes broadened and its half-width is $\Delta E = 2\pi |V|^2$. On the other hand, the theory of transitions yields the zero-order probability of a transition from a discrete level to one of the levels in the continuous spectrum. Equation (9.19) shows that this probability is $p = \frac{4\pi^2}{h} |V|^2 \text{ sec}^{-1}$. The quantity $\tau = 1/p$ is usually called the lifetime of a given discrete state. If we multiply τ by ΔE, we have

$$\Delta E \tau = \frac{h}{2\pi} \tag{10.37}$$

Any process involving a transition from one state to another is unavoidably accompanied by a broadening of the level and the magnitude of the broadening is inversely proportional to the lifetime of the initial state.

The expression given by (10.37) differs from (10.36) by a small factor because the definition of the level width is, to some extent, conventional. In either case the product $\Delta E \tau$ is of the order of Planck's constant.

Similar results hold good for the natural width of levels, which is associated with radiative transitions (Section 14).

All special cases, no matter how different are the relevant interactions, always lead to the same result irrespective of the zero-order level scheme. This suggests that the result is, in fact, quite general.

Uncertainty relation for the energy

The interaction of atoms and molecules with the medium and the associated transfer of energy are unavoidably accompanied by broadening of energy levels. The energy of a particle thus loses its precise meaning. This is a consequence of the basic postulates of quantum mechanics. We shall confine our attention here to a simple analysis using the expression given by (9.15). This can be rewritten in the somewhat simpler form

$$|C_{E^0E}(t)|^2 = 4|V|^2 \frac{\sin^2 \pi \dfrac{E - E^0}{h} t}{(E^0 - E)^2} \tag{10.38}$$

The right-hand side of this expression gives the probability of finding the particle in the level E of the continuous spectrum at time t if at time $t = 0$, when the interaction was 'switched on', it was in the discrete level E^0.

The graph of $|C_{E^0E}|^2$ as a function of E is similar to the curve shown in Fig. 2.4. The maximum probability corresponds to finding the particle in the state with energy $E = E^0$. There is, however, a finite probability that the particle will be found in states with $E \neq E^0$. After the transition the energy is undetermined and individual experiments may yield different results.

The probability $|C_{E^0E}(t)|^2$ has sinusoidal functions approximately from $-\pi/2$ to $+\pi/2$, that is, appreciable values only for $(E - E_0)/h$ between $-\dfrac{\pi}{2t}$ and $+\dfrac{\pi}{2t}$. The energy interval $\Delta \approx \dfrac{\pi h}{t}$, in which the system can be found as a result of the appearance of an interaction continuing for a time t, is very dependent on t. The smaller the time t, i.e. the earlier the state of the system is observed, the larger is the interval ΔE in which the system can be found. All values of the energy are almost equally probable for t close to zero. Conversely the interval ΔE is small for large t, and the energies which are obtained as a result of measurement will approach E^0.

The observed spread $\Delta E \sim \frac{\pi h}{t}$ is independent of the magnitude of the interaction V, and occurs however weak the interaction is between the sub-systems. This is a purely quantum-mechanical result, and it shows that the law of conservation of energy can be verified by two measurements only to within $h/\Delta t$, where Δt is the interval between the two measurements. The relation

$$\Delta E \Delta t \sim h \tag{10.39}$$

is the mathematical expression of the principle of uncertainty for energy.

The formula given by (10.38) is a particular form of the uncertainty relation. In fact, if the state of the system is observed at a time t equal to

$$\tau = \frac{1}{p} = \frac{h}{4\pi^2 |V|^2} \text{ sec}$$

i.e. the mean lifetime of the initial state, it follows from (10.39) that the uncertainty in the energy will be $\Delta E \sim \frac{h}{t}$. Since measurements yield precise values of the energy, the uncertainty in the probability of obtaining a particular experimental result must be due to the uncertainty in the energy of the original state. The smaller the mean lifetime of a given state of the system, the greater will be the mean uncertainty in its energy.

We have regarded the interaction matrix element V as a measure of the interaction between the sub-systems. However, in practice one measures either ΔE or τ. Both of these are directly related to V and unambiguously characterise the magnitude of the interaction.

TABLE 1 Level widths

τ, sec	1	10^{-2}	10^{-4}	10^{-6}	10^{-8}	10^{-10}	10^{-12}	10^{-14}
ΔE, cm^{-1}	$^1/_3 \cdot 10^{-10}$	$^1/_3 \cdot 10^{-8}$	$^1/_3 \cdot 10^{-6}$	$^1/_3 \cdot 10^{-4}$	$^1/_3 \cdot 10^{-2}$	$^1/_3$	33	3300

In Equation (10.39) the energy is expressed in ergs. If it is expressed in cm^{-1}, we have

$$\Delta E \ \tau \sim \frac{1}{c}$$

Table 1 gives the values of ΔE for different τ.

The profile of a system of energy levels

A calculation of the effect of interaction with the medium (continuous spectrum) on a number of discrete levels has been given in papers by Stepanov and Rice [18]. A similar situation is encountered, for example, when one discusses the vibrational or rotational structure of molecules. The problem can be solved subject to the same initial assumptions. It is also postulated that the various discrete levels can interact in different ways with the levels ε_i in the continuous spectrum ($V_n \neq V_m$). We shall only quote the results of these calculations.

By analogy with (10.19), the wave function for the system when the 'interaction' between the levels is allowed for should be sought in the form

$$\begin{aligned}\psi_E = C_E^{(1)} \psi_{E_1} + C_E^{(2)} \psi_{E_2} + \ldots + C_E^{(m)} \psi_{E_m} + \ldots \\ + C_{1E} \psi_{\varepsilon_1} + C_{2E} \psi_{\varepsilon_2} + \ldots\end{aligned} \quad (10.40)$$

where E is the energy of the system as a whole. Substituting (10.40) into the Schroedinger equation (10.14) and solving a system of equations analogous to (10.21) we obtain

$$\sigma(E) = |C_E^{(m)}|^2 = \frac{|V_m|^2}{E_m - E} \; \frac{\dfrac{1}{\pi^2\left[\sum \dfrac{|V_m|^2}{E_m - E}\right]^2}}{1 + \pi^2\left[\sum \dfrac{|V_m|^2}{E_m - E}\right]^2} \quad (10.41)$$

where $C_E^{(m)}$ is defined to within a phase factor. In accordance with (10.23), $\sigma(E)$ as given by (10.41) should be called the profile of the m-th level. It must, however, be noted that it depends not only on $|V_m|$ but also on the position of all the other discrete levels and on the nature of their 'interaction' with the continuous spectrum (i.e. on all the V_n).

The connection between the energy level profiles and the profiles of spectral bands deserves particular attention. They are not identical even when the initial level of a radiative transition is strictly discrete. Let us suppose that there are no transitions to levels in the continuous spectrum on the zero-order approximation. The square of the matrix element for the radiative transition from the state ψ_0 to the state ψ_E given by (10.40) is proportional to the probability of the transition $E_0 \rightarrow E_1$ and is given by

$$|\mathbf{D}_{0E}|^2 = \left|\int \psi_0^* \mathbf{D} \psi_E \, d\tau\right|^2 = \left|\sum_m C^m(E) \mathbf{D}_{0m}\right|^2 \qquad (10.42)$$

where

$$\mathbf{D}_{0m} = \int \psi_0^* \mathbf{D} \psi_m \, d\tau \qquad (10.43)$$

are the matrix elements for the transition $E_0 \rightarrow E_m$ on the zero-order approximation. They are regarded as known.

According to (10.42), a continuous spectrum should be observed both in absorption and in emission. The transition probabilities depend not only on the magnitudes of $\mathbf{D}_{0m}$ and $C^m(E)$ but also on their phase.

Consider the following special cases of (10.42).

1. Distance between the levels very large. We can neglect the small terms for E close to E_m.

$$\frac{1}{E_{m+1} - E}, \quad \frac{1}{E_{m-1} - E}, \ldots$$

and therefore

$$|\mathbf{D}_{0E}|^2 = |\mathbf{D}_{0m}|^2 \frac{|V_m|^2}{\pi^2 |V_m|^4 + (E_m - E)^2}$$

This formula is identical with (10.23), and has already been discussed. Figure 2.15a shows the form of the absorption curve for $2\pi |V_m|^2 = 200$ cm^{-1}.

It is easy to establish the limits of this special case, i.e. the possibility of discussing the broadening of each level (line) independently of all others. The necessary condition is that the half-width $2\pi |V_m|^2$ should be smaller than the separation $E_m - E_{m-1}$ between neighbouring levels.

2. $\mathbf{D}_{0,m} = \mathbf{D}_{0,m+1} = \mathbf{D}_{0,m-1} = \ldots = \mathbf{D}$. The transition probability is the same for all levels and is given by

$$|\mathbf{D}_{0,\mathrm{E}}|^2 = |\mathbf{D}|^2 |\sum_m C^{(m)}(\mathrm{E})|^2 \qquad (10.44)$$

Here, there is a particularly clear distinction between the profile of a level and the profile of a line. The line profile is not formed by superimposing the $|C^{(m)}(\mathrm{E})|^2$. In addition

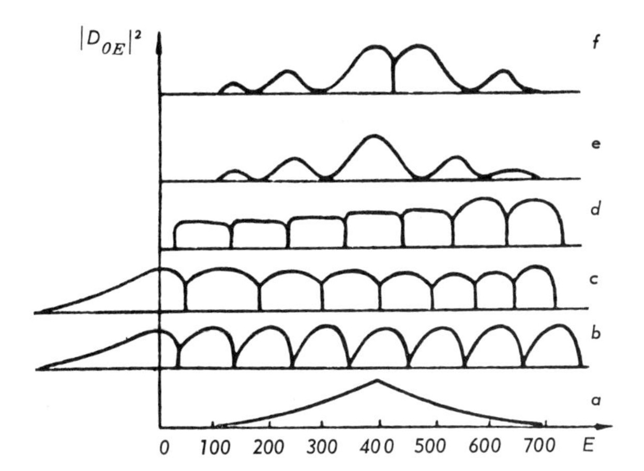

Fig. 2.15 Plots of (10.42) for different special cases

to $\Sigma |C^{(m)}(\mathrm{E})|^2$, the expression given by (10.44) includes cross-terms of the form $C^{(m)}(\mathrm{E})\, C^{(m-1)}(\mathrm{E})$, which are not equal to zero even for equal values of V_m (and equal phases).

Figure 2.15b shows the result of an exact evaluation of (10.44) with the following parameters: $|V_m|^2$ independent of m, $2\pi |V|^2 = 200$ cm^{-1}, levels E_m equidistant from each other ($\mathrm{E}_1 = 0$, $\mathrm{E}_2 = 100$ cm^{-1}, $\mathrm{E}_3 = 200$ cm^{-1}...). At lower values of $|V|^2$ the distribution resembles the broadening of individual lines, while for large values it resembles continuous absorption.

Figure 2.15c shows a graph of (10.44) when the discrete levels are at different distances from each other but V_m is constant, while Fig. 2.15d gives the result for different V_m but constant $\mathrm{E}_m - \mathrm{E}_{m-1}$. The graphs are not drawn to scale.

3. $\mathbf{D}_{0\mathrm{E}} = \mathbf{D}\,\delta_{lm}$. We have

$$|\mathbf{D}_{0\mathrm{E}}|^2 = |\mathbf{D}|^2 |C^{(m)}(\mathrm{E})|^2 \qquad (10.45)$$

which is plotted in Fig. 2.15e using the parameters employed in the preceding special case. It is worth noting that the contour of the absorption spectrum in this case is the same as the contour of the set of discrete levels. The form of the contour is quite unusual, and has a series of secondary maxima.

Figure 2.15f shows a plot of (10.42) when $\mathbf{D}_{0m} = \mathbf{D}_{0,m+1}$. All the remaining matrix elements for the transition are zero. The above special cases show that various secondary spectral formations appear when interactions between molecules and the medium are taken into account, and the interpretation of such details may be quite difficult.

11. MOMENTS OF SPECTRAL BANDS

Definition

The absorption and emission spectrum of any given system exhibits a specific frequency distribution of absorption and emission. The profiles of spectra depend on many parameters and their precise analytical form is unknown as a rule. It is for this reason that the essential result obtained from experiment is the distribution function $\rho(\nu)$ which is usually given in graphical form. Although this method is convenient, it does suffer from serious disadvantages. For example, reproduction of the curves leads to considerable inaccuracies and thus, comparisons between curves obtained at different laboratories is difficult. The most important disadvantage is, however, that with graphical representation of spectral bands it is difficult to investigate the dependence of the spectra on temperature, the frequency of the exciting light, the nature of the solvent, the aggregate state, the structure of the molecules and other parameters at the disposal of the experimenter. Moreover, it is almost impossible to relate the form of a band exhibited graphically to the internal properties of the absorbing and emitting centres.

It is therefore quite usual to augment graphical data on the form of spectral bands by numerical data giving their area, position of maximum and half-width. Such characteristics are unfortunately not unambiguous since the same values of the area, the position of the maximum and the half-width can correspond to very different spectral formations. These difficulties may be overcome with the aid of the method of moments. This method has long been known in the theory of probability but has been introduced only quite recently into spectroscopy.

The moment S_l of order l of a distribution function $\rho(\nu)$ is defined by the integral [19]

$$S_l = \int \nu^l \rho(\nu)\, d\nu \tag{11.1}$$

where the integration is performed over the entire range of ν. If the independent variable is discrete, the integral must be replaced by the sum

$$S_l = \sum_k \nu_k^l \rho_k \tag{11.2}$$

The expression (11.1) may be used to calculate the moments of the spectrum as a whole. If the spectrum can be resolved into a number of independent non-overlapping bands, it is convenient to calculate the moments of each band individually. The moments of the spectrum as a whole will then depend on the moments of the individual bands. It is sufficient to include in the integration only those values of ν which are sufficiently large. If the analytical form of $\rho(\nu)$ is unknown, but is given by a table of numerical values obtained experimentally, the moments (11.1) can easily be evaluated graphically.

The first moments of a spectral band have a simple meaning. The zero-order moment S_0 is equal to the area under the band. The ratio S_1/S_0 gives the centre of gravity of the band, i.e. the mean frequency ν.

In addition to the moments defined by (11.1) it is very convenient to introduce the concept of central moments

$$\bar{S}_l = \int (\nu - \bar{\nu})^l \rho(\nu)\, d\nu \tag{11.3}$$

The central moments $\bar{S}_l$ can be related to the moments defined above:

$$\bar{S}_l = \sum_{k=0}^{l} (-1)^k C_k^l \left(\frac{S_1}{S_0}\right)^k S_{l-k} \tag{11.4}$$

where C_k^l are the coefficients of the binomial expansion. The first few moments are given overleaf.

$$\bar{S}_0 = S_0,\ \bar{S}_1 = 0,\ \bar{S}_2 = S_2 - S_1^2/S_0$$

$$\bar{S}_3 = S_3 - 3S_2\,S_1/S_0 + 2S_1^3/S_0^2$$

The second central moment is related to the bandwidth (root-mean-square deviation from the mean) and the third to the asymmetry of the band. If the function $\rho(\nu)$ is symmetrical with respect to $\bar{\nu}$, all the odd central moments are equal to zero.

Strictly speaking, a band $\rho(\nu)$ is characterised unambiguously by the complete set of moments S_l or $\bar{S}_l$. If the band has a structure, the number of moments necessary for its description is quite large. The absorption and emission bands of complicated systems often resemble the Gaussian curve. When this is so, it is sufficient to determine the first four or five moments in order to specify their form. Structural elements are determined by higher-order moments.

If the moments of a band are known, the band profile can easily be synthesised with the aid of Edgeworth's formula

$$\rho(\nu) = S_0\sigma^{-1}\left[\varphi(\xi) - \frac{\gamma_1}{3!}\varphi^{(3)}(\xi) + \frac{\gamma_2}{4!}\varphi^{(4)}(\xi) + \frac{10\gamma_1^2}{6!}\varphi^{(6)}(\xi) + \dots\right] \tag{11.5}$$

where

$$\xi = \frac{\nu - \bar{\nu}}{\sigma};\ \sigma^2 = S_2/S_0;\ \gamma_1 = \frac{\bar{S}_3}{(\bar{S}_2)^{3/2}}\sqrt{S_0}$$

$$\gamma_2 = \frac{\bar{S}_4}{(\bar{S}_2)^2}S_0 - 3$$

$$\varphi(\xi) = (2\pi)^{-1/2}e^{-1/2\xi^2}$$

and $\varphi^{(n)}(\xi)$ is the n-th derivative of $\varphi(\xi)$.

In Edgeworth's formula, the Gaussian curve $\varphi(\xi)$ is taken as the zero-order approximation. This is very convenient in practice because the functions $\varphi(\xi)$, $\varphi^{(3)}(\xi)$, $\varphi^{(4)}(\xi)$, ... have been tabulated. Other expansions of $\rho(\nu)$ are also used. It is often sufficient to describe experimental curves by the first two terms in (11.5). The first term gives the Gaussian curve while the second represents the asymmetry. The third and fourth terms are even and govern the form of $\rho(\nu)$ near the maximum and in the wings.

If $\rho(\nu)$ is represented by the first two terms of (11.5), it is quite easy to express the frequency ν_{max} corresponding to the maximum of the distribution and the half-width $\Delta\nu$ in terms of the above moments. It can be shown that

$$\nu_{max} = \frac{S_1}{S_0} - \frac{1}{2}\frac{\overline{S}_3}{\overline{S}_2} \tag{11.6}$$

$$\Delta\nu = \sqrt{\frac{\overline{S}_2}{S_0}}\left[2.354 + \left(0.491 - 2.123\frac{1}{\gamma_1^2}\right)^{-1}\right] \tag{11.7}$$

We have used the method of moments to characterise the absorption and emission band profiles. This method can also be used to describe other distribution functions encountered in spectroscopy, for example the distribution of transition probabilities over vibrational levels in excited states, or the distribution of particles over available energy levels.

It is worth noting that not all distribution functions have moments. This is so especially for the dispersion curve

$$\rho(\nu) = \frac{A}{B + C(\nu - \nu_0)^2} \tag{11.8}$$

According to the definition given by (11.1) the second moment of this function is already infinite. This may be avoided by defining (11.8) for a limited frequency range and assuming that elsewhere $\rho(\nu) = 0$.

Examples

As an illustration, let us evaluate the moments of a number of frequently encountered distribution functions. The moments of the Gaussian function normalised to unity with a maximum at $\nu = \nu_0$ i.e.

$$\varphi(\nu - \nu_0) = \sqrt{\frac{\pi}{\alpha}}\, e^{-\alpha(\nu-\nu_0)^2} \tag{11.9}$$

are

$$
\begin{array}{lll}
S_0 = 1, & \overline{S}_0 = 1 & \\
S_1 = \nu_0, & \overline{S}_1 = 0 & \\
& \overline{S}_2 = \dfrac{1}{2\alpha} & \sqrt{\overline{S}_2} = 0.71 \\
& \overline{S}_3 = 0 & \sqrt[3]{\overline{S}_3} = 0 \\
& \overline{S}_4 = \dfrac{\Gamma(2.5)}{\sqrt{\pi\alpha^2}} & \sqrt[4]{\overline{S}_4} = 0.93 \\
& \overline{S}_5 = 0 & \sqrt[5]{\overline{S}_5} = 0 \\
& \overline{S}_6 = \dfrac{\Gamma(3,5)}{\sqrt{\pi\,\alpha^3}} & \sqrt[6]{\overline{S}_6} = 1.11
\end{array}
\qquad (11.10)
$$

where

$$\Gamma(a) = \int_0^\infty a^{-x} x^{a-1}\, dx$$

In addition to the moments, we have also given the values of $\sqrt{\overline{S}_2}$, etc. for $\alpha = 1$. These have immediate physical meaning; they are the root-mean-square deviation from the mean, root-mean-cube deviation from the mean, and so on. They all have the same dimensions, and can easily be compared with each other and exhibited graphically.

All the odd central moments of (11.9) are equal to zero in view of the symmetry of the function with respect to ν_0. The greater the value of α, the smaller the even central moments. Figure 2.16 shows a plot of (11.9) and the values of $\sqrt[l]{\overline{S}_l}$.

The moments of the normalised exponential function

$$\rho(\mathrm{E}) = \frac{1}{kT} e^{-\mathrm{E}/kT} \qquad (11.11)$$

are

$$
\begin{array}{lll}
S_0 = 1, & \overline{S}_0 = 1 & \\
S_1 = kT, & \overline{S}_1 = 0 & \\
S_2 = 2(kT) & \overline{S}_2 = (kT)^2, & \sqrt{\overline{S}_2} = kT \\
S_3 = 3!\,(kT)^3, & \overline{S}_3 = 2(kT)^3, & \sqrt[3]{\overline{S}_3} = \sqrt[3]{2}\; kT \\
S_4 = 4!\,(kT)^4, & \overline{S}_4 = 9\,(kT)^4 & \sqrt{\overline{S}_4} = \sqrt[4]{9}\; kT
\end{array}
\qquad (11.12)
$$

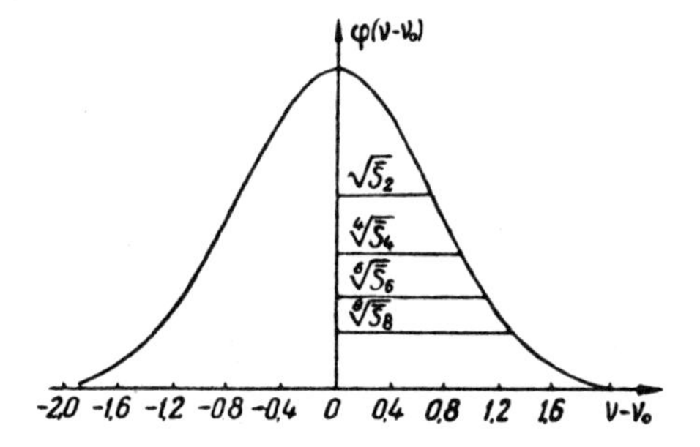

Fig. 2.16 Central moments of the function given by (11.9)

The sharp asymmetry of the function is reflected in the large value of the third central moment. Its positive sign indicates that the function $\rho(\mathrm{E})$ falls off rapidly at large E.

The function

$$\rho(\mathrm{E}) = C(T)\,\mathrm{E}^L e^{-\mathrm{E}/kT} \qquad (11.13)$$

is frequently encountered in spectroscopy. It describes the population of the energy levels of a system of $L-1$ non-interacting classical oscillators. When (11.13) is normalised:

$$C(T) = \frac{1}{\int \mathrm{E}^L\, e^{-\mathrm{E}/kT} d\mathrm{E}} \qquad (11.14)$$

The initial moments of (11.13) are then given by

$$\begin{aligned} S_0 &= 1 \\ S_1 &= kT(L+1) \\ S_2 &= (kT)^2(L+3)(L+2) \qquad (11.15) \\ &\cdots\cdots\cdots \\ S_l &= (kT)^l\,\frac{(L+l)!}{L!} \end{aligned}$$

The expressions for the central moments are more complicated. When $L = 1$ they are

$$\bar{S}_1 = 0,\ \ \bar{S}_2 = 2(kT)^2,\ \ \bar{S}_3 = 4(kT)^3,\ \ \bar{S}_4 = 24(kT)^4$$
$$\bar{S}_5 = 128(kT)^5.$$

A plot of (11.13) for $L = 1$, and the corresponding moments for $kT = 200\ \mathrm{cm}^{-1}$ are shown in Fig. 2.17.

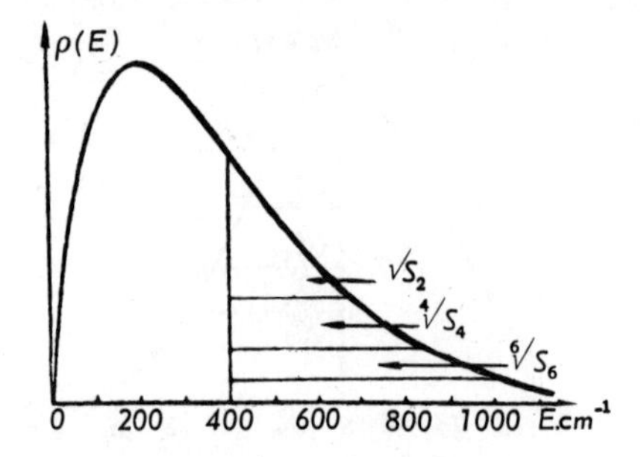

Fig. 2.17 Central moments of the function $\rho(E) = CEe^{-E/kT}$

Figure 2.18 shows the luminescence spectra [20] for 3-aminophthalamide vapour at $t = 265\,^{\circ}C$. The curves are normalised to unity. When the frequency of the exciting radiation is changed, there is a corresponding change in the spectrum. This is not very clear from the graphs.

Table 2 shows the values of S_0, $\bar{S}_1$, $\bar{S}_2$, $\sqrt{\bar{S}_2}$, $\bar{S}_3$, $\bar{S}_4$ and $\sqrt[4]{\bar{S}_4}$ for the three experimental curves. As can be seen, a monotonic change in the frequency of the exciting radiation leads

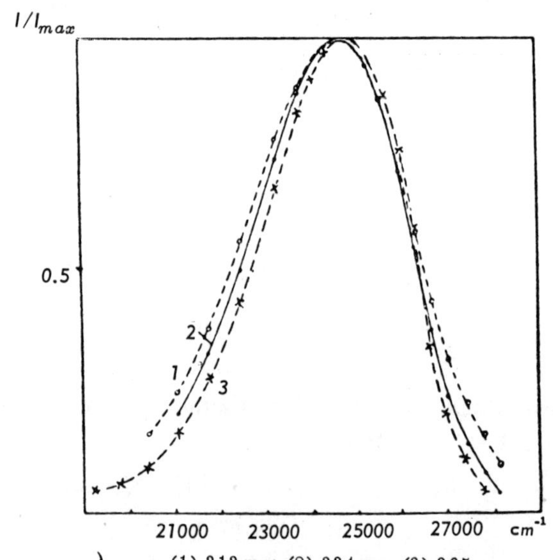

λ_{exc} = (1) 313 mμ; (2) 334 mμ; (3) 365 mμ

Fig. 2.18 Luminescence band of 3-aminophthalamide vapour at T = 265°C

TABLE 2 Moments of the luminescence bands of 3-aminophthalamide

λ_{exc}, mμ	S_0	S_1, cm^{-1}	$\frac{\overline{S_1}}{S_0}$	$\overline{S_2}$, cm^{-2}	$\sqrt{\overline{S_2}/S_0}$, cm^{-1}	$\overline{S_3}$, cm^{-3}	$\overline{S_4}$, cm^{-3}	$\sqrt[4]{\overline{S_4}/S_0}$, cm^{-1}
365	11765	$28120 \cdot 10^4$	23901	$40360 \cdot 10^6$	1850	$-48500 \cdot 10^9$	$44030 \cdot 10^{13}$	2470
334	12985	$30619 \cdot 10^4$	23580	$54200 \cdot 10^6$	2040	$-59500 \cdot 10^9$	$74600 \cdot 10^{13}$	2750
313	14625	$34320 \cdot 10^4$	23466	$82420 \cdot 10^6$	2380	$-139300 \cdot 10^9$	$153650 \cdot 10^{13}$	3200

to a monotonic change in the moments of the luminescence band. As ν_{exc} increases, the first moment decreases, indicating a shift towards lower frequencies.

At the same time, there is an appreciable increase in the second and fourth moments, the fourth moment increasing somewhat more rapidly. The values of $\sqrt{\overline{S}_2}$ and $\sqrt[4]{\overline{S}_4}$ for one of the bands are shown in Fig. 2.19. The negative sign of the third central moment shows that the values of $I(\nu)$

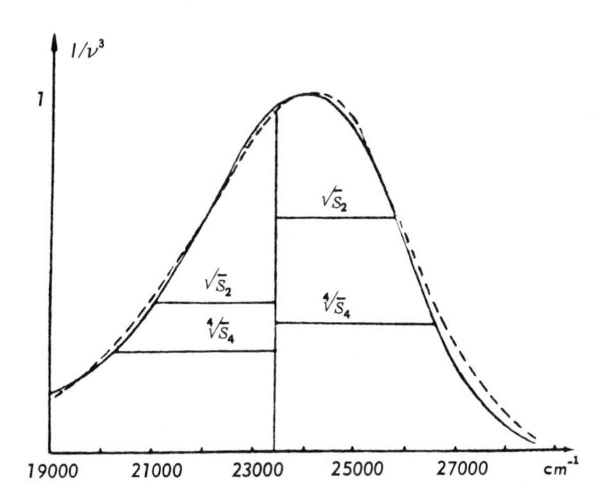

Continuous line - experimental; dashed line - calculated using the first two terms of (11.5)

Fig. 2.19 Central moments of the luminescence band of 3-aminophthalamide vapour

decrease on the right of the maximum more rapidly than on the left $(\bar{\nu} < \nu_{max})$. The increase in the absolute value of the third central moment represents an increase in the asymmetry of the luminescence band.

The broken curve in Fig. 2.19 shows the form of the band calculated using the first two terms in Edgeworth's formula (11.5). The agreement between the experimental and calculated curves is quite good even in the first approximation. When the third term was taken into account the agreement was virtually complete.

By calculating the moments of spectral bands, it is thus possible to obtain a description of the dependence of the band profile on frequency. Similarly, one could investigate the dependence of the moments of other parameters which affect the spectra. Systematic calculations of this kind can serve as a basis for the interpretation of spectra. It is important to note that application of the method of moments, and the determination of the connection between moments and the internal properties of absorbing and emitting objects, require an increase in the accuracy with which the functions $\rho(\nu)$ are measured, particularly in the tails of bands. It is desirable to have experimental determinations of the first and second moments to an accuracy of at least 1% and of the third moment to 5-10%.

Connection between moments of a spectral band and internal properties of absorbing and emitting centres

Consider radiative transitions between vibrational sublevels of two electronic levels. The analysis will be valid for a combination of a fast and slow sub-system. The distribution of transitions from the j-th level of the initial state I to the i-th level of the final electronic state II with the absorption (or emission) of frequency ν_{ij} is determined by the Franck-Condon factor (Section 8)

$$p_{ji} = \text{const} \left| \int \psi_{\mathrm{II}i}(q) P(q) \psi_{\mathrm{I}j}(q)\, dq \right|^2 \qquad (11.16)$$

where $\psi_{\mathrm{I}j}(q)$ and $\psi_{\mathrm{II}i}(q)$ are the corresponding vibrational eigenfunctions and $P(q)$ is the matrix element for the electronic transition. The symbol q represents the vibrational coordinates of the system. The subscript i is looked upon as variable. On the dipole approximation the operator $P(q)$ is given by (see (8.100))

$$P(q) = \int \psi_{II}^{el}(x, q)\, \mathbf{D}_{el}\, \psi_{I}^{el}(x, q)\, dx$$

We have already pointed out that the concept of moments can be used with any distribution function, including of course the distribution function given by (11.16). The initial moment of order l for the distribution (11.16) is given by [21]

$$S_l^{(j)} = \sum_i \nu_{ij}^l\, p_{ji} = \frac{1}{h} \sum_i (E_i - E_j)^l\, p_{ji} \tag{11.17}$$

These formulae hold for discrete vibrational levels, but our subsequent analysis will also be valid for a continuous spectrum. In order to be specific, we shall refer to absorption only.

Consider the quantity

$$\begin{aligned}(E_i - E_j)^l \psi_{IIi}(q) P(q)\, \psi_{Ij}(q) &= \sum_{k=0}^{l} (-1)^k C_l^k\, E_i^{l-k}\, \psi_{IIi}(q)\, P(q) \\ \times E_j^k\, \psi_{Ij}(q) &= \sum_{k=0}^{l} (-1)^k C_l^k\, P(q)\, [\mathbf{H}_I^k \psi_{Ij}(q)]\, [\mathbf{H}_{II}^{*l-k} \psi_{IIi}^*(q)]\end{aligned} \tag{11.18}$$

We have used the Schroedinger equation for vibrational eigenfunctions (see (5.14))

$$\mathbf{H}_I\, \psi_{Ij}(q) = E_j\, \psi_{Ij}(q), \quad \mathbf{H}_{II}\, \psi_{IIi}(q) = E_1\, \psi_{IIi}(q)$$

where $\mathbf{H}_I$ and $\mathbf{H}_{II}$ are the Hamiltonian operators which determine on the adiabatic approximation vibrations of the nuclei in the first and second electronic states $[\mathbf{H} = \mathbf{T}_{vib} + \mathbf{U}(q) = \mathbf{T}_{vib} + E_{el}(q)]$.

Integrating (11.18) with respect to q and using the self-adjoint property of the operators $\mathbf{H}_{II}^{l-k}$, it can readily be shown that

$$\begin{aligned}&(E_i - E_j)^l \int \psi_{IIi}^*(q)\, P(q)\, \psi_{Ij}(q)\, dq \\ &= \sum_{k=0}^{l} (-1)^k C_l^k \int \psi_{IIi}(q)\, \mathbf{H}_{II}^{l-k} P(q)\, \mathbf{H}_I^k\, \psi_{Ij}(q)\, dq\end{aligned} \tag{11.19}$$

If we now multiply (11.19) by

$$\int \psi_{Ij}^{*}(q')\, P^{*}(q')\, \psi_{IIi}(q')\, dq' \tag{11.20}$$

and then sum the resulting expression over all the final states i, we have

$$S_l^{(j)} = \text{const}\, \frac{1}{h} \sum_{k=0}^{l} (-1)^k C_l^k \int \psi_{Ij}^{*}(q)\, P^{*}(q)\, \mathbf{H}_{II}^{l-k} \times P(q)\, \mathbf{H}_I^k\, \psi_{Ij}(q) dq. \tag{11.21}$$

since $\sum_i \psi_i^*(q) \times \psi_i(q') = \delta(q - q')$. In contrast to (11.19), this expression does not contain the eigenfunctions $\psi_{IIi}(q)$ for the final states, and is determined exclusively by the eigenfunctions for the original state and by the form of the operators $\mathbf{H}_I$, $\mathbf{H}_{II}$ and $P(q)$.

The first few moments of the distribution of transition probabilities from level j are

$$S_0^{(j)} = \text{const}\, \frac{1}{h} \int |P(q)|^2 |\psi_{Ij}(q)|^2 dq = |P|^2 \tag{11.22}$$

$$S_1^{(j)} = \text{const}\, \frac{1}{h} \int \psi_{Ij}^{*}(q)\, [P^{*}(q)\, \mathbf{H}_{II}\, P(q) - |P(q)|^2 \mathbf{H}_1]\, \psi_{Ij}(q) dq$$

$$\approx \frac{|P|^2}{h} \int \psi_{Ij}^{*}(q)\, [\mathbf{H}_{II} - \mathbf{H}_I]\, \psi_{Ij}(q)\, dq \tag{11.23}$$

$$S_2^{(j)} = \text{const}\, \frac{1}{h} \int \psi_{Ij}^{*}(q)\, [P^{*}(q)\, \mathbf{H}_{II}^2\, P(q) - 2P(q)\, \mathbf{H}_{II}\, P(q)\, \mathbf{H}_I + |P(q)|^2 \mathbf{H}_I^2]\, \psi_{Ij}(q)\, dq \cong \frac{|P|^2}{h} \int \psi_{Ij}^{*} [\mathrm{H}_{II}^2 - 2\mathrm{H}_{II}\, \mathrm{H}_I + \mathrm{H}_I^2]\, \psi_{Ij}(q)\, dq \tag{11.24}$$

The second parts of these expressions (after the sign $\approx$) give the moments of the distribution in the Franck-Condon approximation (the operator P is independent of the vibrational coordinates).

It is quite easy to transform the moments of the transition probability distribution into the moments of the distribution of absorption or emission probabilities, i.e. into the moments of the spectral bands produced as a result of transitions between electronic levels. Note that in order to obtain the absorption and emission probabilities p, the experimentally determined absorption coefficient k must be divided by ν and the emitted power ν^4 (see Chapters 4 and 5). To do this, the transition probabilities corresponding to frequencies ν must be averaged (with allowance for the population of the levels) over all the initial levels j. The analogous operation is valid for the moments

$$S_l = \sum_j n_j S_l^{(j)} \tag{11.25}$$

If there is an equilibrium distribution over the energy levels before the absorption (emission), the values of n_j given by (11.25) are determined by the Boltzmann formula. The expression given by (11.25) may be reduced to the more convenient form

$$S_0 = \frac{1}{h} |P|^2 n \tag{11.26}$$

$$S_1 = \frac{1}{h} |P|^2 n \int U(q) \rho_T(q)\, dq \tag{11.27}$$

$$S_2 = \frac{1}{h} |P|^2 n \int U^2(q) \rho_T(q)\, dq \tag{11.28}$$

where $n = \Sigma n_j$ and $U(r) = \mathbf{H}_{II} - \mathbf{H}_I = U_{II}(q) - U_I(q)$ is the difference between the potential energy for the upper and lower electronic states (the kinetic energy operators are the same in both states and therefore cancel out). Moreover,

$$\rho_T(q) = \sum_j \frac{n_j}{n} |\psi_{Ij}(q)|^2 \tag{11.29}$$

is the distribution of the system over the coordinates of the nuclei. If the system is in a state of thermodynamic equilibrium (if only over the vibrational degrees of freedom), $\rho_T(q)$ is a function of temperature. $\rho(q)$ is different for each specific system and depends on the vibrating masses. The form of the third initial moment is somewhat more complicated, and will not be given in this instance.

It should be noted that according to (11.26) the area of a band on the Condon approximation is independent of temperature and generally of the distribution function for the vibrational levels.

The moments of bands are thus unambiguously related to the internal properties of matter, i.e. to the operators $\mathbf{H}_I$, $\mathbf{H}_{II}$, $P(q)$ and the population n_j. If these quantities are known, the moments can be calculated. Conversely, if the moments are known from experiment, it is possible to deduce specific information about the properties of absorbing or emitting centres. The calculations are particularly simple for special cases such as the harmonic oscillator, the diatomic molecule and the polyatomic molecule with small anharmonicity. The moments for a harmonic oscillator will be calculated in Chapter 6. In practice, one is often concerned with more complicated objects for which the form of the operators $\mathbf{H}_I$, $\mathbf{H}_{II}$, $P(q)$ is unknown. Further progress may be achieved by assuming a particular model, i.e. by calculating the moments for a number of suitable systems and then comparing the results with experimental data. Work on this very promising procedure is only just beginning. The first interesting results were reported in papers quoted earlier [21]. The Condon approximation has been used to investigate a single harmonic oscillator, a set of independent harmonic oscillators (normal vibrations of a molecule), a single anharmonic oscillator and a system of anharmonic oscillators with a small anharmonicity. A few cases in which the Condon approximation cannot be used have been solved.

Lax's method

The profile of a given band may be written in the form [22]

$$\rho(\nu) = \frac{1}{h}\int I(t)\, e^{-2\pi i \nu t}\, dt \tag{11.30}$$

where

$$I(t) = h\int \rho(\nu)\, e^{2\pi i \nu t}\, d\nu \tag{11.31}$$

is the Fourier transform of $\rho(\nu)$. It is found that for electronic-vibrational transitions the Fourier transform is quite simply related to the internal properties of the absorbing or emitting centre, i.e. to the operators $\mathbf{H}_I$, $\mathbf{H}_{II}$ $P(q)$ and to the population of the vibrational levels of the initial electronic state. In practice, the calculations are based on the expression (11.16) for the probability of transition from a vibrational level j of the first electronic state to the vibrational level i of the second state. By averaging (11.16) over all the initial levels j, with allowance for the population n_j, it is quite easy to show that the probability of transitions from all levels j is given by

$$\rho(\nu) = \sum_j n_j\, p_{ji} = \text{const} \sum n_j \left| \int \psi^*_{Ij}\,(q)\, P(q)\, \psi_{Ii}(q)dq \right|^2 \qquad (11.32)$$

For complicated systems, the vibrational energy levels form a continuous spectrum and therefore, instead of summing over j, we must integrate over E_I.

A detailed comparison of (11.30) with the equivalent expression (11.32) is given in Lax's original paper. It is found that the Fourier transform of $\rho(\nu)$ may be written in the form

$$I(t) = \frac{S_p\left[P^*(q)\, e^{2\pi i\, \mathbf{H}_{II}\, t/h}\, P(q)\, e^{-2\pi i\, \mathbf{H}_I\, t/h}\, e^{-\mathbf{H}_I\, k/T}\right]}{S_p(e^{-\mathbf{H}_I\,/kT})} \qquad (11.33)$$

where the symbols are defined in the preceding section. The symbol S_p represents the trace of the matrix, i.e. the sum of its diagonal elements.

By calculating the Fourier transform $I(t)$ with the aid of (11.30) for a given model of the molecule (impurity centre), it is possible to calculate the appropriate band profile. The initial moments of $\rho(\nu)$ are particularly easy to determine. Expanding the exponential under the integral sign in (11.30) into a series, we have

$$S_l = \int \nu^l \rho(\nu)\, d\nu = \left[\frac{d}{d(2\pi it)}\right]^l_{t=0} I(t) \qquad (11.34)$$

On substituting (11.33) into (11.34) and evaluating the first few elements, it is found that they are the same as the moments (11.21)-(11.24), averaged with the aid of (11.25).

This method has been used in recent years to study the luminescence of complicated molecules of crystals and other systems [23]. It appears to be very promising. It is worth noting that Lax's method can also be used when the moments of $\rho(\nu)$ do not exist (the integrals diverge). In contrast to the moments, the Fourier transform exists for all $\rho(\nu)$.

3

Quantum-electrodynamic Theory of the Interaction of Radiation with Matter

12. QUANTISATION OF THE FREE ELECTROMAGNETIC FIELD

Formulation of the problem

Classical electrodynamics and quantum mechanics each provide a good description of different aspects of the interaction of radiation with matter. However, they do not give a strictly correct and entirely complete representation of this process. Classical theory does not take into account the quantum properties of atoms and molecules, or the quantum properties of radiation. Wave mechanics, on the other hand, takes into account the quantum properties of mechanical systems only. The electromagnetic field itself is not studied in wave mechanics and enters into it only as a perturbation which leads to transitions of the mechanical system from one stationary state to another.

The absorption and emission of radiation, and even more

so, the properties of radiation, can be investigated in quantum mechanics only in an indirect way. Since the electromagnetic field is described by classical quantities, and its specifically quantum-mechanical properties are not taken into account, it follows that the perturbing effect of light on an atom is allowed for only approximately. In particular, the analysis does not include the interaction of matter with zero-point fields, which leads to such important processes as the spontaneous emission of radiation. Quantum electrodynamics is free from these shortcomings. Matter and fields interacting with it are regarded as a single quantum-mechanical system with its own specific wave functions.

The quantum theory of radiation starts with the Schroedinger equation

$$i\frac{h}{2\pi}\frac{\partial\psi}{\partial t} = \mathbf{H}\psi \tag{12.1}$$

where $\mathbf{H}$ is the Hamiltonian of the matter + field system. If we denote by $\mathbf{H}_m$ and $\mathbf{H}_f$ the energy operators for matter and for the field, and by $\mathbf{V}$ the operator for the interaction between them, the resultant Hamiltonian may be written as the sum

$$\mathbf{H} = \mathbf{H}_m + \mathbf{H}_f + \mathbf{V} \tag{12.2}$$

In general, both $\mathbf{H}$ and ψ must be relativistically invariant. However, in the optical region the motion of particles takes place relatively slowly, the energy quanta are small, and therefore relativistic effects are also small and may be disregarded.

We shall regard the operator $\mathbf{H}_m$ and its eigenfunctions as known. Many of the specific forms of these operators are considered in quantum mechanics. The first step must therefore be to find the explicit form for $\mathbf{H}_f$ and to investigate its eigenfunctions. In other words, we must first consider the quantisation of the free electromagnetic field.

Quantisation of the free field

The electromagnetic field equation in a vacuum may be derived from the vector potential, which satisfies the equations

$$\Delta\mathbf{A} + \frac{1}{c^2}\frac{\partial^2\mathbf{A}}{\partial t^2} = 0 \tag{12.3}$$

$$\operatorname{div} \mathbf{A} = 0 \tag{12.4}$$

with the scalar potential φ set equal to zero. It was shown in Section 4 that these two equations follow directly from Maxwell's equations when $\rho = 0$. The relation between the electric and magnetic fields on the one hand, and the vector potential on the other, is given by (1.2). If we confine our attention to a cube of volume $v = L^3$, the solution of (12.3) can be written as the sum of plane monochromatic waves equivalent to the set of harmonic oscillators (Section 4)

$$\mathbf{A} = \sum_{\lambda} [q_{\lambda}(t)\, \mathbf{A}_{\lambda}(\mathbf{r}) + q_{\lambda}^{*}\,(t)\, \mathbf{A}_{\lambda}^{*}\,(\mathbf{r})] \tag{12.5}$$

where

$$\mathbf{A}_{\lambda}(\mathbf{r}) = \sqrt{\frac{4\pi\, c^2}{v}}\, \mathbf{e}_{\lambda} e^{i\,(\varkappa\, \mathbf{r})}\,; \qquad q_{\lambda}(t) = q_{\lambda} e^{-2\pi\, i\, \nu_{\lambda}\, t} \tag{12.6}$$

Summation over λ is equivalent to summation over all the possible values of the wave vector $\varkappa_{\lambda}$ and over the two polarisation directions.

If we substitute

$$Q = q_{\lambda} + q_{\lambda}^{*}\,, \quad P_{\lambda} = -2\pi\, i\, \nu_{\lambda}\, (q_{\lambda} - q_{\lambda}^{*}\,) \tag{12.7}$$

the energy of an individual plane wave will be given by the Hamilton function

$$H_{\lambda} = \frac{1}{2}\, [P_{\lambda}^2 + (2\pi\nu_{\lambda})^2\, Q_{\lambda}^2\,] \tag{12.8}$$

and the total Hamiltonian for the field will be

$$\mathrm{H}_{f} = \sum_{\lambda} \mathrm{H}_{\lambda} = \frac{1}{2} \sum_{\lambda} [P_{\lambda}^2 + (2\pi\nu_{\lambda})^2\, Q_{\lambda}^2\,] \tag{12.9}$$

The expression given by (12.8) is the Hamilton function for harmonic oscillator. In quantum theory Q_{λ} must be looked upon as a time-independent multiplication operator and P_{λ} must be replaced by the momentum operator $\mathbf{P}_{\lambda} = -i\dfrac{h}{2\pi}\dfrac{\partial}{\partial Q_{\lambda}}$. The operators $\mathbf{Q}_{\lambda}$ and $\mathbf{P}_{\lambda}$ are analogous to the position and momentum operators for an ordinary oscillator in that they satisfy the commutation relations

$$[\mathbf{P}_\lambda \mathbf{Q}_\lambda] = \mathbf{P}_\lambda \mathbf{Q}_\lambda - \mathbf{Q}_\lambda \mathbf{P}_\lambda = -i\frac{h}{2\pi}$$

$$[\mathbf{P}_\lambda \mathbf{Q}_\mu] = [\mathbf{P}_\lambda \mathbf{P}_\mu] = [\mathbf{Q}_\lambda \mathbf{Q}_\mu] = 0 \qquad (12.10)$$

The Schroedinger equation for the free electromagnetic field is of the form

$$\mathrm{H}_f \Phi = \frac{1}{2} \sum_\lambda [\mathbf{P}_\lambda^2 + (2\pi\nu_\lambda)^2 \mathbf{Q}_\lambda] \Phi = \mathrm{E}_n \Phi \qquad (12.11)$$

Since the field oscillators do not interact with each other, the function Φ may be sought in the form of the product of wave functions of the individual oscillators:

$$\Phi = \varphi_{\lambda_1} \varphi_{\lambda_2} \varphi_{\lambda_3} \ldots \varphi_{\lambda_N} \ldots = \prod_\lambda \varphi_\lambda \qquad (12.12)$$

The total field energy will then be equal to the sum of the individual energies:

$$\mathrm{E}_f = \sum_\lambda \mathrm{E}_\lambda \qquad (12.13)$$

and (12.11) will become the system of equations for φ_λ and E_λ

$$\frac{1}{2}\left[-\frac{h^2}{4\pi^2}\frac{\partial^2}{\partial Q^2} + (2\pi\nu_\lambda)^2 Q_\lambda^2\right]\varphi_\lambda = \mathrm{E}_\lambda \varphi_\lambda \qquad (12.14)$$

This equation is well known in quantum mechanics (Chapter 6). It describes the properties of a harmonic oscillator of mass $m = 1$, and its solutions are

$$\varphi_{\lambda n}(\xi_\lambda) = e^{-\xi_\lambda^2/2} H_{\lambda n}(\xi_\lambda) \qquad (12.15)$$

$$\mathrm{E}_{\lambda n} = \left(n_\lambda + \frac{1}{2}\right) h\nu_\lambda \qquad (12.16)$$

where $H_{\lambda n}(\xi_\lambda)$ is the Chebyshev-Hermite polynomial of order n and

$$\xi_\lambda = \sqrt{\frac{4\pi^2 \nu_\lambda}{h}}\, Q_\lambda \qquad (12.17)$$

In these expressions ν_λ is the frequency of the λ-th oscillator, which in view of (4.10) is given by

$$\nu_\lambda = \frac{c\,|\varkappa_\lambda|}{2\pi} = \frac{c}{L}\sqrt{n_{\lambda x}^2 + n_{\lambda y}^2 + n_{\lambda z}^2} \tag{12.18}$$

and $n_{\lambda j} = 0, 1, 2, 3, \ldots$.

Substituting (12.15) into (12.12) we obtain the possible total energies of the electromagnetic field:

$$E_f = \sum_\lambda \left(n_\lambda + \frac{1}{2}\right) h\nu_\lambda \tag{12.19}$$

As can be seen from (12.16), the energy of the individual field oscillators is strictly discrete. The energy of the field as a whole is also discrete. According to (4.26) the number of oscillators per unit frequency range is equal to $\frac{8\pi\nu^3}{c^3}\,v$, and therefore as ν and the volume of the cube increase, the energy level density will increase very rapidly. It is for this reason that the energy of the electromagnetic field can frequently be regarded as practically continuous. The lowest state of the field has the energy

$$E_f^0 = \frac{1}{2}\sum_\lambda h\nu_\lambda \tag{12.20}$$

which is known as the zero-point energy. The state of the field with zero-point energy is the ground (unexcited) state of the electromagnetic field. It can only be achieved at the absolute zero of temperature. All other states are excited states and are completely defined by the set of quantum numbers of all the oscillators $n_{\lambda_1}, n_{\lambda_2}, n_{\lambda_3}, \ldots$.

Field quanta

Since the field can communicate to matter, or receive from it, only discrete amounts (quanta) of energy, it may be said that the field itself consists of individual objects, i.e. quanta of radiation or photons. If an oscillator λ is excited to the k-th level, then it is said that the field consists of k photons of energy $h\nu_\lambda$. The zero-point energy is then ignored, since it remains constant.

The concept of a photon as the elementary particle of a

field can also be introduced by considering the momentum of the field. It may be shown that each plane wave has a set of discrete momenta which are proportional to the energy of the wave. The momentum and energy of the wave transform as a 4-vector under the Lorentz transformation [1]. Therefore, each individual plane wave λ occupying a level n_λ behaves as a beam of n_λ free particles of energy $h\nu_\lambda$ each with momentum $\frac{1}{2\pi} h \chi$.

Matrix elements of the operators $\mathbf{q}_\lambda$

After quantisation the vector potential **A**, and therefore the field vectors **E** and **H**, will also transform into operators acting on the field wave function Φ. In further calculations we shall be concerned with the effect of the operators $\mathbf{q}_\lambda$ and $\mathbf{q}^*_\lambda$.

According to the quantum theory of the harmonic oscillator (Chapter 6) the matrix elements of a coordinate Q_λ calculated with the aid of the wave functions (12.15) are

$$Q^{\lambda}_{n,n+1} = Q^{*\lambda}_{n+1,n} = \sqrt{\frac{h(n_\lambda + 1)}{8\pi^2\nu_\lambda}}$$
$$Q^{\lambda}_{n,n'} = 0, \quad \text{if} \quad n'_\lambda \neq n_\lambda \pm 1 \qquad (12.21)$$

Since $P^{\lambda}_{mn} = -2\pi i \nu_{mn} Q^{\lambda}_{mn}$, it is easy to show, using (12.7) and (12.21), that the matrix elements of the operators $\mathbf{q}_\lambda$ and $\mathbf{q}^*_\lambda$ are

$$q^{\lambda}_{n,n+1} = \sqrt{\frac{h(n_\lambda + 1)}{8\pi^2\nu_\lambda}}, \quad q^{*\lambda}_{n+1,n} = \sqrt{\frac{h(n_\lambda + 1)}{8\pi^2\nu_\lambda}}$$
$$q^{\lambda}_{n+1,n} = q^{*\lambda}_{n,n+1} = 0 \qquad (12.22)$$

Applying the operator $\mathbf{q}_\lambda$ to the function $\varphi_{\lambda n}$ we obtain a new function which can be expanded into a series of terms of the eigenfunctions of the operator $\mathbf{q}_\lambda$:

$$\mathbf{q}_\lambda \varphi_{\lambda n} = \sum_m a^{\lambda}_{mn} \varphi_{\lambda n} \qquad (12.23)$$

If we now multiply (12.23) by $\varphi^*_{\lambda k}$ and integrate over all space, we have

$$a^{\lambda}_{kn} = \int \varphi^{*}_{\lambda k}\, q_{\lambda}\, \varphi_{\lambda n}\, d\tau = \sqrt{\frac{h}{8\pi^2 \nu_{\lambda}}}\sqrt{n_{\lambda}}\, \delta_{k,n-1} \tag{12.24}$$

Consequently

$$q_{\lambda}\, \varphi_{\lambda n} = \sqrt{\frac{h}{8\pi^2 \nu_{\lambda}}}\sqrt{n_{\lambda}}\, \varphi_{\lambda, n-1} \tag{12.25}$$

Similarly

$$q^{*}_{\lambda}\, \varphi_{\lambda n} = \sqrt{\frac{h}{8\pi^2 \nu_{\lambda}}}\sqrt{n_{\lambda}+1}\, \varphi_{\lambda, n+1} \tag{12.26}$$

As can be seen from these equations, the operators q_{λ} and q^{*}_{λ} transform the wave function $\varphi_{\lambda n}$ into the wave functions $\varphi_{\lambda,n-1}$ and $\varphi_{\lambda,n+1}$. Since these functions describe the state of the field in which the number of photons of a given sort is fewer or greater by one as compared with the state $\varphi_{\lambda n}$, the operators q_{λ} and q^{*}_{λ} are called the absorption and emission operators respectively.

Equations (12.25) and (12.26) can be used to find the result of the application of the operator $\frac{8\pi^2 \nu_{\lambda}}{h} q^{*}_{\lambda}\, q_{\lambda}$ to the wave function:

$$\frac{8\pi^2 \nu_{\lambda}}{h} q^{*}_{\lambda}\, q_{\lambda}\, \varphi_{\lambda n} = n_{\lambda}\, \varphi_{\lambda n} \tag{12.27}$$

Since the eigenvalues of this operator are equal to the number of quanta, the operator itself represents the number of quanta of a given kind.

Uncertainty relation for field operators

All the physical quantities which describe the electromagnetic field, i.e. vector potential, field strengths, energy, momentum, number of photons, and phase of the wave are replaced by operators in the quantum theory. Some of them, for example the energy and momentum operators, commute with each other, and therefore the corresponding physical quantities can be simultaneously measured with an arbitrary degree of precision. Other field operators do not commute.

It is known from quantum mechanics that if two operators **A** and **B** satisfy the commutation relation

$$\mathbf{AB} - \mathbf{BA} = \mathbf{C} \tag{12.28}$$

then the uncertainties $\Delta \mathbf{A}$ and $\Delta \mathbf{B}$ in a simultaneous determination of the physical quantities **A** and **B** satisfy the inequality

$$\Delta \mathbf{A} \Delta \mathbf{B} \geqslant |\mathbf{C}| \tag{12.29}$$

According to (12.7) and (12.10), the operators $\mathbf{q}_\lambda$ and $\mathbf{q}_\lambda^*$ satisfy the commutation relation

$$\mathbf{q}_{\lambda_1} \mathbf{q}_{\lambda_2}^* - \mathbf{q}_{\lambda_2}^* \mathbf{q}_{\lambda_1} = \frac{h}{8\pi^2 \nu_{\lambda_1}} \delta_{\lambda_1 \lambda_2} \tag{12.30}$$

Hence it is clear that the operator for the number of photons does not commute with the field operators **E** and **H**, i.e. the number of photons (the energy of the field) and the field strengths cannot be simultaneously measured with an arbitrary degree of accuracy. If n_λ is determined precisely, then **E** and **H** are completely undetermined and vice versa. This holds good even when $n_\lambda = 0$.

The uncertainties in the electric and magnetic field strengths are of a fundamental significance, but are not reflected in practical calculations in typical optical problems. In Chapter 1, when we were concerned with these problems, the final results never contained the instantaneous values of the field strength or the coordinates, but always involved averages over the period of the oscillations. A similar averaging procedure is carried out in the quantum theory of radiation.

Quantisation of the field for open systems [24]

All the results in the present section were obtained on the assumption that the Hamiltonian (12.9) does not explicitly depend on time, i.e. that the electromagnetic field forms a closed system. However, in most cases encountered in practice, the field does not form a closed system. As a rule a real experiment will involve not only field and matter, but also sources of radiation (lamps, arcs, sparks, etc) and detectors of radiation (photographic plates, bolometers, etc), which means that the system is not, in fact, closed.

The methods of classical electrodynamics which were described in Chapter 1 do not as a rule require the introduction of the assumption that the system is closed when optical processes are analysed. The methods of quantum electrodynamics can also be applied to real experimental conditions. Analyses of closed systems will, of course, be useful, since they frequently yield good approximation to reality, and the necessary mathematical formalism is much simpler.

To begin with, let us consider the difference between a closed and an open system from the point of view of the classical theory of radiation. A closed system is basically an electromagnetic field in a finite volume bounded by perfectly reflecting walls. The set of field oscillators is determined by the size and form of the volume, and their vibrational state is prescribed by the amplitudes $\mathbf{q}_{\lambda}$ and $\mathbf{q}_{\lambda}^{*}$. The field can be represented in such a system by the sum of plane standing waves, since to each wave with wave vector $\varkappa_{\lambda}$ and amplitude $\mathbf{q}_{\varkappa}$ there corresponds an opposite wave with wave vector $-\varkappa_{\lambda}$ and amplitude $\mathbf{q}_{-\varkappa_{\lambda}} = \mathbf{q}_{\varkappa_{\lambda}}$. The quantities $\mathbf{q}_{\varkappa_{\lambda}}$ may be regarded as time-independent internal parameters of the system.

In open systems the field consists of travelling waves whose parameters can be functions of time. However, if we confine our attention to radiation within a finite volume, for example a cube, the field may be written down as the superposition of plane waves or harmonic oscillators. By choosing suitable dimensions of the cube, we can select only the complete system of orthonormal functions and any vector potential $\mathbf{A}$ can be expanded in terms of them. In contrast to the previous case, the quantities $q_{\varkappa_{\lambda}}$ can be functions of time, and moreover $q_{\varkappa_{\lambda}} \neq q_{-\varkappa_{\lambda}}$. If the field parameters vary slowly enough so that in a single period, $T = 10^{-14}$ sec, they can be regarded as practically constant, then all the formulae which we obtained before will remain valid for open systems also.

The transition to open systems can be performed in a similar way in the quantum theory. The state of the field is prescribed by the number of oscillator quanta n_{λ} (plane waves). In a closed system they are constant, while in an open system they are regarded as functions of time. This fact should not affect the form of either the equations for the free field or the equations for the interaction between light and matter. In point of fact, the possible states of the field are determined in quantum theory by the possible solutions of (12.11) and not by the values of the amplitude. These solutions are quite independent of whether the particular system is open or closed. Consequently, the equations for the interaction between an atom and a field have similar form both for closed and open systems. The difference may be reduced to the fact that in the first case the numbers are the internal parameters of the system, while in the second they are determined by external conditions (method of illumination) and depend explicitly on time.

13. BASIC EQUATIONS OF QUANTUM OPTICS

Expansion of the wave function in terms of the eigenfunctions for matter and field

The electromagnetic field interacting with matter is described by the Schroedinger equation (12.1), which may be written in the form

$$i\frac{h}{2\pi}\frac{\partial}{\partial t}\psi = (\mathbf{H}_0 + \mathbf{V})\psi \tag{13.1}$$

where $\mathbf{H}_0 = \mathbf{H}_m + \mathbf{H}_f$ is the energy operator for matter and field in the absence of interaction between them.

In the preceding section we considered the wave functions for a free field of radiation (see (12.12)). Suppose that we also know the solution of the time-independent Schroedinger equation

$$\mathbf{H}_m \psi_j = E_j \psi_j \tag{13.2}$$

In the absence of interaction the state of the total system consisting of the two sub-systems (matter + field) can be characterised by the wave functions

$$\Psi^0_{j(n_\lambda)} = \psi_j \varphi_{(n_\lambda)}\, e^{-2\pi i E_{i(n_\lambda)} t/h} \tag{13.3}$$

where the subscript j indicates the fact that the matter is in the j-th state and (n_λ) is an abbreviation for the complete set of quantum numbers $n_{\lambda_1}, n_{\lambda_2}, n_{\lambda_3} \dots$ which completely characterise the state of all the field oscillators. The symbol $\varphi_{(n_\lambda)}$ is equivalent to $\prod_\lambda \varphi_{n_\lambda}$.

Let us seek the solution of (12.3) in the form of the sum

$$\Psi = \sum_{i(n_\lambda)} C_{i(n_\lambda)} \psi_i \varphi_{(n_\lambda)} e^{-2\pi i\, E_{i(n_\lambda)} t/h} \tag{13.4}$$

where $E_{i(n_\lambda)} = E_i + E_{(n_\lambda)}$ is the energy of the field and matter. Substituting (13.4) into (13.1), multiplying from the left by $\psi_j^* \varphi^*_{(m_\lambda)}$ and integrating over all the variables on which the wave functions depend, we obtain

$$i \frac{h}{2\pi} \frac{\partial C_{j(m_\lambda)}}{\partial t} = \sum_{i(n_\lambda)} V_{j(m_\lambda)/i(n_\lambda)} C_{i(n_\lambda)} e^{\frac{2\pi i [(E_{j(m_\lambda)} - E_{i(n_\lambda)})] t/h}{h}} \tag{13.5}$$

This equation is analogous to (8.10), which was derived within the framework of quantum mechanics. There is an important difference between them, however. The expansion coefficients C_j in (8.10) refer only to states of the mechanical system; the state of the electromagnetic field is not taken into account at all. In the present case the equation describes the field and matter as a single quantum system. The expansion coefficients $C_{j(m_\lambda)}$ depend on the time and have the meaning of probability amplitudes. Thus, the quantity $|C_{j(m_\lambda)}|^2$ is equal to the probability that the matter is in the j-th state, while the field consists of a set of photons (m_λ). Using (13.5) and the analogous equation for the complex conjugate quantities, it is easy to verify that

$$\frac{d}{dt} \sum_{j(m_\lambda)} |C_{j(m_\lambda)}|^2 = 0 \tag{13.6}$$

i.e. the normalisation of $C_{j(m_\lambda)}$ remains constant.

The set of coefficients $C_{j(m_\lambda)}$ specifies completely the function $\Psi(t)$, and therefore uniquely defines the state of the system.

The number of coefficients $C_{j(m_\lambda)}$, like the number of field oscillators, is infinitely large. It follows that if (13.5) is written out in the explicit form we obtain a complicated system consisting of an infinite number of connected equations. These equations have not so far been solved exactly even for the simplest problems in the theory of radiation. As a rule they are solved by the methods of perturbation theory. This leads to correct results when the interaction operator is in fact small, and the eigenfunctions for the unperturbed system can be taken as the starting point for the application of perturbation theory.

Matrix elements of the interaction operator

In Chapter 2 we derived an expression for the operator representing the interaction of light with matter. For a single optical electron it is of the form

$$\mathbf{V} = -\frac{e}{mc}(\mathbf{pA}) + \frac{e^2}{2mc^2}\mathbf{A}^2 \tag{13.7}$$

where e, m and $\mathbf{p}$ are the charge, mass and momentum of an electron. If $\mathbf{p}$ and $\mathbf{A}$ are defined in the usual way, then $\mathbf{V}$ will be equal to the interaction energy of classical electrodynamics. If, on the other hand, we substitute the momentum operator $\mathbf{p} = -i\dfrac{h}{2\pi}\nabla$ for $\mathbf{p}$ in (13.7), we obtain the quantum mechanical interaction operator. In quantum electrodynamics, both matter and field are quantised, and therefore the vector potential $\mathbf{A}$ is also replaced by an operator. This leads to the following expression

$$\mathbf{V} = \frac{e}{mc}\left[\mathbf{p}\sum_{\lambda}(\mathbf{q}_{\lambda}\mathbf{A}_{\lambda} + \mathbf{q}_{\lambda}^{*}\mathbf{A}_{\lambda}^{*})\right] + \frac{e^2}{2mc^2}\sum_{\lambda}\sum_{\lambda'}(\mathbf{q}_{\lambda}\mathbf{A}_{\lambda} + \mathbf{q}_{\lambda}^{*}\mathbf{A}_{\lambda}^{*})(\mathbf{q}_{\lambda'}\mathbf{A}_{\lambda}^{*} + \mathbf{q}_{\lambda'}^{*}\mathbf{A}_{\lambda'}^{*}) \tag{13.8}$$

where $\mathbf{q}_{\lambda}$ and $\mathbf{q}_{\lambda}^{*}$ are time-independent operators whose effect on the wave functions for the field is determined by (12.25) and (12.26) and $\mathbf{A}_{\lambda}$ is given by (12.6).

Let us consider in detail integrals which are encountered in evaluating the matrix elements of the interaction operator.

1.
$$I_1 = \frac{e}{mc}\int \psi_j \varphi_{(m_\lambda)}\left(\mathbf{p}\sum_\lambda q_\lambda \mathbf{A}_\lambda\right)\psi_i\,\varphi_{(n_{\lambda'})}\,d\tau$$

$$= \frac{e}{m}\sqrt{\frac{4\pi}{c}}\sum_\lambda \int \psi_j(\mathbf{p}\mathbf{e}_\lambda e^{i(\varkappa_\lambda \mathbf{r})})\,\psi_i\,dx \int \varphi_{(m_\lambda)}\,q_\lambda\,\varphi_{(n_{\lambda'})}\,d\tau_1 \quad (13.9)$$

The last expression is based on the fact that the particle momentum operator $\mathbf{p}$ does not act on the field wave functions and q_λ does not act on the particle wave functions. Integrals analogous to the integral with respect to x in (13.9) have already been encountered earlier (see (8.46)) and were discussed in detail there. It will therefore be sufficient to consider the integral with respect to τ_1.

On writing the wave function for the field in the explicit form, we have in view of (12.25),

$$\int \varphi_{(m_\lambda)}\,q_\lambda\,\varphi_{(n_{\lambda'})}\,d\tau_1 = \int \prod_\lambda \varphi_{m_\lambda}\,q_\lambda \prod_\lambda \varphi_{n_{\lambda'}}\,d\tau_1$$

$$= \sqrt{\frac{hn_\lambda}{8\pi^2\nu_\lambda}}\prod_{\lambda'\neq\lambda}\int \varphi_{m_{\lambda'}}\varphi_{n_{\lambda'}}\,d\tau_{\lambda'}\int \varphi_{m_\lambda}\varphi_{n_\lambda - 1}\,d\tau_\lambda \quad (13.10)$$

$$= \sqrt{\frac{hn_\lambda}{8\pi^2\nu_\lambda}}\prod_{\lambda'\neq\lambda}\delta_{m_{\lambda'},\,n_{\lambda'}}\,\delta_{m_\lambda,\,n_\lambda - 1}$$

where the operator q_λ acts only on the wave function of the single oscillator λ. Substituting (13.10) into (13.9) and introducing for the sake of brevity the notation

$$(P_\lambda^+)_{ji} = \frac{e}{m}\sqrt{\frac{h}{2\pi}}\int \psi_j(\mathbf{p}\mathbf{e}_\lambda e^{i(\varkappa \mathbf{r})})\,\psi_i\,dx \quad (13.11)$$

we obtain

$$I_1 = \sum_\lambda (P_\lambda^+)_{ji}\sqrt{\frac{n_\lambda}{v\nu_\lambda}}\prod_{\lambda'\neq\lambda}\delta_{m_{\lambda'},n_{\lambda'}}\,\delta_{m_\lambda,\,n_\lambda - 1} \quad (13.12)$$

It should be noted that the summation over (n_λ) in (13.5) has a different significance compared with the summation over λ

in (13.12). In the first case the summation is carried out over all the possible states of the field, each of which is described by an infinite set of numbers n_{λ_1}, n_{λ_2}, n_{λ_3}, and so on. The sum over λ represents summation over the field oscillators in a particular given state.

$$2.\ I_2 = \frac{e^2}{2mc^2}\int \psi_j\, \varphi_{(m_\lambda)} \left(\sum_\lambda \sum_{\lambda'} q_\lambda \mathbf{A}_\lambda \overset{*}{q}_{\lambda'} \overset{*}{\mathbf{A}}_{\lambda'}\right) \psi_i\, \varphi_{(n_{\lambda''})}\, d\tau$$

$$= \frac{2\pi e^2}{mv} \sum_\lambda \sum_{\lambda'} \int \psi_j\, \mathbf{e}_\lambda \mathbf{e}_{\lambda'}\, e^{i(\boldsymbol{\varkappa}_\lambda - \boldsymbol{\varkappa}_{\lambda'}\cdot \mathbf{r})} \psi_i\, dx$$

$$\times \int \varphi_{(m_\lambda)}\, q_\lambda \overset{*}{q}_{\lambda'} \varphi_{(n_{\lambda''})}\, d\tau_1 \tag{13.13}$$

$$= \frac{1}{2} \sum_\lambda \sum_{\lambda'} (\eta_{ji})_{\lambda\lambda'} \sqrt{\frac{n_\lambda (n_{\lambda'} + 1)}{v^2 \nu_\lambda \nu_{\lambda'}}}$$

$$\times \prod_{\lambda'' \neq \lambda, \lambda'} \delta_{m_{\lambda''} n_{\lambda''}}\, \delta_{m_\lambda n_\lambda - 1}\, \delta_{m_{\lambda'}, n_{\lambda'} + 1}$$

where

$$(\eta_{\lambda\lambda'})_{ji} = \frac{he^2}{2\pi m} \left(\mathbf{e}_\lambda \mathbf{e}_{\lambda'}\, e^{i(\boldsymbol{\varkappa}_\lambda - \boldsymbol{\varkappa}_{\lambda'})\,\mathbf{r}}\right)_{ji} \tag{13.14}$$

The matrix elements of all the other terms in (13.8) can be evaluated in a similar way.

System of equations for the probability amplitudes

After substituting the matrix elements of the operator (13.8) into (13.5) we must carry out the summation over all the states of the field (n_λ). Since the summation sign is followed by products of δ-functions, it follows that only a small number of terms correspond to states of the field which differ from the state $\varphi_{(m_\lambda)}$ by one or two quanta of a particular sort. Next, let us replace m_λ by n_λ so that, using (13.5), we have

$$i\frac{h}{2\pi}\frac{\partial C_{j(n_\lambda)}}{\partial t}=-\sum_i\sum_{\lambda'}\frac{(P^-_{\lambda'})_{ji}\sqrt{n_{\lambda'}}}{\sqrt{v\nu_{\lambda'}}}$$

$$\times C_{i(n_\lambda-\delta_{\lambda\lambda'})}\,e^{2\pi i(\nu_{ji}+\nu_{\lambda'})t}-\sum_i\sum_{\lambda'}\frac{(P^+_{\lambda'})_{ji}\sqrt{n_{\lambda'}+1}}{\sqrt{v\nu_{\lambda'}}}$$

$$\times C_{i(n_\lambda+\delta_{\lambda\lambda'})}\,e^{2\pi i(\nu_{ji}-\nu_{\lambda'})t}+\sum_i\sum_{\lambda'}\sum_{\lambda''}$$

$$\times\frac{(\eta_{\lambda'\lambda''})_{ji}\sqrt{n_{\lambda'}(n_{\lambda''}+1)}}{v\sqrt{\nu_{\lambda'}\nu_{\lambda''}}}C_{i(n_\lambda-\delta_{\lambda\lambda'}+\delta_{\lambda\lambda''})}\,e^{2\pi i(\nu_{ji}+\nu_{\lambda'}-\nu_{\lambda''})t} \tag{13.15}$$

$$+\sum_i\sum_{\lambda'}\sum_{\lambda''}\frac{(d^-_{\lambda'\lambda''})_{ji}\sqrt{n_{\lambda'}n_{\lambda''}}}{v\sqrt{\nu_{\lambda'}\nu_{\lambda''}}}C_{i(n_\lambda-\delta_{\lambda\lambda'},-\delta_{\lambda\lambda''})}$$

$$\times e^{2\pi i(\nu_{ji}+\nu_{\lambda'}+\nu_{\lambda''})t}+\sum_i\sum_{\lambda'}\sum_{\lambda''}\frac{(d^+_{\lambda'\lambda''})\sqrt{(n_{\lambda'}+1)(n_{\lambda''}+1)}}{v\sqrt{\nu_{\lambda'}\nu_{\lambda''}}}$$

$$\times C_{i(n_\lambda+\delta_{\lambda\lambda'}+\delta_{\lambda\lambda''})}\,e^{2\pi i(\nu_{ji}-\nu_{\lambda'}-\nu_{\lambda''})t}$$

where $(\eta_{\lambda'\lambda''})_{ji}$ is given by (13.14), and

$$(P^-_{\lambda'})_{ji}=(P^+_{\lambda'})^*_{ji}=\frac{e}{m}\sqrt{\frac{h}{2\pi}}\left(\mathbf{p}\mathbf{e}_{\lambda'}\,e^{-i(\vec{\varkappa}_{\lambda'},\mathbf{r})}\right)_{ji}$$

$$(d^-_{\lambda'\lambda''})_{ji}=(d^+_{\lambda'\lambda''})^*_{ji}=\frac{e^2h}{2\pi m}\left(\mathbf{e}_{\lambda'}\mathbf{e}_{\lambda''}e^{-i(\varkappa_{\lambda'}+\vec{\varkappa}_{\lambda''})\mathbf{r}}\right)_{ji} \tag{13.16}$$

The amplitudes $C_{i(n_\lambda\pm\delta_{\lambda\lambda'})}$ refer to states which differ from the initial state of the field $\varphi_{(n_\lambda)}$ by one photon belonging to the class λ'. The system of equations given by (13.15) can easily be generalised to the case when the electromagnetic field interacts not with one, but with many optical electrons.

Equation (13.15) for the coefficients $C_{i(n_\lambda)}$ is equivalent to the original wave equation (13.11) and completely describes the interaction of radiation with matter. It contains all the possible special cases. The various atoms and molecules are characterised by different matrix elements $\mathbf{p}_{ji}$, and the different states of the field by the set of quantum numbers n_λ. By solving (13.15) we can obtain a description of all the main experimental facts concerning absorption and emission of radiation. It is for this reason that they may be referred to as the basic equations of the quantum theory of radiation, or better still, of quantum optics.

The first two terms on the right of (13.15) are due to the interaction described by the linear term in the operator (13.7). As a result of this interaction the original state of the field is transformed into a new state which differs from the original state by one photon. A change by two or more photons can only occur in stages through intermediate states. The remaining three terms in (13.15) reflect the interaction of field with matter which is characterised by $\mathbf{A}^2$. In such interactions the state of the field is changed by two photons at once: the two photons λ' and λ'' are simultaneously absorbed or emitted, or the photon λ' disappears from the field and is replaced by the photon λ''.

14. SINGLE-PHOTON RADIATIVE PROCESSES

Absorption

Let us apply (13.15) to the simplest interaction, i.e. to the absorption or emission of a single photon. This is well described by first-order perturbation theory. To simplify the calculations we can neglect the radiation reaction which leads to spectral line broadening. This will not prevent us from reaching the correct general conclusions. Problems connected with line profiles and energy level profiles will be discussed separately.

Suppose that at the initial instant of time the atom is in the lowest level $j = 0$, and the external incident field is specified by the set of protons n_λ^0. Consequently the probability amplitudes at $t = 0$ are

$$C_{0\,(n_\lambda^0)} = 1,\ C_{i\,(n_\lambda^0)} = 0 \text{ when } i \neq 0 \qquad (14.1)$$

At subsequent times the absorption of external radiation will ensure that other $C_{i(n_\lambda^0)}$ are also different from zero. On the first approximation we have from (13.15)

$$i\frac{h}{2\pi}\frac{\partial}{\partial t}C_{i(n_{\lambda}^{0}-\delta_{\lambda\lambda'})}=-\frac{(P_{\lambda'}^{+})_{i0}\sqrt{n_{\lambda'}^{0}}}{\sqrt{v\nu_{\lambda'}}}C_{0(n_{\lambda}^{0})}e^{2\pi i(\nu_{i0}-\nu_{\lambda'})t} \tag{14.2}$$

The coefficients $C_{i(n^{0}+\delta_{\lambda\lambda'})}$ refer to the state of the system in which the atom is in level i and the single photon λ' disappears from the field. Substituting (14.1) into the right-hand side of (14.2) and integrating between 0 and t, we obtain

$$C_{i(n_{\lambda}^{0}-\delta_{\lambda\lambda'})}=\frac{1}{h}\frac{(P_{\lambda'}^{+})_{i0}\sqrt{n_{\lambda'}^{0}}}{\sqrt{v\nu_{\lambda'}}}\frac{1-e^{2\pi i(\nu_{i0}-\nu_{\lambda'})t}}{\nu_{i0}-\nu_{\lambda'}} \tag{14.3}$$

This formula is analogous to (8.19), and therefore all the conclusions which were obtained in the analysis of the frequency and time dependence (8.19) will also hold for equation (14.3). We shall derive the transition probabilities in a somewhat different way, however.

In view of (14.3), the probability of finding the atom in level i at time t as a result of an interaction with a photon λ' is given by

$$|C_{i(n_{\lambda}^{0}-\delta_{\lambda\lambda'})}|^{2}=\frac{2}{h^{2}}\frac{|(P_{\lambda'}^{+})_{i0}|^{2}n_{\lambda}^{0}}{v\nu_{\lambda'}}\frac{1-\cos 2\pi(\nu_{i0}-\nu_{\lambda'})t}{(\nu_{i0}-\nu_{\lambda'})^{2}} \tag{14.4}$$

Since at the initial instant of time $C_{i(n_{\lambda}^{0}-\delta_{\lambda\lambda'})}$ was equal to zero, it follows that the time derivative of (14.4) determines the probability of the transition of the atom from the state 0 to the state i under the action of the photons λ';

$$\frac{d}{dt}|C_{i(n_{\lambda}^{0}-\delta_{\lambda\lambda'})}|^{2}=\frac{4\pi^{2}}{h^{2}}\frac{|(P_{\lambda'}^{+})_{i0}|^{2}n_{\lambda'}^{0}}{v\nu_{\lambda}}\frac{\sin 2\pi(\nu_{i0}-\nu_{\lambda'})t}{\nu_{i0}-\nu_{\lambda'}} \tag{14.5}$$

For large t the last factor has the properties of a δ-function:

$$\lim_{t\to\infty}\frac{\sin\nu t}{\nu}=\pi\delta(\nu) \tag{14.6}$$

In the present case t cannot be regarded as infinitely large. It must be small enough to enable us to use first-order perturbation theory. This approximation is valid provided the conditions given by (14.1) are satisfied. We already know, and we shall prove below, that the mean lifetime of the system in the lowest level is usually very much greater than 10^{-8} sec. Such an interval of time can be approximately regarded as infinitely long in comparison with the period of the oscillations, which is of the order of 10^{-15} -10^{-14} sec. Consequently, we have from (14.5)

$$\frac{d}{dt}|C_i|^2 = \frac{4\pi^2}{h^2}\frac{|(P^{+}_{\lambda'})_{i0}|^2 n^0_{\lambda'}}{v\,\nu_{\lambda'}}\,\delta(\nu_{i0} - \nu_{\lambda'}) \tag{14.6a}$$

This expression gives the probability of absorption of quanta of radiation with strictly defined properties, namely frequency, direction of propagation, and amplitude. In practice, the beam of radiation is usually specified not by the number of photons, but by the energy density $u^{\alpha}(\nu, \Omega)$ per unit frequency interval and unit solid angle and for one of the two independent polarisations ($\alpha = 1, 2$). It is evident that the number of photons and $u^{\alpha}(\nu, \Omega)$ are related by the expression

$$u^{\alpha}(\nu, \Omega)\,d\nu\,d\Omega = \frac{1}{v}\sum_{\varkappa}^{\varkappa + d\varkappa} n_{\varkappa\alpha}\,h\nu_{\varkappa} \tag{14.7}$$

where the numbers of photons carry two subscripts, since (14.7) does not include summation over the polarisations. The summation is carried out only over the wave vector $\varkappa$ where magnitude lies between $2\pi\nu/c$ and $2\pi\nu + d\nu/c$ and whose direction lies between Ω and $\Omega + d\Omega$. If we represent by $n_{\varkappa\lambda}$ the mean number of quanta of a given oscillator of frequency ν, polarisation α and given direction of propagation, and if we take it out from under the summation sign, we obtain

$$\begin{aligned} u^{\alpha}(\nu, \Omega)\,d\nu\,d\Omega &= \bar{n}_{\varkappa,\alpha} h\nu_{\varkappa}\,\frac{1}{v}\sum_{\varkappa}^{\varkappa + d\varkappa} 1 \\ &= \mathrm{n}_{\varkappa\alpha}\,h\nu_{\varkappa}\,\rho_{\alpha}(\nu, \Omega)\,d\nu\,d\Omega \end{aligned} \tag{14.8}$$

where $\rho_{\alpha}(\nu, \Omega)$ is the density of states. Substituting for it from (4.25) into the last expression, we obtain

$$\bar{n}_{\varkappa\alpha} = \frac{c^3}{h\nu^3}\,u^{\alpha}(\nu, \Omega) \tag{14.9}$$

From (14.8) and (4.25) we can obtain the general rule for the replacement of summation over $\varkappa$ by integration with respect to ν and Ω:

$$\sum_{\varkappa} f(\varkappa) \underset{v\to\infty}{\longrightarrow} \frac{v}{c^3} \int_{\nu}\int_{\Omega} f(\nu, \Omega)\, \nu^2 d\nu\, d\Omega \tag{14.10}$$

In order to obtain the total probability p_{0i} of absorption of all the incident photons, we must sum (14.6a) over $\varkappa'$ and α. In view of (14.9) and (14.10) this sum may be written in the form

$$p_{0i} = \sum_{\alpha}\sum_{\varkappa} \frac{d}{dt} |C^{\alpha}_{i(n^0_{\lambda}-\delta_{\lambda\lambda'})}|^2$$

$$= \frac{4\pi^2}{h^3} \sum_{\alpha} \int_{\nu}\int_{\Omega} \frac{|(\mathbf{P}^+_{\varkappa\alpha})_{i0}|^2}{\nu^2_{\varkappa}} u^{\alpha}(\nu, \Omega)\, \delta(\nu_{i0} - \nu_{\varkappa})\, d\nu_{\varkappa}\, d\Omega$$

$$= \frac{4\pi^2}{h^3} \sum_{\alpha} \int_{\Omega} \frac{|(\mathbf{P}^+_{\varkappa\alpha})_{i0}|^2}{\nu^2_{i0}} u^{\alpha}(\nu_{i0}, \Omega)\, d\Omega \tag{14.11}$$

$$= \sum_{\alpha} \int_{\Omega} b^{\alpha}_{0i}(\Omega)\, u^{\alpha}(\nu_{i0}, \Omega)\, d\Omega$$

where

$$b^{\alpha}_{0i}(\Omega) = \frac{4\pi^2}{h^0} \frac{|(P^+_{\varkappa\alpha})_{i0}|^2}{\nu^2_{i0}} \tag{14.12}$$

If we confine our attention to the dipole approximation, we have from (13.21)

$$|(P^+_{\varkappa\alpha})_{i0}|^2 = \frac{e^2 h}{2\pi m} |(\mathbf{p}_{i0}\, \mathbf{e}_{\varkappa\alpha})_{i0}|^2$$

$$= 2\pi e^2 h\, \nu^2_{i0} |(\mathbf{r}\mathbf{e}_{\varkappa\alpha})|^2 = 2\pi \nu^2_{i0} h\, |(\mathbf{D}_{i0}\mathbf{e}_{\varkappa\alpha})|^2 \tag{14.13}$$

Substituting this expression into (14.12) we arrive at the expression for the differential Einstein coefficients (see (8.72))

$$b^{\alpha}_{0i}(\Omega) = \frac{8\pi^3}{h^2} |(\mathbf{D}_{0i}\mathbf{e}_{\varkappa\alpha})|^2 \tag{14.14}$$

The amount of energy absorbed per unit time can be determined with the aid of (14.11) as in quantum mechanics by multiplying the transition probability by the energy of the photon:

$$W_{\text{abs}} = p_{0i} h \nu_{i0} = \sum_{\alpha} \int_{\Omega} b^{\alpha}_{0i}(\Omega)\, u^{\alpha}(\nu_{i0}, \Omega)\, h\nu_{i0}\, d\Omega$$
$$= \sum_{\alpha} \int_{\Omega} W^{\alpha}_{\text{abs}}(\Omega) \qquad (14.15)$$

where

$$W^{\alpha}_{\text{abs}}(\Omega) = b^{\alpha}_{0i}(\Omega)\, u^{\alpha}(\nu_{i0}, \Omega)\, h\nu_{i0} \qquad (14.16)$$

is the amount of absorbed energy of given linear polarisation and given direction of propagation per unit solid angle.

Comparison of these results with the expressions in Section 8 shows that quantum electrodynamics completely confirms the results of quantum mechanics as far as the absorption of radiation is concerned.

Spontaneous and stimulated emission

Suppose that at the initial instant of time the atom is in an excited state i such that

$$C_{i\,(n^0_{\lambda})} = 1, \quad C_{j\,(n^0_{\lambda})} = 0, \quad \text{if} \quad j \neq i$$

Transitions to lower energy levels are then described by

$$i\frac{h}{2\pi}\frac{\partial C_{j\,(n^0_{\lambda}+\delta_{\lambda\lambda'})}}{\partial t} = -\frac{(P_{\overline{\lambda'}})_{ji}\sqrt{n^0_{\lambda'}+1}}{\sqrt{v\nu_{\lambda'}}}$$
$$\times C_{i\,(n^0_{\lambda})}\, e^{2\pi i(\nu_{\lambda'}-\nu_{ij})t} \qquad (14.17)$$

On solving this equation and completing the same operations as for the probability of absorption, we obtain

$$C_{j(n^0_{\lambda}+\delta_{\lambda\lambda'})} = -\frac{1}{h}\frac{(P_{\overline{\lambda'}})_{ji}\sqrt{n^0_{\lambda'}+1}}{\sqrt{v\nu_{\lambda'}}}\frac{1-e^{-2\pi i(\nu_{ij}-\nu_{\lambda'})t}}{\nu_{ij}-\nu_{\lambda'}} \qquad (14.18)$$

$$\frac{d}{dt}|C_j|^2 = \frac{4\pi^2}{h^2}\frac{|(P_{\lambda'}^{-})_{ji}|^2}{v\nu_{\lambda'}}\delta(\nu_{ij}-\nu_{\lambda'})$$

$$+\frac{4\pi^2}{h^2}\frac{|(P_{\lambda'}^{-})_{ji}|^2}{v\nu_{\lambda'}}n_{\lambda}^{0}\delta(\nu_{ij}-\nu_{\lambda'}) \quad (14.19)$$

As can be seen from the latter expression, the probability of emission consists of two parts. The first part is independent of the energy density of the external radiation and is not equal to zero even when ($n_{\lambda}^{0} = 0$). It is therefore called the probability of spontaneous emission.

If we sum the first term in (14.19) over all the emitted quanta, and confine our attention to the dipole approximation, we shall find that

$$A_{ij} = \sum_{\lambda'}\frac{4\pi^2}{h^2}\frac{|(P_{\lambda'}^{-})_{ji}|^2}{v\nu_{\lambda'}}\delta(\nu_{ij}-\nu_{\lambda'})$$

$$= \sum_{\alpha}\int_{\Omega}\frac{8\pi^3\nu_{ij}^3}{hc^3}|(\mathbf{D}_{ij}\mathbf{e}_{\varkappa\alpha})|^2 d\Omega \quad (14.20)$$

$$= \sum_{\alpha}\int_{\Omega} a_{ij}^{\alpha}(\Omega)\, d\Omega = \frac{64\pi^4\nu_{ij}^3}{3hc^3}|\mathbf{D}_{ij}|^2$$

where

$$a_{ij}^{\alpha}(\Omega) = \frac{8\pi^3\nu_{ij}^3}{hc^3}|(\mathbf{D}_{ij}\mathbf{e}_{\varkappa\alpha})|^2 \quad (14.21)$$

is the differential Einstein coefficient for spontaneous emission which was postulated earlier, and A_{ij} is the integral Einstein coefficient.

Calculation of the rate of spontaneous emission per unit solid angle, and the integral emission, again leads to (8.78) and (8.79).

The second term in (14.19) is proportional to the number of incident photons, and gives the probability of stimulated emission which has already been considered within the framework of quantum mechanics. However, in quantum mechanics, it was found from the law of conservation of energy in terms of which only the energy (frequency) of the photon was determined. The direction of propagation and the polarisation of the emitted photon remained unknown. Quantum electrodynamics, on the other hand, resolves this problem and confirms Einstein's elementary theory of emission.

In contrast to the quantum-mechanical expression given by (8.70), which represents only the probability of a change of $h\nu_{kj}$ in the energy of the atom, equation (14.19) gives the probability of emission of photons with strictly defined values of frequency, polarisation and direction of propagation. The probability of stimulated emission of a photon λ' is proportional to $n^0_{\lambda'}$. It follows that for stimulated emission the emitted radiation propagates in the direction of the exciting radiation and has the same frequency and polarisation. On the dipole approximation the differential Einstein coefficient for stimulated emission is given by (see (14.19))

$$b^{\alpha}_{ij}(\Omega) = \frac{8\pi^3}{h^2}\,|(\mathbf{D}_{ij}\mathbf{e}_{\varkappa\alpha})|^2 = b^{\alpha}_{ji}(\Omega) \qquad (14.22)$$

Using (14.21) and (14.22) we can find the relation between the differential Einstein coefficients

$$a^{\alpha}_{ij}/b^{\alpha}_{ij} = \frac{h\nu^3_{ij}}{c^3} \qquad (14.23)$$

which corresponds to

$$A_{ij}/B_{ij} = 8\pi h\nu^3_{ij}/c^3 \qquad (14.24)$$

The last two expressions were derived in Section 7 by Einstein's method from a consideration of the thermodynamic equilibrium of matter and radiation in a closed cavity. Quantum electrodynamics provides a rigorous derivation of these relationships.

Natural profile of energy levels [25]

It was shown in Section 11 that the energy levels of quantised systems broaden as a result of interactions. The level width due to the interaction of an atom with the zero-point electromagnetic field is called the natural width. The atom or molecule can then be regarded as an isolated and closed system whose behaviour is described by the time-independent Schroedinger equation

$$(\mathbf{H}_m + \mathbf{H}_f + \mathbf{V})\,\psi = E\,\psi \qquad (14.25)$$

We shall take the interaction operator in the simplest form

$$\mathbf{V} = -\frac{e}{mc}(\mathbf{pA}) \tag{14.26}$$

If we expand ψ in terms of the eigenfunctions for matter and field and complete operations similar to those leading to (13.15), we shall find that

$$(E_{i(n_\lambda)} - E)\, C_{i(n_\lambda)} = \sum_{j\lambda'} \frac{(P_{\lambda'})_{ij}\sqrt{n_{\lambda'}+1}}{\sqrt{v\nu_{\lambda'}}}\, C_{j(n_{\lambda'}+\delta_{\lambda\lambda'})} + \sum_{j\lambda'} \frac{(P_{\lambda'})_{ij}\sqrt{n_{\lambda'}}}{\sqrt{v\nu_{\lambda'}}}\, C_{j(n_{\lambda'}-\delta_{\lambda\lambda'})}, \tag{14.27}$$

where

$$E_{i(n_\lambda)} = E_i + \sum_\lambda \left(n_\lambda + \frac{1}{2}\right) h\nu_\lambda \tag{14.28}$$

$$(P_{\lambda'})_{ij} = i\sqrt{\frac{h}{2\pi}}\, 2\pi e\, \nu_{ij}\, (\mathbf{r}\mathbf{e}_{\lambda'})_{ij} \tag{14.29}$$

The change in the phase of the wave within the limits of the atom is neglected in these expressions (dipole approximation).

Let us suppose for the sake of simplicity that the atom has two energy levels, and let us solve the system of equations given by (14.27) using the method of successive approximations. As the zero-order approximation we shall take the state in which there are no photons and the atom is in the excited state $E_2 = h\nu_{21}$. For the first approximation we have from (14.27)

$$\begin{gathered}(E_2 - E)C_{2(0)} - \sum_{\lambda'} \frac{(P_{\lambda'})_{21}}{\sqrt{v\nu_{\lambda'}}}\, C_{1(\delta_{\lambda\lambda'})} = 0 \\ (h\nu_{\lambda'} - E)\, C_{1(\delta_{\lambda\lambda'})} - \frac{(P_{\lambda'})_{12}}{\sqrt{v\nu_{\lambda'}}}\, C_{2(0)} = 0\end{gathered} \tag{14.30}$$

This system is equivalent to (10.17) and (10.18) and describes the interaction of an excited level of the molecule with the practically continuous spectrum of possible field states.

The required coefficient $|C_{2(0)}|^2$ can be taken directly from (10.23). As a first step we must find the matrix element per unit energy interval. The number of frequencies in an interval of 1 $\sec^{-1}$ is, according to (4.26), equal to $\frac{8\pi\nu^2}{c^3}v$ and the separation between them is $\frac{c^3}{8\pi\nu^2 v}$, while the separation between neighbouring levels is $\Delta = \frac{hc^3}{8\pi v \nu^2}$. The quantity Δ is independent of frequency. However, since $C_{2(0)}$ decreases rapidly as E departs from E_2, we can take the value of Δ near the level E_2 of the excited molecule and let $h\nu = h\nu_{21}$. Therefore, the value of $|V|^2$ in (10.23) is $|P_{21}|/\Delta$. Using (14.29) we obtain the following final expression for the level profile

$$\sigma(E) = |C_{2(0)}(E)|^2 = \frac{\beta}{\pi} \frac{1}{\beta^2 + (E_2 - E)^2} \tag{14.31}$$

where

$$\beta = \frac{|16\pi^3 e^2 \nu_{21}^3 |r_{21}|^2}{3c^3} \tag{14.32}$$

This profile is due to the unavoidable interaction between the molecule and the electromagnetic field, and is therefore called the natural level profile.

The expression given by (14.31) has a simple physical interpretation. It gives the probability that an atom occupying the second energy level will have the energy E. As can be seen, the most probable energy is E_2. We shall see below that the natural level profile has the same form as the natural profile of a spectral line.

Natural profile of a spectral line

To begin with, consider an atom which can only occupy one of the two energy states E_2 and E_1. If at the initial instant of time the atom is in an excited state and $n_0^\lambda = 0$ (zero number of photons), then according to (13.15), the probability amplitudes are given by

$$i\frac{h}{2\pi}\frac{\partial}{\partial t} C_{2(0)} = -\sum_{\lambda'} \frac{(P_{\lambda'}^{+})_{21}}{\sqrt{v\nu_{\lambda'}}} C_{1(\delta_{\lambda\lambda'})} e^{2\pi i(\nu_{21} - \nu_{\lambda'})t} \tag{14.33}$$

$$i\frac{h}{2\pi}\frac{\partial}{\partial t}C_{1(\delta_{\lambda\lambda'})} = -\frac{(P_{\overline{\lambda'}})_{12}}{\sqrt{v\nu_{\lambda'}}}C_{2(0)}e^{-2\pi i(\nu_{21}-\nu_{\lambda'})t} \tag{14.34}$$

We shall solve these equations on the assumption that the probability of finding the atom in the excited state decreases exponentially with time, i.e.

$$C_{2(0)} = e^{-\gamma t} \tag{14.35}$$

On solving (14.34) subject to (14.35) we obtain

$$C_{1(\delta_{\lambda\lambda'})} = \frac{1}{h}\frac{(P_{\overline{\lambda'}})_{12}}{\sqrt{v\nu_{\lambda'}}}\frac{1-e^{-2\pi i(\nu_{21}-\nu_{\lambda'})t-\gamma t}}{(\nu_{21}-\nu_{\lambda'})+i\gamma'} \tag{14.36}$$

where $\gamma' = \gamma/2\pi$. Substituting (14.35) and (14.36) into (14.33) we obtain

$$2\gamma = i\frac{4\pi}{h^2}\sum_{\lambda'}\frac{|(P_{\lambda'}^{+})_{21}|^2}{v\nu_{\lambda'}}\frac{1-e^{2\pi i(\nu_{21}-\nu_{\lambda'})t+\gamma t}}{(\nu_{21}-\nu_{\lambda'})+i\gamma'} \tag{14.37}$$

The parameter γ will in general be complex. It is quite easy to show from (14.36) that its imaginary part gives the correction for the shift of the natural frequency ν_{21}. This shift is usually very small, and may be ignored on the first approximation. If we confine our attention to the real part, we have from (14.37),

$$2\gamma = \sum_{\lambda'}\frac{2\pi}{h^2}\frac{|(P_{\lambda'})_{21}|^2}{v\nu_{\lambda'}}\delta(\nu_{21}-\nu_{\lambda'}) = A_{21} \tag{14.38}$$

According to (14.20) this gives the total probability of a spontaneous transition of the atom from the excited to the ground state. This was in fact to be expected, since the quantity γ in (14.35) was introduced as the reciprocal of the lifetime of the atom in the excited state.

The line profile is determined by the probability of the final state for $t \gg 1/\gamma$, when the atom has definitely reached the excited state:

$$|C_{1(\delta_{\lambda\lambda'})}|^2 = \frac{1}{h^2}\frac{|(P_{\overline{\lambda'}})|_{12}}{v\nu_{\lambda'}}\frac{1}{(\nu_{21}-\nu_{\lambda'})^2+\gamma'^2} \tag{14.39}$$

The total probability that the atom occupies the lower level, and a photon with arbitrary physical properties appears in the field, is given by the sum

$$\sum_{\lambda} |C_{1(\delta_{\lambda\lambda'})}|^2 = \frac{\gamma'}{\pi}\int \frac{d\nu}{(\nu_{21}-\nu)^2+\gamma'^2} \tag{14.40}$$

which leads directly to the classical formula (2.17) for the natural profile of a spectral line:

$$I(\nu) = \frac{\gamma'}{\pi}\,\frac{1}{(\nu_{21}-\nu)^2+\gamma'^2}$$

$$\text{or } I(\omega) = \frac{\gamma}{\pi}\,\frac{1}{(\omega_{21}-\omega)^2+\gamma^2} \tag{14.41}$$

In evaluating (14.40), the summation was replaced by integration in accordance with (14.10), and use was made of (14.20) and (14.38).

If the atom can occupy a number of energy levels, the spectral line profiles corresponding to the various transitions will be given by formulae analogous to (14.41). However, the quantity $2\gamma_{ij}$ will not be equal to the probability of spontaneous transitions from the upper level i to the lower level j. It will be given by the sum [26]

$$\gamma_{ij} = \gamma_i + \gamma_j \tag{14.42}$$

where $\gamma_k = \sum_l A_{kl}$ is the sum of all the probabilities of spontaneous transition from the level k to the lower lying levels, and is equal to the natural width of the k-th energy level. It follows that the natural width of a spectral line is equal to the sum of the widths of the upper and lower levels (this is valid for all quantum-mechanical systems except for the harmonic oscillator - see Chapter 6). In the quantum theory of radiation there is therefore no unambiguous connection between the width of a line and its intensity. Even very weak lines can have a large width as a result of a considerable broadening of the lower level.

As an example, consider an atom with three energy levels

(Fig. 3.1). Suppose that the probability of a transition from the third to the first and second levels is small, while the probability A_{21} is large. The third level will then be narrow and the second broad. Lines with frequency ν_{32} and ν_{31} will have a low intensity, but ν_{32} will correspond to a broad line

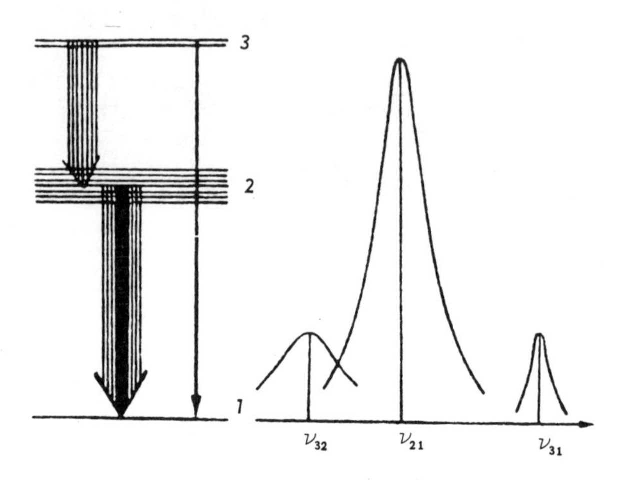

Fig. 3.1 Dependence of the width of a spectral line on the width of the energy levels

and ν_{31} to a narrow line. The line at ν_{21} will be broad and strong.

In contrast to the situation in the classical theory of the harmonic oscillator, the natural line width is always proportional to its intensity. In fact, the energy of free oscillations of an oscillator falls off exponentially with time:

$$E = E_0 e^{-2\gamma t} \tag{14.43}$$

The intensity of a line is equal to the energy lost by the dipole per second, and is proportional to 2γ:

$$I = \left| \frac{dE}{dt} \right| = 2\gamma E \tag{14.44}$$

where 2γ is the natural line width.

15. TWO-PHOTON AND THREE-PHOTON PROCESSES

Time-independent illumination

For the sake of simplicity we shall consider the secondary emission of photons on the assumption that an atom has only three non-degenerate energy levels.

Suppose that an atom is illuminated by a beam of external radiation of arbitrary spectral composition. The physical characteristics of this radiation will be assumed to be time-independent. If the atom and the incident photons did not interact, the state of the system could be described by the amplitude $C_1(n_\lambda^0)$, i.e. the atom would occupy the lowest state and the field would consist of photons emitted by the external source only. The existence of an interaction between the atom and the field leads to the conversion by the atom of some of the incident photons into one or two secondary photons. The atom may or may not undergo a change of state in the process. For example, the disappearance of a primary photon λ' may be accompanied by the transition of the atom from level 1 to level 2, with the result that the system may assume the state $C_2(n_\lambda^0 - \delta'_{\lambda\lambda})$ or $C_3(n_\lambda^0 - \delta'_{\lambda\lambda})$.

If the transformation of the primary photon λ' into the secondary photon λ'' occurs without any change in the atom, the system will enter one of two different states, $C_1(n_\lambda^0 - \delta'_{\lambda\lambda} + \delta''_{\lambda\lambda})$ or $C_1(n_\lambda^0 + \delta''_{\lambda\lambda} - \delta'_{\lambda\lambda})$, corresponding to two different intermediate states. The second of these transformations occurs through a virtual state. Virtual states have the property that their energy is different from both the initial and the final state of the system. For example, if at the initial instant of time the atoms were in the lowest state $E_1 = 0$ and the field did not include photons so that $n_\lambda^0 = 0$, the new state $\Psi^0_{i(n_\lambda)}$ in which $E_i > 0$ and $n_\lambda \neq 0$ will be virtual, since it corresponds to a simultaneous increase in the energy of matter and field. Transitions to virtual states are also referred to as virtual. Virtual transitions involve a formal reversal of the temporal order of physical processes. For example, (13.15) describes not only the process of absorption of the photon λ' and the subsequent emission of the photon λ'', but also the reverse process. It follows from this system of equations that an atom occupying the lowest state can at first emit a photon and undergo a transition to an excited state and then absorb an incident photon and return to the original ground state. The apparent departure from the law of conservation of energy and from the temporal order of processes in virtual transitions is not of profound physical significance; it is associated with the expansion of the wave function into a series and the use of perturbation theory.

Interaction of the field with the atom under consideration may also result in the appearance of the states $C_{2(n_\lambda^0-\delta_{\lambda\lambda'}+\delta_{\lambda\lambda''})}$ and $C_{2(n_\lambda^0+\delta_{\lambda\lambda''}-\delta_{\lambda\lambda'})}$. These are formed as the result of a two-photon process accompanied by a change in the state of the atom (transition 1 $\rightarrow$ 2). The atom may undergo a downward transition from this state by emitting a photon λ''' corresponding to the possible states $C_{1(n_\lambda^0-\delta_{\lambda\lambda'}+\delta_{\lambda\lambda''}+\delta_{\lambda\lambda'''})}$ and $C_{1(n_\lambda^0+\delta_{\lambda\lambda''}-\delta_{\lambda\lambda'}+\delta_{\lambda\lambda'''})}$. The onset of multi-photon processes will lead to the appearance of various other states. However, by analysing the solution of (13.15) it can be shown that different possible states contribute to the true state of the system with quite different weights. States which arise as a result of multi-photon processes are of relatively low probability and in the first approximation may be neglected altogether. If necessary, they can be allowed for by the method of successive approximations.

Using (13.15) it is quite easy to write down the system of equations connecting the probability amplitudes for states which arise as a result of the first possible transformation of a photon λ' into λ'' (through the intermediate states $C_{2(n_\lambda^0-\delta_{\lambda\lambda'})}$ and $C_{3(n_\lambda^0-\delta_{\lambda\lambda'})}$):

$$i\frac{h}{2\pi}\frac{\partial}{\partial t}C_{2(n_\lambda^0-\delta_{\lambda\lambda'})}=-\frac{(P_{\lambda'})_{21}\sqrt{n_{\lambda'}^0}}{\sqrt{v\nu_{\lambda'}}}\times e^{2\pi i(\nu_{21}-\nu_{\lambda'})t}C_{1(n_\lambda^0)}$$

$$-\sum_{\lambda''}\frac{(P_{\lambda''})_{21}\sqrt{n_{\lambda''}^0+1}}{\sqrt{v\nu_{\lambda''}}}\times e^{2\pi i(\nu_{21}-\nu_{\lambda''})t}C_{1(n_\lambda^0-\delta_{\lambda\lambda'}+\delta_{\lambda\lambda''})}$$

$$(15.1)$$

$$i\frac{h}{2\pi}\frac{\partial}{\partial t}C_{3(n_\lambda^0-\delta_{\lambda\lambda'})}=-\frac{(P_{\lambda'})_{31}\sqrt{n_{\lambda'}^0}}{\sqrt{v\nu_{\lambda'}}}\times e^{2\pi i(\nu_{31}-\nu_{\lambda'})t}C_{1(n_\lambda^0)}$$

$$-\sum_{j=1,2}\sum_{\lambda''}\frac{(P_{\lambda''})_{3j}\sqrt{n_{\lambda''}^0+1}}{\sqrt{v\nu_{\lambda''}}}\times e^{2\pi i(\nu_{3j}-\nu_{\lambda''})t}C_{j(n_\lambda^0-\delta_{\lambda\lambda'}+\delta_{\lambda\lambda''})};$$

$$i\frac{h}{2\pi}\frac{\partial}{\partial t}C_{1(n_\lambda^0-\delta_{\lambda\lambda'}+\delta_{\lambda\lambda''})} = -\frac{(P_{\lambda''})_{12}\sqrt{n_{\lambda''}^0+1}}{\sqrt{v\,\nu_{\lambda''}}}$$

$$\times e^{2\pi i(\nu_{12}+\nu_{\lambda''})t}C_{2(n_\lambda^0-\delta_{\lambda\lambda'})};$$

$$i\frac{h}{2\pi}\frac{\partial}{\partial t}C'_{1(n_\lambda^0-\delta_{\lambda\lambda'}+\delta_{\lambda\lambda''})} = -\frac{(P_{\lambda''})_{13}\sqrt{n_{\lambda''}^0+1}}{\sqrt{v\,\nu_{\lambda''}}}$$

$$\times e^{2\pi i(\nu_{13}+\nu_{\lambda''})t}C_{3(n_\lambda^0-\delta_{\lambda\lambda'})};$$

$$i\frac{h}{2\pi}\frac{\partial}{\partial t}C_{2(n_\lambda^0-\delta_{\lambda\lambda'}+\delta_{\lambda\lambda''})} = -\frac{(P_{\lambda''})_{23}\sqrt{n_{\lambda''}^0+1}}{\sqrt{v\,\nu_{\lambda''}}}$$

$$\times e^{2\pi i(\nu_{23}+\nu_{\lambda''})t}C_{3(n_\lambda^0-\delta_{\lambda\lambda'})}-\sum_{\lambda'''}\frac{(P_{\lambda'''})_{21}\sqrt{n_{\lambda'''}^0+1}}{\sqrt{v\,\nu_{\lambda'''}}} \qquad \begin{matrix}(15.1)\\ \text{(cont)}\end{matrix}$$

$$\times e^{2\pi i(\nu_{21}-\nu_{\lambda'''})t}C_{1(n_\lambda^0-\delta_{\lambda\lambda'}+\delta_{\lambda\lambda''}+\delta_{\lambda\lambda'''})}$$

$$i\frac{h}{2\pi}\frac{\partial}{\partial t}C_{1(n_\lambda^0-\delta_{\lambda\lambda'}+\delta_{\lambda\lambda''}+\delta_{\lambda\lambda'''})} = \frac{(P_{\lambda'''})_{12}\sqrt{n_{\lambda'''}^0+1}}{\sqrt{v\,\nu_{\lambda'''}}}$$

$$\times e^{2\pi i(\nu_{12}+\nu_{\lambda'''})t}C'_{2(n_\lambda^0-\delta_{\lambda\lambda'}+\delta_{\lambda\lambda''})}$$

$$i\frac{h}{2\pi}\frac{\partial}{\partial t}C''_{1(n_\lambda^0-\delta_{\lambda\lambda'}+\delta_{\lambda\lambda''})}$$

$$= -\frac{(\eta_{\lambda\lambda''})_{11}\sqrt{n_{\lambda'}^0(n_{\lambda''}^0+1)}}{v\sqrt{\nu_{\lambda'}\nu_{\lambda''}}}e^{2\pi i(\nu_{\lambda''}-\nu_{\lambda'})t}C_{1(n_\lambda^0)}$$

In these expressions $(P_\lambda)_{ji}$ is, as before, given by (13.11), while $(\eta_{\lambda\lambda'})_{ji}$ is the dipole approximation to (13.14) and is given by

$$(\eta_{\lambda\lambda'})_{ji} = \frac{he^2}{2\pi m}(\mathbf{e}_\lambda\,\mathbf{e}_{\lambda'})\,\delta_{ji} \qquad (15.2)$$

The solution of (15.1) is of the form

$$C_{1(n^0_\lambda)} = a_1 \tag{15.3}$$

$$C_{j(n^0_\lambda - \delta_{\lambda\lambda'})} = \frac{(P_{\lambda'})_{j1}\sqrt{n^0_{\lambda'}}}{h\sqrt{v\nu_{\lambda'}}} \frac{a_1 e^{2\pi i(\nu_{j1}-\nu_{\lambda'})t}}{\nu_{j1}-\nu_{\lambda'}-i\gamma'_j} \quad (j=2,3) \tag{15.4}$$

$$C_{1(n^0_\lambda - \delta_{\lambda\lambda'}+\delta_{\lambda\lambda''})} = \frac{(P_{\lambda''})_{12}(P_{\lambda'})_{21}\sqrt{n^0_{\lambda'}(n^0_{\lambda''}+1)}}{h^2 v\sqrt{\nu_{\lambda'}\nu_{\lambda''}}}$$

$$\times \frac{a_1}{\nu_{21}-\nu_{\lambda'}-i\gamma'_2} \frac{e^{2\pi i(\nu_{\lambda''}-\nu_{\lambda'})t}-1}{\nu_{\lambda''}-\nu_{\lambda'}} \tag{15.5}$$

$$C_{1(n^0_\lambda - \delta_{\lambda\lambda'}+\delta_{\lambda\lambda''})} = \frac{(P_{\lambda''})_{13}(P_{\lambda'})_{31}\sqrt{n^0_{\lambda'}(n^0_{\lambda''}+1)}}{h^2 v\sqrt{\nu_{\lambda'}\nu_{\lambda''}}}$$

$$\times \frac{a}{\nu_{31}-\nu_{\lambda'}-i\gamma'_3} \frac{e^{2\pi i(\nu_{21}-\nu_{\lambda'}+\nu_{\lambda''})t}-1)}{\nu_{\lambda''}-\nu_{\lambda'}} \tag{15.6}$$

$$C_{2(n^0_\lambda - \delta_{\lambda\lambda'}+\delta_{\lambda\lambda''})} = \frac{(P_{\lambda''})_{23}(P_{\lambda'})_{31}\sqrt{n^0_{\lambda'}(n^0_{\lambda''}+1)}}{h^2 v\sqrt{\nu_{\lambda'}\nu_{\lambda''}}}$$

$$\times \frac{a_1}{\nu_{31}-\nu_{\lambda'}-i\gamma'_3} \frac{e^{2\pi i(\nu_{21}-\nu_{\lambda'}+\nu_{\lambda''})t}}{\nu_{21}-\nu_{\lambda'}+\nu_{\lambda''}-i\gamma'_2} \tag{15.7}$$

$$C_{1(n^0_\lambda - \delta_{\lambda\lambda'}+\delta_{\lambda\lambda''}+\delta_{\lambda\lambda'''})}$$

$$= \frac{(P_{\lambda''})_{23}(P_{\lambda'})_{31}(P_{\lambda'''})_{12}\sqrt{n^0_{\lambda'}(n^0_{\lambda''}+1)(n^0_{\lambda''}+1)}}{h^3\sqrt{v^3\nu_{\lambda'}\nu_{\lambda''}\nu_{\lambda'''}}}$$

$$\times \frac{1}{\nu_{31}-\nu_{\lambda'}-i\gamma'_3} \frac{a_1}{\nu_{21}-\nu_{\lambda'}+\nu_{\lambda''}-i\gamma_2}$$

$$\times \frac{e^{2\pi i(\nu_{\lambda'''}+\nu_{\lambda''}-\nu_{\lambda'})t}-1}{\nu_{\lambda'''}+\nu_{\lambda''}-\nu_{\lambda'}} \tag{15.8}$$

$$C''_{1(n^0_\lambda-\delta_{\lambda\lambda'}+\delta_{\lambda\lambda''})} = -\frac{(\eta_{\lambda'\lambda''})_{11}\sqrt{n^0_{\lambda'}(n^0_{\lambda''}+1)}}{h^2 v\sqrt{\nu_{\lambda'}\nu_{\lambda''}}} \times a_1 \frac{e^{2\pi i(\nu_{\lambda''}-\nu_{\lambda'})t}-1}{\nu_{\lambda''}-\nu_{\lambda'}} \tag{15.9}$$

where $4\pi\gamma_2'$ and $4\pi\gamma_3'$ are the total probabilities that the atom will leave levels 2 and 3, respectively. In other words, each of these quantities is the sum of the probabilities of spontaneous and stimulated transitions starting from levels 2 or 3.

In order to obtain a complete description of the interaction of the atom with radiation, one must find the probability amplitudes for the states arising as a result of virtual transitions. These amplitudes satisfy a system of equation analogous to (15.1). We shall omit mathematical derivations and will simply quote the final amplitudes which will be necessary in our subsequent calculations:

$$C_{1(n^0_\lambda+\delta_{\lambda\lambda''}-\delta_{\lambda\lambda'})} = \frac{1}{h^2}\sum_{j=2,3}\frac{(P_{\lambda'})_{1j}(P_{\lambda''})_{j1}\sqrt{n^0_{\lambda'}(n^0_{\lambda''}+1)}}{v\sqrt{\nu_{\lambda'}\nu_{\lambda''}}} \times \frac{a_1}{\nu_{j1}+\nu_{\lambda''}}\;\frac{e^{2\pi i(\nu_{\lambda''}-\nu_{\lambda'})t}-1}{\nu_{\lambda''}-\nu_{\lambda'}} \tag{15.10}$$

$$C_{2(n^0_\lambda+\delta_{\lambda\lambda''}-\delta_{\lambda\lambda'})} = \frac{1}{h^2}\,\frac{(P_{\lambda'})_{23}(P_{\lambda''})_{31}\sqrt{n^0_{\lambda'}(n^0_{\lambda''}+1)}}{v\sqrt{\nu_{\lambda'}\nu_{\lambda''}}} \times \frac{a_1}{\nu_{31}+\nu_{\lambda''}}\;\frac{e^{2\pi i(\nu_{21}+\nu_{\lambda''}-\nu_{\lambda'})t}}{\nu_{21}+\nu_{\lambda''}-\nu_{\lambda'}-i\gamma_2'} \tag{15.11}$$

$$C_{1(n^0_\lambda+\delta_{\lambda\lambda''}-\delta_{\lambda\lambda'}+\delta_{\lambda\lambda'''})} = \frac{1}{h^3}\,\frac{(P_{\lambda'''})_{12}(P_{\lambda'})_{23}(P_{\lambda''})_{31}}{\sqrt{v^3\nu_{\lambda'}\nu_{\lambda''}\nu_{\lambda'''}}\,(\nu_{31}+\nu_{\lambda''})} \times \frac{a_1\sqrt{n^0_{\lambda'}(n^0_{\lambda''}+1)(n^0_{\lambda'''}+1)}}{\nu_{21}+\nu_{\lambda''}-\nu_{\lambda'}-i\gamma_2'}\;\frac{e^{2\pi i(\nu_{\lambda'''}+\nu_{\lambda''}-\nu_{\lambda'})t}-1}{\nu_{\lambda'''}+\nu_{\lambda''}-\nu_{\lambda'}} \tag{15.12}$$

The states described by these expressions are physically identical with the states described by (15.5)-(15.8), and therefore in calculating parameters which are of interest in practice, the corresponding amplitudes must be added.

In order to determine the rate of absorption or emission it is necessary to know the radiative transition probabilities. At the same time, the solution of (15.1) yields only the probabilities of the various states of the atom + field system as functions of time if the origin of time is chosen at some arbitrary instant, once the steady state has been reached. For example, the quantity $|C_{2(n_\lambda^0-\delta_{\lambda\lambda'})}|^2$ cannot be regarded as the probability of the transition $C_{1(n_\lambda^0)} \to C_{2(n_\lambda^0-\delta_{\lambda\lambda'})}$ because $|C_{2(n_\lambda^0-\delta_{\lambda\lambda'})}|^2$ is the result of an equilibrium between two opposite processes, one of which takes the system into a given state and the other takes it from that state. The same conclusions hold good for all excited states of an atom, or more precisely, for all states of the system in which the excited atom participates.

The situation is quite different if at the end of the process taking place in the system the atom is in the initial state, i.e. in the state which it occupied at $t = 0$. In this case, the process occurring in the system leads to a change in the properties of the field without affecting the state of the atom, and therefore the probability of the final state of the system may serve as a measure of the probability of the process (disappearance of two photons or appearance of other photons). The quantity

$$|C_{1(n_\lambda^0-\delta_{\lambda\lambda'}+\delta_{\lambda\lambda''})} + C'_{1(n_\lambda^0-\delta_{\lambda\lambda'}+\delta_{\lambda\lambda''})} + C''_{1(n_\lambda^0-\delta_{\lambda\lambda'}+\delta_{\lambda\lambda''})} + C_{1(n_\lambda^0+\delta_{\lambda\lambda''}-\delta_{\lambda\lambda'})}|^2 \tag{15.13}$$

may thus be regarded as the number of photons λ' which are transformed by the atom into photons λ'' between times 0 and t, or as the number of the corresponding elementary transformations of radiation.

It is worth noting that the system under consideration is not closed, and the sum of the probabilities of all the states does not remain constant. This is connected with the accumulation of photons which are produced as a result of the transformation of the radiation emitted by the atoms. For example, the quantity given by (15.13) increases continuously, i.e. there is a continuous accumulation of the transformed radiation quanta. If we take the time derivative of (15.13), we obtain the number of photons transformed per second by the atom.

$$N(\lambda' \to \lambda'') = \frac{d}{dt} \left| C_{1\,(n^0_\lambda - \delta_{\lambda\lambda'} + \delta_{\lambda\lambda''})} + C'_{1\,(n^0_\lambda - \delta_{\lambda\lambda'} + \delta_{\lambda\lambda''})} + C''_{1\,(n^0_\lambda - \delta_{\lambda\lambda'} + \delta_{\lambda\lambda''})} + C_{1\,(n^0_\lambda + \delta_{\lambda\lambda''} - \delta_{\lambda\lambda'})} \right|^2 \quad (15.14)$$

Like any other transition probability, the quantity N has the dimensions of $\sec^{-1}$.

Similarly, the quantity

$$N(\lambda' \to \lambda''\lambda'') = \frac{d}{dt} \left| C_{1\,(n^0_\lambda - \delta_{\lambda\lambda'} + \delta_{\lambda\lambda''} + \delta_{\lambda\lambda'''}} + C_{1\,(n^0_\lambda + \delta_{\lambda\lambda''} - \delta_{\lambda\lambda'} + \delta_{\lambda\lambda'''})} \right|^2 \quad (15.15)$$

may be regarded as the probability of transformation of a photon λ' into two photons λ'' and λ'''. This interpretation of (15.14) and (15.15) is in complete agreement with experimental data and the usual concept of the probability of processes. Equations (15.14) and (15.15) provide us with a means of studying all the properties of secondary radiation under time-independent illumination.

Two-photon processes

Substituting the amplitudes from (15.5), (15.6), (15.9) and (15.10) into (15.14), we obtain the following expression for the rate of the two-photon process

$$N(\lambda' \to \lambda'') = \left| \sum_{j=2,\,3} \left[\frac{(P_{\lambda''})_{1j}(P_{\lambda'})_{j1}}{\nu_{j'} - \nu_{\lambda'} - i\gamma'_2} + \frac{(P_{\lambda'})_{1j}(P_{\lambda''})_{j1}}{\nu_{j1} + \nu_{\lambda''}} \right] + h(\eta_{\lambda'\lambda''})_{11} \right|^2 \frac{4\pi}{h^4} |a_1|^2 \frac{n^0_{\lambda'}(n^0_{\lambda''} + 1)}{v^2 \nu_{\lambda'}\nu_{\lambda''}} \frac{\sin 2\pi(\nu_{\lambda''} - \nu_{\lambda'})t}{\nu_{\lambda''} - \nu_{\lambda'}} \quad (15.16)$$

For the purposes of comparison with experiment, this formula must be rewritten in terms of new variables, i.e. it must be expressed in terms of the radiation density per unit frequency interval per unit solid angle. Using (14.10) and (14.7), we obtain from (15.16) the following expression for the number of photons of frequency $\nu_{\lambda'}$ travelling in the direction Ω' and having a polarisation α', which are transformed per second into photons with parameters λ'', Ω'' and α'':

$$N(\nu', \Omega', \alpha' \to \nu'', \Omega'', \alpha'')$$

$$= \frac{8\pi^3 \nu^3 e^4}{h^3 c^3} \left| \sum_{j=2,3} \left[\frac{(\mathbf{re}_{\alpha''})_{1j} (\mathbf{re}_{\alpha'})_{j1}}{\nu_{j1} - \nu' - i\gamma_j'} + \frac{(\mathbf{re}_{\alpha'})_{1j} (\mathbf{re}_{\alpha''})_{j1}}{\nu_{j1} + \nu''} \right] \right|^2 \times |a_1|^2 u^{\alpha'}(\nu', \Omega'), \delta(\nu'' - \nu') \qquad (15.17)$$

In deriving this formula we used the fact that [28]

$$\frac{h}{2\pi m} (\mathbf{e}_{\alpha'} \mathbf{e}_{\alpha''}) = \frac{i}{m} \sum_k [(\mathbf{pe}_{\alpha'})_{jk} (\mathbf{re}_{\alpha''})_{kj} - (\mathbf{re}_{\alpha''})_{jk} (\mathbf{pe}_{\alpha'})_{kj}] \qquad (15.18)$$

and neglected stimulated emission, since at those densities of the exciting radiation at which the perturbation theory used here is valid, this emission is much smaller in intensity than spontaneous emission $\left(\rho_{\alpha''}(\nu'', \Omega'') \ll \frac{h\nu^3}{c^3}\right)$

The expression given by (15.17) includes a δ-function. This means that in two-photon processes occurring under time-independent conditions the frequencies of the incident transformed photons are equal and the transformation of photons of one frequency into photons of another does not take place. We could arrive at the same result by taking $C_{2(n_\lambda^0)}$ or $C_{3(n_\lambda^0)}$ as the initial state instead of $C_{1(n_\lambda^0)}$. In the classical theory (Chapter 1), we arrived at the same result, namely, the frequency of the secondary radiation was found to be equal to the frequency of the incident radiation.

We shall now continue our analysis of the two-photon process on the assumption that the incident radiation is plane-polarised and propagates in a certain direction Ω_0. If we multiply (15.17) by the energy $h\nu''$ of the transformed photon and integrate with respect to the frequency of the incident photons, we obtain the following expression for the rate of secondary emission

$$W_{\text{scat}}(\nu'', \Omega'', \alpha'') = \frac{16\pi^4 \nu''^4}{h^2 c^3} |a_1|^2 \left| \sum_{j=2,3} \left[\frac{(\mathbf{re}_{\alpha''})_{1j} (\mathbf{re}_{\alpha'})_{j1}}{\nu_{j1} - \nu'' - i\gamma_j} + \frac{(\mathbf{re}_{\alpha'})_{ij} (\mathbf{re}_{\alpha''})_{j1}}{\nu_{j1} + \nu''} \right] \right|^2 u_0(\nu'') \qquad (15.19)$$

This is a generalisation of the well-known Kramers-Heisenberg formula to the case of resonance interaction. Kramers and Heisenberg obtained their formula from the correspondence principle by analysing the dispersion of light.

As in the classical theory, the rate of scattering is proportional to the intensity of the incident radiation and to the fourth power of its frequency.

The expression given by (15.19) may be used to establish the shape of a spectral line in the scattered radiation. It can readily be seen that the line shape is largely determined by two factors, namely, the energy density distribution $u_0(\nu)$ in the primary beam and the resonance factor

$$\left| \sum_{j=2,3} \left[\frac{(\mathbf{re}_{\alpha''})_{1j}\,(\mathbf{re}_{\alpha'})_{j1}}{\nu_{j1} - \nu'' - i\,\gamma_j'} + \frac{(\mathbf{re}_{\alpha'})_{1j}\,(\mathbf{re}_{\alpha''})_{j1}}{\nu_{j1} + \nu''} \right] \right|^2 \tag{15.20}$$

If the incident radiation is strictly monochromatic, the secondary radiation will also be monochromatic. If the atom is illuminated by radiation consisting of a narrow line lying well away from the natural frequencies of the atom ($\nu \gg \nu_{j1}$ or $\nu \ll \nu_{j1}$), then (15.20) is a monotonic function of ν, and therefore the scattered line shape is very approximately the same as the incident line shape.

If the frequency of the incident line or band is close to one of the natural frequencies of the atom, for example, ν_{31}, and the effect of the second excited level of the atom may be neglected, i.e. we may neglect the $j = 2$ term in (15.19), this expression assumes the simpler form

$$W(\nu'', \Omega'', e_{\alpha''}) = \frac{16\,\pi^4\nu^4}{hc^4}\,|a_1|^2\,|(\mathbf{re}_{\alpha''}|^2\,|(\mathbf{re}_{\alpha'})|^2 \frac{u_0(\nu'')}{(\nu_{31}-\nu'')^2 + \gamma_3'^2} \tag{15.21}$$

It can readily be seen that the resonance factor (15.21) is very similar to the resonance factor in the classical theory of scattering (see (2.46)) based on the harmonic oscillator model.

When the width of the incident line is large in comparison with $\gamma_3'(\Delta\nu \gg \gamma_2')$, i.e. when $u_0(\nu)$ varies slowly with frequency, within the natural line of the atom, the scattered line has the usual dispersion profile of width γ_3. The position of the maximum of the line and its shape are independent of the incident radiation and are determined by the internal properties of the atom. The line resembles the natural line of the atom which it will emit on thermal excitation. However, in contrast to thermal luminescence, the radiation scattered by the various atoms is coherent. In the intermediate cases, the shape of the secondary emission line is more complicated.

Three-photon process

Consider now the interaction of an atom with an electromagnetic field, which is accompanied by the disappearance of one primary photon λ' and the creation of secondary photons λ'' and λ'''. This process corresponds to Raman scattering in which the atom undergoes a transition from state 1 to state 2 with the subsequent emission of radiation which brings the atom back to the initial state. The probability of the elementary three-photon process as a result of which the atom returns to the initial state is given by (15.15). We already know that this probability is equal to the number of such events per second.

Substituting (15.8) and (15.12) into (15.15), we obtain

$$N(\lambda' \to \lambda'', \lambda''') = \frac{4\pi^2}{h^6} \left| \frac{(P_{\lambda''})_{23}(P_{\lambda'})_{31}}{\nu_{31} - \nu_{\lambda'} - i\gamma_3'} + \frac{(P_{\lambda'})_{23}(P_{\lambda''})_{31}}{\nu_{31} + \nu_{\lambda''}} \right|^2 |a_1|^2 \frac{n^0_{\lambda'}(n^0_{\lambda''}+1)(n^0_{\lambda'''}+1)}{v^3 \nu_{\lambda'}\nu_{\lambda''}\nu_{\lambda'''}} \times \frac{|(P_{\lambda'''})_{12}|^2}{(\nu_{21}+\nu_{\lambda''}-\nu_{\lambda'})^2 + \gamma_2'^2}\, \delta(\nu_{\lambda'''} + \nu_{\lambda''} - \nu_{\lambda'}) \tag{15.22}$$

The presence of the δ-function in (15.22) means that, as in the two-photon process, we again have the law of conservation of energy $\nu_{\lambda''} + \nu_{\lambda'''} = \nu_{\lambda'}$. The above expression enables us to calculate the rate of emission of photons of both types λ'' and λ'''.

We can now transform from the number of quanta to the radiation densities, integrate over all quanta λ''' emitted by the atom during the $2 \to 1$ transition, and thus find the law of transformation of the incident photons (ν', Ω', α') into the secondary photons (ν'', Ω'', α''):

$$N'(\nu', \Omega', \alpha', \rightarrow \nu'', \Omega'', \alpha'') = \frac{8\pi^2 e^4 \nu''^3}{h^3 c^3} \left| \frac{(\mathbf{re}_{\alpha''})_{23} (\mathbf{re}_{\alpha'})_{31}}{\nu_{31} - \nu' - i\gamma_3'} \cdot \right.$$

$$\left. + \frac{(\mathbf{re}_{\alpha'})_{23} (\mathbf{re}_{\alpha''})_{31}}{\nu_{31} + \nu''} \right|^2 |a_1|^2 \frac{\gamma_2'}{(\nu_{21} + \nu'' - \nu')^2 + \gamma_2'^2} u(\nu') \qquad (15.23)$$

This formula was obtained on the same assumptions as (15.19). It is readily seen that for a given ν', (15.23) gives a dispersion profile with a half-width γ_2' and a maximum at $\nu'' = \nu'' - \nu_{21}$. Accordingly, (15.23) describes Raman scattering of radiation. If the exciting radiation is monochromatic, the spectral distribution in the scattered radiation is determined only by the internal properties of the atom. The profile of the Raman line can then be found by integrating (15.23) with respect to the frequency ν' of the exciting radiation. This is particularly easy in the following special cases.

1. The width of the second level is much smaller than the widths of the third level and of the incident line ($\gamma_2' \ll \gamma_3', \gamma_2' \ll \Delta\nu'$). On multiplying (15.23) by $h\nu''$ and integrating with respect to ν', we have

$$W(\nu'', \Omega'', \alpha'')$$

$$= \frac{16\pi^4 \nu''^4}{h^2 c^3} |a_1|^2 \left| \frac{(\mathbf{re}_{\alpha''})_{23} (\mathbf{re}_{\alpha'})_{31}}{\nu_{32} - \nu'' - i\gamma_3'} + \frac{(\mathbf{re}_{\alpha'}) (\mathbf{re}_{\alpha''})_{31}}{\nu_{31} + \nu''} \right| u_0(\nu'' + \nu_{21}) \qquad (15.24)$$

The Raman and Rayleigh profiles, (15.24) and (15.21), are thus practically identical in this case, but their maxima are shifted by ν_{21}.

2. The atom is excited by a broad line so that $u(\nu_{31})$ may be regarded as constant near the natural frequency. In this case, the Raman profile

$$W(\nu'', \Omega'', \alpha'') = \frac{\gamma_2' + \gamma_3'}{\pi} \frac{C}{(\nu_{32} - \nu'')^2 + (\gamma_2' + \gamma_3')^2} \qquad (15.25)$$

where $C = \text{const}$ is identical with the natural line shape produced as a result of a transition of the atom between two broadened levels.

We have considered one of the forms of radiation which arises in three-photon processes and is described by (15.22).

This emission follows directly the absorption of the external radiation, i.e. there are no intermediate processes between the absorption of the incident photon and the emission of the secondary photon.

On summing (15.22) over the photons λ'', we obtain the probability of emission of the photons λ''' which are not produced immediately after the absorption of the incident photon λ' but only after the transition of the atom from level 3 to level 2:

$$N'(\nu', \Omega', \alpha' \to \nu''', \Omega''', \alpha''') = \frac{8\pi^3 e^4 (\nu' - \nu''')^3}{h^3 c^4} |a_1|^2 \frac{|(P_{\lambda'''})_{12}|^2 \nu'''}{c^3} \times \sum_{\alpha''} \int_{\Omega''} \left| \frac{(\mathbf{re}_{\alpha''})_{23} (\mathbf{re}_{\alpha'})_{31}}{\nu_{31} - \nu' - i\gamma_3'} + \frac{(\mathbf{re}_{\alpha'})_{23} (\mathbf{re}_{\alpha''})}{\nu_{31} + \nu' - \nu'''} \right|^2 d\Omega'' \frac{I_0(\nu')}{(\nu_{21} - \nu''')^2 + \gamma_2'^2} \tag{15.26}$$

To obtain the line profile for this radiation, we must integrate with respect to the exciting frequency ν'. Since (15.26) has a sharp peak at $\nu''' = \nu_{21}$, it follows that, when we integrate with respect to ν', we can replace ν''' by ν_{21} in all the factors except the last. As a result, (15.26) yields the following expression for the rate of emission

$$I'(\nu''', \Omega''', \alpha''') = \frac{\nu'''^2}{h^2 c^3} \left| (P_{\lambda'''})_{12} \right|^2 \frac{C'}{(\nu_{21} - \nu''')^2 + \gamma_2'^2} \tag{15.27}$$

where C' is proportional to the density of the exciting radiation and is independent of frequency.

According to (15.27), both the spectrum and the line profile which arise when the intermediate process participates in the phenomenon are completely determined by the internal properties of the atom and are therefore independent of the properties of the exciting radiation. Such lines are emitted by the atom as a result of thermal and certain other forms of excitation.

As in other forms of secondary emission the line intensity

is proportional to the intensity of the exciting radiation.

Equations (15.16) and (15.22) can be used to calculate not only the rate of secondary emission but also the rate of absorption of the external radiation. This can be done by summing these expressions over all the emitted photons and combining the resulting expressions. Calculations show that under time-independent illumination, the energy yield is equal to unity while the quantum yield is either one or two (two-photon and three-photon processes, respectively).

In addition to the processes discussed above, a number of other similar processes will take place in the system. It is possible, for example, to calculate the conversion processes in which the second and third levels are the initial and final states. The secondary emission is then naturally divided into two parts, one of which arises only when there are intermediate processes between the absorption and emission. Rayleigh and Raman scattering always occur without the intervention of intermediate processes. However, the Raman line in the case of an excited atom occurs at the frequency $\nu'+\nu_{21}$ instead of $\nu'-\nu_{21}$. It follows that when the population of excited levels is increased, for example, by increasing the temperature, there is a parallel increase in the intensity of the Raman line $\nu'+\nu_{21}$.

The interaction of electromagnetic fields with atoms having a large number of energy levels differs from the above simple cases in the great variety of the possible processes, and is described by a relatively complicated system of equations. The secondary emission spectrum for such atoms is very complicated. However, in all cases the position and profile of the lines formed without participation of intermediate processes depend on the properties of the incident radiation, while the frequency and profile of lines which arise with the participation of intermediate processes are determined exclusively by the internal properties of the atom (level widths and separations).

Afterglow

The duration of afterglow, which continues after the exciting light is removed, is an important characteristic of the interaction of light with matter. It can, of course, be studied under time-independent conditions. To determine the properties of afterglow, we must use (13.20) to set up a system of equations for the state amplitudes under the condition that there are no photons in the field. For the purposes of comparison with the time-independent case, the initial conditions will be taken to be the amplitudes at a particular instant under the time-independent conditions. There are no inherent difficulties in these operations but the final result takes the form of very unwieldy formulae [29]. We shall not reproduce them here, especially since their analysis leads to conclusions already familiar from classical electrodynamics. Calculations show that if an atom is first exposed to an electromagnetic field, then whatever the nature of the interaction (Raman scattering, Rayleigh scattering or luminescence), the atom acquires a store of energy which is emitted as afterglow as soon as the excitation ceases. Afterglow always occurs, and its duration is determined exclusively by the internal properties of the atom. It consists of the natural lines of the atom whose positions and profiles are independent of the properties of the radiation used in the preliminary excitation. The spectral composition and intensity of exciting radiation affect only the intensity of the afterglow lines. It is evident that one can select a spectrum for the exciting radiation for which some of the natural lines of the atom will not appear in the afterglow spectrum.

The afterglow spectra and the spectra emitted under time-independent conditions are, in general, different although they do have common lines. The similarities and differences between them may be varied because the position and profile of some of the lines produced under time-independent conditions depend not only on the properties of the atoms but also on the spectral composition of the exciting radiation. It has already been shown (see (15.27)) that even under time-independent illumination the atom can emit its natural lines. They always arise as a result of intermediate atomic processes and are always coincident with the afterglow lines. This is the basic difference between them and the lines in the scattered spectrum. The Raman line, (15.24), disappears immediately after the exciting radiation is cut off. In the case of Rayleigh scattering (15.19) there are a number of possibilities. If the atom is illuminated by radiation consisting of a narrow line whose frequency is not equal to one of the natural frequencies of the atom, the scattered line will not reappear in the afterglow spectrum. In the case of resonance illumination by a broad line, the Rayleigh line coincides with the natural lines of the atom and is present in afterglow.

16. THE PROBABILISTIC METHOD

Principle of the method

Quantum electrodynamics is the most rigorous and fundamental theory of interaction of radiation with matter. However, it is impossible in practice to find exact solutions of the basic equations of quantum optics, while the use of the perturbation theory severely restricts the range of application of the final results. In particular, for high densities of the exciting radiation, the interaction operator is large and perturbation theory is invalid. Other methods of applying quantum electrodynamics to problems of this kind have not as yet been developed. Many optical problems are therefore solved with the aid of the probabilistic method put forward by Einstein in 1917, well before the advent of quantum mechanics. The basic ideas of Einstein have been described already in Section 7, in connection with radiative transitions.

When absorption and emission of radiation are examined by the probabilistic method, it is assumed that the energy levels and transition probabilities of the quantum-mechanical system are known. The problem is then reduced to determination of level populations [30]. Knowing the distribution of transition probabilities, it is easy to find the number of transitions through all the channels, and therefore the rates of absorption and emission.

If we assume that the number of transitions per unit time from level i to level j is proportional to the population of the i-th level and the total $(i \to j)$ transition probability p_{ij}, we then obtain the following set of equations for the distribution function:

$$\frac{dn_i}{dt} = n_i \sum_j p_{ij} + \sum_j n_j p_{ji} \tag{16.1}$$

The time derivative of n_i gives the rate of change in the population of the i-th level, i.e. the rate of change in the number of particles occupying the level i. This rate of change is equal to the number of particles leaving the i-th level to all the remaining levels (negative terms in (16.1)) plus the number of particles arriving at the i-th level from all other levels per unit time. When $i>j$, the total transition probabilities are related to the Einstein coefficients by the following formulae:

$$p_{ij} = A_{ij} + B_{ij} u_{ij} + d_{ij}$$
$$p_{ji} = B_{ji} u_{ij} + d_{ji} \tag{16.2}$$

where A_{ij}, $B_{ij} u_{ij}$ are the probabilities of spontaneous and induced transitions, and d_{ij} and d_{ji} are the probabilities of non-radiative transitions.

The coefficients of A_{ij} and B_{ij} cannot be calculated from Einstein's theory. Originally, they were determined empirically but were subsequently calculated on the basis of quantum mechanics and electrodynamics (see Sections 7, 14). However, by considering the thermodynamic equilibrium Einstein found that (Section 7)

$$g_i B_{ij} = g_j B_{ji}, \; A_{ij}/B_{ij} = \frac{8\pi h \nu_{ij}^3}{c^3} \tag{16.3}$$

where g_i and g_j are the statistical weights of the i-th and j-th levels.

Because of the simplicity and ease of interpretation, the probabilistic method has found wide application. It is used to determine the excitation, quenching, quantum yield and kinetics of luminescence. It has been found to be effective in the solution of fundamental problems in the theory of luminescence, for example, in calculations of the quantum yield with allowance for the thermal background, in proving the possibility of energy yields greater than unity and of the existence of negative luminescence, and so on. The probabilistic method has recently found wide application in the theory of masers and lasers.

Limits of applicability of the probabilistic method

Many results obtained by this method have been confirmed both experimentally and by rigorous quantum-mechanical solutions. Quantum electrodynamics has also confirmed the validity of the relations in (16.3) which were first obtained by the probabilistic method, and Einstein used them in his derivation of Planck's formula which is in agreement with all known experimental data. The probabilistic method leads to correct results for the growth and quenching of luminescence, the quantum yield, and other phenomena.

There is therefore a definite range of optical phenomena which may be investigated by this method. On the other hand, many optical phenomena lie outside the framework of the probabilistic method. Among them are problems associated with the profile of energy levels and spectral lines of simple systems, the shifting and broadening of levels under the action of exciting radiation, non-resonance interactions (scattering), and certain other problems. There have been attempts to find a quantum-electrodynamic justification of the method and to delineate its limits of applicability. Landau [31] and Bloch [32] have shown that in a special case (absence of external radiation and of non-radiative transitions) the balance equations. (16.1) on which the probabilistic method is based are a consequence of the fundamental equations of quantum optics. Recently, Apanasevich [33] investigated the limits of applicability of the probabilistic method for the general case when the quantum system is not only illuminated by external radiation but also interacts with the surrounding medium. Starting with the rigorous quantum-electrodynamic equations for the probability amplitudes $C_{i\alpha(n_\lambda)}$, he found the equations relating the quantities

$$b_i = \frac{n_i}{n} \sum_{\alpha(n_\lambda)} \overline{|C_{i\alpha(n_\lambda)}|^2} \tag{16.4}$$

and their time derivatives. In this expression, n_i is the number of molecules occupying the level i and n is the total number of molecules. The quantity $|C_{i\alpha(n_\lambda)}|^2$ is equal to the probability that a molecule is in the state i, while the surrounding medium is in the state α and the field consists of a set of photons (n_λ). The bar over the modulus represents averaging over the phases. The set of equations for b_i is analogous to the equations for $\overline{n}$.

Equations (16.1) do not generally follow from the equations of quantum electrodynamics. They can only be obtained under certain special assumptions which limit the range of applicability of the probabilistic method. They are valid provided the incident radiation contains broad enough spectral lines whose mean frequencies are equal to the natural frequencies of the medium under consideration. They can always be used in the absence of illumination or when the incident radiation has a broad spectrum, for example white light, and also in the case of illumination with narrow lines if the energy levels are broad (vibrational sub-levels of complicated molecules).

The last conclusion also follows from other considerations. It was shown in Section 7 that if the probability amplitudes C_i cannot be integrated with respect to the frequency of the incident radiation, or the natural frequency of the atom, neither the transition probabilities nor the Einstein coefficients can be introduced, and without them the balance equation loses its physical meaning.

As an example of the application of the probabilistic method, we may mention the propagation of energy along a linear chain (Section 9). It was shown in this case that the probabilistic method leads to results which do not resemble, even qualitatively, the results of exact calculations. In the case of non-resonance excitation by narrow lines, the probabilistic method will not even yield the mean value of the distribution function.

In conclusion it is worth noting that the probabilistic method is apparently inapplicable in the case of a strong interaction of the atom (molecule) with the surrounding medium. However, this problem has not as yet been adequately explored.

4

Absorption

17. ABSORPTION OF INCIDENT RADIATION

Basic spectrophotometric relationships

The basic photometric quantity is the flux of radiant energy. It is defined as the amount of energy flowing through a given surface per unit time. For non-monochromatic radiation the flux $d\Phi_\nu$ must be referred to a frequency interval $d\nu$. If we consider a bundle of rays travelling in a solid angle $d\,\Omega_\nu$,

$$d\,\Phi_\nu = S_\nu d\,\nu \cos\theta\, dS d\,\Omega = S_\nu d\,\nu\, dS_n d\,\Omega \qquad (17.1)$$

where θ is the angle between the normal to the area dS and the axis of the bundle, and $dS_n = dS \cos\theta$. The quantity S_ν is referred to as the intensity of the beam per unit frequency interval. It is a function of frequency and the direction of propagation.

Consider an element of volume, dV, inside the medium. The energy emitted by this volume per unit time in the frequency interval $d\nu$ in all directions is equal to $W_\nu^{\text{em}}\, d\nu\, dV$.

If we surround the volume element by a sphere of arbitrary radius R, we find that the flux $\Delta\Phi_\nu$ through this sphere is

$$\Delta\Phi_\nu = W_\nu^{em}\, d\nu\, dV \tag{17.2}$$

while the flux within the solid angle $d\Omega$ is

$$d\Phi_\nu = \frac{1}{4\pi} W_\nu^{em}(\Omega)\, d\nu\, dV d\Omega \tag{17.3}$$

where

$$W_\nu^{em} = \frac{1}{4\pi}\int W_\nu^{em}(\Omega)\, d\Omega \tag{17.4}$$

The constant W_ν^{em} is called the rate of emission. It is related to the intensity by

$$S_\nu(\Omega) = \frac{1}{4\pi} W_\nu^{em}(\Omega)\, \frac{dV}{dS_n} \tag{17.5}$$

As the radiation propagates in the absorbing medium, the flux gradually decreases. The reduction in the intensity over a path length l is given by

$$dS_\nu = -k_\nu S_\nu dl \tag{17.6}$$

The absorption coefficient k_ν is characteristic of the particular material. In general, it depends on the frequency, the intensity, the coordinates of the absorbing region, the direction of propagation, the temperature and the external conditions. The absorption coefficient of a homogeneous isotropic medium is independent of the angle Ω and the coordinates of the absorbing element.

If the absorption coefficient is independent of the intensity of the beam, S_ν, which is the basic assumption of linear optics, we can integrate (17.6) to yield

$$S_\nu = S_\nu^0\, e^{-k_\nu l} = S_\nu^0\, e^{-D_\nu} \tag{17.7}$$

where S_ν^0 is the intensity at $l = 0$ and S_ν is the intensity after a path length l. The quantity D_ν is called the optical density of a layer of thickness l. According to (17.7), the intensity of a beam of radiation which passes through an absorbing medium decreases exponentially. This is a basic result, known as Bouguer's law (17.29). It is valid in many cases. Departures from this law, which are in fact observed in practice, suggest that the assumptions introduced between (17.6) and (17.7) were inaccurate, and the departures themselves are sources of important additional information about the properties of the medium.

The intensity of a beam can easily be related to the density $u_\nu(\Omega)$ of radiant energy, i.e. the amount of energy per unit volume propagating in a unit solid angle. The amount of energy dE_ν in a volume $dV = dS_n dl$ is $d\Phi_\nu dt$, where $d\Phi_\nu$ is given by (17.3) and dt is the time necessary for the light to traverse the path dl. Since the velocity of propagation is c/n, where n is the refractive index, it follows that $dt = n dl/c$, and hence

$$dE_\nu = u_\nu(\Omega)\, d\nu\, d\Omega\, dV = d\Phi_\nu \frac{dl}{c} n \tag{17.8}$$

or from (17.3)

$$u_\nu(\Omega) = \frac{n}{c} S_\nu(\Omega) \tag{17.9}$$

This expression is valid for anisotropic radiation. In the case of isotropic radiation there is no dependence on the direction of propagation.

If the line has a finite width, and to some extent this is always the case, it is useful to introduce the integral flux, integral intensity and integral rate of emission. These are defined by

$$d\Phi = \int_{\nu=0}^{\infty} d\Phi_\nu \tag{17.10}$$

$$S = \int_{\nu=0}^{\infty} S_\nu d\nu \tag{17.11}$$

$$W^{em} = \int_{\nu=0}^{\infty} W_\nu^{em}\, d\nu \tag{17.12}$$

It is usually sufficient to integrate within the more restricted limits of the real line or band profile. A similar generalisation cannot be introduced for the absorption coefficient k_ν. If k_ν is a function of frequency, Bouguer's law is valid only within a narrow frequency band. The integral flux is reduced not exponentially but in accordance with the expression

$$S = \int S_\nu^0 e^{-k_\nu l} d\nu \tag{17.13}$$

which is sometimes very different from (17.7).

There are a number of related quantities, each of which can be used as a measure of the absorption of radiation in a medium. In the case of an object of finite dimensions, the absorption may be characterised by the absorptive power or simply absorption. If the 'transmission' is defined by

$$P = \frac{S_\nu^{tr}}{S_\nu^0} \tag{17.14}$$

and the reflection factor by

$$R = \frac{S_\nu^{refl}}{S_\nu^0} \tag{17.15}$$

the absorptive power of the object is given by

$$A = 1 - R - P \tag{17.16}$$

The coefficient of absorption can serve as a measure of the absorption of radiation in an element of volume. If the absorption coefficient at all points in a finite object is the same, it can easily be related to the experimentally observed value of P and R. In particular, if the reflection factor is small and may be ignored, the relation between the absorption coefficient and the measured transmission P is given by Bouguer's law (17.17)

$$k_\nu = -\frac{1}{l} \ln P \tag{17.17}$$

In the presence of reflection, particularly multiple reflection within the body under consideration, the relation between P and R is more complicated and Bouguer's law cannot be used for the body as a whole.

Rate of emission

Suppose that the element of volume dV contains $n_j(\Omega_1)\,d\Omega_1$ particles occupying the level j and oriented so that they lie within the solid angle between Ω_1 and $\Omega_1 + d\Omega_1$. The number of radiative transitions from level j to the higher level i in a time dt under the action of radiation of density $u^\alpha(\nu_{ij}, \Omega_2)\, d\Omega_2$, and with polarisation α is, according to (7.3), given by

$$dn_{ji\alpha}(\Omega_1, \Omega_2) = b^\alpha_{ji}(\Omega_1, \Omega_2)\, u^\alpha(\nu_{ij}, \Omega_2)\, d\Omega_2\, n_j(\Omega_1)\, d\Omega_1 dV dt \qquad (17.18)$$

where $b^\alpha_{ji}(\Omega_1, \Omega_2)$ is the differential Einstein coefficient, which is proportional to the square of the modulus of the matrix element of the transition operator. On the dipole approximation, it is determined by (8.74).

The energy absorbed as a result of $j \to i$ transitions is $h\nu_{ij} dn_{ji\alpha}(\Omega_1, \Omega_2)$. The experimentally observed absorption, and the associated reduction in intensity of the incident beam, is always less than this. It was shown in Section 14 that both $j \to i$ and $i \to j$ transitions occur under the influence of the incident radiation. There is a close connection between absorption and stimulated emission. The probability of stimulated emission is equal to $b^\alpha_{ij}(\Omega_1, \Omega_2) \times u^\alpha(\nu_{ij}, \Omega_2)$. For non-degenerate levels, or sub-levels of degenerate levels, the Einstein coefficients $b^\alpha_{ij}(\Omega_1, \Omega_2)$ and $b^\alpha_{ji}(\Omega_1, \Omega_2)$ are equal. For degenerate levels we have the relation given by (7.14).

The most characteristic feature of stimulated emission is that the radiation is emitted in the same direction as the direction of propagation of the incident beam. Some of the radiation propagating in the direction Ω_2 is absorbed in the medium, while a fraction of the absorbed energy is compensated by the energy released by stimulated emission. The phenomena of absorption and stimulated emission co-exist and form two aspects of the same process, i.e. the process of the interaction of radiation with matter. In experiment they are observed at the same time. The actual change in energy per unit time per unit volume is

$$\begin{aligned} dW^{\text{abs}}_{ji} &= (dn_{ji\alpha} - dn_{ij\alpha})\, h\nu_{ij} \\ &= b^\alpha_{ji}(\Omega_1, \Omega_2)\, u^\alpha(\nu_{ij}, \Omega_2)\, h\nu_{ij} d\Omega_2 \left[n_j(\Omega_1) - \frac{g_j}{g_i} n_i(\Omega_1) \right] d\Omega_1 \end{aligned} \qquad (17.19)$$

The quantity W^{abs}_{ji} is the rate of absorption. In general, it depends on the nature of the absorption spectrum, the frequency and orientation of the molecules relative to the direction of propagation and polarisation of the external radiation, the intensity of the incident radiation and the population of the two levels. The expression for the rate of absorption is considerably simplified in the two special cases considered in Section 7. They include isotropic alignment of particles when the incident light is anisotropic, and the effect of an isotropic beam of arbitrarily disposed particles. In either case there is no dependence on the angles or the polarisation of the incident radiation:

$$W_{ji}^{\text{abs}} = B_{ji} u(\nu_{ij}) \left(n_j - \frac{g_j}{g_i} n_i \right) h\nu_{ij} \tag{17.20}$$

The relation between the integral Einstein coefficients and the corresponding differential coefficients is determined by (7.20), (7.21), (7.27), (7.28), (7.36) and (7.37), while the relation between $u^{\alpha}(\nu_{ij}, \Omega_2)$ and the total density $u(\nu_{ij})$ is determined by (7.19). For dipole absorption the relation between B_{ji} and $\mathbf{D}_{ji}$ is given by (8.76).

The rate of absorption (17.20) depends both on the properties of the medium and on the intensity of the external radiation. The ability of matter to absorb radiation is conveniently described by dividing (17.20) or (17.19) by the density of the incident radiation. The resulting quantity is

$$\mathrm{k}_{ji} = \frac{W_{ji}^{\text{abs}}}{u(\nu_{ij})} = B_{ji} \left(n_j - \frac{g_j}{g_i} n_i \right) h\nu_{ij} \tag{17.21}$$

and is commonly referred to as the absorptive power of the medium. It has the dimensions of the reciprocal of time and is generally different from the absorption A, of the body which is the ratio of absorbed to incident fluxes. The new parameter k can be defined as the ratio of the energy $W_{\nu}^{\text{abs}}\, d\nu\, dV$ absorbed in the volume dV per second to the amount of energy $u_{\nu} d\nu\, dV$ present in this volume. The quantity A automatically takes into account all processes such as secondary emission, scattering and reflection within the body, whereas k refers only to the elementary volume.

In most special cases, absorption is studied under the conditions of thermodynamic equilibrium. It is assumed that the incident radiation does not appreciably affect the level population. Substituting the ratio

$$\frac{n_i}{n_j} = \frac{g_i}{g_j} e^{-h\nu_{ij}/kT} \tag{17.22}$$

into (17.20)-(17.21), we obtain

$$\begin{aligned} W_{ji}^{\text{abs}} &= B_{ji}\, u(\nu_{ij})\, n_j (1 - e^{-h\nu_{ij}/kT})\, h\nu_{ij} \\ &= \frac{c^3}{8\pi n^3 \nu_{ij}^2} A_{ij} u(\nu_{ij})\, n_j \frac{g_i}{g_j} (1 - e^{-h\nu_{ij}/kT}) \end{aligned} \tag{17.23}$$

The absorptive power k_{ji} can easily be estimated from the following numerical examples. Suppose that the number of absorbing particles per cubic centimetre is 10^{19}. For electronic levels $n_1 \sim n$, while for vibrational levels it is somewhat smaller (see Section 6). For electronic levels $A_{ij} \sim 10^8$ sec^{-1} m, $\nu \sim 20{,}000$ $\text{cm}^{-1} \times 3 \times 10^{10}$ cm sec^{-1}, and so

$$k_{ji} = \frac{W_{ji}^{\text{abs}}}{u(\nu_{ij})} \approx 10^{27}\ \text{sec}^{-1}$$

This means that if $u_{ij} = 1$, erg cm^{-3}, the amount of energy absorbed per second per unit volume is 10^{27} erg. The amount of energy absorbed per atom is 10^8 erg. All this is valid provided, of course, the absorbed energy is continuously compensated by the arrival of energy from outside, and the density of radiation in the volume under consideration remains constant and equal to 1 erg cm^{-3}. Since the energy of a single quantum is $h\nu \sim 4 \times 10^{-12}$erg, we see that one atom undergoes $B \approx 10^{20}$ transitions per second.

At first sight, these results are paradoxical. This is because the radiation density was assumed to be 1 erg cm^{-3}. In reality, this can only be achieved when the radiation source is at a very high temperature. At room temperature the density of equilibrium radiation is only 4.87×10^{-53}erg cm^{-3}, and therefore the energy absorbed by an atom per second is only 10^{-45} erg.

For vibrational levels the corresponding values are somewhat different. Here $A \sim 100$, $\nu \sim 1{,}000$ $\text{cm}^{-1} \times 3 \times 10^{10}$cm sec^{-1} and therefore $k = 10^{24}$ sec^{-1}. Thus, one molecule absorbs 10^5 erg (10^8 photons) per second.

In the radio-frequency region it is necessary to take into account the population of the upper levels participating in the transitions. Suppose that $\nu = 1$ $\text{cm}^{-1} \times 3 \times 10^{10}$ cm sec^{-1} and that the transitions occur at room temperature. The quantity $(1 - e^{-h\nu_{ij}/kT})$ in (17.23) above is then approximately equal to 1/200. Substituting this into (17.23) and assuming that $A \sim 10^{-11}\text{sec}^{-1}$, we have $k \sim 10^{10}$ sec^{-1}. Thus, one particle absorbs 10^{-9}erg and since the photon energy is 2×10^{-16}erg, the number of photons absorbed per second is $\sim 10^7$.

The absorptive power corresponding to transitions from higher energy levels is as a rule much smaller ($n_j \ll n$). This is not so in the radio-frequency region where the populations of neighbouring levels are practically the same.

In the absence of thermodynamic equilibrium it is necessary to use equations (17.19)-(17.21). The values of n_j and n_i should be determined by exact calculation, for example by the probabilistic method with allowance for the intensity of the external perturbation modifying the equilibrium. A large number of such calculations will be carried out in subsequent chapters for radiative excitation. Equations (17.19) and (17.20) may be used to solve the converse problem; knowing B_{ji}, g_i and g_j, from measured values of the absorptive power, we can determine the difference $n_j - n_i$ and thereby the departure from equilibrium.

Equation (17.20) may be rewritten in a somewhat different form by considering the change in the rate of absorption, or the absorptive power, due to the departure from thermodynamic equilibrium:

$$W_{ji}^{\mathrm{abs}} = W_{ji}^{\mathrm{abs\ eq}} + B_{ji} u(\nu_{ij}) \left[\Delta n_j - \frac{g_j}{g_i} \Delta n_i \right] h \nu_{ij} \quad (17.24)$$

where $\Delta n_j = n_j - n_j^{\mathrm{eq}}$ and $\Delta n_i = n_i - n_i^{\mathrm{eq}}$ are the departures of n_j and n_i from their equilibrium values and can either be positive or negative.

We have assumed for the sake of simplicity that the spectral lines were strictly monochromatic. In reality, all lines have a finite width. Correct results are obtained if ν_{ij} in the formulae given in this section is understood to mean the mean frequency of a narrow spectral line. The quantities B_{ji} and B_{ij} are the integral (with respect to frequency) Einstein coefficients.

If the particles under investigation have a continuous energy spectrum, the rate of absorption must be calculated individually for each frequency. Suppose that $\rho(E)\,dE$ is the probability that the energy of a given particle is between E and $E + dE$, and $\rho(E)\,dE\,n\,dV$ is the total number of such particles in the volume element dV. If we multiply this value by the probability of a transition from the level E to the set of levels between $E + h\nu$ and $E + h\nu + h\,d\nu$ and integrate over all E, we obtain the total number of absorbed photons of frequency ν, which is

$$nu(\nu)\, d\nu\, dV \int B_{\text{abs}}(E, \nu)\, \rho(E)\, dE$$

In order to find the rate of absorption we must take into account stimulated emission and multiply by $h\nu$. Thus

$$W^{\text{abs}}(\nu) = nu(\nu)\, h\nu \int_{E=0}^{\infty} B_{\text{abs}}(E, \nu)\, \rho(E) \times \left[1 - \frac{g(E)}{g(E+h\nu)} \frac{\rho(E+h\nu)}{\rho(E)}\right] dE \tag{17.25}$$

This gives the rate of absorption per unit interval of frequency. The absorptive power is obtained by dividing this expression by $u(\nu)$. In the visible part of the spectrum the second term in (17.25) is as a rule unimportant.

If the electronic levels have a discrete character, while the vibrational levels have a continuous spectrum (Fig. 4.1), the

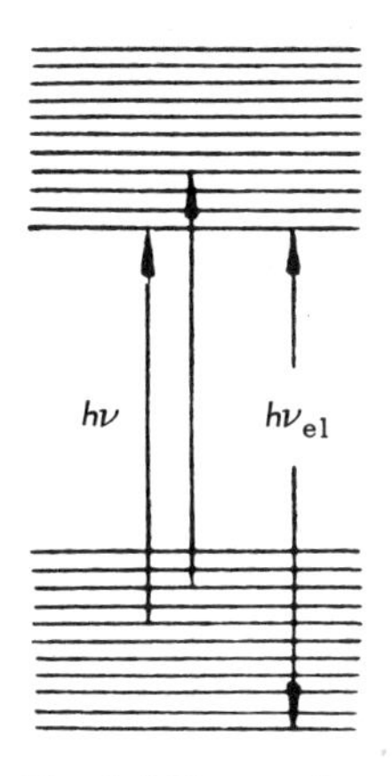

Fig. 4.1 Electronic-vibrational levels

rate of absorption is still given by (17.25) except that the limits of integration must be modified. For frequencies $\nu < \nu_{el}$ the lower limit of integration is zero as before, while for $\nu < \nu_{el}$ it is $E = h(\nu_{el} - \nu)$. .

Absorption coefficient

The absorptive power is not measured directly but is usually calculated from the absorption coefficient. According to

(17.6) the absorption coefficient is the constant of proportionality between the change dS_ν in the intensity of a beam in a path dl. The dimensions of the absorption coefficient are L^{-1} and it is usually expressed in cm^{-1}.

The absorption coefficient can easily be related to transition probabilities between the corresponding energy levels and their population. To calculate k_ν we note that the change in the energy flux in a path dl is equal to the amount of energy absorbed per second in the volume element $dV = dS_n dl$. Equation (17.19) gives the amount of energy absorbed as a result of $j \to i$ transitions per unit volume within the entire line. If the frequency dependence of the Einstein coefficient is taken into account, then the flux change $d\Phi_\nu = S_\nu d\nu\, dS_n d\Omega_2$ in a length dl is given by

$$d\,[d\,\Phi_\nu] = -\,b^{\alpha}_{ji}(\Omega_1, \Omega_2, \nu)\, u^{\alpha}(\nu_{ij}, \Omega_2)\, d\nu\, d\Omega_2\, h\nu_{ij} \times \left[n_j(\Omega_1) - \frac{g_j}{g_i}\, n_i(\Omega_1)\right] d\Omega_1\, dS_n dl \qquad (17.26)$$

According to (17.1) and (17.9), the flux $d\Phi_\nu$ through the area dS_n, expressed in terms of the radiation density is

$$d\,\Phi_\nu = \frac{c}{n}\, u^{\alpha}(\nu_{ij}, \Omega_2)\, d\nu\, d\Omega_2\, dS_n \qquad (17.27)$$

From (17.26), (17.27) and (17.6) we have

$$k^{\alpha}_{\nu}(\Omega_1, \Omega_2)\, d\nu = \frac{n}{c}\, b^{\alpha}_{ji}(\Omega_1, \Omega_2, \nu)\, d\nu \times \left[n_j(\Omega_1) - \frac{g_j}{g_i}\, n_i(\Omega_1)\right] h\nu_{ij} \qquad (17.28)$$

or integrating with respect to frequency within the profile of the line,

$$\int\limits_{\nu_{ij}} k^{\alpha}_{\nu}(\Omega_1, \Omega_2)\, d\nu = \frac{n}{c}\, b^{\alpha}_{ji}(\Omega_1, \Omega_2) \times \left[n_j(\Omega_1) - \frac{g_j}{g_i}\, n_i(\Omega_1)\right] h\nu_{ij} \qquad (17.29)$$

where, in accordance with (7.7a)

$$b^{\alpha}_{ji}(\Omega_1, \Omega_2) = \int\limits_{\nu_{ij}} b^{\alpha}_{ij}(\Omega_1, \Omega_2, \nu)\, d\nu_{ij}$$

is the measured value of the Einstein coefficient for the spectrum line as a whole.

The absorption coefficient in (17.28) and (17.29) refers to molecules whose axes lie within the solid angle $d\Omega_1$ and which absorb radiation with polarisation α propagating within the solid angle $d\Omega_2$. Integration of (17.29) with respect to Ω_1 yields the absorption coefficient for all molecules of arbitrary orientation:

$$\begin{aligned}\int_{\nu_{ij}} k_\nu^\alpha (\Omega_2)\, d\nu_{ij} &= \int_{\Omega_1} d\Omega_1 \int_{\nu_{ij}} k_\nu^\alpha (\Omega_1, \Omega_2)\, d\nu_{ij} \\ &= \frac{\mathrm{n}}{c} h\nu_{ij} \int_{\Omega_1} d\Omega_1\, b_{ji}^\alpha (\Omega_1, \Omega_2) \left[n_j(\Omega_1) - \frac{g_j}{g_i} n_i(\Omega_1) \right]\end{aligned} \qquad (17.30)$$

When the angular distribution of the particles is random, the expression given by (17.30) assumes the simpler form

$$\int_{\nu_{ij}} k_\nu d\nu_{ij} = \frac{\mathrm{n}}{c} h\nu_{ij} B_{ji} \left(n_j - \frac{g_j}{g_i} n_i \right) \qquad (17.31)$$

where B_{ji} is the integral Einstein coefficient. The absorption coefficient is then independent of the direction of the incident beam.

The rate of absorption was introduced above not only for directed but also for isotropic fluxes of radiation. This generalisation has no meaning for the absorption coefficient because it always characterises the attenuation of directed radiation.

When the absorption band is wide enough, the frequency dependence must be taken into account. By analogy with (17.25) we may write for transitions between electronic-vibrational levels

$$\begin{aligned} k(\nu) &= \frac{1}{c/\mathrm{n}} h\nu n \int_E B_{\mathrm{abs}}(\mathrm{E}, \nu)\, \rho(\mathrm{E}) \\ &\times \left[1 - \frac{g(\mathrm{E})}{g(\mathrm{E} + h\nu)} \, \frac{\rho(\mathrm{E} + h\nu)}{\rho(\mathrm{E})} \right] d\mathrm{E} \end{aligned} \qquad (17.32)$$

where $\rho(\mathrm{E})$ and $\rho(\mathrm{E} + h\nu)$ are the distribution functions for the initial and final states. The dimensions of the Einstein coefficient $B_{\mathrm{abs}}(\mathrm{E}, \nu)$ differ from those of B_{ji} in (17.31), since the former is calculated per unit frequency interval. Care must be exercised not to confuse the refractive index in c/n and the total number n of particles per unit volume.

Comparison of (17.31), (17.20) and (17.21) shows that there is a simple relation between the absorption coefficient, the rate of absorption and the absorptive power:

$$\int_{\nu_{ij}} k_\nu d\nu = \frac{\mathrm{n}}{c} \frac{W_{ji}}{u(\nu_{ij})} = \frac{\mathrm{n}}{c} \mathrm{k}_{ji} \quad (17.33)$$

where n is the refractive index. An analogous result holds not only for the integral values (with respect to the solid angles) but also for each direction individually.

The quantities k_ν, W^{abs} and k refer to a particular point in space (element of volume) and may be functions of co-ordinates.

If we substitute the Boltzmann level population into (17.29)-(17.31), we obtain the absorption coefficient for thermodynamic equilibrium.

Let us consider (17.31) in somewhat greater detail. If we express the Einstein coefficient in terms of the parameters of the classical electric dipole and the oscillator strength in accordance with (8.83), we obtain

$$k = \int_{\nu_{ij}} k_\nu d\nu_{ij} = \frac{\mathrm{n}}{c} \frac{\pi e^2}{3m} n f_{ji} \left(\rho_j - \frac{g_j}{g_i} \rho_i \right) \quad (17.34)$$

It has already been pointed out in Section 2 that this integral was obtained by Kravets on the basis of the classical theory as long ago as 1912. The area under the curve of the absorption coefficient plotted as a function of frequency is the best measure of the integral absorption of radiation. It is quite easy, by measuring it experimentally, to calculate the transition probabilities (Einstein coefficients) and oscillator strengths or, knowing these, determine the number of absorbing centres $n\left(\rho_j - \frac{g_j}{g_i}\rho_i\right)$. In the original variant of the theory, (17.34) did not include the factor $(\rho_j - g_j/g_i\rho_i)$. In this form it is valid only for transitions from the lowest level although in this case also there is no dependence of the absorption coefficient on the temperature or on changes in the level population. Moreover, induced emission has no effect on the phenomenon although it is known that under certain conditions it can be very appreciable even in the visible spectrum [34].

In (17.34) the integral on the right is represented by the letter k. It must be emphasised once again that k is quite different from the constant k_ν in Beer's law (17.6) or (17.7) and does not unambiguously determine the rate of attenuation of the intensity of the beam of radiation in the medium. The integral represented by k has different dimensions from those of the absorption coefficient ($cm^{-1} sec^{-1}$ instead of cm^{-1}).

The Kravets integral is convenient for estimating the numerical values of the absorption coefficient. Let us suppose that the oscillator strength is of the order of unity (e and m are the charge and mass of the electron; the Einstein coefficient A is of the order of 10^8), so that $\rho_j \sim 1$, $\rho_i \sim 0$. The integrated value of the absorption coefficient is then $0.03n$, where n is the number of absorbing particles per unit volume. At lower oscillator strengths

$$\int k_\nu d\nu = 3 \cdot 10^{-2} nf \text{ cm}^{-1} \text{ sec}^{-1}$$

For metastable states the oscillator strength is lower by a factor of 10^4-10^8 as compared with the preceding case. When estimating the absorption coefficient for transitions between vibrational levels it is necessary to take into account the increase in the mass of the vibrating nuclei in (17.36) in comparison with the electron mass. This leads to a reduction in the absorption coefficient.

Sections 2 and 10 show that the introduction of frictional forces corresponding to some non-radiative transitions, leads to the broadening of spectral lines and bands, but that the area under them remains constant. The same result follows automatically from (17.34) since the right-hand side does not contain any quantities which depend on the probabilities of non-radiative transitions.

To estimate the numerical value of k_ν at the maximum, the integral (17.34) may be written in the approximate form (see Fig. 4.2)

$$\int k_\nu d\nu \approx k_{max} \Delta\nu$$

where $\Delta\nu$ is the half-width of the band or line. In the absence of non-radiative transitions, k_{max} is determined by the ratio of the Kravets integral to the natural width of the line $\Delta\nu \sim 10^7 sec^{-1}$. For electronic transitions where $f = 1$, $\rho_j = 1$, we have in accordance with the above estimates $k_{max} \sim 10^{-2} \times 10^{-7} n = 10^{-9} n$. If $n \sim 10^{19}$ then $k_{max} \sim 10^{10} cm^{-1}$. This is the maximum possible value of the absorption coefficient. If, on the other hand, there is an increase in the width of the band as a result of non-radiative transitions, it is found that k_{max} decreases. In the case of broad bands ($\Delta\nu \sim 1{,}000$ cm^{-1}

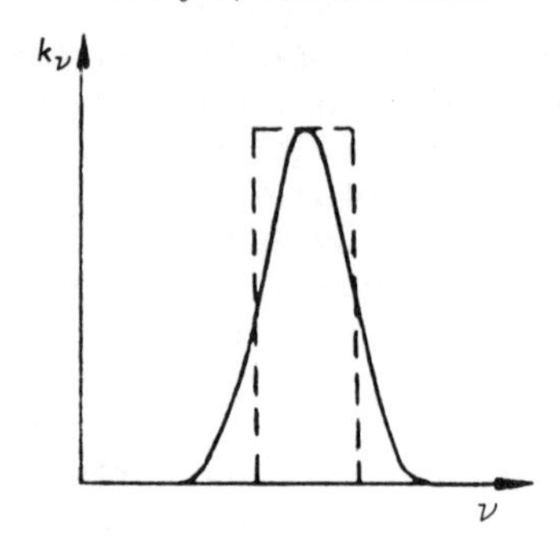

Fig. 4.2 Profile of an absorption band

or 3×10^{13} sec^{-1}) the value of k_{max} is of the order of 10^4.

The sum rule which was derived for dipole transitions in Section 8 may be used to establish certain general properties of the absorption coefficient. Suppose that all the particles occupy the same energy level ($n = n_1$, $n_{i \neq 1} = 0$). If we apply the sum rule (8.88)

$$\sum_i h \nu_{ij} B_{ji} = \frac{\pi e^2}{3m} \sum_i f_{ji} = \frac{\pi e^2}{3m} N \qquad (17.35)$$

to (17.31), where N is the number of optical electrons in the atom or molecule, we have

$$\int_{\nu=0}^{\infty} k_\nu d\nu_{i1} = \frac{1}{c/n} \frac{\pi e^2}{3m} n N \qquad (17.36)$$

The right-hand side of this equation is the integral of the absorption coefficient for transitions from the level $j = 1$ to all the possible energy levels of the given atom or molecule and represents the total absorption by it when the incident radiation has a continuous spectrum.

The sum rule is valid not only for the lowest level but for any level j. It is therefore sometimes written

$$\int_{\nu=0}^{\infty} k_\nu d\nu = \frac{1}{c/n} \frac{\pi e^2}{3m} N \sum_j n_j = \frac{1}{c/n} \frac{\pi e^2}{3m} N n \qquad (17.37)$$

where stimulated emission is neglected and we have used

(17.31). From this it is often concluded that the integral absorption is independent of the distribution over the energy levels, $\rho_i = n_i/n$, and that in thermodynamic equilibrium it is also independent of the temperature. This conclusion is incorrect. In reality, if all the particles occupy the j-th level, the sum

$$\sum_i \int_{\nu_{ij}} k_\nu d_\nu = \frac{1}{c/\mathrm{n}} \sum_i h\nu_{ij} B_{ji} n_j$$

includes both positive $(i>j)$ and negative terms, and the sum of only positive terms is not equal to (17.37).

Very general results may be obtained by considering the Kravets integral for transitions between electronic-vibrational levels of a complicated system. In the section devoted to the Franck-Condon principle (Section 8) it was shown that if the operator for the electronic transitions was independent of the vibrational coordinates (Condon approximation), the sum of the transition probabilities from a given level E in the lowest electronic state to all the vibrational levels of the upper electronic state was constant and independent of the energy of the initial vibrational level. This means that in (17.32)

$$\int B_{\mathrm{abs}}(\mathrm{E}, \nu)\, d\nu = \mathrm{const} \qquad (17.38)$$

for all E. If we neglect stimulated emission in (17.34), which is always possible for electronic spectra, we obtain

$$\int_\nu \frac{k_\nu d\nu}{h\nu} = \frac{1}{c/\mathrm{n}} n \int \mathrm{const}\ \rho(\mathrm{E})\, d\mathrm{E} = \mathrm{const} \frac{1}{c/\mathrm{n}} n \qquad (17.39)$$

where n is the refractive index and n is the number of particles per unit volume. The value of the integral given by (17.39) is determined by the square of the matrix element for transitions between electronic levels and is independent of the distribution function $\rho(\mathrm{E})$; for an equilibrium distribution over the vibrational levels it is also independent of the temperature of the system. A temperature change can lead only to a change in the profile of the band but not in the area under it.

Experiment shows that the integral (17.39) is in fact very frequently constant, e.g. in the spectra of complicated molecules. However, in some cases, the temperature dependence is particularly strong, and this clearly suggests that the Condon approximation is invalid, and the operator for the electronic transition does, in fact, depend on the vibrational coordinates.

Absorption coefficient for mixtures. Lambert-Beer law

The absorption coefficient of a mixture of non-interacting components is equal to the sum of the individual absorption coefficients:

$$k(\nu) = \sum_{\alpha} k^{\alpha}(\nu) \tag{17.40}$$

If the individual spectral lines do not overlap, the discussion in the preceding paragraph will apply. Some new properties arise in the case of mixtures of substances whose spectra contain broad level bands.

If we neglect stimulated emission, then in view of (17.32) we have

$$k(\nu) = \frac{1}{c/n} h\nu \sum_{\alpha} n_{\alpha} \int B^{\alpha}_{\text{abs}}(\mathrm{E}, \nu)\, \rho_{\alpha}(\mathrm{E})\, d\mathrm{E} \tag{17.41}$$

where n_{α} is the number of molecules of the α component per unit volume

$$\left(n = \sum_{\alpha} n_{\alpha}\right)$$

Let

$$\frac{n_{\alpha}}{n} = C_{\alpha} \tag{17.42}$$

$$k_{\alpha} = \frac{1}{c/n} h\nu \int B^{\alpha}_{\text{abs}}(\mathrm{E}, \nu)\, \rho_{\alpha}(\mathrm{E})\, d\mathrm{E} \tag{17.43}$$

where C_{α} is the fraction of molecules of the α component and k_{α} is the corresponding absorption coefficient, i.e. the absorption coefficient which would be observed if the number of molecules of the $\bar{\alpha}$ component were equal to the total number of molecules in the mixture. The expression given by (17.41) can then be written in the form

$$k(\nu) = \sum_{\alpha} C_{\alpha} k_{\alpha}, \; k^{\alpha} = C_{\alpha} k_{\alpha} \qquad (17.44)$$

and this determines the dependence of the absorption coefficient of the mixture on the concentration of the components.

This formula can be used to calculate the absorption by a mixture if the absorption coefficients of the individual components and their concentrations are known. The converse problem is more frequently encountered in molecular spectral analysis: knowing the absorption coefficients of the mixture and its components from experiment, a determination is made of the concentrations C_{α}. The formula given by (17.44) is frequently used for the determination of the absorption coefficient of an unknown material in a mixture.

For a mixture of two substances we have from (17.44)

$$k(\nu) = C_1 k_1(\nu) + C_2 k_2(\nu) = C_1 k_1(\nu) + (1 - C_1) k_2(\nu) \qquad (17.45)$$

If one of the substances, for example, the solvent, is transparent $(k_1 = 0)$

$$k(\nu) = C_2 k_2(\nu) = k^{(2)}(\nu) \qquad (17.46)$$

and all the formulae assume the form characteristic of the one absorbing material.

Let us suppose that the absorption bands $k_1(\nu)$ and $k_2(\nu)$ of the components of a two-component mixture intersect at a single point, i.e. $k_1(\nu') = k_2(\nu')$. At this frequency the absorption coefficient of the mixture is, in view of (17.45), independent of the concentration of the components:

$$k(\nu') = k_1(\nu') = k_2(\nu') \qquad (17.47)$$

When the relative concentration of the components of the mixture is altered, there is a change in the absorption spectrum. If the condition given by (17.47) is satisfied, all the spectra intersect at the point ν' As an illustration, Fig. 4.3 shows the spectra of a mixture of monomers and dimers of rhodamine [35]. The different relative concentrations were obtained by introducing different amounts of rhodamine into the water solution. In this way, a definite ratio of C_1 and $C_2 = 1 - C_1$ was established. Curve 1 corresponds to the spectrum of the monomer and curve 5 to the spectrum of the dimer. All the remaining curves correspond to the spectra of their mixtures. In this case, the condition given by (17.47) is satisfied for two values of ν and therefore

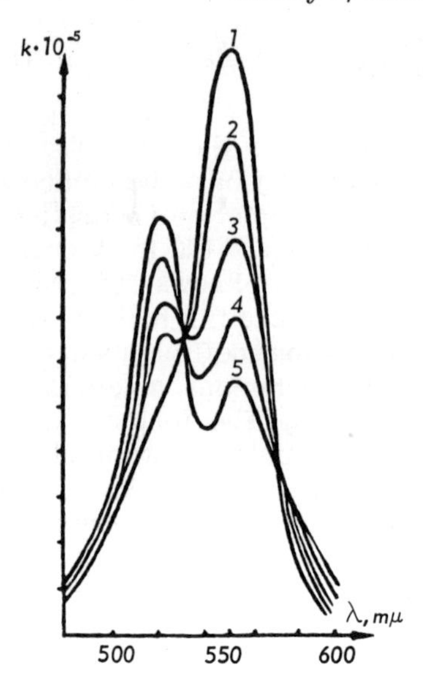

1 – 10^{-6} ml^{-1}; 2 – 10^{-4}; 3 – 4×10^{-4}; 4 – 3×10^{-3}; 5 – dimers

Fig. 4.3 Dependence of the absorption spectrum of a water solution of rhodamine C on the concentration

the curves intersect at two points. The existence of such points unambiguously shows the presence in the mixture of only two absorbing components.

It follows from (17.45) that

$$C_1 = \frac{k(\nu) - k_2(\nu)}{k_1(\nu) - k_2(\nu)} \tag{17.48}$$

If $k_1(\nu)$ and $k_2(\nu)$ are known, the relative concentrations of the components of the mixture can readily be determined.

Let us substitute (17.44) into (17.7) which gives the intensity of the beam on the linear approximation. The result is

$$S_\nu = S_\nu^0\, e^{-\left(\sum\limits_\alpha C_\alpha k_\alpha\right) l} = S^0\, e^{-\left(\sum\limits_\alpha n_\alpha \frac{k_\alpha}{n}\right) l} \tag{17.49}$$

where k_α/n is the absorption coefficient of the α-component per molecule and n_α is the number of molecules of the α-component per unit volume.

The expression given by (17.49) may be transformed by isolating the term characterising the absorption by a particular component of the mixture:

$$S_\nu = L e^{-n_\beta \frac{k_\beta}{n} l} \tag{17.50}$$

If all the remaining components do not absorb radiation, $L = S_\nu^0$. The above formula gives the intensity as a function of the number of absorbing molecules n_β. Knowing l and having determined experimentally the values of S_ν for different concentrations of the absorbing component, it is quite easy to calculate the absorption coefficient for a particular component. The formula (17.50) was verified experimentally by Beer (1852) and was analysed in detail by Lambert (1760). It is frequently referred to in the literature as the Bouguer-Lambert-Beer law.

The Lambert-Beer law is valid only when the introduction of the β component into the mixture does not lead to interaction of these molecules with the solvent or the other molecules present in the mixture and it will not hold if the β molecules either associate or dissociate when they are introduced into the mixture. Any breakdown of the Lambert-Beer law may be regarded as evidence of the appearance of some physico-chemical changes and the formation of new distinct products in the mixture.

It must, however, be emphasised that (17.50) will always be valid for n_β representing the number of molecules of the particular component β which is actually present in the mixture and k_β/n, the absorption coefficient per molecule. When we spoke of the breakdown of the Lambert-Beer law, we had in mind the fact that the measured absorption coefficient $k(\nu)$ did not equal $\sum_\alpha n_\alpha \frac{k_\alpha}{n}$, where n_α and k_α/n refer to the components α introduced into the mixture from outside.

Absorption of light by a plane-parallel layer

So far, we have been concerned with the absorption of light in an element of volume or in a uniform medium. One is frequently concerned, however, with objects having finite linear dimensions so that reflection and transmission at their boundaries must be taken into account. The simplest example of this is the plane-parallel layer. We shall discuss its properties in some detail since plane-parallel layers are widely used in practice as light filters, interferometers, absorbing specimens, and so on. Plane-parallel layers have recently found application as cavity resonators for the generation and amplification of visible radiation.

For thin enough layers, the problem must be solved within the framework of wave optics with allowance for interference. However, in many cases involving thick layers, high absorption or slowly varying thickness, the methods of geometrical optics provide an adequate approximation. We shall employ this approximation here, and will indicate at the end how the results will be modified when interference is allowed for.

Let us suppose, to begin with, that the radiation is incident at right-angles to the plane-parallel layer. The reflection coefficient r and the absorption coefficient k will, in general, depend on the frequency of the incident radiation. The reflection coefficient is given by the Fresnel formula which for normal incidence assumes the form

$$r = \frac{(n - n_0)^2 + \varkappa^2}{(n + n_0)^2 + \varkappa^2} \tag{17.51}$$

where n and n_0 are the refractive indices of the layer and of the surrounding medium respectively, and $\varkappa = k\lambda/4\pi$.

It is evident from Fig. 4.4 that the reflected and transmitted radiation is made up of a large number of individual rays which are multiply reflected in the layer and consequently traverse different paths (for the purpose of illustration, the incident beam is shown at an angle to the normal).

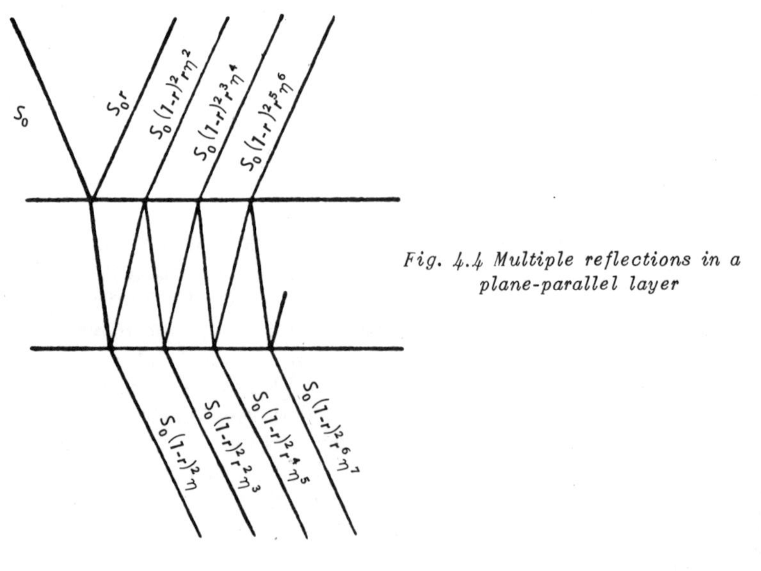

Fig. 4.4 Multiple reflections in a plane-parallel layer

If the intensity of this incident beam is S_0, the intensity of the reflected beam will be $S_0 r$. The second beam is formed by transmission at the first boundary, where its intensity is reduced by a factor of $(1 - r)$, followed by transmission through the layer (thickness l) in which it is attenuated by a factor of $\eta = e^{-kl}$ in accordance with Bouguer's law. On reaching the second surface, it is partly reflected and partly transmitted, and so on (Fig. 4.4).

On the geometrical optics approximation we can add the intensities of the various beams propagating in a particular direction so that the total intensity is given by

$$S_{refl} = S_0 \{r + (1 - r)^2 r \eta^2 [1 + r^2 \eta^2 + \dots]\} \tag{17.52}$$

The number of beams which have to be added is infinite. Since $r^2\eta^2$ is very small, it follows that

$$S_{refl} = S_0 \left\{ r + \frac{(1 - r)^2 r \eta^2}{1 - r^2 \eta^2} \right\} \tag{17.53}$$

The ratio

$$R = \frac{S_{refl}}{S_0} = r + \frac{(1 - r)^2 r \eta^2}{1 - r^2 \eta^2} \tag{17.54}$$

is known as the reflection factor of the layer. Similarly, the transmission coefficient of the layer is given by

$$P = \frac{S_{tr}}{S_0} = (1 - r)^2 \eta (1 + r^2 \eta^2 + \dots) = \frac{(1 - r)^2 \eta}{1 - r^2 \eta^2} \tag{17.55}$$

The absorptive power of a plane parallel layer is, by definition, given by

$$A = 1 - R - P = \frac{(1 - r)(1 - \eta)}{1 - r\eta} \tag{17.56}$$

When $\eta = 1$, i.e, when the layer is transparent, we have $A = 0$. If $r = 0$ then $A = 1 - e^{-kl}$. It is sometimes necessary to know the intensity of the beam inside the layer. Let $\eta_x = e^{-kx}$, $\eta_{l-x} = e^{-k(l-x)}$, where x is the distance from the first boundary. Simple calculations will show that for a beam travelling from the first boundary to the second

$$S_x^{(1,2)} = \frac{(1-r)\eta_x}{1-r^2\eta^2} S_0 \qquad (17.57)$$

while for a beam travelling from the second to the first

$$S_x^{(2,1)} = \frac{(1-r)r\eta\eta_{l-x}}{1-r^2\eta^2} S_0 \qquad (17.58)$$

Consider the formula (17.55). The transmission of radiation by an absorbing layer is not described by Bouguer's law which is valid only for propagation in a uniform medium. The exponential reduction in the intensity of a beam in a layer of thickness l can occur only for small coefficients of reflection at the surface. If r^2 can be neglected in comparison with unity, then

$$P = (1-2r)e^{-kl}$$

Figures 4.5-4.7 show plots of the functions $R(k)$, $P(k)$, and $A(k)$. It is evident that in the case of transmission by a layer it is necessary to allow for the reflection at the layer-medium boundary, while in the case of reflection the absorption within the layer must be allowed for.

It is important to note that for large r, multiple reflections may give rise to an apparent change in the spectral band profile.

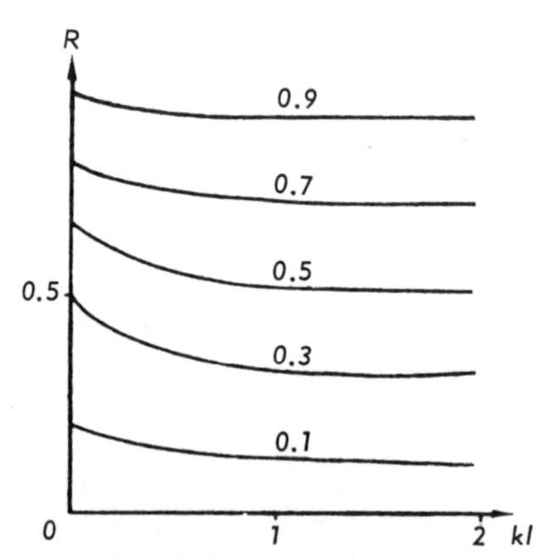

Fig. 4.5 Reflection by a plane-parallel layer as a function of the optical density for different values of the reflection coefficient r

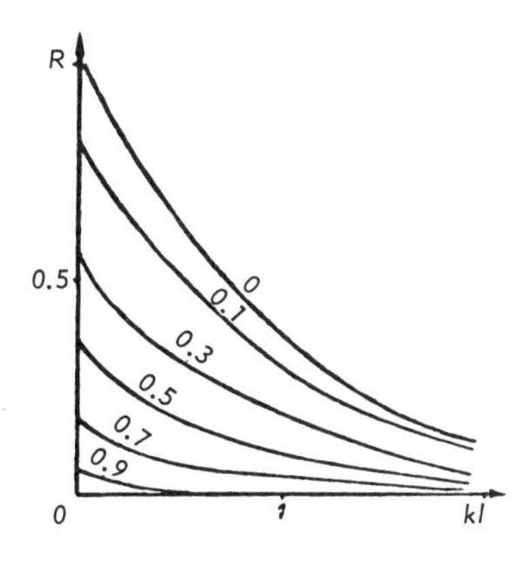

Fig. 4.6 Dependence of the transmission by a plane-parallel layer on the optical density for different values of reflection coefficient r

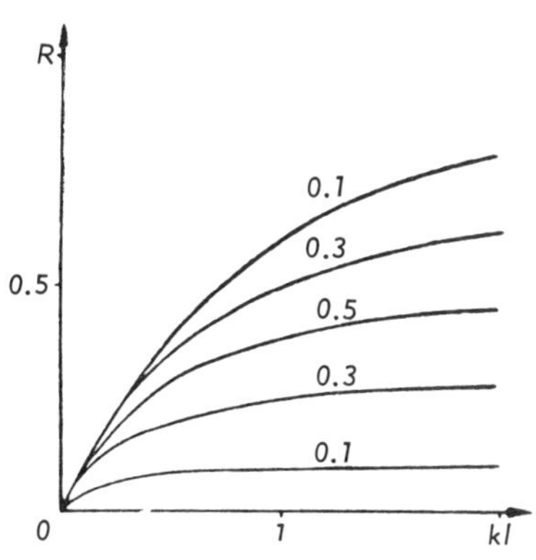

Fig. 4.7 Dependence of the absorption by a plane-parallel layer on the optical density for different values of reflection coefficent r

In contrast to the above case, the number of beams which have to be added at oblique incidence is finite, so that

$$
\begin{aligned}
R &= r + (1-r)^2 r \eta^2 [1 + r^2 \eta^2 + \dots (r^2 \eta^2)^{n-1}] \\
&= r + \frac{(1-r)^2 r \eta^2}{1 - r^2 \eta^2} [1 - (r^2 \eta^2)^n]
\end{aligned}
\tag{17.59}
$$

where n is the number of beams. Similarly,

$$
\begin{aligned}
P &= (1-r)^2 \eta [1 + r^2 \eta^2 + \dots + (r^2 \eta^2)^{n-1}] \\
&= \frac{(1-r)^2 \eta}{1 - r^2 \eta^2} [1 - (r^2 \eta^2)^n]
\end{aligned}
\tag{17.60}
$$

where $\eta = e^{-kl/\cos\psi}$ and ψ is the angle of refraction.

In contrast to the geometrical-optics approximation the exact wave-optics solution of the problem requires the addition of the amplitudes of the partial waves rather than the intensities. Here, we shall only quote without proof the final result which is obtained for normal incidence by solving Maxwell's equation subject to the appropriate boundary conditions [36]. If the plane-parallel layer is surrounded by a non-absorbing medium,

$$R = \frac{r\left[(1-\eta)^2 + 4\eta \sin^2 \frac{2\pi l}{\lambda}\right]}{(1-r\eta)^2 + 4r\eta \sin^2\left(\frac{2\pi}{\lambda} l - \delta\right)} \tag{17.61}$$

$$P = \left(1 + \frac{\varkappa^2}{n^2}\right) \frac{(1-r)^2 \eta}{(1-r\eta)^2 + 4r\eta \sin^2\left(\frac{2\pi}{\lambda} l - \delta\right)} \tag{17.62}$$

where $\varkappa$ is the imaginary part of the complex refractive index $\left(\varkappa = \frac{\lambda}{4\pi} k\right)$, r is the reflection coefficient at the surface of the layer which is given by (17.51), n is the refractive index of the layer and

$$\eta = e^{-\frac{4\pi}{\lambda} l\varkappa} = e^{-kl} \tag{17.63}$$

$$\delta = \tan^{-1} \frac{2n\varkappa}{n_0^2 - n^2 - \varkappa^2} \tag{17.64}$$

In view of the presence in (17.61) and (17.62) of the sine function, the values of R and P are periodic functions of the thickness of the layer. Thus, the maxima of P occur at

$$\frac{2\pi l}{\lambda} - \delta = \frac{2\pi n l \nu}{c} - \delta = s\pi, \quad s = 0, 1, 2 \tag{17.65}$$

while the minima occur at

$$\frac{2\pi l}{\lambda} - \delta = \frac{2\pi n l \nu}{c} - \delta = (2s+1)\frac{\pi}{2} \tag{17.66}$$

The maxima and minima are more difficult to find in the case of reflection since it is then necessary to allow for the periodic nature of both the numerator and denominator in (17.61). For a transparent layer ($\eta = 1$) $R_{min} = 0$ when (17.65) is satisfied.

According to (17.62), a plane-parallel plate of given optical thickness nl will transmit light easily only at certain wavelengths, i.e. those which satisfy (17.65). The remaining wavelengths are removed not simply by absorption within the layer, but also as a result of reflection at this surface.

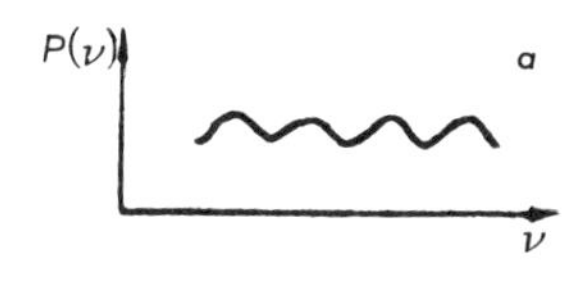

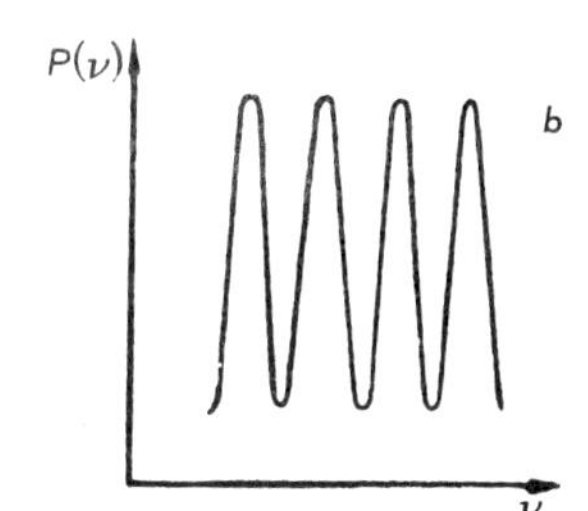

a – $4r\eta < (1 - r\eta)^2$; b – $4r\eta > (1 - r\eta)^2$

Fig. 4.8 Transmission by a plane-parallel layer

Figure 4.8 shows the form of $P(\nu)$ for $4\,r\eta < (1-r\eta)^2$ and $4\,r\eta > (1-r\eta)^2$. Case (a) corresponds to $r\eta$ approaching zero; case (b) corresponds to $r\eta$ approaching unity. The maxima correspond to a sequence of integral values of s. The separation between the peaks is

$$\nu_{s+1} - \nu_s = \frac{c}{2\,ln}$$

and is a function of only the optical thickness nl. The peak-to-valley ratio is

$$\frac{P_{max}}{P_{min}} = \left(\frac{1+r\eta}{1-r\eta}\right)^2$$

For small $r\eta$ the change in $P(\nu)$ in the interval $\nu_{s+1} - \nu_s$ is small, while for $4r\eta = 0$ (strongly absorbing thick layers, or weakly reflecting materials) there is no change at all. For such layers interference phenomena can be neglected in practice, and the frequency dependence of the transmission coefficient for the layer is then largely determined by the exponential factor $\eta = e^{-k(\nu)l}$ in the numerator of (17.62), i.e. the transmission is determined by the ordinary Bouguer law.

In the opposite case, i.e. when the reflection coefficient is large while the absorption coefficient is small (Fig. 4.8b), the dependence of the transmission on frequency is determined largely by interference phenomena. The spectrum of the transmitted radiation consists of individual sharp peaks each of which has a definite half-width, which we shall call the transmission band, near the frequency ν_s. Within the band the transmission is higher; elsewhere it is low. The half-width is given by

$$\Delta\nu = \frac{c}{\pi nl} \sin^{-1} \frac{1 - r\eta}{2\sqrt{r\eta}}. \qquad (17.67)$$

This expression is valid only when $1 - r\eta \leqslant 2\sqrt{r\eta}$. When $1 - r\eta = 2\sqrt{r\eta}$ the half-width is equal to $c/2nl$, i.e. to the distance between the peaks. The half-width decreases rapidly with decreasing $(1 - r\eta)$. For small values of the argument

$$\Delta\nu = \frac{c}{2\pi nl} \frac{1 - r\eta}{\sqrt{r\eta}} \qquad (17.68)$$

If, for example, $nl = 5$cm, $\eta = 1$ (transparent material) and $r = 0.95$, $\Delta\nu = 0.5 \times 10^8 \sec^{-1}$. This is comparable with the natural width of spectral lines. If the incident radiation has a broad spectral distribution, the radiation leaving the layer will be highly monochromatic; only individual, very narrow lines will be transmitted.

From this point of view, a plane-parallel layer is equivalent to a cavity resonator which transmits only narrow spectral ranges. The Q-factor (see Section 2) is then given by

$$Q \cong \frac{\nu}{\Delta\nu} = \nu \frac{\pi nl}{c} \frac{1}{\sin^{-1} \dfrac{1 - r\eta}{2\sqrt{r\eta}}} \approx \frac{2\pi nl\nu}{c} \frac{\sqrt{r\eta}}{1 - r\eta} \qquad (17.69)$$

The latter approximate equation is valid for $r\eta$ approaching unity. The Q-factor of a plane-parallel layer may be very high. When $\nu \sim 10^{15} \sec^{-1}$ and $\Delta\nu \sim 10^8 \sec^{-1}$, it is equal to 10^7. It is worth emphasising that the Q-factor is very dependent on frequency.

The properties of a plane-parallel layer are of particular importance for a further reason. When a light beam is allowed to fall on such a layer, the steady-state conditions are reached only after a certain interval of time. This is associated with the fact that there is a gradual accumulation of energy within the layer. Similarly, when the incident radiation is switched off, a certain amount of time is necessary before the radiation is damped out. This retention of radiation within the layer is, as a rule, not very marked. However, it can frequently affect the width of spectral lines transmitted through the layer and is particularly important in the case of generation of radiation.

This problem can often be solved on the geometrical-optics approximation. Suppose that at time $t = 0$ a beam of intensity S_0 is incident normally on the plane-parallel layer. The reflected intensity will be $S_0 r$. The second partial intensity $S_0(1-r)^2 r \eta^2$ will leave the layer at time $2nl/c$, where c/n is the velocity of propagation, the third beam whose intensity will be $S_0(1-r)^2 r^3 \eta^4$ will leave the layer at time $2\dfrac{2nl}{c}$, and generally the $(i+1)$-th beam whose intensity will be $S_0(1-r)^2 r \eta^2 (r^2\eta^2)^i$ will leave it at time $i\,2nl/c$. At time $t = \dfrac{i\,2nl}{c}$, the total reflected intensity will be

$$
\begin{aligned}
S_{\text{refl}} &= S_0 \{r + (1-r)^2 r \eta^2 [1 + r^2\eta^2 + \dots (r^2\eta^2)^{i-1}]\} \\
&= S_0 \left\{ r + \frac{(1-r)^2 r \eta^2}{1 - r^2\eta^2} [1 - (r^2\eta^2)^{\frac{ct}{2nl}}] \right\}
\end{aligned}
\tag{17.70}
$$

and therefore

$$
R(t) = r + \frac{(1-r)^2 r \eta^2}{1 - r^2\eta^2} [1 - e^{-t/\tau}] \tag{17.71}
$$

$$
\tau = -\frac{nl}{c} \frac{1}{\ln r\eta} \tag{17.72}
$$

Similarly, for the transmitted radiation we have

$$
P(t) = \frac{(1-r)^2 \eta}{1 - r^2\eta^2} [1 - r\eta e^{-t/\tau}] \tag{17.73}
$$

For transparent and absorbing media $r\eta$ is less than unity and therefore the exponential factor falls off rapidly with time. When $t = \infty$, (17.71) and (17.73) become identical with (17.54) and (17.55) respectively. It is natural to take the time τ defined by (17.72) as a measure of the rate at which steady-state conditions are established. When $r = 0$ and $\eta = 0$ $(k = \infty)$, the time τ is zero and the steady-state conditions are reached instantaneously. When $\eta = 1$ (transparent medium) and $r \to 1$ the value of τ tends to infinity, i.e. the steady-state conditions are never reached. If, for example, $nl = 5$ cm, and $r = 0.95$, then $\tau = 0.33 \times 10^{-8}$ sec. When $r = 0.99$, we have $\tau = 1.6 \times 10^{-8}$ sec. These times are comparable with the lifetimes of the excited states of atoms, or the emission decay time of a classical oscillator.

A definite store of radiant energy accumulates within the layer while it is exposed to the incident radiation. When the incident radiation is switched off, this energy is re-emitted and the original state of the layer is gradually re-established. The rate at which the relaxation process proceeds can readily be found with the aid of (17.57) and (17.58). Consider, for example, the flux through the second boundary. For times between 0 and nl/c the intensity of the beam is determined by (17.57) with $x=l$ multiplied by $1-r$. The next part of the beam to emerge is that given by (17.58). This is followed by the twice-reflected and twice-transmitted rays, and so on. The flux leaving the second boundary at time t is thus given by

$$S_2 = S_0 \frac{(1-r)^2 \eta}{1-r^2 \eta^2} (r^2 \eta^2)^{\frac{c}{2nl} t} = S_0 \frac{(1-r)^2 \eta}{1-r^2 \eta^2} e^{-t/\tau} \quad (17.73a)$$

where S_0 is the constant intensity incident on the layer for a long time (up to $t=0$).

The mean luminescence decay time can be determined as before from (17.72). The rate at which the steady-state conditions are reached and the rate of relaxation are equal. When $r\eta=1$, the afterglow process will continue indefinitely, while for $r=0$, $\eta=0$ it takes place instantaneously (no storage of energy).

Dichroism

We shall now consider the dependence of the absorption coefficient on the orientation of the particles and the polarisation of the absorbed light. In experimental studies of dichroism (Section 2) it is usual to measure the absorption coefficients for two plane-polarised rays with perpendicular electric vectors. In order to be specific suppose that the incident beams are parallel to the y axis and that the electric vector of one ray is parallel to the z axis while that of the other is parallel to the x axis. If we confine our attention to the absorption coefficient at a particular frequency ν_{ij}, there is no need to prescribe the orientation of the molecule in space. A more general analysis will be given in Chapter 8, but it is sufficient at present to define the matrix element of the dipole moment for a particular transition with the aid of the angles ϑ and φ.

It then follows from (8.72), (8.76) and (17.29) that

$$k_\nu^z(\vartheta, \varphi) = 3\frac{n}{c} B_{ji} \cos^2\vartheta \left[n_j(\vartheta, \varphi) - \frac{g_j}{g_i} n_i(\vartheta, \varphi) \right] \quad (17.74)$$

and

$$k_\nu^x(\vartheta, \varphi) = 3\frac{n}{c} B_{ji} \sin^2\vartheta \cos^2\varphi \left[n_j(\vartheta, \varphi) - \frac{g_j}{g_i} n_i(\vartheta, \varphi) \right] \quad (17.75)$$

These expressions describe absorption by particles of given orientation (per unit solid angle). In order to determine k_ν^z and k_ν^x for the entire ensemble of molecules we must integrate with respect to the two angles:

$$k_\nu^z = 3\frac{n}{c} B_{ji} \int_0^{2\pi} d\varphi \int_0^{\pi} \cos^2\vartheta \times \left[n_j(\vartheta, \varphi) - \frac{g_j}{g_i} n_i(\vartheta, \varphi) \right] \sin\vartheta \, d\vartheta \quad (17.76)$$

and

$$k_\nu^x = 3\frac{n}{c} B_{ji} \int_0^{2\pi} \cos^2\varphi \, d\varphi \int_0^{\pi} \sin^2\vartheta \times \left[n_j(\vartheta, \varphi) - \frac{g_j}{g_i} n_i(\vartheta, \varphi) \right] \sin\vartheta \, d\vartheta \quad (17.77)$$

The distribution function can frequently be regarded as independent of the intensity of the incident radiation. In practice the intensity is arranged to be so low that the angular anisotropy which it introduces into excited particle distribution has the minimum possible effect on the final results of measurements.

In the case of natural dichroism the particles will not interact with any external radiation other than Planck radiation. It follows that

$$\frac{1}{g_i} n_i(\vartheta, \varphi) = \frac{1}{g_j} n_j(\vartheta, \varphi) e^{-h\nu_{ij}/kT}$$

and the formulae given by (17.76) and (17.77) become

$$k_\nu^z = 3\frac{n}{c} B_{ji}\left[1 - e^{-h\nu_{ij}/kT}\right] \times \int_0^{2\pi} d\varphi \int_0^{\pi} n_j(\vartheta, \varphi)\cos^2\vartheta \sin\vartheta \, d\vartheta \tag{17.78}$$

and

$$k_\nu^x = 3\frac{n}{c} B_{ji}\left[1 - e^{-h\nu_{ij}/kT}\right] \times \int_0^{2\pi} \cos^2\varphi \, d\varphi \int_0^{\pi} n_j(\vartheta, \varphi)\sin^3\vartheta \, d\vartheta \tag{17.79}$$

Consider the following three special angular distributions.

1. Isotropic distribution: $n_j(\vartheta, \varphi) = \frac{1}{4\pi} n_j$. Taking n_j from under the integral sign we have

$$\frac{1}{4\pi}\int_0^{2\pi} d\varphi \int_0^{\pi} \cos^2\vartheta \sin\vartheta \, d\vartheta = \frac{1}{4\pi}\int_0^{2\pi} \cos^2\varphi \, d\varphi \int_0^{\pi} \sin^3\vartheta \, d\vartheta = \frac{1}{3}$$

Hence $k_\nu^z = k_\nu^x$ and there is no dichroism:

$$D = \frac{k_\nu^z - k_\nu^x}{k_\nu^z + k_\nu^x} = 0$$

2. Dipole moments of all particles parallel to the z axis: $n_j(\vartheta, \varphi) = n_j \delta(\Omega)$. In this case

$$k_\nu^z \sim \int_\Omega \cos^2\vartheta\delta(\Omega)\, d\Omega = 1, \quad k_\nu^x \sim \int \cos^2\vartheta \sin^2\vartheta \, \delta(\Omega)\, d\Omega = 0$$

which corresponds to maximum dichroism ($D = 1$).

3. $n_j(\vartheta, \varphi) = n_j \cos^2\vartheta$. Substituting into (17.78) and (17.79) and integrating with respect to the angles we have $D = 1/2$.

Induced dichroism is discussed in detail in Chapters 6 to 8.

18. THE EFFECT OF THERMAL BACKGROUND ON THE ABSORPTION PROCESS

Kirchhoff's law for finite bodies

Approximately 100 years ago, Kirchhoff formulated the two basic laws of thermal emission and absorption. The first of these is concerned with elementary volumes of matter and will be discussed later in Section 21. The second holds for objects of finite linear dimensions. Both laws have purely thermodynamic foundations and are among the most general laws of nature.

Consider an arbitrary body in thermodynamic equilibrium with the radiation surrounding it (Fig. 4.9). The flux of

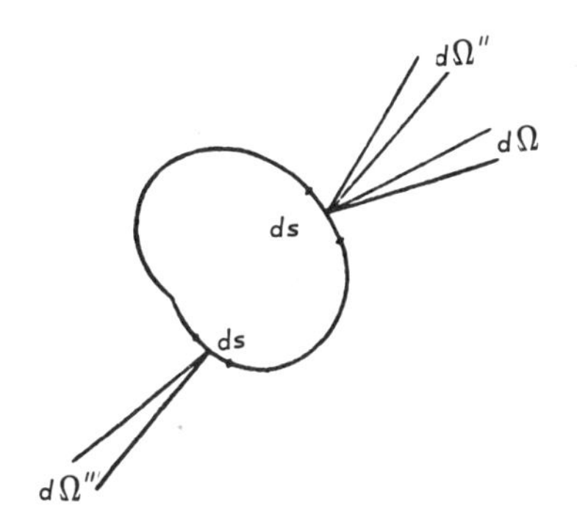

Fig. 4.9 Derivation of Kirchhoff's law

equilibrium radiation falling on an area ds from a solid angle $d\,\Omega$ in the frequency range $d\nu$ is given by

$$d\Phi_\nu = S_\nu\,(\Omega, s)\,d\nu \cos\vartheta\, ds d\,\Omega \tag{18.1}$$

where ϑ is the angle between the normal to ds and the direction of the beam. A part $A_\nu\,(r,\Omega)\,d\Phi_\nu$ of this flux enters the body, is multiply reflected from the walls, and is absorbed. A part $R_\nu\,(r,\Omega,-\Omega')d\Phi_\nu$ is reflected into the solid angle $d(-\Omega')$, while the flux

$$P_\nu\,(r,\Omega)\,d\Phi_\nu = \int\limits_{s'',\,\Omega''} P_\nu\,(r,\Omega, r'' - \Omega'')\,d\,(-\Omega'')\,d\Phi_\nu \tag{18.2}$$

enters the body and after multiple reflections leaves it at

different points on the surface and in different directions (the letter r represents the coordinates of the surface element ds at which the original radiation enters, while the symbol r'' represents the coordinates of the points at which the radiation leaves the body). Positive values of Ω refer to fluxes falling on the body, while negative values refer to the opposite directions. In view of the law of conservation of energy

$$d\Phi_\nu = A_\nu (r, \Omega)\, d\Phi_\nu + P_\nu (r, \Omega)\, d\Phi_\nu + R (r, \Omega)\, d\Phi_\nu \qquad (18.3)$$

or

$$A_\nu (r, \Omega) + P_\nu (r, \Omega) + R_\nu (r, \Omega) = 1 \qquad (18.4)$$

which holds for all r, Ω and ν.

In thermodynamic equilibrium, the flux emitted by ds is exactly equal to that given by (18.1) except that it travels in the opposite direction. It consists of three parts: the thermal emission

$$\varepsilon_\nu (r, -\Omega)\, ds \cos\vartheta\, d(-\Omega)\, d\nu$$

where $\varepsilon_\nu (r, -\Omega)$ is a coefficient of proportionality which is known as the emissive power of the surface element ds within the solid angle $d(-\Omega)$; the reflected flux

$$R'_\nu (r, \Omega', -\Omega)\, S'_\nu\, ds_n\, d\Omega'\, d\nu$$

where S'_ν is the intensity reaching the area ds within the solid angle $d\Omega'$ together with parts of the fluxes of equilibrium radiation $d\Phi''_\nu (r'', \Omega'')$ which penetrate the body from all directions and leave it after multiple reflections in the direction of $d(-\Omega)$ through the surface element ds; and finally, the third part

$$\int_{s''\,\Omega''} P''_\nu (r'', \Omega'', r, -\Omega)\, S''_\nu\, d\nu\, ds''_n\, d\Omega''\, ds_n\, d(-\Omega) \qquad (18.5)$$

Therefore

$$\begin{aligned} d\Phi_\nu (-\Omega) = {} & \varepsilon_\nu (r, -\Omega)\, ds_n\, d(-\Omega)\, d\nu + R'_\nu (r, \Omega', -\Omega) \\ & \times S'_\nu\, ds_n\, d\Omega'\, d\nu + \int_{s''\,\Omega''} P''_\nu (r'', \Omega'', r, -\Omega) \\ & \times S''_\nu\, d\nu\, ds''_n\, d\Omega''\, ds_n\, d(-\Omega) \end{aligned} \qquad (18.6)$$

In view of the principle of detailed balancing, the reflected flux $R_\nu(r, \Omega, -\Omega') S_\nu\, ds_n\, d\Omega\, d\nu$ is equal in magnitude and opposite in direction to the reflected flux $R'_\nu(r, \Omega', -\Omega) \times S'_\nu\, ds_n\, d\Omega'\, d\nu$ whereas the transmitted flux $P_\nu(r, \Omega, r'', -\Omega'') \times S_\nu\, ds_n\, d\Omega\, d\nu\, ds''_n\, d(-\Omega'')$ is equal in magnitude and opposite in direction to $P''_\nu(r'', \Omega'', r, -\Omega) S''_\nu\, ds''_n\, d\Omega''\, d\nu\, ds_n\, d(-\Omega)$.

Hence it follows ($d\Omega = d\Omega'$) that

$$\begin{aligned} R'_\nu(r, \Omega', -\Omega) S'_\nu &= R_\nu(r, \Omega, -\Omega') S_\nu, \\ P''_\nu(r'', \Omega'', r, -\Omega) S''_\nu &= P_\nu(r, \Omega, r'', -\Omega'') S_\nu \end{aligned} \tag{18.7}$$

Since in thermodynamic equilibrium the intensity of Planck radiation is the same in all directions ($S''_\nu = S'_\nu = S_\nu$) it follows that

$$R'_\nu(r, \Omega', -\Omega) = R_\nu(r, \Omega, -\Omega') \tag{18.8}$$

and

$$P''_\nu(r'', \Omega'', r, -\Omega) = P_\nu(r, \Omega, r'' -\Omega'') \tag{18.9}$$

These equations form the so-called principle of reciprocity which is widely used in theoretical optics.

If we equate (18.3) and (18.6) and make use of (18.7), (18.8) and (18.9), we can readily show that

$$\frac{\varepsilon_\nu(r, -\Omega)}{A_\nu(r, \Omega)} = S_\nu^{eq}(\Omega) = \frac{c}{n}\frac{u_\nu^{eq}}{4\pi} \tag{18.10}$$

where we have used (17.9) and the fact that for isotropic radiation

$$u_\nu(\Omega) = \frac{u_\nu^{eq}}{4\pi} \tag{18.11}$$

The formula (18.10) expresses the Kirchhoff law which states that the ratio of the emissive and absorptive powers of a given element is equal to the intensity of the equilibrium radiation. This holds true for any direction of propagation. Kirchhoff's law is universal. It must be emphasised once again that the quantities A_ν and ε_ν are individually different for different points on the surface of the body, different directions of propagation and different frequencies.

Kirchhoff's law leads to a number of important consequencies. Transparent bodies ($A_\nu = 0$) do not emit thermoradiation ($\varepsilon_\nu = 0$) while the emissive power of a black body ($A_\nu = 1$) is given by

$$\varepsilon_\nu^{\text{b.b.}} = \frac{c}{n}\frac{u_\nu^{\text{eq}}}{4\pi} = \frac{2h\nu^3}{c^2/n^2}\frac{1}{e^{h\nu/kT} - 1} \tag{18.12}$$

The expression (18.10) is strictly valid only if the body under consideration is in equilibrium with the surrounding medium, i.e. when the temperatures of the body and of the medium are equal, and also if the body is in equilibrium with the surrounding radiation. However, (18.10) becomes heuristically significant when there is in fact a departure from equilibrium. Let us suppose that the temperature of the medium is reduced. If this does not produce appreciable changes in the state of the body (either because the time necessary for equilibrium to be reached is very large, or because there are sources of energy within the body which maintain the temperature of the body), the emissive and absorptive powers remain constant. By measuring the energy emitted by the body into the surrounding medium we can find its emissive power and then calculate the absorptive power. This method is widely used in determinations of the absorptive power of thermal sources of radiation.

Thermodynamic equilibrium can also be upset by directing a flux of radiation in excess of the equilibrium flux on to a body in thermal equilibrium with the surrounding medium. It can frequently be assumed, however, that the departure from equilibrium is small and the external flux does not appreciably change equilibrium energy level distribution, so that the absorptive power of the body is only slightly modified (the emissive power may undergo a much greater change leading to the appearance of luminescence).

By measuring the absorptive power it is possible to determine the thermal emissive power from Kirchhoff's law. Experimental determinations of absorption coefficients are usually based on this idea. It must, however, be noted that such calculations are only approximate, and when there is a large departure from equilibrium due to changes in the properties of the body, or due to illumination by high-intensity radiation, the emission of radiation is no longer described by Kirchhoff's law.

Kirchhoff's law is quite general and can be applied to all

bodies, including detectors of radiation, i.e. objects capable of recording incident radiation. They not only receive the incident flux of radiation but are themselves sources of radiation. This must be taken into account in measurements, since otherwise incorrect results may be obtained.

When the detector is in thermodynamic equilibrium with the surrounding medium, the energy flux absorbed by it is equal to the emitted flux. This means that the detector in principle will not register a flux of radiation corresponding to its own temperature. It will register only after its thermodynamic equilibrium with the medium has been upset, and its indications will as a rule be proportional to the difference between the absorbed and emitted fluxes:

$$Y = a(S_{\mathrm{d}}^{\mathrm{abs}} - S_{\mathrm{d}}^{\mathrm{em}}) \tag{18.13}$$

where a is a constant for each particular receiver.

The effect of thermal radiation background on the readings of radiation detectors

So far we have discussed absorption without specifying the experimental conditions and without allowing for the thermal radiation background. A commonly used arrangement for the investigation of absorption spectra (Fig. 4.10) consists of a source of radiation, a plane-parallel layer of the material

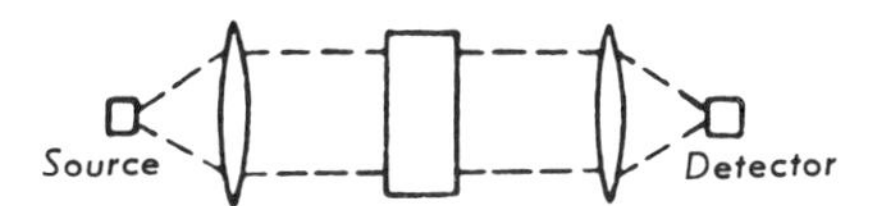

Fig. 4.10 Basic arrangement for studying absorption spectra

under investigation, and a detector. Each element of the system may be at its own temperature. A non-equilibrium source of radiation is usually employed, and the specimen itself may not be in a state of equilibrium either. The system as a whole is located in the medium of temperature T_{m}. The readings of the detector depend on all these factors, and the experimental data must be corrected for the effects of the source, the detector and the medium, before the properties of the absorbing material itself can be found.

The second possible method of measuring the absorptive power of a body is based on Kirchhoff's law. In this method it is sufficient to measure the emissive power and calculate A_ν from (18.10). The experimental arrangement is simpler (in Fig. 4.10 the plane-parallel layer is removed and the object under investigation is used as the source of radiation). However, in this case also it is important to take into account the properties of the detector and the temperature of the medium.

Consider the last case first. The flux of energy leaving the source in the direction of the detector is

$$\alpha[\varepsilon_s + R_s S_0(T_m) + P_s S_0(T_m)] \qquad (18.14)$$

where α is a constant determined by the geometry of the system, ε_s is the flux emitted by the source (for the sake of brevity, the dependence on frequency is not indicated by a subscript), R_s and P_s are the reflection and transmission factors of the source, $S_0(T_m)$ is the flux of equilibrium radiation of density u_0 falling on the source from the medium, and $\alpha R_s S_0(T_m)$ and $\alpha P_s S_0(T_m)$ are the fluxes of equilibrium radiation from the medium which are reflected and transmitted by the source of radiation in the direction of the detector respectively.

The flux of radiation absorbed by the detector is

$$\begin{aligned} S_d^{abs} &= A_d \alpha[\varepsilon_s + R_s S_0(T_m) + P_s S_0(T_m)] \\ &= A_d \alpha[\varepsilon_s + (1 - A_s) S_0(T_m)] \end{aligned} \qquad (18.15)$$

where A_d and A_s are the absorptive powers of the detector and source respectively, which are related to the reflection and transmission factors by (18.4).

In general, the receiver and the source cannot be regarded as black-body sources; for black bodies $A = 1$ and all the formulae become very much simpler.

The flux of radiation emitted by the detector in the direction of the source is determined by Kirchhoff's law:

$$\alpha' S_d^{em} = \alpha' A_d S_0(T_d) \qquad (18.16)$$

where α' is a constant which depends on the geometry of the apparatus. Substituting (18.15) and (18.16) into (18.13) we obtain the reading of the detector:

$$Y = A_d\, a\alpha\, [\varepsilon_s + (1 - A_s)\, S_0(T_m)] - A_d\, a\alpha'\, S_0(T_d) \quad (18.17)$$

This holds for all states of the source and any medium and detector temperatures. When $T_s = T_m$ and at the same time $T_d = T_m$, i.e. all the elements of the system are in thermodynamic equilibrium, we have $Y = 0$. Since for thermal emission $\varepsilon_s\,(T_s) = A_s\; S_0(T_s)$ it follows that by setting the right-hand side of (18.17) equal to zero, we have

$$\alpha = \alpha' \quad (18.18)$$

When this result is introduced into (18.17) we obtain

$$\begin{aligned} Y = {} & A_d\, a\alpha\, [\varepsilon_s - A_s\; S_0(T_m)] \\ & + A_d\, a\alpha\, [S_0(T_m) - S_0(T_d)] \end{aligned} \quad (18.19)$$

The product $a\alpha$ depends on the sensitivity of the detector, the dimensions of the source, the working area of the detector, the mutual disposition of the elements of the system and certain other factors including the slit width. The recorded flux of radiation

$$\frac{Y}{A_d\, a} = \alpha\, [\varepsilon_s - A_s\; S_0(T_m)] + \alpha\, [S_0(T_m) - S_0(T_d)] \quad (18.20)$$

is independent of the properties of the detector (other than its dimensions) and is equal to the difference of two fluxes, namely, the flux given by (18.14) which reaches the detector after the thermodynamic equilibrium has been upset, and the flux $\alpha S_0(T_d)$ which reaches it when $T_s = T_d = T_m$ (i.e. the flux emitted by the detector in the direction of the source). The quantity given by (18.20) consists of two independent components, namely the flux $\alpha\,[\varepsilon_s - A_s S_0(T_m)]$ which depends on the state of the source, and the flux $\alpha\,[S_0(T_m) - S_0(T_d)]$ which is due to the temperature difference between the medium and the detector.

Let us now list the consequences of (18.19) for the most important special cases:

1. In thermodynamic equilibrium the measured flux, and therefore the readings of the detector, are zero.

2. The temperature of the detector differs from the temperature of the medium and there are no sources of radiation ($T_s = T_m$ and therefore $\varepsilon_s = A_s(T_m)$):

$$Y = A_d\, a\alpha [S_0 (T_m) - S_0 (T_d)] \tag{18.21}$$

This shows that when $T_m > T_d$, the detector readings are positive, and vice-versa. Phenomena which are associated with negative detector readings will be discussed in the next section. A reduction in the temperature of the detector is equivalent to an increase in the temperature of the medium and vice versa. The parameters a and A_d may also be functions of temperature. The spectral dependence of the product aA_m may be investigated directly by surrounding the detector with an equilibrium radiation background and using (18.21).

3. If the temperatures of the detector and of the medium are equal ($T_d = T_m$) and there is a source of radiation in the space surrounding the detector, we have

$$Y = A_d\, a\alpha [\varepsilon_s - A_s\, S_0 (T_m)] \tag{18.22}$$

This is the most frequently encountered case. The flux of radiation from the source

$$\frac{Y}{A_d\ a} = \alpha [\varepsilon_s - A_s\, S_0 (T_m)]$$

which is directly recorded by the detector is equal to the difference between the fluxes emitted and absorbed by the source, and is in general independent of the properties of the detector. Under normal conditions $\varepsilon_s > A_s S_0(T_m)$ and consequently $Y > 0$.

The formula (18.22) is valid for any source of radiation (both equilibrium and non-equilibrium). For thermal sources Kirchhoff's law yields $\varepsilon_s = A_s S_0(T_s)$ and (18.22) assumes the simpler form

$$Y = A_d\, a\alpha A_s [S_0 (T_s) - S_0 (T_d)] \tag{18.23}$$

It follows from (18.22) and (18.23) that when A_s and ε_s are calculated from experimental data, it is important to take into account both the temperature of the source and the temperature of the detector (medium), especially when they do not differ greatly from each other.

The detector reading in this scheme is in general given by (18.19). This relation may be modified because in practice one is usually interested not in the absolute detector reading but only in the change in this reading when the light source is switched on. In the absence of the source, or, which amounts to the same thing, when $T_s = T_m$, the detector reading is

$$Y_0 = A_d\, a\alpha\, [S_0(T_m) - S_0(T_d)] \tag{18.24}$$

The change in this due to the introduction of the source is

$$Y' = Y - Y_0 = A_d\, a\alpha\, [\varepsilon_s - A_s\, S_0(T_m)] \tag{18.25}$$

which shows that Y' is independent of the detector's temperature. A departure from equilibrium between the material (source) under investigation and the surrounding medium is a necessary condition for the determination of the emissive (absorptive) power. From the above point of view, a reduction in the temperature of the detector has no special advantages for studies of the optical properties of matter.

The formula (18.23) may be used as a basis for measurements of absorptive power. This method requires (1) a direct measurement of Y at all frequencies; (2) a determination (or elimination) of $A_d\, a\alpha$ which represents the properties of the system; and (3) a calculation of $[S_0(T_s) - S_0(T_d)]$ from Planck's formula. The quantity $A_d\, a\alpha$ may be eliminated through additional measurements in which the object under investigation is replaced by some other object of known emissive power, e.g. a black body. In the latter case

$$Y^{\text{b.b.}} = A_d\, a\alpha\, [S_0(T_s) - S_0(T_d)] \tag{18.26}$$

On dividing (18.23) by (18.26) we obtain

$$\frac{Y}{Y^{\text{b.b.}}} = A_s \tag{18.27}$$

Knowing A_s we can calculate ε_s from Kirchhoff's law.

It is evident from the basic formula (18.19) that the emissive (absorptive) power of a body can be measured only when the equilibrium is disturbed ($T_s \neq T_m$). When $T_s = T_m$ the detector records the quantity Y_0 given by (18.24) which is independent of the properties of the source. A measurement of the detector temperature cannot in principle yield any information since it leads only to a change in Y_0. To measure A_s one must change either the temperature of the body itself or of the medium. To measure A_s at room temperature, one must use the second method.

As an illustration, Fig. 4.11 shows the spectra recorded with an infrared spectrophotometer. They were used to determine the emissive and absorptive powers of a thin film of oil when $T = 230°C$. Curve b shows the results for

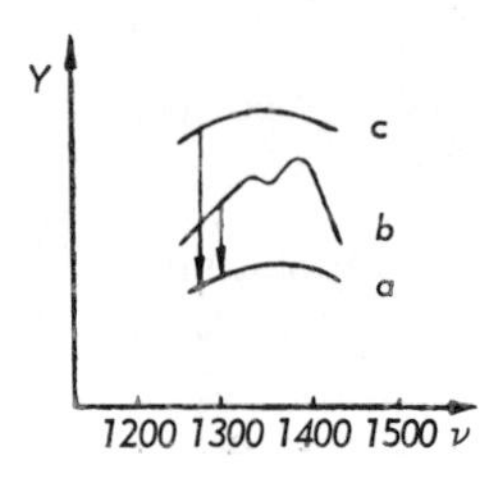

Fig. 4.11 Thermal emission by diffusion pump oil

the layer under investigation, while curve c represents the emissive power of a thick opaque layer of the same oil. If the amount of radiation reflected by the layer is assumed to be small, the layer may be regarded as approximating to a black body. Curve a shows the thermal emission of the heated container which acts as the second independent source of radiation. To determine Y we must subtract Y_a from Y_b and Y_c. From (18.27) we then have

$$\frac{Y_b - Y_a}{Y_c - Y_a} = A_s$$

Absorption spectra may be determined from the thermal emission spectra by using very thin layers of the material under investigation, because as the thickness increases, the properties of the layer approach those of a black body and the band structure is lost.

In equilibrium ($T_s = T_l = T_d = T_m$), the detector intercepts the flux $\alpha S_0(T_d)$. When the equilibrium is disturbed the incident flux is equal to

$$\alpha[\varepsilon_s + R_s S_0(T_m) + P_s S_0(T_m)] P_l + \alpha[\varepsilon_l + R_l S_0(T_m)] \tag{18.28}$$

where ε_l is the emissive power of the layer and P_l and R_l are its transmission and reflection factors. The first term represents the flux (18.14) from the source of radiation attenuated in the layer under investigation and then incident on the receiver. The second term in (18.28) gives the independent flux which leaves the layer in the direction of the detector. The detector records the difference between (18.28) and $\alpha S_0(T_d)$, i.e.

$$Y^* = A_d a \Delta S = A_d a \alpha [\varepsilon_s - A_s S_0(T_m)](1 - A_l - R_l)$$

$$+ A_d a \alpha [\varepsilon_l - A_l S_0(T_m)] + A_d a \alpha [S_0(T_m) - S_0(T_d)] \quad (18.29)$$

where we have used the relation $A_l + P_l + R_l = 1$. The detector readings may be positive or negative depending on the temperature of the detector and the medium and also on the state of the source and the layer under investigation.

The most important special case of (18.29) corresponds to measurements of the absorption coefficient at room temperature when $T_l = T_d = T_m$. This yields

$$Y^* = A_d a \alpha [\varepsilon_s - A_s S_0(T_m)] P_l \quad (18.30)$$

where the asterisk represents the readings of the detector in the system illustrated in Fig. 4.10. P_l can be determined by measuring Y^*. The simplest method is to perform additional measurements of Y in the absence of the absorbing layer. The corresponding values of Y are then given by (18.22). On dividing (18.30) by (18.22) we obtain

$$\frac{Y^*}{Y} = P_l \quad (18.31)$$

In contrast to (18.30), the ratio Y^*/Y is independent of the state of the source and of the temperature of the medium.

We have already seen in Section 17 that the transmission factor of a plane-parallel layer is unambiguously related to the absorption coefficient of the material of the layer and the reflection coefficient at the boundary with the surrounding medium. k and r cannot be determined simply from a knowledge of P_l. This requires a second measurement, for example a measurement of P_l, for a different thickness of the layer, or of R_l.

At temperatures well in excess of the room temperature,

the simple formula of (18.30) cannot be used because the thermal emission of the layer must then be allowed for. If we substitute $T_d = T_m$ into (18.29) and assume that $T_1 \neq T_m$ we have

$$Y^* = A_d a \alpha \{[\varepsilon_s - A_s S_0(T_m)] P_1 + \varepsilon_1 - A_1 S_0(T_m)\} \quad (18.32)$$

In order to eliminate the thermal emission by the layer we must carry out additional measurements of Y^* with and without the source (respectively $Y' = A_d a \alpha [\varepsilon_1 - A_1 S_0(T_m)]$ and $Y = A_d a \alpha [\varepsilon_s - A_s S_0(T_m)]$. These three measurements yield

$$P_1 = \frac{Y^* - Y'}{Y} \quad (18.33)$$

A number of examples of this procedure will be given in the next section. More detailed information will be found in original papers [37].

Negative fluxes of radiant energy

The flux of radiation emitted by a system in thermodynamic equilibrium is equal to the flux of the ambient thermal radiation which is absorbed by the system. This ensures the continuation of the state of equilibrium. For any pair of levels of the system $(E_i > E_j)$, the number of downward transitions $(A_{ij} + B_{ij} u_0) n_i$ is equal to the number of upward transitions $B_{ji} u_0 n_j$. It is only in this sense that one can say that the system is not subject to the effect of external radiation. External (exciting) radiation is usually defined as necessarily giving rise to a departure from thermodynamic equilibrium. The external flux S acting on the system is equal to the difference between the total flux of radiation $S_0 + S$ penetrating the system and the flux of thermal radiation S_0 intercepted by the system before the source was switched on in complete thermodynamic equilibrium with the medium (the representation of the total flux as the sum of external and equilibrium radiation is not possible in non-linear optics). Similarly, the density of radiation u incident on the particles is equal to the difference between the total density $u_0 + u$ and the equilibrium density u_0.

In the great majority of cases encountered in practice, the

temperature of the source is higher than the temperature of the object under investigation. The total flux of radiant energy incident on the object is therefore greater than the flux of equilibrium radiation, so that the exciting flux is positive and the number of upward transitions is greater than the number of downward transitions. The absorption of external radiation for $u > 0$ is usually represented by an arrow pointing upwards (Fig. 4.12). Absorption of radiation

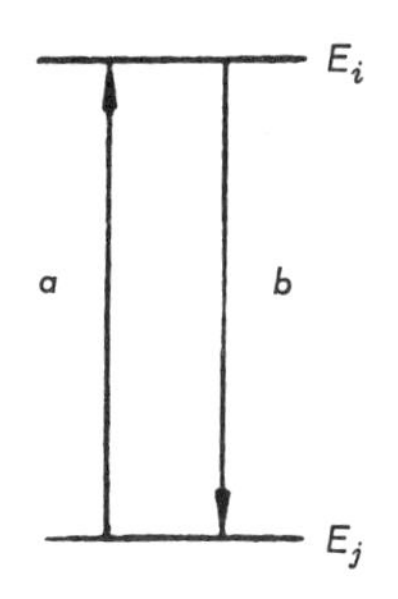

Fig. 4.12 Absorption of positive (a) and negative (b) radiation fluxes

gives rise to a redistribution of the particles over the energy levels. The number of particles occupying level n_i increases ($n_i > n_i^{eq}$) whereas the number of particles occupying level j is reduced ($n_j < n_j^{eq}$). The mean energy of the system will also increase and the system as a whole will leave the state of equilibrium, i.e. will become excited.

The same system of particles may, however, undergo an opposite departure from the state of thermodynamic equilibrium. Suppose that a body whose temperature is lower than the temperatures of system and medium is introduced into the space surrounding the system. This will reduce the flux of radiation incident on the system and the flux S_0+S will therefore become smaller than the equilibrium value S_0. As a result, the equilibrium distribution over the energy levels will be disturbed, and the number of $j \rightarrow i$ transitions will become smaller than the number of $i \rightarrow j$ transitions, so that there will be a reduction in the number of particles occupying level i and an increase in the number of particles occupying level j. The mean energy of the system will decrease. In this case the system may also leave the state of equilibrium, but in the opposite direction. The reason for this negative excitation is the illumination of the system by a negative flux of radiation (negative S and u). In contrast to the more usual case, absorption of a negative flux of radiation must be represented by an upward arrow (Fig. 4.12).

The expressions for the rate of absorption and the absorption coefficient introduced at the beginning of the preceding section are valid for both positive and negative fluxes. If the density of the incident radiation is negative, then, according to (17.20), the rate of absorption is also negative when $n_j > n_i$, and the system loses some of its energy. In contrast to the rate of absorption, the absorption coefficient given by (17.31) is independent of the sign of the radiation incident on the material.

In accordance with Bouguer's law (17.7), negative fluxes are absorbed exponentially. As l increases, there is a reduction in the absolute magnitude of S but its sign remains the same. For large l, the magnitude of $|S|$ tends to zero and the radiation approaches equilibrium.

There is an important, but not fundamental, difference between positive and negative exciting fluxes of radiation. At a given temperature of the system, the magnitude of a positive flux is not limited, and may lie between zero and infinity. Negative fluxes on the other hand can assume values only within certain limits. The minimum absolute magnitude of a negative flux is zero, while the maximum is equal to the intensity of equilibrium radiation at the particular temperature of the medium. This maximum value can only be obtained in the limit by completely surrounding the illuminated system by a body (light source) at a temperature of $T=0$. This is not an absolute difference between negative and positive fluxes. If the temperature of the object under investigation is continuously increased at a constant source temperature, the absolute magnitude of the negative flux will become very large. Stars, for example, are continuously illuminated by exceedingly strong negative fluxes from outer space.

The magnitude of negative fluxes of radiation which can be obtained experimentally is very dependent on the temperature of the object under investigation and the working spectral region. In the visible and ultraviolet parts of the spectrum and at room temperature, the negative fluxes are exceedingly low. They may be large in this region only at ultra-high temperatures. On the other hand, in the infrared

u_0 is quite high and negative fluxes are easily detected by normal radiation detectors.

A rather special situation arises at radio frequencies. There the density of equilibrium radiation, u_0, is low, but the energy levels are very closely spaced ($h\nu/kT \ll 1$) and therefore the populations of upper and lower levels are practically the same. Excitation by positive fluxes of any intensity is ineffective and can only lead to a complete equalisation of level populations. However, illumination by negative fluxes can, in principle, lead to a reduction in the population of the upper level.

It might appear at first sight that the introduction of the concept of negative fluxes is due to an arbitrariness in the choice of the direction of propagation. This is not so. Negative fluxes will always arise when the density of radiation incident on a body is lower than the density of equilibrium radiation u_0. We shall see later that most of the formulae of theoretical spectroscopy are valid for both negative and positive fluxes. The concept of negative fluxes extends the limits of their applicability and enables us to describe a large number of new experimental facts.

The direction of the flux is also an important question. The mathematical solution of the problem may be obtained for any set of coordinate axes, but not all such sets correspond to the physical processes which occur in the object under consideration. In the above case of a system of particles and an external source of radiation, we are concerned with the optical properties of the system and not of the source, which may be chosen in a number of ways. Our system is an indicator of the radiation incident upon it, and can receive and variously react to both negative and positive fluxes.

It has already been mentioned that any body whose temperature is lower than the temperature of the system under investigation will be a source of negative flux. The properties of such sources can readily be investigated with the aid of a simple scheme consisting of a source, a detector and various focusing attachments. According to (18.19), when $T_d = T_m$ and $\varepsilon_s < A_s \times S_0(T_d)$, the flux of radiation recorded by the detector is less than zero, and the detector readings are negative. The maximum negative flux which is absorbed by the source for $\varepsilon_s = 0$ is equal to $\alpha A_s \times S_0(T_d)$.

Figure 4.13 shows the detector output as a function of wavelength when it is exposed to a cold source (a piece of

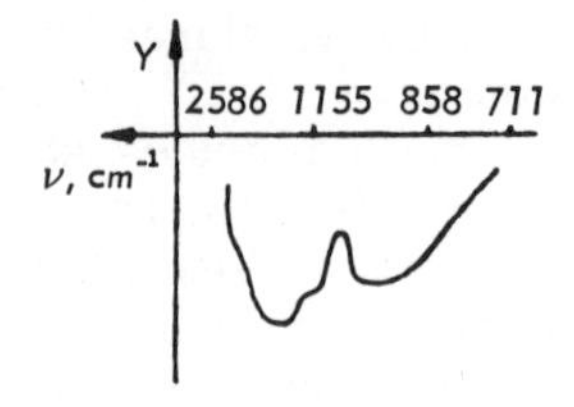

Fig. 4.13 Negative emissive power of aluminium (see text)

aluminium at $T = -140\,°C$). The detector was at room temperature. The spectrum was obtained with an infrared spectrometer between the source and the detector. When the metal was heated, the detector gave positive readings. Each metal has its own specific 'negative emission spectrum'. The absorptive and emissive powers can readily be calculated from (18.23) by dividing the detector reading by $A_d a\alpha[S_0(T_s) - S_0(T_m)]$. The procedure for the elimination of $A^p a\alpha$ is the same as in the case of positive fluxes. All the formulae in the preceding section are valid for cold sources of radiation.

Figure 4.14 shows the detector readings for a non-equilibrium source of radiation. At some frequencies the

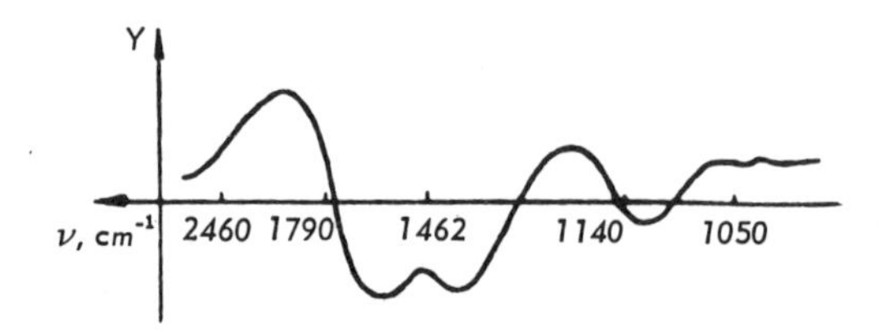

Fig. 4.14 Dependence of positive and negative detector readings on frequency for a non-equilibrium source

detector recorded positive fluxes, while at others it indicated negative fluxes. In the former case the density of radiation incident on the detector was greater than the density of equilibrium radiation $u_0(T_d)$, while in the second case it was smaller. The non-equilibrium source consisted of a combination of two bodies at different temperatures (electrical furnace and a container with cooled toluene).

Absorption of negative fluxes

The properties of negative fluxes do not differ fundamentally from those of positive fluxes. Negative fluxes can be absorbed, reflected or scattered in a medium, they can give rise to luminescence, undergo double refraction in anisotropic media, have definite polarisation and so on. Let us consider the absorption of negative fluxes in an experiment of the kind illustrated in Fig. 4.10. All the necessary formulae have already been given. In particular, it follows from (18.30) that the transmission factor of a layer can equally well be determined with positive or negative fluxes. The accuracy of the measurements will depend on the intensity of the beam but not on its sign.

Figure 4.15 shows the absorption spectra of cyclohexanone. Both the specimen and the detector were at room temperature. The horizontal line corresponds to the absence of the source. The two upper curves with positive values of Y correspond to a heated source, the two lower ones to a cooled source. The source was a blackened cavity in a brass cylinder maintained at $T = 80\,^{\circ}\mathrm{C}$ in the first case, and at $T = -180\,^{\circ}\mathrm{C}$ in the second. Curves a and d show the detector readings in the absence of the cyclohexanone layer,

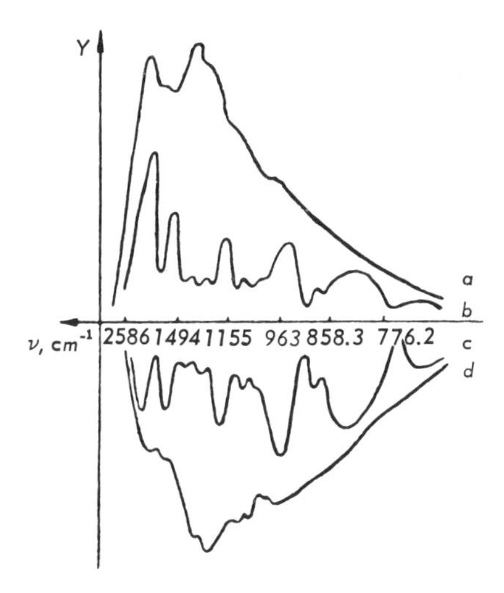

Fig. 4.15 Absorption spectra of cyclohexanone for heated and cooled sources

and characterise the emissive power of the source. Curves b and c show the detector readings in the presence of the cyclohexanone layer. The spectra are symmetrical about the zero line, but the symmetry is approximate and occurs only so long as there is a symmetry in the emission spectra of heated and cooled sources, i.e. curves a and d. The spectrum $P_1(\nu)$ calculated from these experimental data using (18.31) were found to be the same for both the heated and the cooled source. Bouguer's law was found to be strictly valid for negative fluxes.

Figure 4.16 shows the absorption spectra of cyclohexanone when the temperature of the layer was different from the temperature of the medium and of the detector ($T_s = 240\,°C$, $T_1 = -140\,°C$). The correct interpretation of these curves can be obtained with the aid of (18.32), except that it is necessary to allow for the thermal emission of the windows of the container, which are rarely perfectly transparent. Curve b in Fig. 4.16 shows the spectrum of the source with the container empty. Curves d, c and e respectively show the spectra in the presence of the cyclohexanone layer, the thermal emission spectrum of the container filled with cyclohexanone and the thermal emission spectrum of the empty container. In the latter two cases the source was

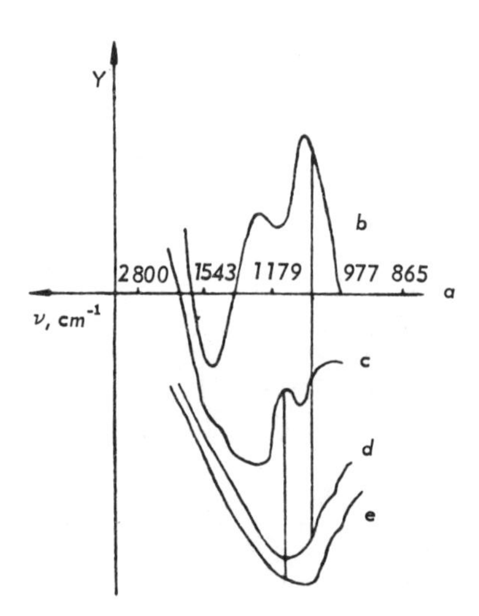

Fig. 4.16 Detector readings for a hot source and a cooled layer of cyclohexanone

removed. The difference $Y_b - Y_c$ enables us to eliminate the thermal emission of the empty container, and is equal to $[\varepsilon_s - A_s S_0(T_m)] P_{cont}$. The difference $Y_d - Y_e$ is equal to $[\varepsilon_s \; A_s . S_0(T_m)] P_1 P_{cont}$ and does not contain the thermal emission of the layer and the container. These differences are represented in Fig. 4.16 by the vertical lines. Their ratio is equal to the transmission factor of the layer which is under investigation for $T_1 \neq T_m$:

$$\frac{Y_d - Y_e}{Y_b - Y_c} = P_1$$

Four measurements are necessary (at each frequency) before P_1 can be determined. This is the only way in which the effects of the source, the detector, the windows of the container and the thermal emission of the layer itself can be eliminated.

In Fig. 4.16 curve b cuts the abscissa axis. At those frequencies at which the windows absorbed an appreciable proportion of the external flux, the detector recorded a negative flux due to the thermal emission by the windows of the container. At other frequencies the positive flux from the source predominated over the negative flux from the container. The readings of the detector were always negative when the cyclohexanone was introduced into the container.

Negative fluxes have recently found interesting applications in connection with the negative optical-acoustic effect. They have been used in measurements of the emissive power of a number of gases at room temperature, and in the development of effective methods for quantitative spectral gas analysis.

19. DEPENDENCE OF THE ABSORPTION COEFFICIENT AND RATE OF ABSORPTION ON THE DISTRIBUTION OF PARTICLES OVER THE ENERGY LEVELS

Equilibrium distribution

The absorption coefficient and the rate of absorption of an elementary volume integrated over the profile of a line are given by (17.31) and (17.20). It follows from these expressions that the rate of absorption within a line ν_{ij} is very dependent on the level populations n_j and n_i. The latter in their turn depend on the number of molecules per unit volume, the temperature of the medium, the nature of the external perturbation (in particular the intensity of the incident radiation) and the probabilities of radiative and non-radiative transitions between levels i and j and between all other levels.

When the system of particles and the surrounding medium are in thermodynamic equilibrium the formulae given by (17.20) and (17.31) assume the simpler forms

$$k_{ji} = \frac{B_{ji} h \nu_{ij}}{c/n} n_j (1 - e^{-h \nu_{ij}/kT}) \tag{19.1}$$

$$W_{ji} = B_{ji} u_{ij} h \nu_{ij} n_j (1 - e^{-h \nu_{ij}/kT}) \tag{19.2}$$

or

$$k_{ji} = \frac{B_{ji} h \nu_{ij}}{c/n} C(T) g_j n \left[e^{-E_j/kT} - e^{-E_i/kT}\right] \tag{19.3}$$

$$W_{ji} = B_{ji} u_{ij} h \nu_{ij} C(T) g_j n \left[e^{-E_j/kT} - e^{-E_i/kT}\right] \tag{19.4}$$

when $C(T)$ is the normalisation factor (Section 6). These formulae are valid if the external flux density u_{ij} used to measure k_{ji} or W_{ji} is small enough and cannot modify the equilibrium distribution over the energy levels. They are also valid for large u_{ij}, but only for systems for which the probabilities of non-radiative transitions are large (which ensures the return to equilibrium). These cases are frequently realised in practice and most experimental data refer to the equilibrium values of k_{ij} and W_{ji}. At very high temperatures, departures from thermodynamic equilibrium are in general difficult to achieve.

When (19.3) and (19.4) are valid, the absorption coefficient depends only on the properties of the molecules under investigation, the refractive index n, the temperature and the total number of particles per unit volume n. In the present case we are interested only in the temperature dependence which determines the distribution of the particles over the energy levels.

The maximum values of the absorption coefficient and the rate of absorption for transitions from the lowest level

$j=1$ ($E_1=0$) to some other level i occur at the absolute zero of temperature ($T=0$)

$$k_{1i}(T=0)=\frac{B_{1i}h\nu_{i1}}{c/\text{n}}g_1 n \tag{19.5}$$

$$W_{1i}(T=0)=B_{1i}u_{i1}h\nu_{i1}g_1 n \tag{19.6}$$

These formulae refer to transitions from a sub-level of a degenerate level. The statistical weight g_1 disappears after summation over all levels (Section 7).

At $T=0$ there is no absorption as a result of transitions from excited levels, and $k_{ji}(E_j\neq 0)$ and W_{ij} are both zero. This was to be expected since $n_j=0$ for $T=0$.

As the temperature increases, the absorption coefficient for $1\rightarrow i$ transitions begins to decrease as a result of the reduction in n_1 (first term in (19.3)) and the appearance of and increase in stimulated emission as the population of level i increases (second term in (19.3)). The dependence k_{1i} on temperature is monotonic. At very high temperatures the level populations become equal, so that the absorption coefficient tends to zero. At infinite temperatures, stimulated emission completely compensates for absorption, and the material becomes transparent.

Absorption of external radiation at frequencies corresponding to $j\rightarrow i$ transitions from excited levels j has a different temperature dependence. When $T=0$ the absorption coefficient is zero. As the temperature increases, there is an increase in the population of the initial level, n_j, and a gradual increase in absorption. At high enough temperatures, the absorption coefficient begins to decrease and tends to zero as $T\rightarrow\infty$. This is associated with the more rapid increase in stimulated emission at high temperatures.

The variation of the absorption coefficient with temperature is very dependent on the frequency of the absorbed radiation. In (19.3) and (19.4) the temperature enters as part of the ratio $h\nu/kT$. This means that if the frequency is increased by a factor of 2, the temperature must be increased by an equal factor in order to achieve the same effect. In the visible region $e^{-h\nu/kT}$ is very small, even at high temperatures. It follows that the absorption coefficient for the lower level is determined by (19.5), while the absorption coefficient for transitions from excited levels is zero. Temperature effects become appreciable at temperatures $T \sim h\nu/k$, i.e. $T \sim$ 30,000°K when $\nu \sim$ 20,000 cm^{-1} (λ = 5,000 Å). In the infrared region, an appreciable temperature dependence sets in earlier. Thus $T \sim$ 450°C when $\nu \sim$ 500 cm^{-1} ($\lambda = 20\mu$).

In the infrared region one frequently observes the overlap of spectral lines due to transitions from different excited levels and therefore the temperature dependence of the observed band may be different. For the harmonic oscillator (see Chapter 6) the absorption coefficient is independent of temperature by definition.

At radio frequencies $h\nu \ll kT$ and therefore the temperature dependence is very different. In this region the values of kT are very large in comparison with the energies of the absorbed quanta, even at the temperatures of liquid hydrogen. Instead of (19.3) and (19.4) we now have

$$k_{ji} = \frac{B_{ji} h\nu_{ij}}{c/n} C(T) g_j n \frac{h\nu_{ij}}{kT} = \frac{B_{ji}(h\nu_{ij})^2}{c/n} C(T) g_j n \frac{1}{kT} \tag{19.7}$$

$$W_{ji} = B_{ji} u_{ij} (h\nu_{ij})^2 C(T) g_j n \frac{1}{kT} \tag{19.8}$$

When $h\nu/kT \ll 1$, the normalisation factor $C(T)$ is not very different from the constant $1/\Sigma g_j$ and therefore the temperature dependence is determined mainly by the factor $1/kT$, which represents the relation between $j \to i$ and $i \to j$ transitions, i.e. the effect of stimulated emission. The temperature at which maximum absorption coefficient is reached is approximately $h\nu_{ij}/k$, i.e. for $h\nu_{ij} \sim 1$ cm the temperature is approximately 1.5 °K. It follows that at normal temperatures the absorption coefficient falls off in accordance with the simple hyperbolic law.

It should be noted that in accordance with (19.7) the absorption coefficient decreases with decreasing frequency and tends to zero in the limit of very long wavelengths.

All the above considerations apply also to the temperature dependence of the rate of absorption. The rate of absorption of positive fluxes is positive, while the rate of absorption of negative fluxes ($u < 0$) is negative. The absorption coefficient is independent of the sign of the incident radiation since it is calculated per unit density of the external flux.

Dependence of the absorption coefficient and rate of absorption on density of the incident radiation. Non-linear effects

A detailed analysis of this problem can more conveniently be carried out for specific systems with two, three or more energy levels. It is then possible to obtain exact solutions and exhibit successively more general features (see Chapters 6-8). In this section we shall confine our attention to the general formulation of the problem and will summarise the results which are valid for all systems.

In thermodynamic equilibrium, the number of particles occupying a particular level is independent of time; the number of transitions from level i to any other level $j<i$ is equal to the number of reverse transitions. In accordance with the balance equations (16.1), which are valid within the framework of the probabilistic method, we may write

$$\frac{dn_i}{dt} = [-(A_{ij} + B_{ij} u_{ij}^{\text{eq}}) n_i^{\text{eq}} + B_{ji} u_{ij}^{\text{eq}} \cdot n_j^{\text{eq}}] + [-d_{ij} n_i^{\text{eq}} + d_{ji} n_j^{\text{eq}}] + \left[-\sum_{k \neq j} p_{ik} n_i^{\text{eq}} + \sum_{k \neq j} p_{ki} n_k^{\text{eq}}\right] = 0 \tag{19.9}$$

In view of the principle of detailed balancing, each of the square brackets in this expression must be individually equal to zero. The notation in (19.9) is the same as before. As can be seen, transitions from level i to level j have been written out separately. The last bracket contains transitions from level i to all other levels $k \neq j$ and the reverse transitions from all levels $k \neq j$ to level i. The symbols p_{ik} and p_{ki} represent total probabilities of the corresponding transitions, including both radiative and non-radiative transitions.

Let us suppose now that the medium is illuminated by external radiation of frequency ν_{ij} and density u_{ij}. This leads to a departure from the thermodynamic equilibrium, so that the number of particles in levels i, j, and in all other levels k will undergo a change. After a certain time has elapsed, steady-state conditions will be established and the numbers of particles occupying the various levels will become constant. When this state is reached, we have

$$\left[A_{ij} + B_{ij}(u_{ij}^{eq} + u_{ij}) + d_{ij} + \sum_{k \neq j} p_{ik}\right] n_i + \left[B_{ji}(u_{ij}^{eq} + u_{ij}) n_j + d_{ji} n_j + \sum_{k \neq j} p_{ki} n_k\right] = 0 \qquad (19.10)$$

The term in the first square bracket in (19.10) gives the number of transitions from level i to all other levels, while that in the second bracket represents the number of reverse transitions to level i. Although the numerical values of the two terms are equal as before, the number of direct and reverse transitions in each individual channel is different. Radiative $j \rightarrow i$ transitions due to absorption of the external radiation u_{ij} may be compensated by radiative transitions from level i to any other levels k, or by non-radiative transitions between levels i, j, or levels i, and other levels k.

The higher the density of the incident radiation u_{ij}, the greater the departure from thermodynamic equilibrium. In the limit of infinite incident radiation densities, the last two terms in (19.10) $B_{ij} u_{ij} n_i$ and $B_{ji} u_{ij} n_j$ will predominate, so that, in the absence of degeneracy, the populations of levels i and j will become equal and the state of saturation will be reached. A sketch of the functions $n_i(u_{ij})$ and $n_j(u_{ij})$ is shown in Fig. 4.17.

The degree of departure from thermodynamic equilibrium depends not only on $B_{ij} u_{ij} n_i$ and $B_{ji} n_{ij} n_j$, which give rise to

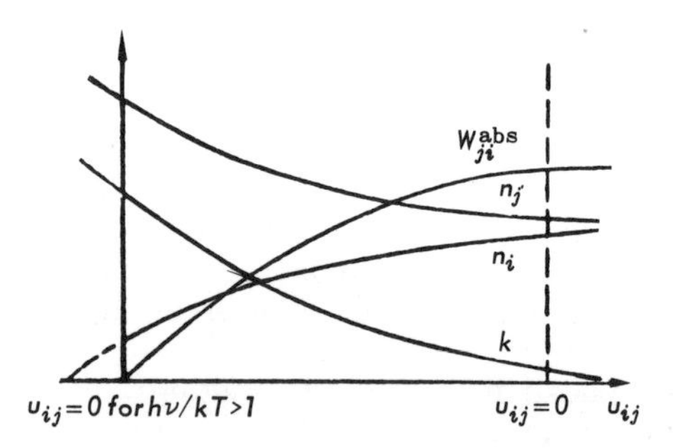

Fig. 4.17 Dependence of the populations n_i *and* n_j $(E_i > E_j)$, *the rate of absorption* W_{ji}^{abs} *and the absorption coefficient* k_{ji} *on the density of radiation. The scale along the ordinate axis is different for the different curves*

this departure, but also on all other properties of the system which tend to restore it to its original state. Thus, if the temperature and therefore the density of equilibrium radiation, u_{ij}^{eq}, are very large, moderate fluxes of radiation will not give rise to appreciable changes in the populations: $B_{ij}u_{ij}n_i$ and $B_{ji}u_{ij}n_j$ will be small in comparison with $B_{ij}u_{ij}^{eq}n_i$ and $B_{ji}u_{ij}^{eq}n_j$. At high temperatures, a departure from thermodynamic equilibrium is much more difficult to achieve.

Similarly, if the probabilities of non-radiative transitions, d_{ij} and d_{ji}, are large in comparison with the radiative transition probabilities $B_{ij}u_{ij}$ and $B_{ji}u_{ij}$, a small change in the radiation density will not lead to an appreciable effect and the population of the levels will remain as before. Transitions between the levels i and j and other levels k have a similar effect. If, for example, the Einstein coefficients B_{ij} are small, the level population is determined mainly by the transitions $j \leftrightarrows k$ and $i \leftrightarrows k$, and are very slow functions of the intensity of the incident flux of frequency ν_{ij}.

If the dependence of the population on u_{ij} is substituted into the expression (17.20) for the rate of absorption, we obtain

$$W_{ji}^{abs} = B_{ij}u_{ij}h\nu_{ij}\left[n_j(u_{ij}) - \frac{g_j}{g_i}n_i(u_{ij})\right] \qquad (19.11)$$

On the first approximation, when n_j and n_i change only slightly, the rate of absorption increases linearly with u_{ij}. This is the approximation of linear optics. Classical theory of the interaction of radiation with matter leads to precisely the same results (see (2.38)). The quantum theory, which takes into account the dependence of n_j and n_i on u_{ij}, enables us, in addition, to describe non-linear phenomena. For large u_{ij} the dependence of the rate of absorption on the density of radiation is non-linear. As u_{ij} increases, the rate of absorption increases more and more slowly and tends to a limit as $u_{ij} \to \infty$. Its magnitude depends on the form of the functions $n_j(u_{ij})$ and $n_i(u_{ij})$ and can only be calculated for a specific system.

If we take into account the functions $n_j(u_{ij})$ and $n_i(u_{ij})$ in the expression for the absorption coefficient, we find from (17.31) that as $u_{ij} \to \infty$, the absorption coefficient, i.e. the rate of absorption per unit intensity, will tend to zero. In accordance with the usual form of Bouguer's law, this means that the material becomes transparent, i.e. the incident radiation is not attenuated.

It must, however, be emphasised at once that in non-linear optics Bouguer's law is valid only in the differential form (17.6):

$$dS_\nu = -k_\nu S_\nu dl \tag{19.12}$$

Integration of this expression does not lead to an exponential relation between S and the thickness:

$$l = -\int_{S_{l=0}}^{S} \frac{dS_\nu}{k_\nu(S_\nu)S_\nu} = -\int_{S_{l=0}}^{S} \frac{dS_\nu}{W(S_\nu)} \tag{19.13}$$

The usual form of Bouguer's law given by (17.7) is valid only in linear optics when $k = \text{const}$, i.e. for relatively low incident intensities.

It will be shown in Chapters 7 and 8 that in many cases, at least within the framework of the probabilistic method, the dependence of the absorption coefficient on the density of the incident radiation may be described by a formula of the form

$$k(u) = \frac{k_0}{1+\alpha u} = \frac{k_0}{1+\alpha' S} \quad (\alpha > 0) \tag{19.14}$$

and therefore

$$W^{\text{abs}} = ku\,c/n = \frac{c}{n}\,\frac{k_0 u}{1+\alpha u} \tag{19.15}$$

Equation (19.12) can then be readily integrated to yield

$$\ln\frac{S}{S_0} - \alpha'(S_0 - S) = -k_0 l \tag{19.16}$$

where S_0 is the intensity for $l = 0$. When $\alpha' = 0$ this result becomes identical with (17.7). The dependence of k and W^{abs} on u_{ij} for $u_{ij} > 0$ is illustrated in Fig. 4.17. The broken curves in this figure show the dependence of n_j, n_i, W_{ji}^{abs} and k_{ii} on u_{ij} for negative u_{ij}. At relatively low temperatures ($h\nu/kT \gg 1$), the range of negative values of u_{ij} is small, so that there is no need to take into account the changes in n_j, n_i,

W_{ji}^{abs} and k_{ji}. If, however, $h\nu \ll kT$, most of the changes in these quantities occur at negative values of u_{ij}. When $u_{ij} = 0$, the values of n_j and n_i approach each other, W_{ji}^{abs} is virtually equal to its limiting value and k_{ji} approaches zero. The form of the curves can then be obtained from the same figure by moving the vertical axis to the position indicated by the broken vertical line and measuring u_{ij} from the new origin. Similar situations occur at radio-frequencies.

It is usually assumed that non-linear effects are very small and can only be observed at very high incident intensities. In reality, this is only true for transitions between the first and second electronic states which are observed in the visible and ultraviolet parts of the spectrum. If the system has metastable energy levels, non-linear phenomena may play an important role even at normal intensities. As can be seen from Fig. 4.17, the conditions for the appearance of non-linear effects in the visible and radio-frequency regions of the spectrum are also quite different. They assume particular importance in the theory of masers and lasers.

Negative absorption coefficients

Any departure from thermodynamic equilibrium due to an external disturbance leads to a change in the absorption coefficient. If the excitation takes place at frequency ν_{ij}, the limiting (minimum) value of k_{ji} is zero. At other frequencies the situation may be different. In some cases the number of particles in the higher level may become larger than the number in the lower state, and when the condition

$$\frac{n_i}{g_i} > \frac{n_j}{g_j} \tag{19.17}$$

is satisfied, the absorption coefficient will become negative in accordance with (17.31). This situation may also arise in other cases of a considerable departure from thermodynamic equilibrium, for example, in chemical reactions, in a gas discharge, under the action of strong electric and magnetic fields and so on.

When the absorption coefficient at the frequency ν_{ij} is negative, (17.20) shows that the rate of absorption by an elementary volume of the medium will also be negative when it is illuminated by a positive flux. This means that there is a liberation rather than an absorption of energy in the volume element. The energy emitted as a result of $i \to j$ transitions exceeds the energy absorbed from the incident flux as a result of $j \to i$ transitions.

This phenomenon was first predicted by Fabrikant in a series of papers published between 1930 and 1940 and was again emphasised in a patent application by Fabrikant, Vudynskii and Butaeva in 1959 [38]. The practical realisation at microwave frequencies is due to Basov and Prokhorov [39]. Recently this phenomenon has been systematically investigated and used as a basis for masers and lasers.

When a beam of radiation is allowed to pass through a medium having a negative absorption coefficient, the intensity of the beam should increase continuously in accordance with Bouguer's law, i.e. there is an amplification of the intensity of the beam. However, Bouguer's law (17.7) is now valid only provided the absorption coefficient is independent of the intensity. For very large path lengths in the medium, the intensity of the beam should, according to (17.7) become infinitely large, which is, of course, impossible. The amplification process within the system is accompanied by the appearance of factors which eventually limit the amplification. As we have seen, the process is accompanied by the liberation of energy within the system and transitions from upper level i to lower level j. At low intensities of radiation of frequency ν_{ij} this process may be compensated by external sources which give rise to a departure from equilibrium and the appearance of a negative absorption coefficient. However, at high intensities, the number of $i \to j$ transitions is very large and the values of n_i and n_j tend to become equal, while the absorption coefficient rapidly decreases in accordance with (19.14).

It is important to note that the condition given by (19.17) is not always sufficient for amplification to set in for a pair of levels i and j. If there are other pairs of levels in the system with energy differences equal to $h\nu_{ij}$, the amplification process may be reduced or even eliminated altogether as a result of transitions between these levels and the absorption of radiation.

Methods of producing negative absorption coefficients

There are many ways of producing negative absorption coefficients. All the methods which have been put forward so far depend on a strong departure from thermodynamic equilibrium. Historically, the first method which was realised at microwave frequencies [40] was based on the separation of excited and unexcited particles in a highly non-uniform electric field. The working substance was ammonia ($\lambda=1.27$ cm). When the electric field was switched on, there was an increase in the energy of the upper level and a reduction in the energy of the lower level. The molecules occupying the upper and lower states were thus subjected to opposite forces; those with the lower energy were drawn into the region of maximum field, while those with higher energy were drawn into the region of minimum field. The non-uniform field was produced by a cylindrical electrostatic quadrupole. The region near the axis of the quadrupole was occupied only by the excited molecules and this ensured a high negative absorption coefficient.

In the experiments of Butaeva and Fabrikant [41] the negative absorption coefficient was obtained in a glow discharge in mercury vapour. Figure 4.18 shows the energy level diagram of the mercury atom. The spectral lines 5,460, 4,358 and 4,046 Å produced as a result of $6\,^3P_{0,\,1,\,2}7^3S_1$ transitions were investigated. The number of atoms in the lowest level of the triplet was reduced by the addition of hydrogen and helium gas which gave rise to non-radiative transitions from the $6\,^3P_{0,\,1,\,2}$ level.

A similar idea was used by Abrekov, Pesin and Fabelinskii [42]. The gas discharge was produced in a mixture of mercury vapour and a small amount of zinc vapour. Amplification was achieved as a result of $4^1D_2 \rightarrow 4^1P_1$ transitions in the

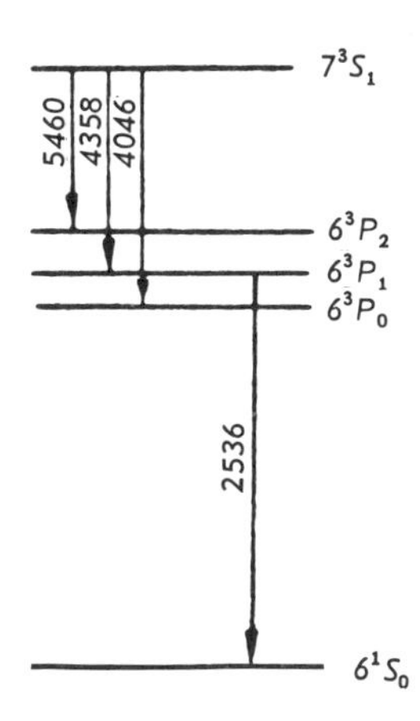

Fig. 4.18 Energy level diagram for the mercury atom

zinc atoms (λ=6,362 Å). The zinc upper level is very close to one of the mercury levels. Collisions between mercury and zinc atoms lead to a transfer of energy from the excited mercury atoms to zinc atoms (collisions of the second kind) and this leads to an accumulation of particles in the 4^1D_2 level.

Many other methods of producing negative absorption coefficients in gas discharges and in semiconductors have been suggested in the literature. One method is to introduce a sudden change in the direction of an external magnetic field in a paramagnetic crystal. This is accompanied by the excitation of the crystal ions, and before equilibrium is re-established, the stimulated emission may be greater than absorption.

We shall only consider the optical methods of producing negative values of k. In a system of particles with two energy levels, intensive excitation can only lead to a reduction of the absorption coefficient to zero (see Chapter 7). In the case of the harmonic oscillator (see Chapter 6), the absorption coefficient is independent of the method and rate of excitation. The simplest system which can have a negative absorption coefficient is a system with three energy levels. In some cases it is convenient to use transitions involving four energy levels. A detailed analysis of the necessary conditions will be given in Chapter 8.

Let us consider the possible transitions in a system of particles with three energy levels. When the system is illuminated by a high-intensity beam of frequency ν_{31}, absorption of this radiation leads to an accumulation of the particles in level 3. Calculations show, however, that n_3 is always less than n_1. A negative absorption coefficient can arise only at other frequencies, namely, ν_{21} or ν_{32}. Suppose, for example, that for one reason or another, the probability of the non-radiative transition $3\rightarrow 2$ is high, while the probability of the $2\rightarrow 1$ transition is low. Under these conditions, there will be a large number of particles occupying level 2 (when $A_{21} + d_{21} \cong 0$ and $d_{23} = 0$ up to $n_2 = n$) which will exceed n_1. As a result, the absorption coefficient will become negative.

This method has been used in the optical region by Maiman [43], who employed a ruby crystal, i.e. Al_2O_3 with a Cr_2O_3 impurity. The active substance was the Cr^{+++} ion. The energy level diagram for this ion is shown in Fig. 4.19. If the ruby crystal is illuminated with high-intensity green

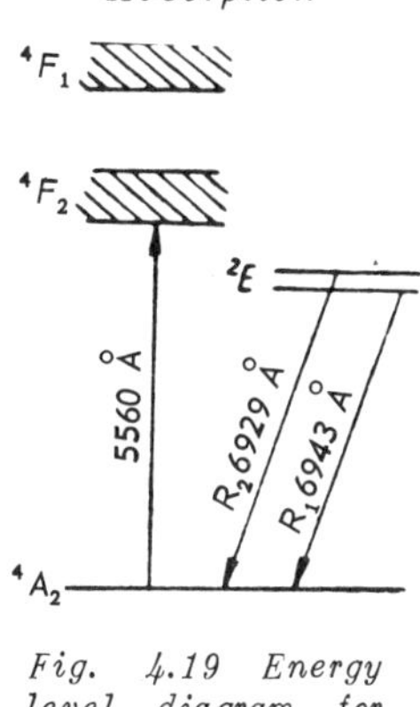

Fig. 4.19 Energy level diagram for the chromium ion in ruby

light ($\lambda = 5{,}500$ - $5{,}600$ Å), some of the atoms will undergo transitions to the $^4F_2$ state. After transferring their surplus energies to the crystal lattice, they are found in the ^2E level. The probability of luminescence as a result of the $^4F_2 \rightarrow {}^4A_2$ transitions is much lower. As the intensity of the exciting light increases, the stimulated emission in the R_1 line begins to exceed absorption.

A negative absorption coefficient in a system of particles with three energy levels can also be obtained for the frequency ν_{32}. In this case, there should be a very high probability of the $2 \rightarrow 1$ transition (largely non-radiative) with the result that the number of particles in level 2 will not be very different from the equilibrium number. To reduce n_2 still further, it is necessary to reduce the temperature. When the excitation is carried out at frequency ν_{31}, the number of particles in level 3 will exceed the number of particles in level 2.

This method has been used with a CaF_2 crystal and a trivalent uranium impurity. Figure 4.20 shows the corresponding energy level diagram. A fourth level was used to enhance the effect. The flux of radiation of frequency ν_{41} in the visible region was absorbed by the uranium ions. After non-radiative transitions these atoms are found to occupy level 3. Up to this stage the number of particles n_3 is small and the emission spectrum contains infrared luminescence lines. Further increase in n_3 due to an increase in the intensity of the exciting radiation leads to an excess in the number of particles in the third level as compared to the number

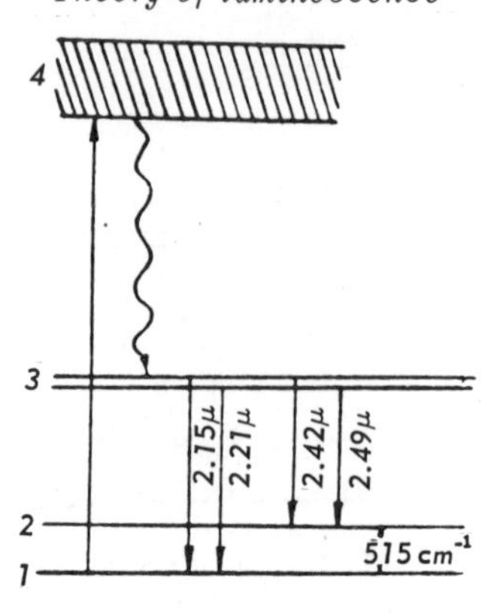

Fig. 4.20 Energy level diagram for trivalent uranium in calcium fluoride

in the second level. The absorption coefficient k_{23} will therefore become negative. A similar effect may be obtained with radiation of frequency ν_{31}, although the rate of absorption of this radiation is small.

The three-level scheme has also been used to obtain negative absorption coefficients in the microwave range [45].

20. AMPLIFICATION AND GENERATION OF RADIATION

Departure from Bouguer's law in media with negative absorption coefficients

The rate of absorption in media with negative absorption coefficients is also negative. This is a direct consequence of (17.33) which shows that

$$W_{ij}^{\text{abs}} = k_{ij} u_{ij} \frac{c}{n} = k_{ij} S_{ij}$$

The negative sign of the rate of absorption (for $u_{ij} > 0$) means that the radiation is amplified and not absorbed in the elementary volume, i.e. there is a liberation of radiant energy of frequency ν_{ij}. This is, of course, consistent with the law of conservation of energy since the production of a medium with a negative absorption coefficient requires the expenditure of energy in some other channels.

According to Bouguer's law (17.7), an absorption coefficient $k_{ij} < 0$ will ensure that the intensity of a beam of radiation of frequency ν_{ij} will become amplified on passing through the medium. The formula (17.7) was obtained within the framework of linear optics, i.e. on the assumption that the absorption coefficient was independent of S_{ij}. On this approximation the variation in S with increasing l is exponential. This approximation cannot be used for media with negative absorption coefficients since as S increases, the value of $|k|$ should decrease.

In non-linear optics the dependence of the intensity on l is given by (19.16). When $\alpha \neq 0$, but S is small, we have

$$S = S_0 \left(1 - \frac{lk_0}{1 + \alpha \dfrac{n}{c} S_0} \right) \tag{20.1}$$

Since $k_0 < 0$, $\alpha > 0$, it follows that S will increase with increasing l. For large l we can neglect the first term on the left-hand side of (19.16), and this yields

$$S = S_0 - \frac{lk_0 c}{\alpha n} \tag{20.2}$$

As l increases further, the quantity $\frac{lk_0 c}{\alpha n}$ becomes greater than S_0 and therefore the intensity of the beam at a depth l becomes independent of the intensity of the primary beam. The intensity within the medium is then wholly determined by its internal properties. A medium with a negative absorption coefficient and large enough linear dimensions will act as a source (generator) of radiation and not merely an amplifier of the incident radiation.

Substituting (20.2) into (19.14) we have

$$k = \frac{k_0}{1 + \dfrac{\alpha n}{c} S_0 - k_0 l} \tag{20.3}$$

For very large l we have $k = 1/l$, and the optical thickness kl remains constant. The rate of negative absorption, taken with the minus sign, is in view of (17.33) given by

$$W^{em} = -W^{abs} = \frac{k_0\left(\frac{\alpha n}{c} S_0 - k_0 l\right)}{\alpha\left(\frac{\alpha n}{c} S_0 - k_0 l + 1\right)} \tag{20.4}$$

For small l the rate of emission depends on the intensity of the primary beam S_0 and the depth l. For large l this is no longer so; each volume element emits the same amount of energy k_0/α. The greater the primary intensity S_0, the more quickly is this limiting value reached. The greater the magnitude of l, the smaller the dependence of W^{em} on S_0.

The limiting value of the rate of emission depends not on k but on the ratio k_0/α. A higher rate of emission can be achieved with a slightly non-linear absorption coefficient.

Properties of a plane-parallel layer on the linear approximation

The absorption of radiation in a plane-parallel layer was discussed in detail in Section 17. All the calculations were carried out on the linear-optics approximation. The results are also valid for layers with negative absorption coefficients. More accurate results can be obtained by taking into account the dependence of the absorption coefficient on the density of the radiation inside the layer (see next section).

Figures 4.21 and 4.22 show the transmission and reflection factors of a plane-parallel layer of this kind as a function of kl and r. The curves were computed from (17.54) and (17.55) which were obtained on the geometrical-optics approximation. In contrast to the usual case ($k > 0$; see Figs. 4.5-4.7), in the case of negative k it is possible to have $P > 1$ and $R > 1$, i.e. the incident radiation is amplified. The sum $P + R$ is always greater than unity, and when

$$e^{-kl} \geqslant \frac{1}{r} \tag{20.5}$$

both P and R increase without limit.

This means that the radiation is produced within the plane-parallel layer, however small is the primary intensity S_0. The layer therefore becomes a generator of radiation. Infinite values of P and R cannot be achieved in practice, that is, there must be some non-linear effects.

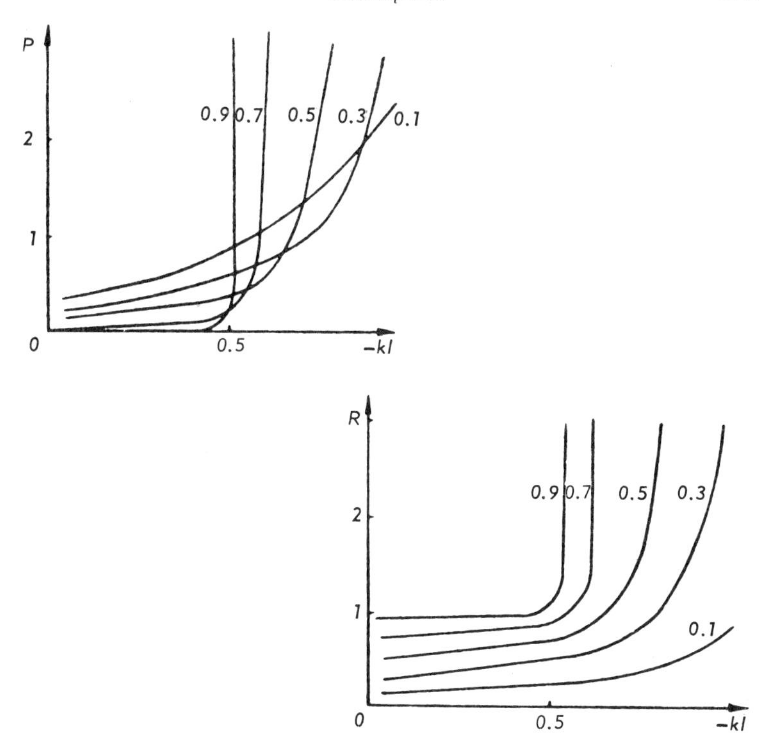

Fig. 4.21 Reflection and transmission of a plane-parallel layer with a negative absorption coefficient as a function of $|k|$ *on the geometrical optics approximation (numbers on the curves indicate the reflection coefficient for the layer-medium boundary)*

The generation condition given by (20.5) has a simple energy interpretation. As it passes from one surface to the other, the radiation is amplified by a factor of e^{-kl}. Some of this radiation leaves the layer and the remainder is reflected to the second surface. If the fraction of reflected radiation, e^{-kl}, is greater than unity, then as a result of multiple reflections there will be a continuous accumulation of energy within the layer. For small reflection coefficients, generation of radiation can occur only for large values of $|k|$ (for example, $|kl| = 1.61$ when $r = 0.2$). For large r, the generation condition is satisfied much more readily ($|kl| = 0.11$ for $r = 0.9$).

Figures 4.21 and 4.22 show that the region of amplification ($R > 1$ or $P > 1$) for r close to unity is very limited,

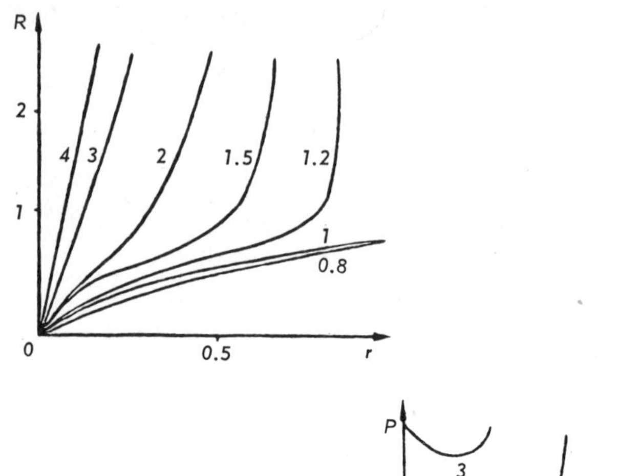

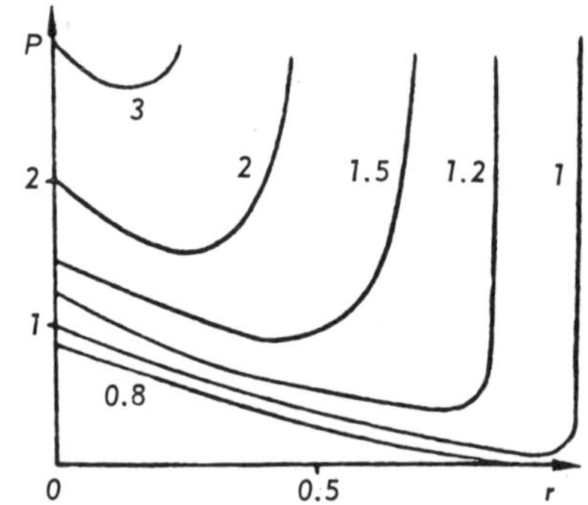

Fig. 4.22 Reflection and transmission of a plane-parallel layer with a negative absorption coefficient as functions of r *(numbers against curves represent the values of* $\eta = e^{-kl}$

since the functions $R(-kl)$ and $P(-kl)$ vary very rapidly, and very small changes in k may lead to transition from amplification to generation.

Equations (17.70)-(17.73) may also be used to estimate the time dependence of the radiation produced by a plane-parallel layer with negative k. They determine the form of $R(t)$ and $P(t)$ at the beginning of the illumination process and after the external radiation has been removed. Equation (17.72) gives the mean time necessary to reach the steady-state and afterglow. It shows, in particular, that τ rapidly decreases as re^{-kl} approaches unity, i.e. as $|kl|$ increases. Layers with negative absorption coefficients may exhibit very considerable inertia and retain the radiation within them for considerable lengths of time. When the condition given by (20.5) is satisfied, equation (17.72) loses its significance. A more rigorous calculation based on the non-linear approximation is then necessary.

The usual geometrical-optics approximation cannot be used for a strictly plane-parallel layer. The transmission and reflection factors of such a layer, obtained by solving Maxwell's equations, lead to equations (17.61) and (17.62), from which it follows that the optical properties of the layer depend on the ratio of the optical thickness to the wavelength. The values of R and P remain finite even when $re^{-kl} = 1$, provided the second generation condition

$$\frac{2\pi}{\lambda} l - \delta = \pi s \tag{20.6}$$

is not satisfied at the same time.

The values of P and R are very sensitive to any departure from (20.6). If, for example, the sinusoidal function in the denominators of (17.61) and (17.62) is equal to unity rather than zero, the transmission and reflection factors will be very small, and can be greater than unity only for very small r and (at the same time), very large $|kl|$.

Figures 4.23 and 4.24 show R and P as functions of $|kl|$ and r. They were computed from (17.61) and (17.62) subject to the condition given by (20.6). It follows from these graphs that R and P will only tend to infinity when $re^{-kl} \to 1$, and will not do so when $re^{-kl} \gg 1$. R and P are particularly rapidly varying functions of kl near the generation point.

It follows from (17.67), which determines the width of the transmission band for the plane-parallel layer, that $\Delta\nu \to 0$ when $re^{-kl} \to 1$. Under generation conditions the layer can emit only narrow monochromatic lines [46].

Properties of a plane-parallel layer on the non-linear approximation

The theory of a plane-parallel layer with a negative absorption coefficient given in the preceding section shows that the linear optics approximation is unsatisfactory. Although it does yield the conditions which are necessary for the intrinsic emission by the layer to occur, it leads to physically incorrect values for R, P, τ, and $\Delta\nu$. This is connected with the initial assumption that the absorption coefficient is independent of the density of radiation inside the layer.

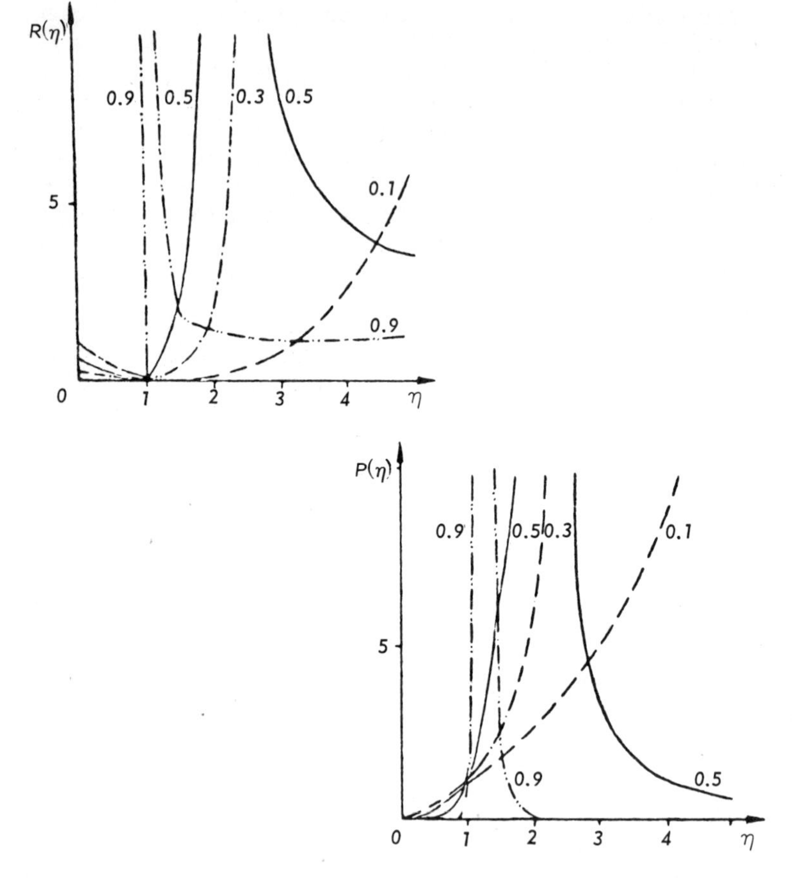

Fig. 4.23 Dependence of R *and* P *on* η *on the wave optics approximation*

It has already been pointed out in connection with Bouguer's law that more accurate calculations may be carried out on the basis of (19.14). This shows that as the density of radiation inside the layer increases, there is a continuous decrease in the absorption coefficient, so that even in the limit of $u \to \infty$, which is never achieved in practice, the intrinsic emission by the layer remains finite. In fact, if $u \to \infty$, then $k \to 0$, and according to (17.33), the negative rate of absorption tends to the finite value

$$W^{\mathrm{em}} = -W^{\mathrm{abs}} = -\frac{c}{n} uk = -\frac{c}{n}\frac{k_0 - k}{\alpha} \qquad (20.7)$$

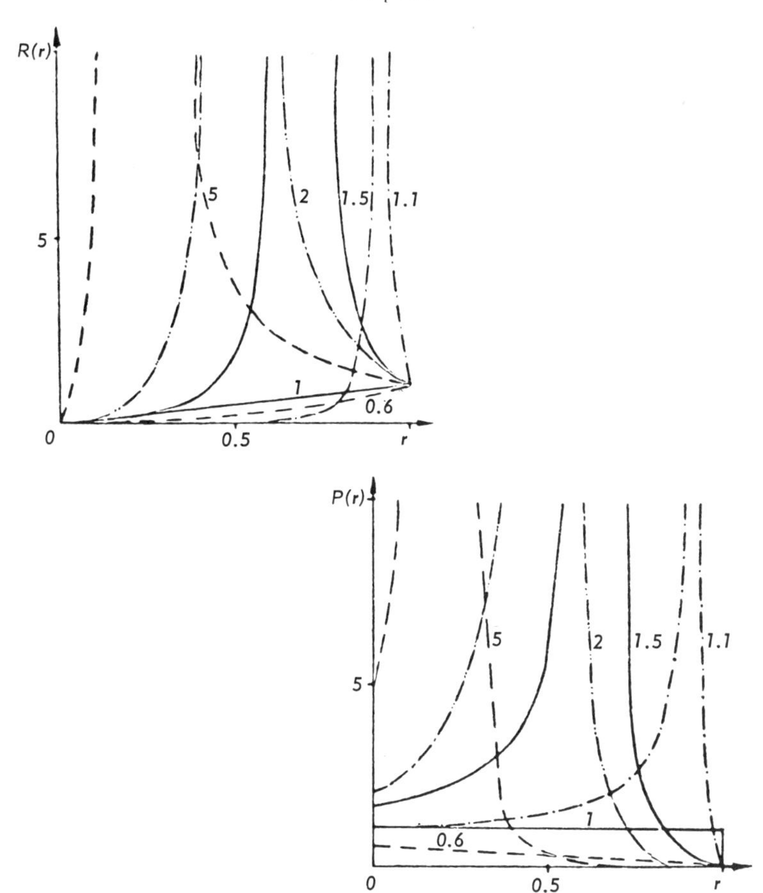

Fig. 4.24 Dependence of R *and* P *on* r *on the wave optics approximation*

This gives the rate of absorption per unit volume. The total intrinsic emission by the layer through both sides is equal to lW^{em}.

The maximum value W^{em} may be found by returning to the formulae of geometrical optics, (17.54) and (17.55), which describe the reflection and transmission by a plane-parallel layer. It was assumed in the preceding section that these formulae were valid only within the framework of the linear approximation, since for given values of the absorption coefficient and with (20.5) satisfied these formulae lead to infinite R and P. In reality (17.61) and (17.62) are valid on the non-linear approximation provided the absorption coefficient depends not only on the properties of the medium but also on the density u of radiation inside the layer. A change in this coefficient as u increases automatically leads to finite P and R. Hence it follows that

$$re^{-kl} \leqslant 1, \ |k| \leqslant \frac{1}{l} \ln \frac{1}{r} \tag{20.8}$$

If $S_0 \neq 0$ then the equality sign cannot apply. When $S_0 = 0$ the outgoing fluxes $S_{\text{refl}} = S_0 R$. $S_d = S_0 P$ can be finite if

$$-k = \frac{1}{l} \ln \frac{1}{r} = k^{\text{loss}} \tag{20.9}$$

The existence of outgoing fluxes in the absence of external radiation shows that radiation is being generated by the layer. The amount of radiation escaping per unit length from the layer is determined by the quantity $(1/l) \ln(1/r)$. The rate of emission by the layer satisfies the inequality

$$W^{\text{em}} = -W^{\text{abs}} \leqslant -\frac{c}{n} \frac{k_0 - \frac{1}{l} \ln \frac{1}{r}}{\alpha} \tag{20.10}$$

The equality sign yields the rate of generation. The generation condition is

$$|k_0| - \frac{1}{l} \ln \frac{1}{r} = |k_0| - k^{\text{loss}} \geqslant 0 \tag{20.11}$$

If this condition is not satisfied, radiation is not generated and emission by the layer leads to the amplification of the incident radiation.

The true values of the negative absorption coefficient within the layer can be calculated through the simultaneous solution of

$$u = \frac{k_0 - k}{\alpha k} \tag{20.12}$$

and

$$u = u_0 \left[\frac{(1-r) e^{-kx} + (1-r) r e^{-kl} e^{-kx}}{1 - r^2 e^{-2kx}} \right] \tag{20.13}$$

The first of these equations follows from (19.14) and determines the non-linearity of the absorption coefficient; the second was obtained by summing over (17.57) and (17.58) and represents the density of radiation within the layer when a flux $S_0 = u_0 c/n$ is incident on one of its faces.

Analysis of (20.12) and (20.13) shows that the absolute magnitude of the absorption coefficient increases with decreasing non-linearity parameter, the density of the internal radiation u_0, the thickness of the layer and the reflection coefficient r, and with increasing absolute magnitude of k_0.

Quantum generators (masers)

Media with negative absorption coefficients have been used to develop new sources of radiation. These are all based on the assumption that the stimulated emission at frequency ν_{ij} will be greater than absorption. Since the direction of propagation of the emitted and incident radiation is the same, and stimulated emission results in coherent radiation, such sources can be made to produce beams of exceedingly small divergence and wavelength spread. The width of a line which may be produced in this way may be much smaller than the natural line width.

The first quantum generators (masers) were successfully developed for the microwave range. The first generator for the visible part of the spectrum (the laser) was developed in 1960. It was based on a synthetic ruby crystal with 0.05-0.5% of chromium oxide. The crystal was in the form of a rod with a diameter of half an inch. The two plane-parallel ends of the rod were coated with silver in order to produce a high reflection coefficient. A negative absorption coefficient was achieved with the aid of a pulsed neon-krypton lamp, which produced high-intensity pulses ($\lambda = 5{,}600$ Å). As a result of multiple reflections between the two parallel faces of the crystal, a pulse of red coherent radiation was obtained (R_1 in Fig. 4.19). The angular divergence of the beam was 0.01°, the wavelength spread was 1 Å the peak power output of the order of 1 kW, and the efficiency was in the region of one per cent. At about the same time, a coherent generator of infrared radiation using a CaF_2 crystal with a uranium impurity was also developed. Quantum-mechanical generators will undoubtedly find extensive applications in science and technology [47].

5

Luminescence

21. RATE OF LUMINESCENCE

Rate of emission

Suppose that in an elementary volume dV there are $n_i(\Omega_1) \times d\Omega_1\, dV$ particles occupying a level i and having a special axis lying within the solid angle $d\Omega_1$. The number of radiative transitions in a time dt from level i to a lower level j with the emission of quanta with polarisation α into the solid angle $d\Omega_2$ is, according to (7.6), given by

$$dn_{ij\alpha}(\Omega_1, \Omega_2) = a_{ij}^{\alpha}(\Omega_1, \Omega_2)\, n_i(\Omega_1)\, d\Omega_1\, d\Omega_2\, dV dt \qquad (21.1)$$

where $a_{ij}^{\alpha}(\Omega_1, \Omega_2)$ is the differential Einstein coefficient. We have taken only the spontaneous transitions into account because stimulated emission is inseparable from the process of absorption and merely compensates the proportion of energy which is absorbed as a result of $j \rightarrow i$ transitions (see Chapter 4). The probability $a_{ij}^{\alpha}(\Omega_1, \Omega_2)$ is proportional to the square of the modulus of the matrix element of the operator representing the interaction between radiation and matter. In the dipole approximation it is given by (8.75).

The rate of spontaneous emission, i.e. the energy emitted by all the particles as a result of $i \to j$ transitions per unit time per unit volume is given by

$$W_{ij\alpha}^{em}(\Omega_2)\,d\Omega_2 = h\nu_{ij}\,d\Omega_2 \int\limits_{\Omega_1} a_{ij}^{\alpha}(\Omega_1, \Omega_2)\,n_i(\Omega_1)\,d\Omega_1 \qquad (21.2)$$

It depends on the nature of the emitting particles, the frequency, the orientation of the particles and the population of the energy levels. In the case of an anisotropic orientation of the particles, the emitted radiation is also isotropic.

If all the particles are distributed randomly, then as was shown in Section 7, the result of integration with respect to Ω_1 is independent of Ω_2 and α. The emission is equally likely in all directions and is completely depolarised. The total rate of emission (after integration with respect to Ω_2) is then given by

$$W_{ij}^{em} = A_{ij} n_i h\nu_{ij} \qquad (21.3)$$

where A_{ij} is the integral Einstein coefficient which is related to B_{ij} by the usual formula

$$A_{ij} = B_{ij}\frac{8\pi h\nu_{ij}^3}{(c/n)^3} \qquad (21.4)$$

The expression given by (21.3) is valid for a discrete spectrum and the value of A_{ij} corresponds to the integral over the profile of the emitted line. If the energy levels form a continuum, then (21.3) may be replaced by

$$W^{em}(\nu) = nh\nu \int A(E^*, \nu)\rho(E^*)\,dE^* \qquad (21.5)$$

where $n\rho(E^*)dE^*$, is the number of molecules in the energy interval between E^* and $E^* + dE^*$, $\rho(E^*)$ is the distribution function over the excited states and $A(E^*, \nu)$ is the probability of spontaneous transition from level E^* to level E ($h\nu = E^* - E$) per unit frequency interval. The dimensions of A_{ij} in (21.3) and $A(E^*, \nu)$ in (21.5) are, of course, different. The limits of integration in (21.5) are different for different specific cases. If the energy spectrum is completely continuous, the integration with respect to E^* is carried out between $h\nu = E$ and infinity. For transitions between different vibrational sub-levels of two electronic levels, the integration must be carried out between $E^* = 0$ and infinity for $h\nu < h\nu_{el}$, and between $E^* = h\nu - h\nu_{el}$ and infinity for $h\nu < h\nu_{el}$.

Rate of emission in thermal equilibrium. Kirchhoff's law for an elementary volume

The rate of thermal emission is given by (21.3) when the level population is replaced by the Boltzmann distribution:

$$W_{ij}^{\text{th.em}} = A_{ij}\, n_i^{\text{eq}}\, h\nu_{ij} = \frac{A_{ij} g_i e^{-h\nu_{ij}/kT}\, h\nu_{ij}}{\Sigma\, g_i e^{-h\nu_{ij}/kT}}\, n \qquad (21.6)$$

where n is the total number of particles per unit volume. An analogous formula may be obtained for anisotropic emission in the presence of particle alignment.

In thermodynamic equilibrium, the rate of spontaneous emission is equal to the rate of absorption of Planck radiation. By equating (21.6) and (19.2) it is easy to obtain the following equilibrium condition

$$A_{ij} n_i^{\text{eq}} = B_{ji}\, u_{ij}^0\, n_j^{\text{eq}}\, (1 - e^{-h\nu_{ij}/kT}) \qquad (21.7)$$

By rearranging this it may be shown that the relations connecting the Einstein coefficients and the density of equilibrium radiation are

$$B_{ji} u_{ij}^0 = \frac{A_{ij} e^{-h\nu_{ij}/kT}}{1 - e^{-h\nu_{ij}/kT}}\, \frac{g_i}{g_j} \qquad (21.8)$$

$$A_{ij} + \frac{g_j}{g_i} B_{ji}\, u_{ij}^0 = \frac{A_{ij}}{1 - e^{-h\nu_{ij}/kT}} \qquad (21.9)$$

We have already noted that the absorptive power is commonly defined as the ratio of the rate of absorption to the density of the incident radiation, i.e. the rate of absorption per unit density. Consequently, if we divide the rate of thermal emission by the absorptive power of matter in equilibrium with radiation we obtain

$$\frac{W_{ij}^{\text{th.em}}}{k_{ji}^{\text{eq}}} = \frac{8\pi h \nu_{ij}^3}{c^3} \frac{1}{e^{h\nu_{ij}/kT} - 1} = u_{ij}^0 \qquad (21.10)$$

This expresses one of the most general laws of physics, namely, Kirchhoff's law. The ratio of the rate of spontaneous emission of any medium in thermal equilibrium with radiation to its absorptive power is equal to the density of the equilibrium radiation. This law is different from that considered in Section 18 in that it is concerned with an elementary volume rather than with emission by bodies of finite dimensions. In the form given by (21.10), Kirchhoff's law is obtained only when stimulated emission is taken into account in the derivation of the expression for the absorptive power. In the absence of stimulated emission the ratio of the rate of emission to the absorptive power would not be given by Planck's formula but by Wien's formula

$$\frac{W_{ij}^{\text{th.em}}}{k_{ji}} = \frac{8\pi h \nu_{ij}^3}{c^3} e^{-h\nu_{ij}/kT} \qquad (21.11)$$

When $h\nu_{ij} \gg kT$, stimulated emission is relatively unimportant, and therefore Wien's formula presents a good approximation to Planck's function. Conversely, at low frequencies and high temperatures ($h\nu_{ij} \ll kT$) the role of stimulated emission becomes appreciable, so that Planck's formula must be employed. At very high temperatures, stimulated emission resulting from $i \to j$ transitions completely balances the absorption of radiation due to $j \to i$ transitions, the absorptive power tends to zero, and therefore, in accordance with Planck's formula, the right-hand side of the expression given by (21.10) tends to infinity. Note that the absorptive power of a harmonic oscillator is independent of temperature. The value of $W^{\text{th.em}}/k$ increases with increasing T only as a result of an increase in the rate of thermal emissions (see Chapter 6).

Rate of luminescence

The thermodynamic equilibrium prevailing in a system may be disturbed by an external agent, for example a beam of radiation from an external source. As a result, the system undergoes a transition to an excited state, and processes develop within it which tend to return it to its original state. Luminescence is one such process.

If the external disturbance affecting thermodynamic equilibrium persists for a sufficient length of time, a steady state is eventually reached. When the excitation is switched off, there is a gradual return to equilibrium.

Let us consider radiative transitions between two energy levels of a system of particles taken out of thermal equilibrium. If the excitation is radiative, the process occurring in the system is simple absorption of the incident radiation. At the same time there is absorption of thermal radiation reaching the system from the surrounding medium. The rate of absorption of thermal radiation is given by

$$W_{ji}^{\text{th.em}} = (B_{ji} u_{ij}^{0} n_j - B_{ij} u_{ij}^{0} n_i) h\nu_{ij} = B_{ji} u_{ij}^{0} h\nu_{ij} \left(n_j - \frac{g_j}{g_i} n_i \right) \tag{21.12}$$

which is a consequence of (17.20).

It was seen above that, in thermodynamic equilibrium, the absorption of thermal radiation within the system is compensated by spontaneous emission (21.6), and therefore the latter must be completely identified with thermal emission. Any departure from equilibrium leads to a change in the rate of spontaneous emission and the rate of absorption of equilibrium radiation. Only a part of the spontaneous emission can then be identified with thermal emission, the remainder being non-equilibrium emission, i.e. luminescence.

In order to determine the rate of thermal emission, it is sufficient to recall that it must compensate the absorption of equilibrium radiation incident on the system from the ambient medium. It is only in the case of such compensation that the density of equilibrium radiation in the surrounding space will remain unaltered and departures from equilibrium within the system will produce no changes in the state of the medium. The rate of luminescence should therefore be defined as the difference between the total rate of spontaneous emission and the rate of absorption of thermal radiation given by (21.12), i.e.

$$W_{ij}^{\text{lum}} = \left[A_{ij} n_i - B_{ji} u_{ij}^{0} \left(n_j - \frac{g_j}{g_i} n_i \right) \right] h\nu_{ij} \tag{21.13}$$

It would be incorrect to define the rate of luminescence as the difference between the rate of spontaneous emission by a system and its thermal emission prior to the departure from thermodynamic equilibrium. This takes no account of the change in the absorption of equilibrium radiation and may lead to important errors, especially when the departure from equilibrium is considerable [48].

From (17.31) and (21.13) it follows that

$$W_{ij}^{\mathrm{lum}} = A_{ij} n_i h\nu_{ij} - k_{ji} u_{ij}^0 c/\mathrm{n} \tag{21.14}$$

In many cases it is more convenient to use

$$W_{ij}^{\mathrm{lum}} = \frac{A_{ij} h\nu_{ij}}{1 - e^{-h\nu_{ij}/kT}} \left(n_i - \frac{g_i}{g_j} n_j e^{-h\nu_{ij}/kT} \right) \tag{21.15}$$

If the system is in a state of thermodynamic equilibrium, i.e. if

$$\frac{n_i}{n_j} = \frac{g_i}{g_j} e^{-h\nu_{ij}/kT}$$

the rate of luminescence is zero. The intensity of luminescence is specified by the degree of departure from the equilibrium distribution over the energy levels, i.e. the quantities

$$\Delta n_i = n_i - n_i^{\mathrm{eq}} \tag{21.16}$$

and

$$\Delta n_j = n_j - n_j^{\mathrm{eq}} \tag{21.17}$$

Substituting (21.16) and (21.17) into (21.15), we have

$$W_{ij}^{\mathrm{lum}} = \left(\Delta n_i - \frac{g_i}{g_j} \Delta n_j e^{-h\nu_{ij}/kT} \right) \frac{A_{ij} h\nu_{ij}}{1 - e^{-h\nu_{ij}/kT}} \tag{21.18}$$

This shows that the appearance of luminescence may be associated with a change in the population of both the upper level i and the lower level j. All the formulae in this section were obtained for luminescence emitted by randomly oriented molecules. Under these conditions the luminescence is isotropic and unpolarised.

Positive and negative luminescence

If, in the presence of a departure from thermodynamic equilibrium we have

$$\frac{n_i}{g_i} > \frac{n_j}{g_j} e^{-h\nu_{ij}/kT} \tag{21.19}$$

then according to (21.15), the rate of luminescence is positive. Spontaneous emission then exceeds absorption of Planck radiation, and the system becomes a source of positive flux of frequency ν_{ij}. Positive luminescence has been extensively investigated and can readily be produced in practice. If, however, there is an opposite departure from equilibrium

$$\frac{n_i}{g_i} < \frac{n_j}{g_j} e^{-h\nu_{ij}/kT} \tag{21.20}$$

then the rate of luminescence is negative, and therefore the total emission by the system is less than its thermal emission, and the system becomes a source of negative flux [49].

The intensity of luminescence is, according to (21.18), determined by the degree of departure from the equilibrium distribution, i.e. the quantities Δn_i and Δn_j. It follows from (21.18) that the appearance of positive luminescence may be associated not only with an increase in the population of the upper level but also with a reduction in the population of the lower level. On the other hand, the appearance of negative luminescence is associated with a reduction in the population of the upper level or an increase in the population of the lower level.

Changes in the population of levels i and j (in comparison with their values in equilibrium) have different effects on the rate of luminescence. Suppose, for example, that during the excitation process the population of the upper level i changes by $\Delta n_i = a$, while the population of the lower level remains unaltered ($\Delta n_j = 0$). The rate of the resulting positive luminescence of frequency ν_{ij} is then equal to

$$\frac{A_{ij} h\nu_{ij}}{1 - e^{-h\nu_{ij}/kT}} a \tag{21.21}$$

If there is a reverse change in the population ($\Delta n_j = a$, $\Delta n_i = 0$), negative luminescence is produced at a rate which is lower than (21.21) by a factor of

$$\frac{g_i}{g_j} e^{-h\nu_{ij}/kT}$$

It is precisely for this reason that negative luminescence is as a rule more difficult to observe in practice.

This result is valid when the departure from equilibrium is associated with an excess of particles in one of the levels ($\Delta n_i = a$, or $\Delta n_j = a$ are positive). If, on the other hand, there is a negative excitation ($\Delta n_i = -a$ or $\Delta n_j = -a$), then the inverse relationship prevails between the rates of emission of positive and negative luminescence. Strictly speaking, the probability of appearance of positive and negative luminescence is the same. The difference is only that at normal temperatures, and in the visible part of the spectrum, it is easier to obtain positive values of Δn_i or Δn_j by taking a proportion of the particles from the lowest to the most populated level. At high temperatures, or low frequencies ($h\nu < kT$), it is quite possible to achieve the inverse situation with a high intensity of negative luminescence. It should also be noted that when $h\nu_{ij} \ll kT$ the intensities of positive and negative luminescence are of the same order of magnitude.

The concept of negative luminescence can most simply be explained by analysing (21.14) and (21.15), which are valid for an elementary volume, together with (18.20), which gives the flux of radiation recorded by a detector. This flux is equal to

$$\frac{Y}{A_d a} = \alpha [\varepsilon_s - A_s S_0(T_m)] + \alpha [S_0(T_m) - S_0(T_d)] \quad (21.22)$$

where S_0 is the flux of equilibrium radiation which is proportional to its density u, ε_s is the emissive power of the source of radiation, A is its absorptive power and α is a constant of proportionality which is determined by the geometry of the apparatus. In contrast to (21.14), the expression given by (21.22) does not refer to any particular point inside the luminescing object, but to the flux of radiation produced and absorbed within a given solid angle throughout the body. The first term in (21.22) is equivalent to (21.14) and represents the difference between the total emission by the body and its thermal emission, i.e. the luminescence of the object under investigation. The second term characterises the detector readings in the absence of the luminescing body which are due to the flux of equilibrium radiation from the surrounding medium. This flux is independent of the properties of the source of luminescence.

Let us suppose that the detector used to analyse the spectral composition of the luminescence is at a temperature of absolute zero. If the system under investigation is in a state of thermodynamic equilibrium with the medium, then $\varepsilon_s = A_s\ S_0\,(T_m)$, the first term in (21.22) is zero, and the detector reading Y corresponds to the usual equilibrium radiation given by Planck's function with $T = T_m$.

When the excitation is switched on, the thermal equilibrium will be disturbed, luminescence will appear, and the system of particles will become a source of radiation. The detector will then register the total rate of emission including the thermal background. At some frequencies, the total rate of emission may be greater than the thermal background, in other cases it may be smaller. This is illustrated in Fig. 5.1a for a medium with a discrete spectrum (the spectral lines were assumed to have a finite width). Positive luminescence is observed at ν_1, ν_3 and ν_4, while negative luminescence is observed at ν_2 and ν_5.

It is evident from Fig. 5.1a that negative luminescence will be appreciable only when the equilibrium emission corresponding to ν_2 and ν_5 is high enough. The maximum rate of emission of negative luminescence cannot exceed the rate of thermal emission. It is precisely for this reason that it is practically impossible to observe negative luminescence in the visible part of the spectrum at room

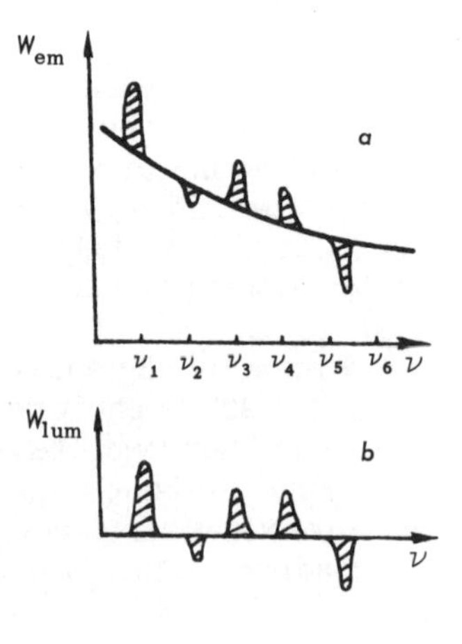

a – $T_d = 0$; b – $T_d = T_m$

Fig. 5.1 Positive and negative luminescence

temperature. We have already noted that it can only become appreciable at high temperatures or low frequencies. At the same time, the rate of positive luminescence may be very appreciable for large $h\nu/kT$. Figure 5.1b shows a plot of the rate of luminescence recorded by a detector at the same temperature as the medium. Prior to excitation the detector readings are zero. As soon as luminescence appears the detector records positive signals for ν_1, ν_3 and ν_4 and negative signals for ν_2 and ν_5. The detector reading is proportional to the rate of luminescence as given by (21.22) with $S_0(T_{\bar{m}}) = S_0(T_d)$.

We have already noted that luminescence is one of the non-equilibrium processes tending to restore the system to thermodynamic equilibrium. Depending on the nature of the departure from equilibrium, the luminescence process may be accompanied either by a reduction or an increase in the energy of the system. In the former case, luminescence is positive, in the latter negative. When $n_i > n_i^{eq}$, the luminescence will result in a reduction of the excess number of particles in level i and vice versa. The luminescence process will terminate only after the equilibrium distribution over the energy levels is restored.

Figure 5.2 shows the directions of luminescence for all the possible radiative transitions in a system of particles with four energy levels when $n_1 \approx n_1^{eq}$, $n_2 < n_2^{eq}$, $n_3 > n_3^{eq}$ and $n_4 \approx n_4^{eq}$. This distribution over the energy levels may occur in particular in the case of excitation by radiation of frequency ν_{32} (dashed arrow). The downward and upward arrows represent positive and negative luminescence respectively. The arrows point in both directions from the

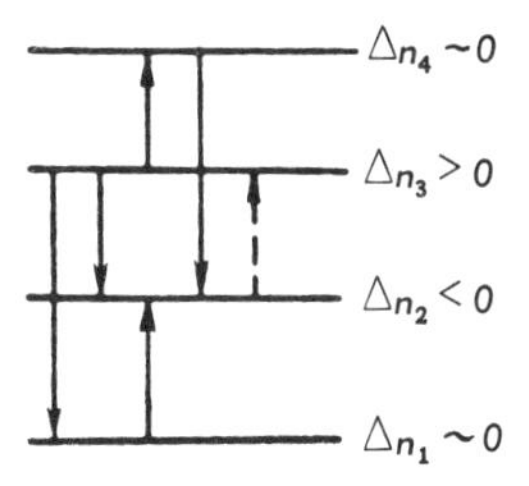

Fig. 5.2 Direction of transitions between energy levels in the case of positive and negative luminescence

level with a surplus of particles and end on the level with a deficiency of particles. In general, the direction of luminescence for any particular pair of levels can be found from the direction which brings the system closer to equilibrium.

It follows from the above discussion that negative luminescence is a common, though not easily noticeable, phenomenon, and arises for any type of departure from equilibrium. The only exception occurs for systems which interact with the medium at absolute zero, (according to (2.14) when $T_m = 0$ and therefore $u^0_{ij} = 0$, the rate of luminescence is equal to the rate of spontaneous emission, which is always positive).

If the system is illuminated by a negative flux of frequency ν_{32}, both positive and negative luminescence will arise, but with opposite signs (positive for ν_{21}, ν_{43} and negative for ν_{31}, ν_{32}, ν_{42}).

It is worthwhile noting a particular property of the harmonic oscillator (Chapter 6). Negative luminescence may arise in a given system only for illumination by negative fluxes. When the harmonic oscillator is placed in a normal flux, the luminescence is always positive.

Negative luminescence exhibits all the main features of ordinary luminescence which we shall discuss below. It may be characterised by definite duration, polarisation, quenching and so on.

Excitation and quenching of luminescence

The departure of a system from equilibrium may be due to one of many different causes. Photoluminescence (fluorescence and phosphorescence) arises as a result of excitation by radiation, whereas cathodoluminescence is produced through excitation by a beam of electrons. Again, the system may be found in a non-equilibrium state as a result of a chemical reaction, in which case one speaks of chemiluminescence. The excitation of luminescence may be associated with the effect of a beam of heavy ions, neutral particles, or various mechanical, electrical and magnetic phenomena. The penetration of an electric field into a phosphor may give rise to electroluminescence.

The intensity of luminescence is directly related to the magnitude of the external effect. For example, in the case of excitation by radiation, it depends on the magnitude of the luminous flux. It also depends on the degree of interaction between the system and the incident radiation. If the radiation is poorly absorbed, departure from equilibrium is negligible and the luminescence weak.

Any departure from the equilibrium distribution over the

energy levels within a system will inevitably give rise to processes which tend to restore the state of equilibrium. Apart from luminescence, these include de-activation by collision, transfer of excitation energy to a solvent, transformation of electronic energy into vibrational energy, and so on. In the steady state, the rate of luminescence depends not only on the intensity of the external agents which bring about the departure from equilibrium, but also on all the internal processes in the system which tend to restore equilibrium.

Let us suppose that a time-independent non-equilibrium distribution over the energy levels has been established in the system of particles as a result of the action of various external and internal agents. The form of this distribution is determined by the solution of an equation such as (19.10). This distribution will correspond to a definite rate of luminescence at all possible frequencies, given by (21.14). Let us suppose further that the experimenter can vary the rate of the processes tending to restore thermodynamic equilibrium, for example, by introducing a foreign substance into the luminescing solution which gives rise to new non-radiative $i \rightleftarrows j$ transitions. The degree of departure of n_i/n_j from $(g_i/g_j)\, e^{-h\nu_{ij}/kT}$ can thus be regulated, and by achieving small changes of this kind we can quench (suppress) luminescence of frequency ν_{ij}. The quenching of luminescence essentially involves the restoration of the equilibrium distribution of the particles over the energy levels. All other methods of quenching involve the destruction of the luminescing particles themselves. Thus, the disappearance of luminescence may be associated with an increase in the probability of photochemical disintegration of the molecules on collision. An increase in the concentration of molecules frequently leads to associations and a reduction in the number of luminescing centres.

Quenching can be either general or selective. In the former case, the external agent gives rise to a simultaneous reduction in luminescence at all frequencies. This can be achieved, for example, by increasing the temperature of the system, leading to a reduction in the effect of the external excitation which tends to introduce a departure from equilibrium. Thus, it follows from (19.10) that as the temperature increases, and therefore the density of equilibrium radiation increases, the effect of the external radiation may become insufficient to bring about an appreciable departure from equilibrium, and this will reduce the rate of luminescence.

A very large increase in temperature is required to produce quenching of this kind in the visible region. The mechanism of the more appreciable temperature quenching in complex systems (molecules) is somewhat different. The increase in the temperature leads to an increase in the store of vibrational energy of the molecules which is associated with an increase in the probability of transformation of electrical energy into vibrational energy and the reduction in a number of excited molecules.

It is also possible to achieve general quenching of luminescence (at all frequencies) by increasing the strength of the interaction between the molecules. In infrared studies, one frequently encounters systems in which the static intermolecular interactions are very considerable, and there are no known ways in which a departure from the equilibrium distribution over the vibrational levels of the lowest electronic state can be effectively produced. It is precisely for this reason that luminescence is not observed in the infrared. Attempts to remove the system from the state of equilibrium most frequently lead to an increase in its temperature, i.e. a new equilibrium distribution at a higher temperature, and the emission of thermal radiation in accordance with Kirchhoff's law.

Selective quenching of luminescence at a single frequency ν_{ij} can only be achieved when non-radiative transitions have a non-resonance nature and produce $i \rightleftarrows j$ transitions. As we have seen in Section 9, the probabilities of non-radiative transitions $i \rightarrow j$ and $j \rightarrow i$ are frequently related by (see (9.11))

$$\frac{d_{ij}}{d_{ji}} = \frac{g_j}{g_i} e^{-h\nu_{ij}/kT}$$

If the probabilities d_{ij} and d_{ji}, due to an extraneous quencher are very high, and play a dominating role in (19.10) in comparison with probabilities of all other transitions, then it follows from (19.10) that

$$d_{ij}n_i = d_{ji}n_j$$

From (9.11) we obtain

$$\frac{n_i}{n_j} = \frac{g_i}{g_j} e^{-h\nu_{ij}/kT}$$

The ratio of the populations of levels i and j is then determined by Boltzmann's formula, and in accordance with (21.15) there is no luminescence at frequency ν_{ij}. This does not exclude a non-equilibrium distribution over all other pairs of levels, or luminescence at other frequencies.

A detailed phenomenological description of excitation and quenching of luminescence can only be given for specific systems (see Chapters 7 and 8). The number of possible types of quenching is quite large, and the quenching mechanism must be discussed separately for each molecule.

Stokes' rule

The absorption and subsequent emission of radiation is usually associated with a change in its spectral composition; fluorescence frequencies differ from absorption (excitation) frequencies. They are equal only in the special case of the resonance fluorescence of atoms.

Stokes' rule was for many years regarded as the basic law determining the transformation of the spectral composition of radiation. It states that the frequencies of the exciting radiation are always greater than the frequencies of the fluorescence spectrum:

$$\nu_{\text{exc}} \geqslant \nu_{\text{em}} \tag{21.23}$$

This rule reflects only the overall features of the phenomenon and is not always valid. Stokes' rule has a particularly clear interpretation in the case of atomic spectra. Let us suppose that the luminescence spectrum is emitted as a result of absorption of radiation of frequency ν_{exc} (Fig. 5.3) and is accompanied by a transition to one of the higher

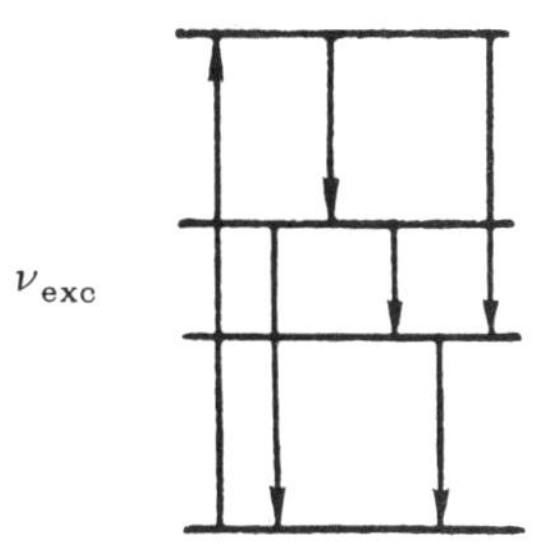

Fig. 5.3 Stokes luminescence

energy levels. The emission spectrum then exhibits a set of lines of different frequencies. They all arise as a result of transitions between the lower energy levels. One of them is the same as the frequency of the incident radiation, but the remainder are all lower.

Departures from Stokes' rule can also be easily demonstrated in the case of atoms. Suppose that the frequency of the exciting light is ν_{32} (Fig. 5.4). If all the atoms are in the electronic ground state, there is no absorption. At temperatures other than zero, a small proportion of the atoms will occupy level 2 and will undergo transitions to level 3

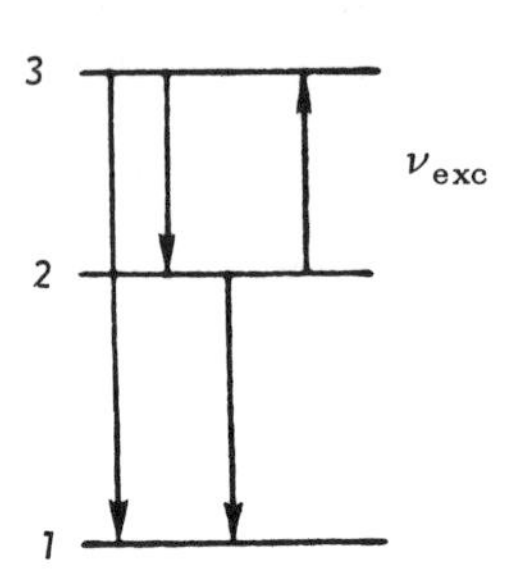

Fig. 5.4 Origin of anti-Stokes luminescence

as a result of absorption of radiation of frequency ν_{32}. They can then undergo transitions to state 2 with the emission of quanta of frequency equal to that of the absorbed radiation. They can also undergo transitions to state 1 with the emission of frequency $\nu_{31} > \nu_{exc}$. The higher the temperature, the smaller the difference $E_2 - E_1$, and the more easy it is to observe departures from Stokes' rule.

In complex systems with a continuous spectrum of energy levels, departures from Stokes' rule are much more probable and are in fact frequently observed. Figure 5.5 shows

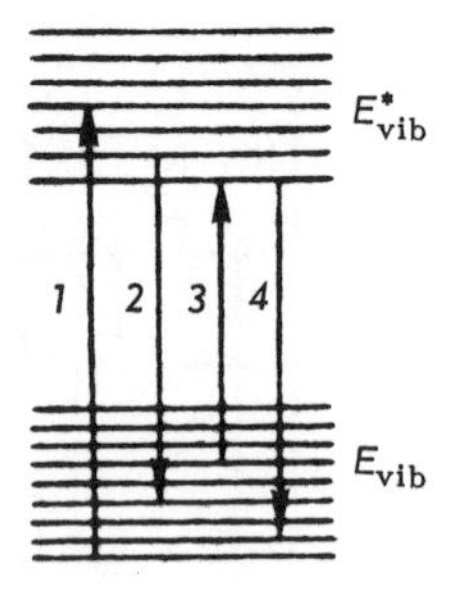

Fig. 5.5 Stokes and anti-Stokes luminescence due to transitions between electronic-vibrational levels of molecules

transitions between vibrational sub-levels of two electronic levels. Arrows 1 and 2 represent transitions involving absorption and emission in accordance with Stokes' rule, whereas arrows 3 and 4 represent exceptions to Stokes' rule. Since the frequency of the initial level for transition 3 is small, the number of particles occupying this level is considerable, and the rate of absorption is sufficient for the appearance of appreciable anti-Stokes luminescence.

As can be seen from Fig. 5.5, the process of absorption and subsequent Stokes fluorescence is always accompanied by the liberation of some vibrational, i.e. thermal, energy. In the case of a departure from Stokes' rule, the reverse situation is observed; the store of vibrational energy of a molecule is reduced, and some of it is transformed into radiation.

Figure 5.6 shows typical luminescence and absorption spectra of complex molecules. The line of approximate symmetry of the spectra corresponds to the frequency of the purely electronic transition. Experiments show that the form of a luminescence band of complex molecules is independent of the exciting frequency for $\nu_{exc} > \nu_{el}$, and is only occasionally slightly transformed for $\nu_{exc} < \nu_{el}$. In the latter case a considerable proportion of the observed luminescence is therefore, in fact, anti-Stokes luminescence.

Universal relationships between absorption and emission spectra of complex molecules

The problem considered in this section is concerned only with complex systems consisting of two sub-systems. For complex molecules, the analysis can be most simply carried out by considering transitions between the vibrational sub-levels of the lowest and the first excited electronic states. Stimulated emission will not be taken into account.

A considerable number of experimental results, including in particular the fact that the luminescence band profile is independent of the exciting frequency, show that a very rapid redistribution of energy over the vibrational degrees of freedom takes place between the absorption and emission of radiation. As a result, the act of emission is preceded by the establishment of a temperature distribution over the vibrational levels of the excited molecules. At the same time, the system does not exhibit complete equilibrium; there is an appreciable surplus of molecules in the excited electronic state. An equilibrium distribution over the vibrational energy levels in the excited electronic state is characteristic for complex molecules and other complex systems, and is associated with the interaction between the vibrational degrees of freedoms. In simple systems this does not occur.

Let us consider again the transition scheme between electronic-vibrational levels of a molecule (see Fig. 5.5). The rate of emission of frequency ν (per unit frequency range) is given by

$$W_\nu^{em} = nh\nu \int A(E^*_{vib}, \nu)\, \rho(E^*_{vib})\, dE^*_{vib} \tag{21.5}$$

where $n\rho(E^*_{vib})$ is the number of excited molecules per unit energy range, $A(E^*_{vib}, \nu)$ is the probability of spontaneous transition from vibrational level E^*_{vib} of the upper electronic state to the vibrational level $E_{vib} = E^*_{vib} - h\nu + h\nu_{el}$ of the lower electronic state. The integration is carried out over all the values of the vibrational energy of the excited state.

Equation (21.5) is valid for all distribution functions. If a complete thermodynamic equilibrium exists in the system, then

$$\rho(E^*_{vib}) = C(T)\, g^*(E^*_{vib})\, e^{-\frac{h\nu_{el}+E^*_{vib}}{kT}} \tag{21.24}$$

where $g^*(E^*_{vib})$ is the statistical weight of the vibrational levels of the excited electronic state, $h\nu_{el} + E^*_{vib}$ is the total energy of the given vibrational level of the excited state and $C(T)$ is a normalising factor.

Substituting (21.24) into (21.5), we obtain an expression for the thermal emission rate. In accordance with Kirchhoff's law (21.10) this emission is related to the absorptive power in equilibrium by the equation

$$W_\nu^{th.em} = u_\nu^0 k_\nu^{th.em} = nh\nu C(T)\, e^{-h\nu_{el}/kT} \int g^*(E^*_{vib}) \times A(E^*_{vib}, \nu)\, e^{-E^*_{vib}/kT}\, dE^*_{vib} \tag{21.25}$$

where we have taken into account the fact that $h\nu \gg kT$.

In the visible region, the rate of thermal emission is very small and difficult to measure. It can readily be calculated from (21.25) if the absorptive power is known from experiment. For our purposes, it is particularly important to know the frequency dependence of $W_\nu^{\text{th.em}}$, i.e. the form of the thermal emission band of a complex molecule. A very convenient description of this can be obtained as follows.

According to (17.33), the absorptive power k_ν is proportional to the absorption coefficient $k_\nu\,(\mathrm{k}_\nu = k_\nu c/\mathrm{n})$. For many complex molecules the frequency dependence of k_ν (profile of absorption line) is well known, and in the long-wave region is of the form shown in Fig. 5.7 by curve 1. Curve 2 in

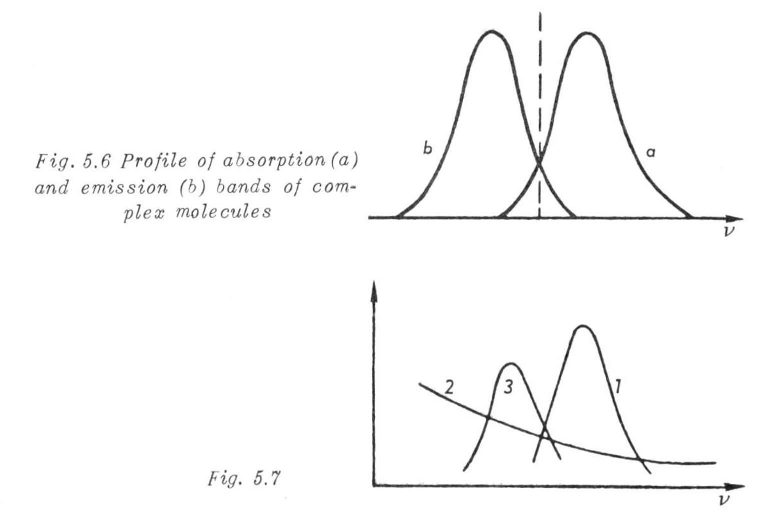

Fig. 5.6 Profile of absorption (a) and emission (b) bands of complex molecules

Fig. 5.7

this figure shows the approximate form of Planck's function in the visible region. The scale is arbitrary. For the purposes of our analysis, it is only important to know that as ν decreases there is a monotonic increase in u_ν^0. Curve 3 (which is also plotted on an arbitrary scale) represents the multiplication of k_ν and u_ν^0, i.e. the profile of the thermal emission band for the particular material. The maximum of the thermal emission band is shifted towards lower frequencies relative to the maximum of the absorption band.

Let us now find the expression for the luminescence rate. According to our original assumption, the act of emission is preceded by the establishment of an equilibrium distribution of excited molecules over the vibrational energy levels. At the same time, absorption of external radiation ensures that the total number of excited molecules is much greater than their number in complete thermodynamic equilibrium. The number of excited molecules with energies between E^*_{vib} and $E^*_{vib} + dE^*_{vib}$ is

$$n^* (E^*_{vib})\, d E^*_{vib} = C^* (T)\, n^* g^* (E^*)\, e^{-E^*_{vib}/kT} dE^*_{vib} \qquad (21.26)$$

where n^*_{vib} is the number of excited molecules which depends on the type of excitation (intensity of the incident radiation, the presence of quenchers, the temperature and the properties of the molecule itself). The statistical weight $g^* (E^*_{vib})$ has the same meaning as in (21.24) and depends only on the properties of the molecule. The temperature T is determined by the temperature of the medium at which the transfer of vibrational energy occurs. The normalising factor $C^* (T)$ differs from $C(T)$ and is determined from

$$\int n^* (E^*_{vib})\, dE^*_{vib} = n^*$$

Substituting $n^* (E^*_{vib})$ from (21.26) for $n \rho (E^*_{vib})$ in (21.5), we obtain

$$W_{\nu}^{lum} = h \nu C^* (T)\, n^* \int \rho^* (E^*_{vib})\, A(E^*_{vib}, \nu) e^{-E^*_{vib}/kT} dE^*_{vib} \qquad (21.27)$$

Comparison of this with (21.25) yields

$$W^{lum} = \frac{C^* (T)\, n^*}{C (T)\, n e^{-h \nu_{el}/kT}} W^{th.em} \qquad (21.28)$$

and therefore in complex molecules with a rapid redistribution of energy over the vibrational degrees of freedom, the rate of luminescence is proportional to the rate of thermal emission. The coefficient of proportionality depends only on the temperature of the medium. The profiles of the thermal emission and luminescence bands are the same.

From Kirchhoff's law we have finally

$$\frac{W_{\nu}^{lum}}{k_{\nu}} = d (T)\, \nu^3 e^{-h \nu/kT} \qquad (21.29)$$

where

$$d (T) = \frac{8\pi h}{c^2} \frac{n^*}{n} \frac{C^* (T)}{C (T)} e^{h \nu_{el}/kT} \qquad (21.30)$$

The rate of luminescence at frequency ν is, therefore, unambiguously related to the absorption coefficient at this frequency. Knowing the frequency dependence of k_ν we can easily find the frequency dependence of W_ν^{lum}. Equation (21.29) is valid for all systems for which the distribution over the sub-levels of the excited state is independent of the frequency of the exciting radiation and, in general, of the method of excitation. In addition, (21.29) will only be valid provided the system does not contain absorbing, but non-luminescing, impurities, there is no non-exciting absorption, and so on. These conditions are obvious and should be taken into account in experimental work [50]. (21.29) can be written as

$$\ln \frac{W_\nu^{lum}}{k_\nu} - 3 \ln \nu \tag{21.31}$$

$$= \text{const} - \frac{h\nu}{kT}$$

and is thus readily amenable to experimental verification.

It has been shown that in the case of solutions of complex molecules the universal relation given by (21.29) is in fact valid, and this confirms the correctness of the original assumption about the redistribution of vibrational energy [51].

For vapours of complex molecules, the distribution over the vibrational levels of the excited molecule depends not only on the temperature of the medium, but also on the frequency of the exciting radiation. In the case of Stokes excitation ($\nu_{exc} > \nu_{\overline{el}}$), the excited molecule has a certain surplus of vibrational energy and a higher effective temperature. It is only for $\nu_{exc} = \nu_{el}$ that the surplus is zero and (21.29) is no longer valid [52].

A universal relationship such as (21.29) is valid not only for electronic-vibrational spectra of complex molecules, but also for other systems consisting of two sub-systems, one fast and one slow. It is only necessary for the energy redistribution time of the slow sub-system to be much greater than the lifetime of the excited state of the fast sub-system. The relation given by (21.29) is also valid for narrow atomic lines, provided the line profile is connected with the interaction with the medium. It shows in particular that a luminescence line is never strictly coincident with an absorption line, but is always shifted towards longer wavelengths, and has a slightly different profile.

22. LUMINESCENCE YIELD

Energy yield

An important characteristic of the luminescence properties of a system is its energy yield, i.e. the ratio of the rate of luminescence to the rate of absorption. The concept of the luminescence yield was first introduced by Vavilov.

It was considered for a long time that the energy absorbed by luminescing objects was largely converted to heat and only a very small proportion was re-emitted in the form of luminescence. This is so for most natural objects. There are however, many cases when losses of radiant energy are relatively small and the energy yield can be quite high. This was first demonstrated as far back as 1924 by Vavilov who studied the properties of fluorescein in a number of solvents. He showed that in the case of excitation by a continuous spectrum, the fluorescence energy yield was 0.71. Subsequent more accurate measurements led to similar results. In many cases the energy yield approached unity.

The energy yield of a particular system is of decisive importance in technological problems concerned with practical applications of luminescence. Studies of the energy yield as a function of the conditions of illumination, the properties of the medium and the system itself are very important from the point of view of the theory of interaction of radiation with matter and the processes which take place in the system after excitation.

Detailed calculations of the luminescence energy yield can only be carried out for specific systems under given conditions of excitation. This will be done in the next three chapters. Here, we shall confine our attention to some general remarks.

An accurate determination of the energy yield will involve measurement of the total amount of energy absorbed by the system while it is being illuminated by the exciting agent, and the total energy of luminescence from the onset of excitation to its complete removal (between $t = 0$ and $t = \infty$). The ratio of these two quantities is then calculated. Most frequently, it is sufficient that we should divide the rate of luminescence by the rate of absorption under stationary conditions. With the exception of special cases, the two results will be identical [53].

By definition, the energy yield of photoluminescence for a system with a discrete spectrum of energy levels is

$$\gamma_{en} = \frac{W^{lum}}{W^{abs}} = \frac{\sum_{i>j} W_{ij}^{lum}}{\sum_{j<i} W_{ji}^{abs}}$$

$$= \frac{\sum_{i,j}\left[A_{ij}n_i - B_{ji}u_{ij}^0\left(n_j - \frac{g_j}{g_i}n_i\right)h\nu_{ij}\right]}{\sum_{i,j} B_{ji}u_{ij}\left(n_j - \frac{g_j}{g_i}n_i\right)h\nu_{ij}} \qquad (22.1)$$

The numerator in this formula is written in accordance with (21.14), and the denominator in accordance with (17.20). Summation is carried out over all absorption and emission frequencies. In (22.1), u_{ij} represents the density of the external radiation of frequency ν_{ij}, i.e. the difference between the total emission and the thermal emission u_{ij}^0. The denominator takes into account only the absorption of the external flux producing the measured luminescence. There are analogous expressions for the luminescence energy yield of systems with a continuous spectrum of energy levels. They can be obtained by replacing the summation sign in (22.1) by integration with respect to ν and over the vibrational levels of the upper electronic state.

Equation (22.1) requires a knowledge of all the Einstein coefficients and all the level populations. To determine n_i and n_j, it is necessary to solve the balance equation for the time-independent state (see Section 16). In general, the population of the i-th level will depend on all the probabilities of radiative and non-radiative transitions between different pairs of levels, the density of external radiation and the temperature of the system.

If excitation is produced by monochromatic radiation of frequency ν_{ij}, it is sufficient to consider transitions between a limited number of energy levels since most of the higher levels will not be affected by excitation and will retain their original populations. It is frequently sufficient (for excitation at frequency ν_{21}) to consider absorption and emission as a result of transitions between two levels only, in which case equation (22.1) assumes the simpler form

$$\gamma_{en} = \frac{A_{21}n_2 - B_{12}u^0_{21}(n_1 - n_2)}{B_{12}u_{21}(n_1 - n_2)} \tag{22.2}$$

Possible values of the energy yield

In the subsequent chapters we shall calculate the energy yield for a number of specific systems and conditions of excitation. Certain general conclusions can, however, be made without detailed calculation.

If there are no non-radiative transitions, the luminescence energy yield is identically equal to one. This follows from the law of conservation of energy. When $d_{ji} = 0$ (i and j arbitrary), there are no processes which transform absorbed radiant energy into other forms of energy.

The appearance of non-radiative transitions leads as a rule to the quenching (attenuation) of luminescence and therefore to a reduction in the energy yield. This is so only for intramolecular non-radiative transitions, or non-radiative transitions connected with the interaction with the medium. They are statistical in nature, tending to restore the system to thermodynamic equilibrium. Such non-radiative transitions are related by (9.11), where the probabilities of downward transitions are always greater than the probabilities of upward transitions.

Non-radiative transitions are frequently induced by a directional external agent which disturbs the equilibrium of the system and the transitions themselves act as a source of luminescence (for example, in the case of excitation by a beam of electrons).

If the probabilities d_{ij} of irreversible transitions which satisfy (9.11) are much greater than the probabilities of radiative transitions $A_{ij} + B_{ij}\, u_{ij}$, the energy yield approaches zero and the system remains in equilibrium, as before. The great majority of specific systems do not luminesce precisely for this reason, and the number of systems capable of emitting luminescence is comparatively small. Most frequently, the phenomenon of luminescence is exhibited by simple systems in which intramolecular non-radiative transitions are of relatively low probability.

It follows from the above considerations that for most

luminescing objects the energy yield is greater than zero and less than unity:

$$0 < \gamma_{en} < 1 \tag{22.3}$$

It is however possible, in principle, to have the two anomalous cases

$$\gamma_{en} < 0 \text{ and } \gamma_{en} > 1$$

The energy yield may assume a negative value if the numerator and the denominator in (22.1) have different signs.

When the system is excited by a positive flux of radiation, the rate of absorption is positive. The necessary condition for the appearance of a negative energy yield is that the numerator in (22.1) should be negative. As we have seen in the preceding paragraph, negative luminescence is a common phenomenon (see Fig. 5.1) although its intensity is usually low. Negative luminescence at a particular frequency is always accompanied by positive luminescence. For $\gamma_{en} < 0$ it is necessary that positive luminescence should be largely quenched. Selective quenching of this kind is possible in principle.

When $\gamma_{en} < 0$ for excitation by a positive flux, this means that both the absorbed energy from the external source and a proportion of the absorbed Planck radiation are transformed into other forms of energy. Negative values of γ_{en} may also arise for negative exciting fluxes, in which case negative luminescence must be quenched.

Under certain conditions, the energy yield may be greater than unity. To be specific, let us consider excitation by a positive flux. The luminescence will then be positive for some frequencies and negative for others (see Fig. 5.2). If all the d_{ij} are negative, the yield is equal to unity, i.e. the numerator in (22.1) is equal to the denominator. It follows that for $\gamma_{en} > 1$, the sum of all the positive terms in the numerator of (22.1) will be greater than the denominator. If now we selectively quench the negative luminescence in some way, the numerator in (22.1) will become greater than the denominator and the yield will be greater than one.

If in a given system $\gamma_{en} \geq 1$, then under the action of an external agent which disturbs the state of equilibrium, the system may convert into radiation a proportion of its intrinsic thermal energy. This is not in conflict with the second law of thermodynamics, since the cooling of a luminescing body is not accompanied by the transfer of energy to the exciting source of radiation which is at a higher temperature. The energy of the exciting radiation and the thermal energy of the luminescing body are both transferred to the surrounding objects whose temperature is lower than the temperature of the source. This is quite similar to the operation of conventional refrigerators which work off external sources of energy.

We need also to consider the temperature dependence of the luminescence energy yield. At very low temperatures $u_{ij}^0 = 0$, so that the second term in the numerator of (22.1) is zero, i.e.

$$\gamma_{en} = \frac{\sum\limits_{i,j} A_{ij} n_i h \nu_{ij}}{\sum\limits_{i,j} B_{ji} u_{ij} \left(n_j - \frac{g_j}{g_i} n_i \right) h \nu_{ij}} \tag{22.4}$$

Negative luminescence is then impossible, the store of thermal energy is negligible, and $\gamma_{en} \leqslant 1$. Anomalous values of the yield ($\gamma_{en} > 1$ and $\gamma_{en} < 0$) are connected with the presence of a thermal background.

At very high temperatures the distribution of particles over the energy levels is practically constant (if $g_i = g_j = 1$), and the external radiation is insufficient for an appreciable departure from the state of equilibrium. Absorption of external radiation and of luminescence is then small. These considerations are valid only at very high temperatures. For small T, the temperature dependence of γ_{en} may be due to the dependence of d_{ji} on T. In complex molecules this may lead to a reduction in W^{lum} at constant W^{abs}, i.e. to a reduction in γ_{en}. This phenomenon is usually referred to as temperature quenching of luminescence. According to (9.11), when $T \to \infty$ the probabilities of non-radiative transitions, d_{ij}, are equal to the reverse probabilities d_{ji}, and the transformation of radiant energy into heat is completely compensated by the reverse process. It follows that there is no quenching of luminescence and therefore at very high temperatures there is an increase in the energy yield which becomes equal to unity in the limit as $T \to \infty$.

Quantum yield of luminescence

The quantum yield of photoluminescence is defined as the ratio of the number of emitted photons to the number of absorbed photons (under time-independent illumination). The quantum yield of simple systems is frequently much greater than unity and is determined, to a considerable extent, by the number of energy levels. For example, in the case of the scheme illustrated in Fig. 5.3, the absorption of a photon of frequency ν_{41} can be followed by the emission of photons with frequencies $\nu_{41}, \nu_{42}, \nu_{43}, \nu_{32}, \nu_{31}, \nu_{21}$. Under these conditions the concept of the quantum yield loses its usefulness and is not employed.

The concept of the quantum yield of photoluminescence is heuristically useful only when the emission spectrum is independent of the frequency of the exciting radiation. We have encountered a similar situation in the case of complex systems for which the emission spectrum is produced only as a result of transitions from the lowest of the excited electronic states, but is independent of the store of vibrational energy received on excitation. In such systems the probabilities of intramolecular non-radiative transitions ($d_{42}, d_{32},\ldots$) are much greater than the probabilities of the corresponding radiative transitions, the process of intramolecular redistribution of energy takes place very rapidly, and the molecule is in state 2 before the act of emission (independently of the method of excitation). The quantum yield is then largely determined by the relation between the probabilities of radiative (A_{21}) and non-radiative (d_{21}) transitions.

When the emission spectrum is independent of ν_{exc}, the quantum yield is unambiguously related to the energy yield

$$\gamma_{en} = \frac{W^{lum}}{W^{abs}} = \frac{N^{lum}}{N^{abs}} \frac{h\nu_{em}}{h\nu_{exc}} = \gamma_q \frac{\nu_{em}}{\nu_{abs}} \tag{22.5}$$

where N^{lum} and N^{abs} are the numbers of emitted and absorbed quanta respectively. When $\nu_{exc} > \nu_{em}$, the quantum yield is always less than the energy yield. The case of anti-Stokes fluorescence must be considered separately.

Secondary luminescence

Suppose that the plane-parallel layer under investigation is illuminated by an external flux S_0 which is attenuated as it penetrates the specimen. Luminescence is produced under the action of the external radiation, and in the absence of non-linear effects, the rate of luminescence is proportional to the density of the exciting flux and falls off in accordance with Bouguer's law

$$W^{\text{lum}} = aS_0 e^{-kx} \tag{22.6}$$

Experiment shows that in many cases the rate of luminescence does not depend on x in the way indicated by (22.6). This may be due to various causes, above all the appearance of secondary luminescence.

The relation given by (22.6) is valid only provided the source of luminescence is primary radiation reaching the layer from outside. This luminescence can conveniently be called primary. However, in reality, luminescence is produced by all the radiation at a particular point, including primary luminescence. Luminescence produced as a result of absorption of additional radiation from primary luminescence is called secondary luminescence. Luminescence of higher orders can be defined in a similar way. Occasionally, secondary luminescence is understood to mean all luminescence apart from primary luminescence.

Figure 5.8 shows the experimental data for rhodamine C [54]. The ratio of the rate of secondary luminescence to the rate of primary luminescence is plotted in this figure as a function of the optical thickness. In all cases it increases rapidly with depth x. The effect is particularly appreciable for large concentrations of the dye. Secondary absorption and emission of radiation has recently been subjected to systematic experimental [55] and theoretical [56] studies. The theory of these phenomena is based on the theory of transfer of radiant energy and is connected with the luminescence of bodies having finite linear dimensions. The theory is particularly easy to develop in the case of resonance luminescence when absorption and emission frequencies are equal. Corrections for the form of absorption and emission bands complicate the theory but modify only slightly the final results.

The equation for the number n_2 of excited particles at a

particular point r in a system of particles with two energy levels is of the form

$$\frac{dn_2(\bar{r},t)}{dt} = -n_2(\bar{r},t)[A_{21} + d_{21} + B_{21}(u_1(\bar{r}) + u_2(\bar{r}) + u_3(\bar{r}))] + [n_1(\bar{r},t)B_{12}(u_1(\bar{r}) + u_2(\bar{r}) + u_3(\bar{r})) + d_{12}] \tag{22.7}$$

where $u_1(\bar{r})$ is the density of the external equilibrium radiation at the point $\bar{r}$, which is attenuated as a result of absorption within the material, $u_2(\bar{r})$ is the density of radiation due

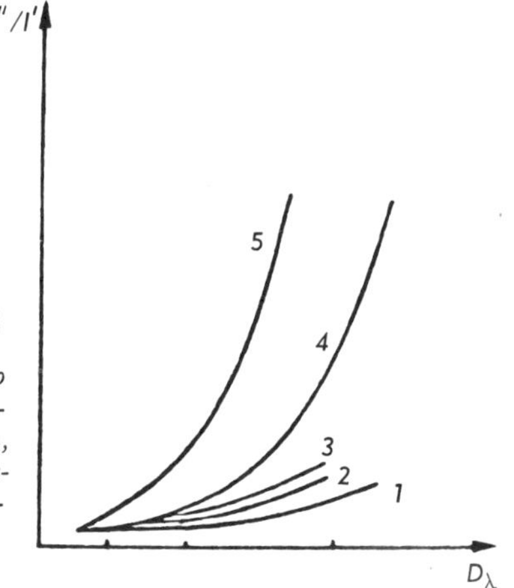

$1-5\times10^{-6}$; $2-3\times10^{-5}$; $3-4\times10^{-5}$; $4-1\times10^{-4}$; $5-3\times10^{-4}$ g/cm³

Fig. 5.8 Ratio of secondary to primary luminescence in a solution of rhodamine C in glycerin, as a function of the optical thickness D = kx *for different concentrations*

to external sources and $u_3(\bar{r})$ is the density of radiation at the particular point due to emission by all elements of the material itself.

When the complicated dependence of u_1, u_2 and especially u_3 on the coordinates is taken into account, (22.7) becomes a non-linear integro-differential equation. When both the stimulated emission and the effect of Planck radiation are neglected, which is practically always valid in the visible and ultraviolet regions, the solution of this equation is considerably simplified. For the stationary case it may be written in the form

$$n_2^{st}(\bar{r}) = \frac{Bn\tau_0}{c} S_{ext}[u^0 + \gamma u^{(1)} + \gamma^2 u^{(2)} + \ldots] \tag{22.8}$$

where $\gamma = \frac{A}{A+d}$, $\tau_0 = \frac{1}{A+d}$ and S_{ext} is the flux due to external sources at the surface of the body. The functions $u^{(i)}(r)$ are proportional to the density of luminescence of various orders at the point $\bar{r}$:

$$u^{(i)}(\bar{r}) = \frac{k}{4\pi} \int u^{(i-1)}(\bar{r}') \frac{e^{-k(r-r')}}{(r-r')^2} dV' \qquad (22.9)$$

$$u^0(\bar{r}) = \frac{cu_1}{S_{ext}} \qquad (22.10)$$

The first term in (22.8) is proportional to the density of radiation reaching the point $\bar{r}$ from outside, the second represents the density of primary luminescence, the third gives the density of secondary luminescence and so on.

The general solution of (22.7) is quite difficult to obtain. In a number of cases the solution has been found by numerical integration in an infinite plane-parallel layer. In some cases the equation can be solved approximately. Calculations reported in the papers quoted show that for large absorption coefficients and high enough values of the quantum yield, the rate of low-order luminescence may become much smaller than the rate of high-order luminescence. The presence of secondary luminescence leads to an increase in the observed quantum yield and in the duration of afterglow. It is often impossible to obtain a correct estimate of the optical properties of an elementary volume, or to determine the probabilities of radiative and non-radiative transitions from experimental data unless secondary luminescence is taken into account.

23. DURATION OF LUMINESCENCE

Dependence of the intensity of luminescence on time

When external radiation is allowed to fall on the specimen for a sufficient length of time, the rate of luminescence becomes independent of time, i.e. a steady state is eventually reached. Studies of time-dependent phenomena reveal many of the characteristic features of luminescence, and may be used to determine the internal properties of matter from experimental data. However, this source of information is not entirely adequate. Additional information may be obtained by the investigation of the properties of time-independent luminescence.

Any change in the conditions of excitation, for example, the intensity of incident radiation, additional illumination at other frequencies and interactions with other quenching objects, will lead to a change in the rate of luminescence. This change is never instantaneous because the transition probabilities between levels are finite and the luminescence exhibits considerable inertia. The most convenient method we have of studying the time characteristics of luminescence is the investigation of the laws of excitation and attenuation of luminescence. The excitation of luminescence begins immediately after the source of excitation is switched on, and continues until the stationary state is reached. Attenuation of luminescence, or what is usually referred to as afterglow, begins immediately after the exciting radiation is switched off.

The decay and rise of luminescence depend on the properties of the luminescing system and can, in general, be quite complicated. The rate of emission and the spectrum are frequently functions of time. For some systems the laws of excitation depend on the intensity of excitation.

The most frequently measured parameter is the duration τ of luminescence. If the law of decay of afterglow is simple and can be described by an exponential function, τ will unambiguously define the form of the attenuation curve. In more complicated cases, the time characteristics of luminescence cannot be represented by only one or two parameters.

The mean duration of afterglow, or simply the duration of luminescence, is usually defined by

$$\tau = \frac{\int_0^\infty t W^{\text{lum}}(t)\,dt}{\int_0^\infty W^{\text{lum}}(t)\,dt} \tag{23.1}$$

where the integration is carried out between the instant at which the exciting radiation is switched off and the complete disappearance of luminescence. This definition is not suitable in all cases because τ may be found to be infinite.

It is sometimes useful to investigate the time dependence of luminescence under pulsed illumination. The source of radiation is switched on for a very short period and is switched off before steady-state conditions set in. The pulse of radiation is followed by a dark pause during which the rate of emission of luminescence decreases. The process is then periodically repeated. A detailed calculation corresponding to this case is given in Section 31.

By comparing the intensity of luminescence as a function of time with the corresponding formulae obtained by the probabilistic method for a particular model of the medium, it is possible to find the transition probabilities characteristic of the given system. A systematic study of the intensity of luminescence as a function of time can only be carried out for specific systems and given conditions of excitation. Analysis of the properties of complex molecules, crystals, and particularly crystal phosphors, requires special calculations which take into account the processes which take place in the molecule between the acts of absorption and emission of radiation. In this chapter we shall discuss only some general problems.

Lifetime of the excited state

Consider the variation in the level population after the excitation is switched off. We shall suppose that at time $t=0$, level i is occupied by $n_i(0)$ particles. We shall also suppose that the population of the higher-lying levels is zero, and therefore there are no $l \to i$ transitions ($E_l > E_i$). After the excitation has been switched off, the number of particles in level i begins to decrease systematically. The change in the number of particles in a time interval between t and $t+dt$ can be found by the probabilistic method (Section 16) and is given by

$$dn_i = -(\sum_j A_{ij})\, n_i dt - (\sum_j d_{ij})\, n_i dt \tag{23.2}$$

where A_{ij} and d_{ij} are the probabilities of spontaneous and non-radiative transitions $i \to j$. The summation is carried out over all transitions to the lowest levels j.

If we integrate Equation (23.2) subject to the initial conditions, we then obtain the following equation.

$$n_i(t) = n_i(0)\, e^{-(\sum_j A_{ij} + \sum_j d_{ij})t}$$
$$= n_i(0) e^{-t/\tau} \tag{23.3}$$

The population of the i-th level falls off exponentially and tends to zero as $t \to \infty$. The rate of decrease is characterised in this case by the single parameter

$$\tau = \frac{1}{\sum_j (A_{ij} + d_{ij})} \tag{23.4}$$

which is called the lifetime of the excited state.

The lifetime of the excited state is numerically equal to the time interval during which the population of the particular level decreases by a factor of e. It is also equal to the mean interval of time during which the molecules are in the excited state. In point of fact, the quantity

$$dn_i = \frac{1}{\tau} n_i(0)\, e^{-t/\tau} dt \tag{23.5}$$

is equal to the number of molecules which undergo transitions to lower levels between t and $t+dt$, and therefore have occupied this level during the time interval between 0 and t. The mean value theorem then yields

$$\overline{t} = \frac{\int_0^\infty t dn_i}{\int_0^\infty dn_i} = \frac{1}{n_i(0)} \int_0^\infty t dn_i = \tau \tag{23.6}$$

According to (23.4), the lifetime of the excited state is inversely proportional to the sum of transition probabilities from level i to the lower-lying levels j. The greater these probabilities, the smaller the lifetime τ. If there are no non-radiative transitions, the lifetime of the excited state is determined only by spontaneous transitions. The appearance of non-radiative transitions, i.e. the quenching of luminescence, leads to a reduction in the lifetime. It is precisely for this reason that determinations of τ for a particular system under different external excitations is a reliable means of establishing the presence of quenching and the magnitude of non-radiative transition probabilities.

For dipole transitions between electronic levels, the quantity A_{ij} is of the order of 10^8 sec^{-1}, and therefore the lifetime of the excited state in the absence of quenching is of the order of 10^{-8}sec. It has already been noted that levels having lifetimes of this order are referred to as labile.

The dipole moment of $i \rightarrow j$ transitions is frequently equal to zero. Transitions can arise only as a result of changes in the quadrupole moment, magnetic dipole moment and higher-order moments. In such cases, the transition probability is much smaller and the lifetime of the excited state is greater by several orders of magnitude. Such levels are commonly referred to as metastable. Very frequently, the metastability of levels is associated with a forbidden transition between levels of different multiplicity, for example, between excited triplet and unexcited singlet levels.

The rate of luminescence at frequency ν_{ij} during the process of afterglow is given by

$$\begin{aligned} W_{ij}^{\text{lum}}(t) &= A_{ij} n_i(t) h \nu_{ij} \\ &= A_{ij} n_i(0) h \nu_{ij} e^{-t/\tau} \\ &= W_{ij}^{\text{lum}}(0) e^{-t/\tau} \end{aligned} \tag{23.7}$$

The quenching of luminescence has the same time-dependence as the variation in the population of the initial level. Substituting (23.7) into (23.1), we find that the mean duration of afterglow is equal to the lifetime of the excited state.

According to (23.7), the duration of afterglow is the same for all frequencies ν_{ij} which are emitted as a result of transitions from the i-th level. The value of τ for luminescence of frequency ν_{ij} depends not only on the probability of the $i \rightarrow j$ transition, but also on the probabilities of all other transitions from the initial level.

Equations (23.3), (23.4) and (23.7) were derived without taking into account the thermal radiation background, and on the assumption that the probabilities of non-radiative $j \rightarrow i$ transitions are zero. They may be valid at low temperatures or high frequencies (in the visible or ultraviolet parts of the spectrum). In the case of values of $h\nu$ which are comparable with kT the thermal radiation background must be taken into account, and therefore the functions $n_i(t)$ and $W_{ij}^{\text{lum}}(t)$ depend on the temperature.

Variation in the level population and rate of luminescence during the afterglow process

In the preceding section we considered the simplest case, when variation in the population during afterglow is associated with transitions from level i to lower-lying levels. The refilling of level i with particles arriving from other, for example higher-lying, levels was assumed not to take place. It is precisely for this reason that we succeeded in deriving a simple exponential law for the variation of the population with time, so that it was possible to introduce the concept of the lifetime of the excited state, which is related to the transition probabilities A_{ij} and d_{ij} as indicated by (23.4).

In the great majority of real cases, the time dependence of the level population is considerably more complicated, particularly when interaction with Planck radiation and non-radiative $j \to i$ transitions is taken into account.

The afterglow process essentially consists of a transition from a stationary distribution over the energy levels to the state of thermal equilibrium with the medium and the disappearance of luminescence. The time necessary to approach the state of equilibrium, i.e. the relaxation time, depends on the probabilities of all radiative and non-radiative transitions between all pairs of levels, and therefore on the properties of the molecule, the degree of departure from equilibrium in the preceding conditions of illumination and the temperature. The concept of the relaxation time is valid for the system as a whole but not each level individually. It is only in some special cases that one can speak of the lifetime of the excited state of an individual level, or occasionally a set of levels.

Let us suppose that when the excitation is switched off, the distribution of the particles over the energy levels is prescribed by the function $n_i(0)$. The values of n_i at any subsequent time may be found by solving the system of equations obtained by the probabilistic method (Section 16):

$$\begin{aligned}
dn_1/dt &= -(p_{12} + p_{13} + \dots + p_{1m})\,n_1 + p_{21}n_2 + \dots + p_{m1}n_m \\
dn_2/dt &= -(p_{21} + p_{23} + \dots + p_{2m})\,n_2 + p_{12}n_1 + \dots + p_{m2}n_m \\
&\dots\dots\dots\dots\dots\dots\dots\dots\dots \\
dn_m/dt &= -(p_{m1} + p_{m2} + \dots)\,n_m + p_{1m}n_1 + \dots\, p_{m-1,\,m}n_{m-1}
\end{aligned} \tag{23.8}$$

$$\Sigma n_i = n$$

where m is the total number of particle energy levels, which is, strictly speaking, infinite, and

$$p_{ij} = A_{ij} + B_{ij}u^{\circ}_{ij} + d_{ji} = \frac{A_{ij}}{1 - e^{-h\nu_{ij}/kT}} + d_{ij} \qquad (23.9)$$

$$p_{ji} = B_{ij}u^{\circ}_{ij} + d_{ji} = \frac{A_{ji}e^{-h\nu_{ij}/kT}}{1 - e^{-h\nu_{ij}/kT}} + d_{ij}e^{-h\nu_{ij}/kT} \qquad (23.10)$$

The equations given by (23.8) may be transformed to a more convenient form. If all the n_i correspond to the state of equilibrium we have

$$n_i^{\text{eq}} = n_1^{\text{eq}} \; e^{-h\nu_{1i}/kT} \qquad (23.11)$$

and the right-hand sides of all the equations in (23.8) are equal to zero. Substituting

$$n_i = n_i^{\text{eq}} + \Delta n_i \qquad (23.12)$$

into (23.8), we obtain

$$\frac{d(\Delta n_1)}{dt} = -(p_{12} + p_{13} + \dots)\,\Delta n_1 + p_{21}\,\Delta n_2 + \dots + p_{m1}\,\Delta n_m$$

$$\dots\dots\dots\dots\dots \qquad (23.13)$$

$$\frac{d(\Delta n_m)}{dt} = -(p_{m1} + p_{m2} + \dots)\Delta n_m + p_{1m}\,\Delta n_1 + \dots + p_{m-1,n}\,\Delta n_{m-1}$$

$$\sum_i \Delta n_i = 0$$

The values of Δn_i are positive for some levels and negative for others.

The solution of (23.13) is

$$\Delta n_i = D_{1i}e^{-\lambda_1 t} + D_{2i}e^{-\lambda_2 t} + \dots + D_{m-1,i}\,e^{-\lambda_{m-1} t} \qquad (23.14)$$
$$i = 1, 2, 3 \;.\;.\;.\; m.$$

where λ_1, λ_2, ..., λ_{m-1} are constants which depend only on the properties of the system of particles and on the temperature, and D_{kl} are constants which in addition depend on the initial conditions, i.e. the method of excitation. The total number of the constants λ_k is equal to $(m - 1)$. They can be complex, and their real part is always positive, since $\Delta n_i \to 0$ as $t \to \infty$.

It follows from (23.14) that the change in level population

as the system approaches the state of equilibrium (after the excitation has been switched off) occurs in accordance with a complicated non-exponential law. For many levels the values of Δn_i can not only decrease, but also increase and pass through zero many times.

The rate of change in the quantities Δn_i as the state of equilibrium is approached may be characterised by a mean time during which this surplus (or deficit) of particles exists in the particular level i. By analogy with (23.6), this time is defined by

$$\overline{t_i} = \frac{\int\limits_{t=0}^{\infty} td\,(\Delta n_i)}{\int\limits_{t=0}^{\infty} d\,(\Delta n_i)} = \frac{1}{\Delta n_i^0}\int\limits_{t=0}^{\infty} td\,(\Delta n_i) \qquad (23.15)$$

Using (23.14) and integrating (23.15) it can readily be shown that

$$\overline{t_i} = \frac{\sum\limits_{k=1}^{m-1} D_{ki}/\lambda_k}{\sum\limits_{k=1}^{m-1} D_{ki}} \qquad (23.16)$$

The values of $\overline{t_i}$ are different for different levels. It follows from (23.16) that they characterise not only the properties of the given i-th level, but also those of all the other levels, i.e. the system as a whole. Moreover, the values of $\overline{t_i}$ depend on not only the transition probabilities and the temperature, but also the initial conditions, i.e. the method of excitation.

Similar calculations may be carried out for the process leading to stationary conditions and the rise of luminescence. In this case the expressions given by (23.10) for the transition probabilities will depend on the density of the exciting radiation. The rate of afterglow at frequency ν_{ij} can readily be found with the aid of (23.14).

As an illustration of the above discussion, we may consider a system of particles with three energy levels excited by a positive flux of radiation of frequency ν_{32}. The necessary formulae will be given in Chapter 8. Figure 5.9 shows graphs of $n_1(t)$ $n_2(t)$ and $n_3(t)$ after the excitation has been switched off for some specific assumptions about the properties of the probabilities (23.9) and (23.10). None of the

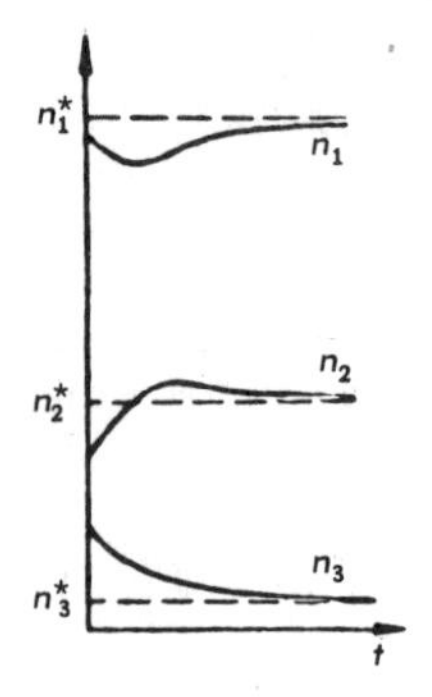

Fig. 5.9 Variation of the population of levels during afterglow for a system of particles with three energy levels excited by radiation of frequency ν_{32}

three quantities are exponential in form and the usual concept of the lifetime of the excited state does not apply. It is only useful in the case of the highest excited states.

The curves shown in Fig. 5.9 are plotted for those values of p_{ij} for which λ is real. They may turn out to be complex, in which case $n_1(t)$, $n_2(t)$ and $n_3(t)$ will continuously pulsate and approach the equilibrium values n_i^{eq}.

Redistribution of vibrational energy within a complex molecule

Complex molecules and other condensed systems have a very large number of vibrational sub-levels which participate in absorption and emission processes. Experiment shows, however, that quenching of the luminescence emitted by solutions of complex molecules of the uranyl compounds and many other substances is strictly exponential and is the same for all frequencies within a broad electron-vibrational luminescence band.

To explain this fact, consider the initial equations given by (23.8). Suppose that n_1 is the number of particles in all the vibrational sub-levels of the first electronic level, while n_2, n_3, ... are the numbers of molecules in various vibrational sub-levels of the upper electronic state. The probabilities p_{1i}^0 of transitions from the lower electronic level which are associated with the presence of the thermal radiation background are very small at room temperature and may be neglected. We shall also suppose that the transition probabilities p_{ij} for $i>1$ and $j>1$ are higher by two or three orders of magnitude than the p_{i1}, this being valid when radiative and largely non-radiative transitions between vibrational sub-levels are much more probable than transitions to the lower electronic state. This relationship between the transition probabilities is characteristic for complex molecules and can even be used as a measure of the degree of complexity of the molecule. Under these conditions, redistribution of the excitation energy of the electronic state over the vibrational levels becomes distinct from the luminescence process. Redistribution ceases after a time interval of the order of 10^{-10} to 10^{-11} sec, i.e. a time usually beyond the control of the experimenter.

If it is assumed that during the short interval of 10^{-10} to 10^{-11} sec there are no transitions to the lowest electronic state, the redistribution of the particles over the vibrational levels of the excited state is described by

$$\left.\begin{aligned} \frac{dn_2}{dt} &= -(p_{23}+p_{24}+\dots)\,n_2 + p_{32}n_3 + \dots + p_{m2}n_m \\ &\dots\dots\dots\dots\dots\dots \\ \frac{dn_m}{dt} &= -(p_{m2}+p_{m3}+\dots)\,n_m + p_{2m}n_2 + \\ &\quad + \dots + p_{m-1,\,m}\,n_{m-1} \end{aligned}\right\} \tag{23.17}$$

These equations were obtained from (23.8) by removing all the transition probabilities p_{i1} (and p_{1i}) to the lower electronic state. The solution of (23.17) is

$$\frac{n_i}{n_0^*} = D_i + D_{i1}e^{-\lambda_1 t} + D_{i2}e^{-\lambda_2 t} + \dots + D_{i,\,m-1}e^{-\lambda_{m-1} t} \tag{23.18}$$

where $n_0^* = n_2 + n_3 + \dots + n_m$ is the total number of particles in the excited electronic state at time $t = 0$. For large t all the exponentials tend to zero and n_i tends to the constant value $n_0^* D_i$. Time intervals of the order of 10^{-9}sec can in this case be regarded as long. A definite distribution of particles over the vibrational levels of the excited state is therefore established in 10^{-9}sec:

$$n_i = n_0^* D_i, \quad \sum_{j=2}^{m} D_i = 1 \tag{23.19}$$

The constant D_i is here identical with the statistical weight of the particular vibrational sub-level.

If there were no transitions from levels $i = 2, 3, \dots$ to level 1, then the values of n_i which are equal to $D_i n_0^*$ would remain constant. In reality, such transitions do take place, and the number of particles in the excited electronic state continuously decreases. However, $i \to 1$ transitions proceed relatively slowly and do not modify the form of the distribution over the vibrational sub-levels. This follows from the fact that (23.17) is very similar to (23.8) apart from the first equation. The probability p_{21} may be neglected in comparison with $p_{23} + p_{24} + \ldots + p_{2m}$, the probability p_{31} may be neglected in comparison with $p_{32} + \ldots + p_{3m}$ and so on.

The populations of the vibrational sub-levels of the excited state at any given time t (after the excitation has been switched off) are thus given by

$$n_i = n_0^* D_i f(t) \tag{23.20}$$

where $f(t)$ is a function of time which is the same for all i. The form of this function, and therefore the form of $n_i(t)$ may be found by solving the first of the equations in (23.8) subject to (23.20). If we neglect the very small terms $(p_{12} + p_{13} + \ldots + p_{1m}) n_1$ and substitute

$$\sum_{i=2}^{m} n_i(t) = n - n_1(t) \tag{23.21}$$

this equation assumes the form

$$-\frac{d}{dt}\left[\sum_{i=2}^{m} n_i(t)\right] = \sum_{i=2}^{m} p_{i1} n_i(t) \tag{23.22}$$

If we now substitute for n_i from (23.20) and use (23.19), we obtain the equation for $f(t)$

$$\left[\sum_{i=2}^{m} D_i\right] \frac{df}{dt} = \frac{df}{dt} = -\left[\sum_{i=2}^{m} p_{i1} D_i\right] f(t) \tag{23.23}$$

It follows that

$$f(t) = f(0)\, e^{-\left[\sum_i p_{i1} D_i\right] t} = e^{-\left[\sum p_{i1} D_i\right] t} = e^{-pt} \tag{23.24}$$

or

$$n_i = n_0^* D_i e^{-pt} = n_0^* D_i e^{-\frac{t}{\tau}} \tag{23.25}$$

The variation in population of all the vibrational sub-levels of the excited electronic state is exponential. The rate of decrease in the number of particles is given by the constant

$$p = \frac{1}{\tau} = \sum_{i=2}^{m} p_{i1} D_i \tag{23.26}$$

i.e. the mean of the transition probabilities p_{i1} from the individual vibrational levels. For a rapid redistribution over the vibrational sub-levels, the individuality of their optical properties is lost. The lifetime of the excited electronic state is inversely proportional to the probability p. It follows that the exponential variation in the number of excited particles, and hence the exponential quenching of luminescence, may be observed for systems characterised by a rapid redistribution of the constituent particles over the various sub-levels of the excited state.

24. POLARISATION OF LUMINESCENCE

It is a direct consequence of both the classical and the quantum-mechanical theory of interaction of radiation with matter that the emission by elementary objects such as atoms and molecules is anisotropic, and that the emitted radiation must be polarised. However, when excited particles are randomly oriented, all the directions become equivalent, and the emitted radiation is unpolarised. Polarised luminescence can only be observed in those media which exhibit a natural or induced anisotropy in the distribution of the constituent particles. In crystals, the degree of polarisation is connected with the disposition of the molecules in the unit cell, while in media with induced anisotropy, it is connected with the nature of the external excitation.

In the following paragraphs we shall be mainly concerned with those media which are completely isotropic prior to the external excitation, and the only reason for the appearance of the anisotropy is the incident radiation. Since electromagnetic oscillations are always transverse, a directed beam of radiation is thus always anisotropic, even if it is unpolarised. When the system is illuminated by isotropic radiation, the resulting luminescence will be unpolarised.

Since the elementary absorbing and emitting centres are anisotropic, the exciting radiation interacts preferentially with those centres which have the appropriate orientation relative to the electric vector. As a result, therefore, only those particles are found in the excited state which have a certain preferred orientation, and the luminescence becomes polarised. The degree of the polarisation is different for different directions of observation.

The anisotropy of luminescence depends not only on the direction of the electric vector in the incident wave and the nature of the elementary emitter, but also on the nature of the processes which proceed in the system while the particles are in the excited state. The latter include above all the interaction with the medium, which reduces the anisotropy in the orientation of the excited particles and thereby the degree of polarisation of the emitted luminescence.

In luminescing gases, collisions between the particles act as a depolarisation process. The longer the lifetime of the excited state, the greater the depolarisation. Depolarisation may also be associated with the rotation of the molecules if the rotational period is comparable with the lifetime of the excited state. In liquid and solid solutions, depolarisation is connected with Brownian motion and depends on the temperature and viscosity of the medium, the dimensions and shape of the molecules and the lifetime of the excited state.

Resonance transport of energy between excited molecules is also a depolarising factor. We have considered similar processes in Section 9, when we were concerned with the transport of energy in linear chains. If, however, the particles interact with each other, the molecule which has absorbed a photon may transmit its excitation energy to some other unexcited molecule. When the interaction is strong enough, this process may occur repeatedly until emission or non-radiative deactivation takes place. Since the molecules participating in the resonance transport of energy may have different orientations, the result of a repeated process of this kind is that the initial angular distribution of the excited molecules is completely suppressed [57]. The degree of depolarisation of luminescence decreases also as a result of the appearance of secondary luminescence (Section 22). The anisotropy of primary luminescence is much less than that of exciting radiation. It follows that the polarisation of secondary luminescence is lower than that of the primary luminescence, and similarly for higher-order luminescence. Since, under certain conditions, the relative fraction of secondary luminescence is quite appreciable, this leads to a general increase in the degree of isotropy of the emitted radiation.

The methods of classical theory can frequently be used in quantitative calculations of the degree of polarisation of luminescence in the linear optics approximation. They were discussed in detail in Section 3. Quantum theory enables us to estimate the limits of applicability of the classical representation, and to elucidate many of the characteristic features of luminescence which are associated with the properties of excited molecules, the transitions between different energy levels and the necessity of taking non-linear corrections into account.

In Section 21 we used equation (21.14) for the rate of luminescence of frequency ν_{ij} averaged over all the orientations of molecules. If we take the anisotropy in the angular distribution of the excited particles into account, we can readily show with the aid of (21.2) and (17.19) that

$$W_{ij\alpha}^{\text{lum}}(\Omega_2)\,d\Omega_2 = h\nu_{ij}d\Omega_2 \int_{\Omega_1} \Big\{ a_{ij}^{\alpha}(\Omega_1, \Omega_2)\, n_i\,(\Omega_1) - b_{ji}^{\alpha}(\Omega_1, \Omega_2)\, u_0^{\alpha}(\nu_{ij}, \Omega_2) \left[n_j(\Omega_1) - \frac{g_j}{g_i}\, n_i(\Omega_1) \right] \Big\} d\Omega_1 \tag{24.1}$$

where $W_{ij\alpha}^{\text{lum}}(\Omega_2)\,d\Omega_2$ is the rate of luminescence of polarisation α which is emitted into the solid angle $d\Omega_2$ by all molecules in a unit volume, $n_i(\Omega_1)\,d\Omega_1$ and $n_j(\Omega_1)\,d\Omega_1$ are the numbers of excited and unexcited molecules with a special axis lying within the solid angle $d\Omega_1$, $a_{jj}^{\alpha}(\Omega_1, \Omega_2)$ is the probability of spontaneous emission of radiation of polarisation α in the direction Ω_2 by a molecule oriented in the direction Ω_1, and $b_{ji}^{\alpha}(\Omega_1, \Omega_2)\, u_0^{\alpha}(\nu_{ij}, \Omega_2)$ is the analogous probability of stimulated transitions under the action of equilibrium radiation. In equation (24.1) we have taken into account the thermal radiation background. It follows from this equation that negative luminescence is also polarised. In the visible region, the second component in (24.1) is as a rule small, and need only be taken into account at very high intensities and temperatures.

According to (24.1), the polarisation characteristics of luminescence depend upon the properties of the molecule itself, the direction, intensity and frequency of the exciting radiation, and on all the depolarising factors. Calculations of $n_i(\Omega_1)$ and $n_j(\Omega_1)$ are in general very complicated, and can only be carried out for specific systems (see Chapters 6-8).

Most frequently, one has to restrict the analysis to the limiting polarisation, i.e. the polarisation of luminescence in the complete absence of all depolarising factors. In practice, depolarisation is reduced by reducing the temperature and increasing the viscosity of the solvent. Calculations become even more complicated if various non-radiative transitions, for example, redistribution of vibrational energy within a complex molecule, take place between absorption and emission [58]. Knowing $W^{\text{lum}}_{ij\alpha}(\Omega_2)$, it is easy to find the degree of polarisation of luminescence in a given direction from the formulae of Section 6.

Absorption and emission by the same oscillator is frequently discussed in classical theory. In quantum theory this corresponds to a transition between the same pair of energy levels, including an unexcited level. We encountered a similar situation in the case of resonance emission by atoms, or the excitation of the luminescence of complex molecules in the long-wave absorption band. In practice, the absorption band frequently corresponds to one electronic transition, while the luminescence band corresponds to another of lower frequency. A radiative or non-radiative transition from one level to the other occurs between absorption and emission. As a result, the polarisation of luminescence of given frequency is very dependent on the frequency of the exciting radiation and has a polarisation spectrum. In quantum theory this is allowed for by taking into account the dependence of $n_i(\Omega_1)$ and $n_j(\Omega_1)$ on the frequency of the exciting radiation. The classical theory assigns to each transition of frequency ν_j its particular elementary oscillator. These are rigidly attached to the molecule and are at definite angles to each other. Knowing the angles between the absorbing and emitting oscillators, it is possible to find the polarisation of luminescence as a function of the frequency of the absorbing oscillator.

25. CLASSIFICATION OF TYPES OF SECONDARY EMISSION

Equilibrium and non-equilibrium emission

The limiting and the most thoroughly investigated case of emission by a real assembly of atoms and molecules is the equilibrium thermal emission of a black body. This is a very general type of emission which masks all the specific properties of the emitting particles, although the elementary processes which give rise to the black-body radiation are no different than the elementary processes in other radiating systems. The total loss of individual particle properties which occurs under the conditions of thermal equilibrium is intimately connected with the strong interaction between elementary processes. Any black body which is in thermal equilibrium at a particular temperature is characterised by its specific equilibrium emissive power. The equilibrium emission of grey bodies of finite dimensions exhibits, if only slightly, the individual properties of the particular material. The loss of individual particle properties is connected not only with their mutual interactions, but also with the repeated occurrence of absorption, emission and scattering, intermolecular transport of excitation energy, reflection at the walls, interference phenomena and so on.

In addition to equilibrium systems there is an infinite number of non-equilibrium systems. They appear as a result of departure from thermodynamic equilibrium due to some external excitation. Emission by non-equilibrium systems depends not only on the nature and the degree of the external excitation, but also on the properties of the system itself. Moreover, the effect of the external excitation will depend both on the properties of the medium surrounding the system, and above all its temperature, and on the rate of exchange of energy between the medium and the system. If, for example, the temperatures of the medium and of the system are very high, small external disturbances will not succeed in modifying the state of equilibrium, and the emission by the system will as before be purely thermal. If, on the other hand, the temperature of the medium is low, external excitation will modify the equilibrium distribution over the energy levels, with the result that non-equilibrium emission will appear against the background of thermal emission. The resultant emission of any system must in some measure include the ordinary thermal emission. This important idea was introduced as far back as 1888 by Wiedemann who introduced the term 'luminescence' to characterise non-equilibrium processes of emission of radiation. According to Wiedemann the chief characteristic of luminescence is the difference between the emissive power of a body and its thermal emissive power (in the particular spectral interval) as can be seen from Equation (25.1).

$$W^{\mathrm{lum}}(\nu) = W^{\mathrm{em}}(\nu) - W^{\mathrm{th.em}}(\nu) \tag{25.1}$$

Wiedemann's definition of luminescence correctly represents some of the characteristic features of the luminescence process, and serves as a basis for the introduction of the concept of negative luminescence. We have used it in deriving the fundamental formula (29.14) for the rate of luminescence. This definition, however, is incomplete. In a series of papers, Vavilov showed that the physical significance of the concept of luminescence as used in modern physics is much narrower than the general concept of non-equilibrium emission. Non-equilibrium emission includes not only luminescence, but also Rayleigh scattering (including reflection), combination (Raman) scattering, and also Vavilov-Cherenkov radiation and bremsstrahlung. A further important type of non-equilibrium emission was discovered in recent years, namely, the emission by media with negative absorption coefficients, i.e. stimulated emission, which again cannot be classified as luminescence.

The experimental separation of the various types of non-equilibrium emission is not an easy task, and cannot always be achieved. It is particularly difficult to separate scattering from photoluminescence, since both are common types of secondary emission produced as a result of a departure from equilibrium due to the incidence of the primary radiation.

By analysing the various types of non-equilibrium emission, Vavilov showed that one of the most characteristic features of luminescence is the presence of an afterglow with decay time longer than the period of the electromagnetic vibrations after the exciting agent has been removed. This is intimately connected with the possibility of quenching luminescence as a result of an increase in the non-radiative transition rate. A systematic application of this criterion enabled Vavilov to develop a correct interpretation of the many types of emission observed in practice, including the Vavilov-Cherenkov emission. Vavilov's criterion is almost entirely valid whenever luminescence arises as a result of non-radiative methods of excitation (thermal excitation, excitation by electron impact, and so on). There are, however, residual difficulties connected with the necessity of separating luminescence from the radiation emitted by quantum generators (masers). A negative absorption coefficient may be produced by non-radiative excitation, and the emission may exhibit afterglow, and, moreover, the generation of radiation may be quenched by impurities, or occasionally by increased temperature.

Whenever secondary non-equilibrium emission is connected with the incidence of a primary radiation, the use of the above criterion is even more difficult, particularly in the case of resonance emission. According to this criterion, resonance emission under the action of an external flux of radiation must be classified as photoluminescence, since it exhibits the characteristic afterglow and is easily quenched. At the same time, experiment shows that when the density of the emitting vapour is increased, the resonance emission is transformed to specular reflection, showing that it is coherent with the incident radiation. Therefore, some of the properties of resonance emission cannot be fitted into the concept of photoluminescence; resonance emission resembles ordinary scattering in many respects. This uncertainty is reflected in some measure of confusion in the terminology. Some authors refer to it as resonance fluorescence, others as resonance scattering, still others, following Wood, call it resonance emission.

The facts considered above show that the single experimental criterion involving the presence of afterglow and quenching is not sufficient for the classification of various types of secondary emission. In his earlier work Vavilov showed that other experimental criteria are also inadequate for reliable classification. For example, one cannot use the coherence criterion, since coherence is absent not only from luminescence, but also from combination (Raman) scattering.

We thus see that the classification of various types of secondary non-equilibrium emission cannot be based on its external characteristics, and it is essential to consider the mechanism responsible for the processes occurring in the system under the action of the external radiation, i.e. to analyse the transformation of radiation within the material.

In concluding the present section we must emphasise that the division of the resultant emission of a system into equilibrium and non-equilibrium emission is possible only within the framework of linear optics, i.e. at low densities of the emitted radiation. This is a direct consequence of the general theory of transport of radiant energy within regions of finite linear dimensions. The integro-differential transport equation given by (22.7), which we introduced in connection with secondary processes, is in general non-linear. It determines the total density of excited particles n_2 at each point. If there is no radiation from external sources ($u_2 = 0$), and as is usual $d_{12} = d_{21} e^{-h\nu/kT}$, (22.7) leads to the usual equilibrium formula

$$n_2^{eq} = \frac{e^{-h\nu/kT}}{1 + e^{-h\nu/kT}} n$$

Any non-zero intensity of exciting radiation leads to a departure from the equilibrium distribution. In accordance with Wiedemann's definition, it seems useful to seek the solution of (22.7) in the form

$$\Delta n_2(\bar{r}) = n_2(\bar{r}) - n_2^{eq}(\bar{r}) \tag{25.2}$$

However, in view of the non-linearity of (22.7), it cannot be separated into two independent equations, one in $n_2^{eq}(\bar{r})$ and the other in $n_2(\bar{r})$, and therefore the separation of the resultant emission into equilibrium and non-equilibrium contributions cannot, in general, be carried out.

Equation (22.7) is linear only under certain special conditions. These include the many special cases when one can neglect stimulated emission, or when the density of the external radiation is small and so is the departure from equilibrium, i.e.

$$n_2(\bar{r}) - n_2^{eq}(\bar{r}) \ll n_2^{eq}(\bar{r})$$

These conditions are usually satisfied in luminescence studies, and therefore (25.2) is valid. However, in the special case of the generation of radiation within a plane-parallel layer with a negative absorption coefficient it is not possible to isolate the equilibrium radiation.

Scattering, combination (Raman) scattering and luminescence

The rigorous criteria which are necessary for the classification of various types of secondary emission can only be established on the basis of quantum electrodynamics [59]. Classical theory does not in general consider stationary luminescence, while quantum mechanics does not describe spontaneous emission. However, the results of classical theory are very illustrative, and therefore it will be convenient to summarise their main consequences. In the case of scattering, they will lead to the same results as quantum electrodynamics, and therefore correctly reflect the true state of affairs.

Absorption and induced emission by a classical dipole has been discussed in detail in Chapter 1 (Section 2). The rate of emission is given by (2.46), and the rate of absorption by (2.35). The emission by a dipole is coherent with the incident radiation and is therefore regarded as scattered radiation. The secondary emission energy yield is

$$\gamma_{\text{эн}} = \frac{W^{\text{em}}(\nu)}{W^{\text{abs}}(\nu)} = \frac{\gamma_{\text{em}}}{\gamma_{\text{em}} + \gamma_{\text{fr}}} \tag{25.3}$$

It is equal to unity only for processes accompanied by the outflow of energy into the surrounding medium (when $\gamma_{\text{fr}} = 0$). Any increase in γ_{fr}, for example, by the addition of gaseous impurities, leads to a reduction in the yield, and therefore to the quenching of the scattering process. The latter undoubtedly exists but is usually masked. Figure 5.10 shows the rate

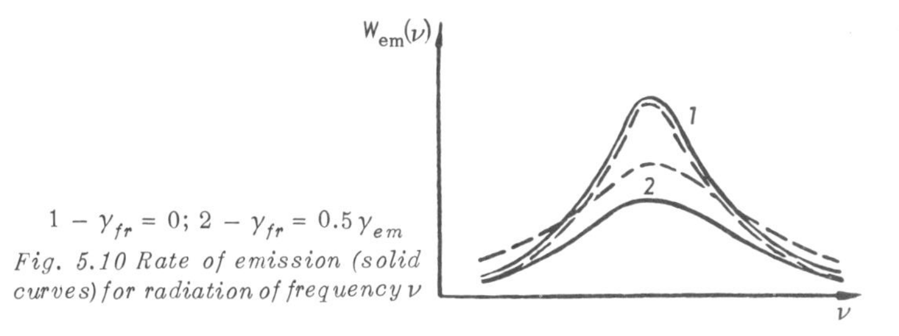

$1 - \gamma_{fr} = 0$; $2 - \gamma_{fr} = 0.5\gamma_{em}$

Fig. 5.10 Rate of emission (solid curves) for radiation of frequency ν

of emission as a function of frequency of the incident radiation for two values of γ_{fr}. When the frequency of the incident radiation is very different from the natural frequency ν_0 of the dipole, the intensity of the scattered light is independent of γ_{fr}. The reduction in the rate of emission with increasing γ_{fr} is associated not with a reduction in the intensity of the emission, but with an increase in the absorption due to the broadening of the absorption line. In the region where $\nu \sim \nu_0$, i.e. for resonance absorption, an increase in γ_{fr} is accompanied by a decrease in the intensity of the emission and in its yield.

It is most important to note that, according to classical theory, the emission spectrum under stationary conditions (scattered spectrum) is very dependent on the spectral composition of the exciting radiation only when the incident radiation contains narrow spectral lines. The profiles of the emission line and the incident line are then identical. Different scattered lines will, of course, have different intensities. If, on the other hand, the incident radiation has a continuous spectrum, then in accordance with (2.46), when E_0=const, the secondary emission spectrum contains a line or a band whose profile is determined exclusively by the properties of the dipole, and is described by the dispersion formula. However, the character of the transformation of light in the two cases is the same. By dividing the incident light into narrow bands, one can see that each of them is transformed without change in profile and frequency (but with a change in intensity).

According to classical theory, afterglow is always present after the removal of illumination, and the afterglow intensity is a function of frequency and of time. For time-integrated quantities the intensity distribution is given by (2.17). The duration of this afterglow is of the order of the corresponding quantity for luminescence (when $\gamma_{fr}=0$, $\tau \sim 10^{-8}$sec). The afterglow spectrum is in general quite different from the spectrum of the incident radiation; it consists of the natural line of the dipole with a profile given by (2.17). At incident frequencies well away from the resonance frequency, there is a rapid change in the emission spectrum as soon as excitation is removed, and this might be interpreted as the absence of afterglow. The spectral composition of the secondary emission under stationary conditions is the same as the spectral composition of afterglow only in the case of resonance illumination by a continuous spectrum. In this case the properties of the secondary emission resemble those of fluorescence, although in fact it does not differ in any way from the secondary emission in the absence of resonance, which is naturally regarded as scattering.

Thus if we consider that the classical theory of induced emission by a dipole correctly reflects the basic properties of scattering and resonance emission of radiation, then the criteria involving the duration and the ability to quench cannot, in fact, be used for the classification of secondary emission; scattering, similarly to luminescence, is quenched and is accompanied by afterglow.

The basic information which is necessary for the classification of secondary emission by the methods of quantum electrodynamics is discussed in detail in Section 15 (Chapter 3). From the point of view of quantum electrodynamics the transformation of light by matter involves the disappearance of primary and the appearance of secondary photons. In this sense there is no difference between scattering, resonance emission and luminescence. The only distinction may lie in whether or not the disappearance and appearance of photons occurs in direct succession, or whether they are separated by intermediate processes. Among such intermediate processes are radiative or non-radiative transitions between energy levels. In complicated molecules they may include redistribution of vibrational energy over the degrees of freedom.

The transformation of radiation without the participation of intermediate processes is described by (15.17) and (15.19) for a two-photon process. In the analysis of three-photon processes we introduced the formula given by (15.22) which describes the appearance of two types of photon. Photons belonging to the first type have frequency, say ν'', and are produced without the participation of intermediate processes. Photons of the second kind are of frequency ν''' say, and appear after the completion of the first process which may be regarded as intermediate. Averaging of (15.22) over all ν''' yields the law of transformation of the incident photons ν' into the secondary photons ν'' as determined by (15.23)-(15.25). Averaging of (15.22) over all photons ν'' yields (15.27) for the $\nu' \to \nu'''$ transformation with the participation of a single intermediate process, in this case, an optical process. A detailed analysis of the spectral properties of such radiations is given in Section 15.

In complicated systems, intermediate processes may be very different with different conditions. The rate of emission of a particular part of secondary emission produced without the participation of intermediate processes is given by

$$W_{i\to j}(\nu' \to \nu'') = \frac{\nu''^2 u_0(\nu')}{2\pi h^2 c^4 \nu'^2} \left| \sum_k \left[\frac{(P_{\lambda'}^{-})_{jk}(P_{\lambda''}^{+})_{ki}}{\nu_{ki} - \nu' - i\gamma_k'} + \frac{(P_{\lambda'}^{+})_{jk}(P_{\lambda''}^{-})_{ki}}{\nu_{ki} + \nu''} \right] \right|^2 |a_i|^2 \delta(\nu_{ij} - \nu' + \nu'') \tag{25.4}$$

This is obtained by averaging over all intermediate radiative and non-radiative processes [60]. In this expression,

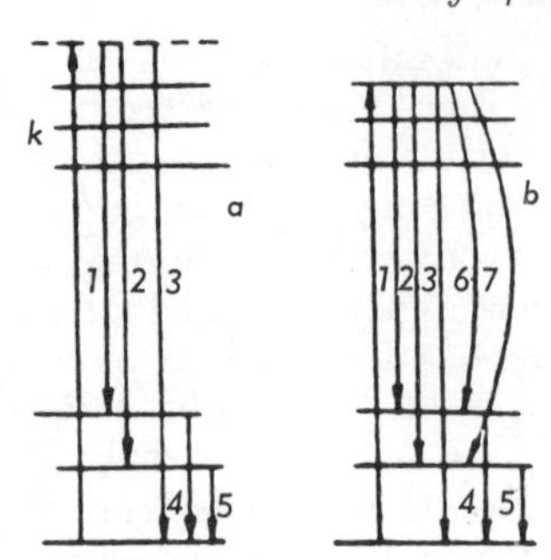

1, 2 - Raman scattering; 3 - Rayleigh scattering; 4, 5 - luminescence; 6, 7 - non-radiative intermediate transitions

Fig. 5.11 Various types of secondary emission in the case of resonance (b) and non-resonance (a) excitation

$|a_i|^2$ is the probability of finding the system in the i-th level and $4\pi\gamma_k'$ is the probability that the system transforming the radiation will leave the intermediate state k (Fig. 5.11). It consists of the probability of transitions with the emission of radiation, $2\gamma_k^{em}$ (spontaneous emission), the probability of non-radiative transitions, $2\gamma_k^{non}$ to lower states, and the probability of redistribution processes $2\gamma_k^{re}$. The sum $2\gamma_k^{non}+2\gamma_k^{re}$ equals the total probability of non-radiative transitions. The remaining symbols in (25.4) have the same meaning as before.

If the frequency of the exciting radiation is not the same as one of the frequencies of the system (Fig. 5.11a), then there is a non-resonance transformation of radiation. When $\nu' = \nu_{ki}$ the transformation is resonant in nature (Fig. 5.11b). Summation is carried out over all intermediate levels k.

When $i = j$, the expression given by (25.4) describes the transformation of radiation without change in frequency ($\nu' = \nu''$). This secondary emission has the properties considered in the case of the classical dipole. The expression given by (25.4) differs from the corresponding classical expression only in the model adopted for the medium. In the case of a harmonic oscillator ($\gamma_k^{non} = \gamma_{fr}$, $\gamma^{re} = 0$) they are identical. Emission by a classical dipole, which is given by (25.4), must therefore be regarded as scattering.

When $i \neq j$, the secondary emission (25.4) contains the combination frequencies $\nu'' = \nu' - \nu_{ij}$. It consists of lines whose position and profiles depend on the spectral composition of the incident radiation and is called combination or Raman scattering.

The energy yield of this transformation process can easily be found by first calculating the number of disappearing photons. The result is

$$\gamma_{en} \sim \frac{\gamma_{em}}{\gamma_{em} + \gamma_{non} + \gamma_{re}} \tag{25.5}$$

where γ^{em}, γ^{non}, γ^{re} are the probabilities γ_k^{em}, γ_k^{non} and γ_k^{re} averaged over all k.

The transformation of radiation which occurs with the participation of only one intermediate process is described by (15.27) and has the energy yield

$$\gamma_{en} = \frac{\gamma_{re}}{\gamma_{em} + \gamma_{non} + \gamma_{re}} \frac{\gamma_{em}}{\gamma_{em} + \gamma_{non} + \gamma_{re}} \tag{25.6}$$

It follows from (15.27) that this part of the emission consists of a set of characteristic lines of the material, whose positions and profiles are independent of the conditions of excitation. The only quantities which depend on the frequency and intensity of the exciting radiation are the C_k', which are closely connected with the population of level k.

In contrast to the emission described by (25.4) for which, well away from resonance, a reduction in the yield is not accompanied by a change in the intensity, a change in the yield of the emission given by (15.27) is always accompanied by a corresponding change in the integral intensity. Emissions which appear as a result of the transformation of radiation with the participation of two, three or a greater number of intermediate redistributions of energy, consist of the characteristic lines of the medium with the same position in the spectrum and the same profile. The intensity of lines corresponding to processes with a different number of intermediate processes may be different, depending on the values of C_k'. All such lines are superimposed upon each other, and cannot be separated experimentally. They must therefore be considered together as a single line.

The quantum yield of the transformation of radiation with the participation of all intermediate processes is of the order of

$$\gamma_{en} \sim \frac{\gamma_{re}}{\gamma_{em} + \gamma_{re} + \gamma_{non}} \frac{\gamma_{em}}{\gamma_{em} + \gamma_{non}} \tag{25.7}$$

In complicated systems the redistributions proceed very rapidly and most of the radiation is emitted after a large number of redistributions so that the band profile is independent of the conditions of illumination. It is evident from

(25.5) that in such systems the transformation of radiation without the participation of redistributions, i.e. scattering, is almost entirely quenched by the redistributions which do take place. All the emission occurs with the participation of redistributions and its yield is entirely determined by the usual formula

$$\gamma_{en} = \frac{\gamma_{em}}{\gamma_{em} + \gamma_{non}} \tag{25.8}$$

which is identical with (25.7) for very large γ_{re}.

The basic equations of quantum optics (Sections 13 and 15) can be used without much difficulty to determine the properties of the emission which continues after the illumination ceases. Calculations show that any stationary emission ends with afterglow whose profile and line positions are independent of the conditions of excitation. The entire afterglow consists of characteristic lines, and therefore its spectrum is the same as the stationary spectrum appearing after the redistributions, and differs from the secondary emission spectrum which is treated as scattering.

From the point of view of quantum electrodynamics, secondary emission may therefore be successively separated into two parts which differ from each other by the mechanism of formation and have quite different properties. One of these is the emission produced with the participation of intermediate transitions. It is characterised by (1) profiles and positions of its constituent lines which are independent of the properties of the incident radiation, (2) a reduction (sometimes an increase) in the quantum yield is always accompanied by a corresponding change in the intensity of the emission, and (3) the stationary spectrum is the same as the afterglow spectrum. All forms of emission which are usually looked upon as photoluminescence have similar properties. It is precisely for this reason that this part of secondary emission is usually treated as photoluminescence. It may be noted, however, that the above properties are clearly exhibited only by line spectra. In other cases, when the lines overlap, the profiles and positions of the bands depend both on the properties of the medium and of the incident radiation.

The second part consists of emission produced without the participation of intermediate processes. In classical theories it is regarded as emission due to forced vibrations of a dipole. This part of the emission has the following characteristic properties: (1) the profiles and positions of the lines depend on the spectral composition of the incident radiation, (2) a reduction in the yield is not always accompanied by a reduction in the intensity of the emission, and (3) the stationary spectrum is not, in general, the same as the corresponding afterglow spectrum. It is only in one special case, i.e. in the case of illumination by a broad resonance line, that the emission has properties that appear to be the same as the properties of emission with the participation of redistributions.

We thus see that any multi-photon emission process which occurs without the participation of redistributions must be regarded as Rayleigh (arrow 3 in Fig. 5.11a) and combination (Raman) scattering (arrows 1 and 2). From this point of view resonance emission by low-density atomic and simple molecular vapours must be regarded as resonance scattering (in Fig. 5.11b Rayleigh resonance scattering is represented by arrow 3, while Raman scattering is represented by arrows 1 and 2). The emission produced after the completion of the intermediate process indicated in Fig. 5.11 by arrows 4 and 5 is classified as photoluminescence. Among the intermediate processes there are also the non-radiative transitions 6 and 7.

The optical properties of specific models of matter, i.e. systems of particles with two or more levels and also of the harmonic oscillator will be considered later. In most cases we shall be concerned with luminescence of such systems rather than with resonance or Raman scattering. Even the emission by a system of particles with two energy levels is classified as luminescence. This is possible if it is assumed that some intermediate processes leading to a change in the phase of the secondary emission occur between absorption and emission. The intensities due to the individual atoms can then be added, since the waves are not coherent.

It follows from (25.5) and (25.7) that the ratio of the photoluminescence to the scattering yield is

$$\frac{\gamma_{en}^{lum}}{\gamma_{en}^{scat}} \sim \frac{\gamma_{re}}{\gamma_{em} + \gamma_{non}} \tag{25.9}$$

With increasing probability of non-radiative transitions,

γ_{en}^{lum} falls off more rapidly than γ_{en}^{scat}. It follows that by using different quenchers, it is possible to suppress completely luminescence without altering scattering. This explains the major practical significance of the quenching criterion for the experimental separation of scattering from photoluminescence.

Media with negative absorption coefficients

In Sections 19 and 20 we considered the special kind of emission which is associated with negative absorption coefficients in highly non-equilibrium media. This cannot be classified as either scattering or luminescence, although it appears when an external electromagnetic field is applied to the medium.

As we have seen, scattering and luminescence are associated with the disappearance of primary and the appearance of secondary photons, and may consist of two-photon or multi-photon processes. The emission by media with negative absorption coefficients, which is much simpler and arises as a result of single-photon processes, was discussed in detail in Section 14.

In the next approximation of quantum electrodynamics, i.e. when two-photon processes are considered, there are further terms which characterise radiation with similar properties. This arises as a result of an excess population in the upper level due to the absorption of the primary photon. We shall not discuss this in detail because it can usually be neglected.

The distinction between media with positive and negative absorption coefficients is not fundamental, from the point of view of the optical properties of volume elements. In the first case, the interaction of light with the medium leads to an attenuation of the incident radiation, whereas in the second, it leads to its amplification. The attenuation is accompanied by an increase in the energy of the absorbing particles, whereas amplification of the beam is accompanied by the transfer of energy from the medium to the field. The mechanism of the two interaction processes is identical, except that the distribution over the energy levels is different. The two processes are mutually symmetrical and differ only in direction.

In the case of stimulated emission by media with negative absorption coefficients the incident photons do not disappear but simply stimulate the emission of further photons. In the terminology which we have adopted, they cannot therefore be referred to as secondary.

Emission by media with negative absorption coefficients is quite different from luminescence. The latter involves the spontaneous emission of a photon, i.e. emission due to the interaction with the zero-point electromagnetic field. A luminescing atom is the simplest generator of radiation. Emission by individual non-interacting atoms is not coherent, and the phases of the emitted waves are random. Emission by media with negative absorption coefficients arises only during interactions with radiation. The individual excited particles do not generate radiation, but simply modify the incident radiation. Photons which are produced as a result of the interaction with the field are coherent both with each other and with the incident photons.

In media with negative absorption coefficients luminescence (spontaneous emission) and stimulated emission always occur side by side. The main experimental criteria which can be used to separate them are (1) coherence of the emitted photons and (2) differences between distribution functions for different directions of propagation. The directions of propagation of the stimulated and incident radiations are the same. Photons produced as a result of spontaneous emission propagate in different directions with a probability which is determined by the properties and orientation of the elementary sources.

Objects having finite linear dimensions are of particular importance in the study of the optical properties of media with negative absorption coefficients. It is then important to take into account not only the amplification of radiation reaching the system from outside, but also the amplification of waves produced in the system itself. Reflection by the walls (see Section 20) can be allowed for at the same time. As a result of the continued repetition of these processes, the object as a whole may become a generator of radiation, even in the absence of external stimulation. Generation will always set in when losses associated with the escape of radiation from the system are compensated by freshly produced radiation.

It was noted at the beginning of this section that the transfer theory is in general non-linear. Linear approximations are admissible only when we can ignore the effect of radiation on the absorption coefficient, i.e. on the distribution of the particles over the energy levels. In media with negative absorption coefficients (finite volumes) this condition is frequently unfulfilled, so that the theory of such media can only be constructed within the framework of non-linear optics. The separation of the emission into equilibrium and non-equilibrium components is then meaningless. In particular, one cannot use formulae such as (21.14) which equate the rate of non-equilibrium emission to the difference between total and thermal emission.

The division of total emission into equilibrium and non-equilibrium components is meaningless not only for the finite volume as a whole, but also for each elementary volume within it. At the same time we can, as before, separate the spontaneous emission of intensity $A_{ij}n_i h\nu_{ij}$ from the stimulated emission of intensity $B_{ij}(n_j - n_i)h\nu_{ij}$ in each elementary volume. In the object as a whole, the directed stimulated emission is usually predominant, while spontaneous emission acts as background noise.

6

Optical Properties of the Harmonic Oscillator

26. QUANTUM-MECHANICAL PROPERTIES OF THE HARMONIC OSCILLATOR

Wave functions and energy levels

The harmonic oscillator is the basic model of matter in classical electrodynamics and retains its importance in quantum theory. It is used not only as a model of matter, but also as a model of the electromagnetic field, which can be represented by a set of such oscillators.

The development of quantum theory, beginning with its original form and ending with the modern quantum field theory, depends to a considerable extent on the properties of this model. The oscillator plays a particularly important role in the theory of emission by quantum systems where it is the basis for the analysis of many complicated problems.

In quantum mechanics, a harmonic oscillator is defined as a system whose Hamiltonian is of the form

$$\mathrm{H} = \frac{p_x^2}{2m} + 2\pi^2 m \nu_0^2 x^2 \tag{26.1}$$

where $\nu_0 = \frac{1}{2\pi}\sqrt{\frac{k}{m}}$ is the natural frequency, and k and m are the quasi-elastic constant and the mass of the oscillator respectively. It is assumed for the sake of simplicity that the problem is one-dimensional, i.e. the oscillations of the particle occur along the x axis. The energy eigenvalues and the wave functions which satisfy the Schroedinger equation for the oscillator

$$-\frac{h^2}{8\pi^2 m}\frac{d^2\psi}{dt^2} + 2\pi^2 m \nu_0^2 \psi = \mathrm{E}\psi \tag{26.2}$$

are given by

$$\mathrm{E}_v = (v + 1/2)h\nu_0, \quad v = 0,\ 1,\ 2,\ \ldots \tag{26.3}$$

$$\psi_v(\xi) = \mathrm{M}_v e^{-\xi^2/2}\mathrm{H}_v(\xi) \tag{26.4}$$

where $\mathrm{M}_v = (2^v v!\sqrt{\pi})^{-1/2}$ is the normalising factor

$$\xi = \frac{x}{x_0} = \sqrt{\frac{4\pi^2 m\nu_0}{h}}\,x$$

$\mathrm{H}_v(\xi)$ is the Chebyshev-Hermite polynomial of order v which is given by

$$\mathrm{H}_v(\xi) = (-1)^v e^{\xi^2}\frac{d^v e^{-\xi^2}}{d\xi^v} = \sum_{k=0}^{\left[\frac{v}{2}\right]} \frac{(-1)^k v!}{k!(v-2k)!}(2\xi)^{v-2k} \tag{26.5}$$

and $\left[\frac{v}{2}\right]$ is the integral part of $v/2$. According to (26.3) the oscillator has an infinite set of equidistant energy levels. The lowest energy is not zero, as predicted by the classical theory, but is $1/2h\nu_0$.

The wave functions for the first three states can readily be written out using (26.4) and (26.5):

$$\psi_0(x) = \frac{1}{\sqrt{x_0\sqrt{\pi}}}\,e^{-x^2/2x_0^2} \tag{26.6a}$$

$$\psi_1(x) = \frac{1}{\sqrt{2x_0\sqrt{\pi}}} e^{-x^2/2x_0} 2 \frac{x}{x_0} \tag{26.6b}$$

$$\psi_2(x) = \frac{1}{\sqrt{2^2 \cdot 2x_0\sqrt{\pi}}} e^{-x^2/2x_0^2} \left(4 \frac{x^2}{x_0^2} - 2\right) \tag{26.6c}$$

The first of these has a maximum at $x = 0$, and does not vanish anywhere apart from infinitely distant points. The second vanishes at $x = 0$ and reaches its maxima at $x = \pm x_0$. In other words, it has one node and two maxima. The third function has two nodes (at $x = \pm x_0\sqrt{2}$). It may be shown that the number of nodes of an eigenfunction is always equal to the quantum number v [61].

Figure 6.1 shows plots of the functions given by (26.6),

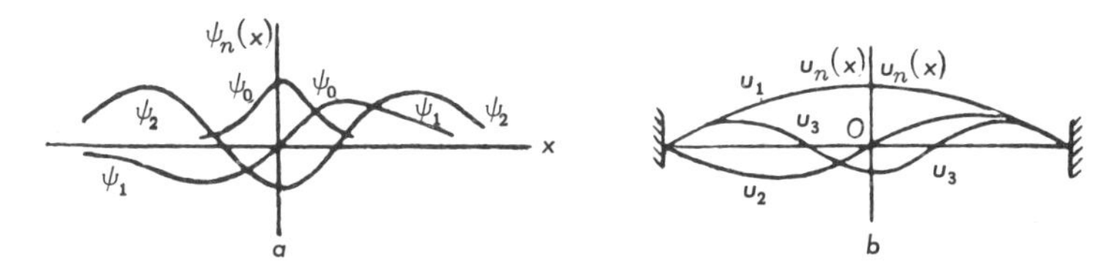

U_1 - fundamental, U_2, U_3 - first two harmonics

Fig. 6.1 Wave functions of (a) harmonic oscillator with v = 0, 1, 2 *and (b) oscillations of a stretched string*

together with graphs of functions describing the vibrations of a string with fixed ends (the fundamental and the first two harmonics). As can be seen, there is a considerable similarity between these graphs.

Probability of localisation of charge

It is well known that a classical operator having the lowest energy $E = 0$ (amplitude $a = 0$) is located precisely at the point of equilibrium at $x = 0$. When $E \neq 0$, the probability of finding it between x and $x + dx$ is

$$W_{cl}(x)dx = \frac{1}{\pi a} \frac{dx}{\left(1 - \frac{x^2}{a^2}\right)^{1/2}} \tag{26.7}$$

The quantum-mechanical probability $W_q(x)dx$ of finding the oscillator of energy E_v between x and $x + dx$ is determined by the square of the modulus of the corresponding wave function:

$$W_q(x)dx = |\psi(x)|^2 dx \qquad (26.8)$$

As can be seen from Fig. 6.2, which shows plots of the classical and quantum-mechanical probabilities, $W_{cl}(x)$ has a maximum at the turning points of the path of the oscillator. The quantum-mechanical probability is a much more complicated function of x, although with increasing level number,

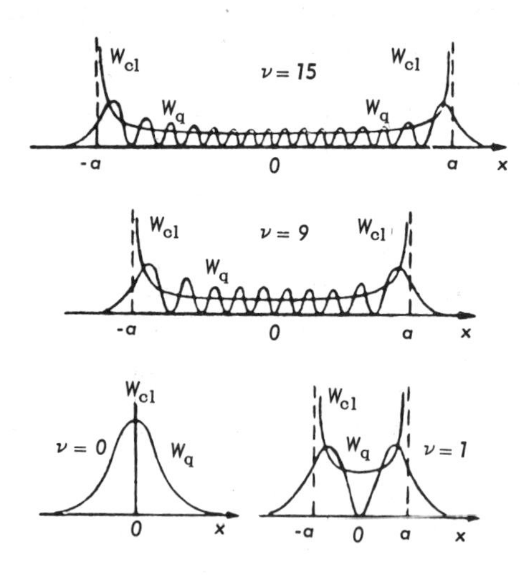

Fig. 6.2 Classical and quantum-mechanical position probabilities for an oscillating charge. The energy of the oscillator at rest is assumed to be zero

it increases near the turning points and assumes a form similar to that of $W_{cl}(x)$. This result is expected from the correspondence principle. It also follows from the elementary criterion of the applicability of classical mechanics given by Fock [62]. In the case of the harmonic oscillator this criterion is

$$v + {}^1/_2 \gg 1 \qquad (26.9)$$

Classical and quantum mechanics of the harmonic oscillator lead to results which become more and more similar as the energy of the oscillator increases.

Mean position

Let us suppose that the motion of the oscillator is described by the wave function

$$\psi(x, t) = \sum_{v} C_v(t)\, \psi_v(x) e^{-2\pi i E_v t/h} \tag{26.10}$$

The mean value of x is then given by

$$\begin{aligned}\bar{x}(t) &= \int_{-\infty}^{\infty} \psi^*(x, t) x \psi(x, t) dx = \sum_{v'}\sum_{v} C^*_{v'}(t) x_{v', v} C_v(t) \\ &= \sum_{v'}\sum_{v} C^*_{v'}(0) x_{v'v}(t) C_v(0)\end{aligned} \tag{26.11}$$

where

$$\begin{aligned}x_{v', v}(t) = &\sqrt{\frac{h}{4\pi^2 m \nu_0}}\left(\sqrt{\frac{v}{2}}\,\delta_{v-1, v}\right. \\ &\left.+\sqrt{\frac{v+1}{2}}\,\delta_{v+1, v}\right) e^{-2\pi i \nu_{vv'} t}\end{aligned} \tag{26.12}$$

are the matrix elements of the position coordinate operator. According to the last expression, $x_{v', v}$ differ from zero only when $v' = v \pm 1$. All the $x_{v', v}$ varv with the same frequency ν_0 and therefore (26.11) may be written in the form

$$\bar{x}(t) = a \cos(2\pi \nu_0 t + \varphi) \tag{26.13}$$

where a and φ are the amplitude and initial phase respectively. The expression given by (26.13) can also be obtained from Ehrenfest's theorem, which states that the mean values of physical quantities obey the classical law of motion

$$m \frac{d^2\bar{x}}{dt^2} = -\frac{\overline{\partial V}}{\partial x}$$

For the oscillator

$$\frac{\overline{\partial V}}{\partial x} = k \int \psi^* x \psi\, d\bar{x} = k\bar{x} \tag{26.14}$$

and therefore

$$m\frac{d^2\bar{x}}{dt^2} + k\bar{x} = 0 \qquad (26.15)$$

This is satisfied by (26.13). The equations of classical mechanics hold rigorously for the mean value $\bar{x}$. This is connected with the special form of the potential function for the harmonic oscillator which is quadratic in x. Even for a small anharmonic oscillator, whose potential function contains higher powers of x, we have

$$\overline{\frac{\partial V}{\partial x}} \neq \frac{\partial V}{\partial \bar{x}}$$

and Ehrenfest's theorem does not lead to the classical equation of motion. The identity of the results of classical and quantum mechanics for the mean value $\bar{x}$ for the harmonic oscillator is, of course, the basic reason for the identity of the results of the classical and quantum theories of emission by the oscillator.

Oscillating wave packet

Consider the time-dependent Schroedinger equation for the harmonic oscillator:

$$i\frac{h}{2\pi}\frac{\partial \Psi}{\partial t} = \left[-\frac{h^2}{8\pi^2 m}\frac{\partial^2}{\partial x^2} + 2\pi^2 m\,\nu_0^2\right]\Psi \qquad (26.16)$$

We shall seek the solution $\Psi(x, t)$ in the form of the superposition of eigenfunctions

$$\Psi(x, t) = \sum_v C_v \psi_v^0 e^{-2\pi iE_v t/h} \qquad (26.17)$$

Suppose that, at the initial time, $\Psi(x, 0)$ describes the wave packet shown in Fig. 6.2 (see the plot of W_q for $v = 0$). The 'centre of gravity' of the packet is at a distance a from the position of equilibrium $x = 0$, so that

$$\Psi(x, 0) = \sum_v C_v \psi_c^0 = \frac{1}{\sqrt{x_0\sqrt{\pi}}} e^{-(x-a)^2/2x_0^2} \qquad (26.18)$$

If we now multiply this expression by $\psi_m^{0*\,1}$ and integrate with respect to x between $-\infty$ and $+\infty$, we have, using (26.4)

$$C_v = M_v \frac{1}{\sqrt{x_0\sqrt{\pi}}} \int\limits_{-\infty}^{\infty} H_v(\xi) e^{-\xi^2/2} e^{-(\xi-\xi_0)^2/2} d\xi$$
$$= \left(\frac{a}{x_0}\right)^v \frac{e^{-a^2/4x_0^2}}{\sqrt{x_0 2^v v!}} \tag{26.19}$$

where $\xi_0 = a/x_0$. This integral can be evaluated with the aid of the generating function for the Chebyshev-Hermite polynomials [63].

Substituting (26.19) into (26.17) we have, after some transformations

$$\Psi(x, t) = \frac{1}{\sqrt{x_0\sqrt{\pi}}} \exp\left\{-\frac{1}{2}[\xi - \xi_0 \cos(2\pi i \nu_0 t)]^2\right.$$
$$\left. - i\left[\pi\nu_0 t + \frac{\xi_0^2}{4}\sin(4\pi\nu_0 t) - \xi\xi_0 \sin(2\pi\nu_0 t)\right]\right\} \tag{26.20}$$

The probability of finding the particle at a particular point is given by

$$|\Psi(x, t)|^2 = \frac{1}{\sqrt{x_0\sqrt{\pi}}} e^{-(x - a\cos 2\pi\nu_0 t)^2/2x_0^2} \tag{26.21}$$

It will readily be seen that $\Psi(x, t)$ describes the wave packet executing harmonic oscillations at the natural frequency of the oscillator. The form of the packet remains constant, and as $a \to 0$, the wave function tends to the eigenfunction for the zero state (26.6a).

Probability of radiative transitions

The quantum-mechanical properties of the oscillator described above are not connected with the presence of its electric charge. Such properties are exhibited by any mechanical system for which the Hamiltonian is of the form given by (26.1). We shall however, only be interested in an optical oscillator constituting a model of an electron in an atom. It has a variable dipole moment and is capable of absorption and emission of electromagnetic waves. We shall confine our attention to the dipole approximation and will calculate Einstein's coefficients for stimulated and spontaneous transitions.

Substituting (26.12) into (8.76) and (8.77) we find that

$$B_{v,\,v+1} = B_{v+1,\,v} = (v+1)\,B_0 = (v+1)\frac{\pi e^2}{3mh\nu_0} \tag{26.22}$$

$$A_{v+1,\,v} = (v+1)A_0 = (v+1)\,\frac{8\pi^2 e^2 \nu_0^2}{3mc^3} = (v+1)\,2\gamma_0 \tag{26.23}$$

where γ_0 is the classical damping constant (see (2.9)). All the remaining transition probabilities are zero, so that on the dipole approximation the oscillator can only undergo transitions between neighbouring levels. Since the levels are equidistant, it follows that all transitions involve the absorption or emission of the proper frequency ν_0.

27. ABSORPTION AND LUMINESCENCE

Absorption of isotropic radiation

Consider n non-interacting harmonic oscillators illuminated by an external isotropic radiation of frequency ν_0 and density u. The dipoles are located in a thermal radiation field of density u_0. In general, the rate of absorption can, in view of (17.19), be represented by the sum

$$W_{\text{abs}} = \sum_{v=0}^{\infty} [n_v(u,\ T,\ t)\,B_{v,\,v+1}\,u - n_{v+1}(u,\ T,\ t)\,B_{v+1,\,v}\,u]\,h\nu_0 \tag{27.1}$$

where $n_v(u,\ T,\ t)$ is the number of oscillators occupying level v. It depends on the density of the exciting radiation. the temperature of the medium and the time t. Stimulated emission (second term under the summation sign) will, as before, be looked upon as negative absorption.

For other quantum-mechanical systems we must calculate the distribution function before we can determine the rate of emission. In the present case this will involve considerable mathematical difficulties. In point of fact, in order to find $n_v(u, T, t)$, we would have to solve an infinite set of differential equations. In some special cases, however, the determination of this function does not involve laborious calculations, and we shall discuss them below.

In view of the specific properties of the oscillator, the sum (27.1) can in general be evaluated without any knowledge of the explicit form of the distribution function. Assuming that coefficients $B_{v,v-1}$ with negative subscripts are zero, we can write (27.1) in the form

$$W_{\text{abs}} = \sum_{v=0}^{\infty} n_v(u, T, t)\,(B_{v,\,v+1} - B_{v,\,v-1})\,uh\nu_0 \qquad (27.1\text{a})$$

Since, according to (26.22), $B_{v,\,v+1} - B_{v,\,v-1} = B_0$ for all v, it follows from (27.1a) that

$$W_{\text{abs}} = Buh\nu_0 \sum_{v=0}^{\infty} n_v(u, T, t) = nB_0 uh\nu_0 = n\frac{\pi e^2}{3m}u \qquad (27.2)$$

This expression is valid both for stationary and non-stationary excitation. It is quite independent of the distribution function, and therefore of the method of excitation and the temperature of the medium. In other words, the absorptive power of the oscillator is independent of whether it is in a normal or an excited state. This follows directly from an analysis of the expression under the summation sign in (27.1a). It gives the observed rate of absorption by the oscillators occupying the v-th level. It is easy to see that for any v the rate of absorption per oscillator is $B_0 uh\nu_0$. When the oscillator is illuminated by a negative flux, the rate of absorption (27.2) is also negative.

Stimulated emission is occasionally neglected to simplify the mathematics. However, inclusion of stimulated emission in the case of the harmonic oscillator not only does not complicate the calculation of the absorbing power, but will also actually help us reduce the number of intermediate steps. and will lead directly to the final result.

From (27.2) and (2.39) it follows that

$$k = \int_0^{\infty} k(\nu)\,d\nu = n\frac{\pi e^2}{3mc} \qquad (27.3)$$

This shows that the absorption coefficient is independent

of the density of the exciting radiation and the temperature and is always positive. For a harmonic oscillator Bouguer's law is always rigorously satisfied.

In the literature, the absorption coefficient is sometimes calculated without allowance for stimulated emission. This leads to the erroneous conclusion that it should increase with temperature. The analysis given here was first published by the authors elsewhere [64].

Equation (27.3) is identical with the Kravets integral (2.45) deduced from classical electrodynamics.

Luminescence under stationary illumination

In view of (21.14), the rate of luminescence by oscillators with allowance for the thermal radiation background can be written in the form

$$W_{\text{lum}} = \sum_{v=1}^{\infty} \{n_v(u, T, t) A_{v, v-1} - [n_{v-1}(u, T, t) - n_v(u, T, t)] B_{v-1, v} u_0\} h\nu_0 \tag{27.4}$$

This sum cannot be evaluated without the use of the properties of the distribution function. In general, the numbers $n_v(u, T, t)$ satisfy the following system of differential equations (see (16.1))

$$\frac{d}{dt} n_v(u, T, t) = -n_v(u, T, t)[A_{v, v-1} + B_{v, v-1}(u_0 + u) + B_{v, v+1}(u_0 + u)] + n_{v+1}(u, T, t)[A_{v+1, v} + B_{v+1, v}(u_0 + u)] + n_{v-1}(u, T, t) B_{v-1, v}(u_0 + u) \tag{27.5}$$

We shall not consider here the interaction of the oscillators with the medium, and therefore the probabilities of non-radiative transitions are absent. The first term on the right-hand side of (27.5) is equal to the number of oscillators which leave level v to levels $v+1$ and $v-1$ per unit time, while the second and third terms correspond to the number of particles arriving from levels $v+1$ and $v-1$ at the level v.

To begin with, let us evaluate W_{lum} under constant illumination. In this case n_v is time-independent, and (27.5) becomes a system of linear algebraic equations

$$
\begin{aligned}
&-n_0B_0(u_0+u)+n_1[A_0+B_0(u_0+u)]=0\\
&-n_1[A_0+B_0(u_0+u)+B_{12}(u_0+u)]+n_0B_0(u_0+u)\\
&\qquad +n_2[A_{21}+B_{21}(u_0+u)]=0\\
&\cdots\cdots\cdots\cdots\cdots\cdots\\
&-n_v[A_{v,\,v-1}+B_{v,\,v-1}(u_0+u)+B_{v+1,\,v}(u_0+u)]\\
&+n_{v-1}B_{v-1,\,v}(u_0+u)+n_{v+1}[A_{v+1,\,v}+B_{v+1,\,v}(u_0+u)]\\
&\qquad =0\\
&\cdots\cdots\cdots\cdots\cdots\cdots
\end{aligned}
\tag{27.6}
$$

From the first equation in this system it follows that

$$n_1=n_0\frac{B_0(u_0+u)}{A_0+B_0(u_0+u)} \tag{27.7}$$

Substituting this into the second equation in (27.6) we have, in view of (26.22) and (26.23)

$$n_2=n_0\left[\frac{B_0(u_0+u)}{A_0+B_0(u_0+u)}\right]^2 \tag{27.8}$$

Similarly

$$n_v=n_0\left[\frac{B_0(u_0+u)}{A_0+B_0(u_0+u)}\right]^v \tag{27.9}$$

If we determine n_0 from the normalisation condition

$$\sum_{v=0}^{\infty} n_v=n \tag{27.10}$$

we obtain the following final expression for the distribution function

$$n_v=n\frac{A_0}{A_0+B_0(u_0+u)}\left[\frac{B_0(u_0+u)}{A_0+B_0(u_0+u)}\right]^v \tag{27.11}$$

When $u\to 0$, (27.9) yields the Boltzmann distribution over the energy levels. Thus, since $(A_0+B_0u_0)e^{-h\nu_0/kT}=B_0u_0$ we have, on letting $u=0$,

$$n_v=n_0e^{-\frac{vh\nu_0}{kT}}=n_0e^{-\frac{E_v-E_0}{kT}} \tag{27.12}$$

When $u \neq 0$, the number $n_v(u, T)$ may be regarded as the v-th term of a geometric progression whose ratio q is given by

$$q = \frac{B_0(u_0 + u)}{A_0 + B_0(u_0 + u)}$$

When $u \to \infty$, $q \to 1$, the distribution becomes uniform and $n_v \to 0$ for any v.

Substituting (27.12) into (27.4), we obtain, after simple transformations, the following expression for the rate of luminescence under constant illumination:

$$W_{\text{lum}}^{\text{st}} = nB_0uh\nu_0 = n\frac{\pi e^2}{3m}u \tag{27.13}$$

This is identical with (27.2) and the corresponding classical formula (see (2.48) for $\gamma = \gamma_{\text{em}}$ and $\overline{\cos^2\theta} = 1/3$).

Growth of luminescence

To calculate luminescence during the growth process, we shall rewrite (27.4) in the somewhat different form

$$W_{\text{lum}} = \sum_{v=1}^{\infty} n_v(u, T, t)A_{v,\,v-1}h\nu_0 - \sum_{v=0}^{\infty}[n_v(u, T, t)B_{v,\,v+1}u_0$$
$$- n_v(u, T, t)B_{v+1,\,v}u_0]\,h\nu_0\,. \tag{27.4a}$$

The second sum in this expression is identical in form with (27.2), and represents the rate of absorption of Planck radiation. We thus have from (27.4a)

$$W_{\text{lum}} = \sum_{v=1}^{\infty} n_v(u, T, t)A_{v,\,v-1}h\nu_0 - nB_0u_0h\nu_0 \tag{27.14}$$

This is a particularly useful formula. It follows from it that (27.4) is entirely consistent with the definition of luminescence as the excess above the thermal background.

In order to avoid complicated mathematical analysis in the evaluation of $n_v(u, T, t)$, we shall employ the following device. Differentiating (27.14) with respect to time t, we obtain

$$\frac{d}{dt}W_{\text{lum}} = h\nu_0\sum_{v=1}^{\infty} A_{v,\,v-1}\frac{d}{dt}n_v(u, T, t) \tag{27.15}$$

and substituting (27.5) into this expression we obtain, after simple transformations, using (26.22) and (26.23)

$$\frac{d}{dt} W_{\text{lum}} = -A_0 \left[\sum_{v=1}^{\infty} n_v(u, T, t) A_{v, v-1} h\nu_0 - nB_0 u_0 h\nu_0 \right] + A_0 n B_0 u h\nu_0 \tag{27.16}$$

Since, according to (27.4), the expression in square brackets is equal to the rate of luminescence, equation (27.16) may be written in the form of a linear differential equation for W_{lum}:

$$\frac{d}{dt} W_{\text{lum}} + A_0 W_{\text{lum}} = A_0 n B u h\nu_0 \tag{27.17}$$

Thus, instead of solving the infinite system of (27.5), and evaluating the sum (27.4), it is sufficient to solve the simple equation (27.17).

As an example, consider the solutions of (27.17) for three different methods of excitation.

1. Suppose that external excitation was absent ($u = 0$) for all times up to $t_0 = 0$ and that the external illumination is switched on and a constant intensity maintained. In classical theory the excitation of the oscillator is calculated under similar conditions. For $u = \text{const}$ the solution of (27.17) is of the form

$$W_{\text{lum}}(t) = Ce^{-A_0 t} + nB_0 u h\nu_0 \tag{27.18}$$

If we determine C from the initial condition ($W_{\text{lum}}(0) = 0$) we obtain

$$W_{\text{lum}}(t) = nB_0 u h\nu_0 (1 - e^{-A_0 t}) = n \frac{\pi e^2}{3m} (1 - e^{-2\gamma_0 t}) \tag{27.19}$$

For $t \gg \frac{1}{A}$, (27.19) becomes identical with (27.13). Comparison of classical and quantum formulae, (2.49) and (27.19), will show that the two theories lead to the same growth law.

2. Let us suppose now that after the source is switched on the intensity of the exciting radiation increases continuously up to the maximum value u. If u increases exponentially,

(27.17) becomes

$$\frac{d}{dt} W_{\text{lum}}(t) + A_0 W_{\text{lum}} = A_0 n B_0 u' h \nu_0 (1 - e^{-\beta t}) \tag{27.20}$$

where β is a parameter representing the rate of increase of u. The solution of (27.20) is

$$W_{\text{lum}}(t) = C_1 e^{-A_0 t} + nB_0 u' \left(\frac{A_0}{\beta - A_0} e^{-\beta t} + 1 \right) \tag{27.21}$$

Having determined C_1 from the initial conditions we obtain the final solution

$$W_{\text{lum}}(t) = nB_0 u' h \nu_0 \times \left[\frac{A_0}{\beta - A_0} e^{-\beta t} - \frac{\beta}{\beta - A_0} e^{-A_0 t} + 1 \right] \tag{27.22}$$

This expression shows that the dependence of u on time modifies the excitation of the oscillator. The modification is determined by the ratio β/A_0. For a Kerr cell, the maximum β is of the order of $10^{10}\,\text{sec}^{-1}$ and A_0 is of the order of $10^8\,\text{sec}^{-1}$ in the visible region of the spectrum. Under these conditions, the error in the growth law is of the order of 1%. Transition to higher frequencies leads to an increase in A_0 (which increases as ν^2) but there is also an increase in contributions due to operation of the cell. When $\beta/A \ll 1$, the growth of the luminescence completely reproduces the form of the increase in u.

3. Consider finally the case of pulsed excitation when an alternating component $\varepsilon u_1 \cos mt\,(\varepsilon < 1)$ is superimposed on a constant intensity u_1. Equation (27.14) now becomes

$$\frac{d}{dt} W_{\text{lum}} + A_0 W_{\text{lum}} = A_0 n B_0 u_1 h \nu_0 (1 + \varepsilon \cos mt) \tag{27.23}$$

Subject to the initial condition $W_{\text{lum}}(0) = 0$ the solution is

$$W_{\text{lum}} = nB_0 u_1 h \nu_0 \Big\{ (1 - e^{-A_0 t}) + \frac{\varepsilon A_0}{\sqrt{m^2 + A_0^2}} [\cos(mt - \varphi) - e^{-A_0 t}] \Big\} \tag{27.24}$$

where $\varphi = \tan^{-1} \frac{m}{A_0}$.

It is evident from this solution that the growth law may differ very substantially from the exponential law. The effect of pulsations becomes negligible as $\frac{A_0}{m} \to 0$. For large t (greater than $\frac{1}{A_0}$) a steady state is reached in which the rate of luminescence is given by

$$W_{\text{lum}} = nB_0 u_1 h \nu_0 \left[1 + \frac{\varepsilon A_0}{\sqrt{m^2 + A_0^2}} \cos(mt - \varphi) \right] \tag{27.25}$$

Decay of luminescence

The law of decay of luminescence may be most simply established from (27.17). In the absence of excitation it assumes the simpler form

$$\frac{d}{dt} W_{\text{lum}} + A_0 W_{\text{lum}} = 0 \tag{27.26}$$

It should be noted that (27.17) is valid only for radiative excitation. With other methods of excitation, for example electron impact, W_{lum} may not satisfy (27.17). However, once the excitation is switched off, the rate of luminescence is always determined by (27.26). Let W^0_{lum} denote the value W_{lum} at $t = 0$ so that

$$W^{\text{dec}}_{\text{lum}}(t) = W^0_{\text{lum}} e^{-A_0 t} = W^0_{\text{lum}} e^{-2\gamma_0 t} \tag{27.27}$$

It follows that according to quantum theory the law of attenuation of the radiation emitted by an oscillator is independent of the temperature of the medium, the method of excitation and the distribution function at $t = 0$. It is identical with the law of decay given by classical theory (see (2.50)).

As has already been pointed out, the distribution of the

oscillators over the energy levels is, in general, a complicated function of time. The fact that this is not reflected in the time dependence of the rate of luminescence is one of the characteristic properties of the harmonic oscillator. We must consider this problem in greater detail. Let us determine the numbers n_v for the decay process in the special case when the temperature is low enough for the thermal background to be negligible. Knowing the explicit form of the distribution function it is possible to calculate the decay law directly from (27.4) and to elucidate a number of other problems.

Let us suppose that the oscillator has j levels [65]. The system of equations (27.5) now becomes

$$\begin{aligned} \frac{d}{dt} n_j(t) &= -jA_0 n_j(t) \\ \frac{d}{dt} n_{j-1}(t) &= -(j-1) A_0 n_{j-1}(t) + jA_0 n_j(t) \\ &\cdots\cdots\cdots \\ \frac{d}{dt} n_{j-i}(t) &= -(j-1) A_0 n_{j-i}(t) + (j-i+1) A_0 n_{j-i+1}(t) \\ &\cdots\cdots\cdots \\ \frac{d}{dt} n_0(t) &= A_0 n_1(t) \end{aligned} \tag{27.28}$$

Integration of the first equation yields

$$n_j(t) = n_j^0 e^{-jA_0 t} \tag{27.29}$$

where n_j^0 is the value of $n_j(t)$ at $t = 0$. Substituting this expression into the second equation in (27.28), we obtain the inhomogeneous equation

$$\frac{d}{dt} n_{j-1}(t) + (j-1) A_0 n_{j-1}(t) = jA_0 n_j^0 e^{-jA_0 t} \tag{27.30}$$

whose solution is

$$n_{j-1}(t) = -jn_j^0 e^{-jA_0 t} + (n_{j-1}^0 + jn_j^0) e^{-(j-1)A_0 t} \tag{27.31}$$

The functions n_{j-2}, n_{j-3} can be found in a similar way.

In general,

$$n_{j-i}(t) = \sum_{k=0}^{i}\sum_{l=0}^{k} \frac{(-1)^{i-k}(j+l-k)!}{(i-k)!\,(j-i)!\;l!} \times n^0_{j+l-k}\, e^{-(j-k)A_0 t} \tag{27.32}$$

Since $j-i$ is the number level v, we have

$$n_v(t) = \sum_{k=0}^{j-v}\sum_{l=0}^{k} \frac{(-1)^{j-v-k}(j+l-k)!}{(j-v-k)!\,v!\,l!} \times n^0_{j+l-k}\, e^{-(j-k)A_0} \tag{27.33}$$

If we substitute $l' = j-k$ and $l'' = l+j-k$ we obtain

$$n_v(t) = \sum_{l''=l'}^{j}\sum_{l'=v}^{l''} \frac{(-1)^{l'-v}\,l''!}{(l'-v)!\,v!\,(l''-l')!}\, n^0_{l''}\, e^{-l'A_0 t} \tag{27.34}$$

Since the factorial of a negative integer is infinite, the lower limit of summation with respect to l'' may be extended to zero and the upper limit reduces to l'', so that

$$n_v(t) = \sum_{l''=0}^{j}\sum_{l'=v}^{l''} \frac{(-1)^{l'-v}\,l''!}{(l'-v)!\,v!\,(l''-l')!}\cdot n^0_{l''}\, e^{-l'A_0 t} \tag{27.35}$$

Since j is not subject to any limitations, we may let it tend to infinity. This yields the final expression for the distribution function during the decay of a harmonic oscillator:

$$n_v(t) = \sum_{l''=0}^{\infty}\sum_{l'=v}^{l''} \frac{(-1)^{l'-v}\,l''!}{(l'-v)!\,v!\,(l''-l')!}\, n^0_{l''}\, e^{-l'A_0 t} \tag{27.36}$$

Direct substitution will show that this expression satisfies (27.5) for $u_0 = u = 0$ and the initial conditions $n_v(t=0) = n^0_v$. As $t \to \infty$, all the $n_v(t)$ except $n_0(t)$ tend to zero, whilst $n_0(t)$ itself tends to n.

The rate of luminescence in the absence of the thermal

background is given by

$$W^{\text{dec}}_{\text{lum}}(t) = \sum_{v=1}^{\infty} n_v(t) A_{v\ v-1} h\nu_0$$

$$= A_0 h\nu_0 \sum_{v=1}^{\infty} \sum_{l''=0}^{\infty} \sum_{l'=v}^{l''} v \frac{(-1)^{l'-v} l''!}{(l'-v)!\, v!\, (l''-l')!} \times n^0_{l''} e^{-l' A_0 t} \tag{27.37}$$

Since $\frac{1}{(l'-v)!} = 0$ for $l' < v$, summation with respect to l' may be extended to zero:

$$W^{\text{dec}}_{\text{lum}}(t) = A_0 h\nu_0 \sum_{l''=0}^{\infty} \sum_{l'=0}^{l''} \frac{(-1)^{l'} l''!}{(l''-l')!} \times n^0_{l''} e^{-l' A_0 t} \sum_{v=1}^{\infty} \frac{(-1)^v}{(l'-v)!\,(v-1)!} \tag{27.38}$$

The last sum in this expression is not zero only for $l' = 1$, and therefore

$$W^{\text{dec}}_{\text{lum}}(t) = A_0 \left(\sum_{v=1}^{\infty} n^0_v v h\nu_0 \right) e^{-A_0 t} = W^0_{\text{lum}} e^{-2\gamma_0 t} \tag{27.39}$$

In fact,

$$\sum_{v=1}^{\infty} \frac{(-1)^v}{(l'-v)!\,(v-1)!} = -\frac{1}{(l'-1)!}(1-1)^{l'-1} = \begin{cases} -1 & \text{when } l' = 1 \\ 0 & \text{when } l' \neq 1 \end{cases} \tag{27.40}$$

Comparison of (27.27) and (27.29) shows that (27.4) and (27.17) lead to identical results. In spite of the fact that $n_v(t)$ is a complicated function of time, the form of the function $W^{\text{dec}}_{\text{lum}}(t)$ is exponential.

Equation (27.39) may be written in the somewhat different form

$$W^{\text{dec}}_{\text{lum}}(t) = A_0 E' \tag{27.41}$$

where $E' = \sum_{v=1}^{\infty} n_v^0 vh\nu_0 e^{-A_0 t}$ is the part of the energy which can be wholly transferred to the electromagnetic field.

In classical theory, the rate of emission is also given by (27.40). The difference is that in classical theory the energy of an oscillator at rest is supposed to be zero, and therefore E′ is equal to the total energy of the oscillator. Its optical properties are not, however, affected when a constant E′ is added to E. It is thus clear that in classical theory the formula for the rate of emission includes only the variable part of the oscillator energy. This must be remembered when comparing results of classical and quantum theories of radiation since otherwise an erroneous conclusion may be reached.

As an example, consider the simple calculation which is usually introduced to illustrate the correspondence principle. Suppose the oscillator occupies level v at time $t_0 = 0$. For t near t_0, the rate of emission is

$$W_{\text{em}} = A_{v,\,v-1} h\nu_0 \tag{27.42}$$

If we multiply and divide this expression by the energy of the oscillator $E_v = (v + 1/2) h\nu_0$ and remember that $A_{v,\,v-1} = v\, 2\gamma_0$, we obtain

$$W_{\text{em}} = 2\gamma_0 E \frac{v}{v + 1/2} \tag{27.43}$$

It is concluded from this equation that the results of classical and quantum theories are identical when the oscillator is excited to high-energy states, i.e. when $\frac{v}{v+1/2} \to 1$. The mistake in this calculation lies in the fact that the total energy of the oscillator is introduced into the formula for the emission, whereas E_{em} depends only on the variable part of the energy $E = vh\nu_0$. It follows that (27.42) must be multiplied and divided by $E' = vh\nu_0$ and not by

$$E = \left(v + \frac{1}{2}\right) h\nu_0$$

The result, whatever the value of v, is

$$W_{\text{em}} = E' \frac{A_{v,\,v-1} h\nu_0}{vh\nu_0} = 2\gamma_0 E' \tag{27.44}$$

and this is identical with the classical formula.

Moments of the transition probability distribution

In Section 11 we established the expressions for the moments of an arbitrary function. From (11.17) we have that the l-th moment of the distribution of the probability of stimulated transitions of an oscillator between levels j and i is

$$S_l^{(j)} = \sum_i (E_i - E_j)^l p_{ji} = \sum_i (E_i - E_j)^l B_{ji} u \tag{27.45}$$

where B_{ji} is Einstein's coefficient for stimulated transitions and u is the intensity of the exciting radiation. Since transitions between neighbouring states are allowed in a harmonic oscillator, the last expression may be rewritten in the form

$$S_l^{(j)} = [B_{j,\,j+1} + (-1)^l B_{j,\,j-1}]\, u (h\nu_0)^l \tag{27.46}$$

If we allow l to assume the values 0, 1, 2, ... and recall (27.22), we find that

$$\begin{aligned} S_0^{(j)} &= B_{j,\,j+1} u + B_{j,\,j-1} u \\ S_1^{(j)} &= (B_{j,\,j+1} - B_{j,\,j-1}) uh\nu = B_0 uh\nu = \frac{\pi e^2}{3m} u \\ S_2^{(j)} &= (B_{j,\,j+1} u + B_{j,\,j-1} u)(h\nu_0)^2 \end{aligned} \tag{27.47}$$

It follows that the zero-order moment is equal to the sum of all the transition probabilities between level j and other levels. The first moment gives the rate of absorption of the incident radiation. This enables us to calculate the rate of absorption of an oscillator and other systems without finding the wave functions and matrix elements. In fact, if we replace in (27.45) the coefficient B_{ji} by the square of the matrix element of the dipole moment $\mathbf{D}_{ji}$ in accordance with (8.76), and repeat the steps leading to (11.22), we find that

$$S_1^{(j)} = \frac{8\pi^3}{3h^2} u \int \psi_j^* \mathbf{D}(\mathbf{H}\mathbf{D} - \mathbf{D}\mathbf{H}) \psi_j \, dx \tag{27.48}$$

where the operator $\mathbf{H}$ is given by (26.1) and the dipole moment of a one-dimensional operator is $D = ex$. Since

$$p^2x - xp^2 = -\frac{h^2}{2\pi^2}\frac{\partial}{\partial x} \quad (27.49)$$

where $\mathbf{p}$ is the momentum operator, we can integrate (27.48) by parts and find that the rate of absorption is

$$S_1^{(j)} = \frac{\pi e^2}{3m} u \quad (27.50)$$

It may be shown that the quadrupole part of absorption by an oscillator is given by the third moment of the distribution of quadrupole transition probabilities and can also be easily calculated without solving Schroedinger's equation [66].

28. POLARISATION OF LUMINESCENCE AND INDUCED DICHROISM

The oscillator energy distribution for excitation by plane-polarised light

Consider a set of n harmonic oscillators which are randomly distributed in space. Suppose further that the directions of their dipole moments are fixed in space and the harmonic oscillators are illuminated by plane-polarised light propagating along the x axis. The electric vector in the incident radiation lies along the z axis and the luminescence is observed along the y axis. The oscillator energy distribution will depend not only on the temperature of the ambient medium and the density of exciting radiation, but also on the directions of the oscillator dipole moments which are defined by angles θ and φ.

Under constant illumination the number of particles $n_v(u, T, \theta)$ satisfies the balance equation

$$-n_v(u,T,\Omega)[A_{v,v-1} + B_{v,v-1}u_0 + B_{v,v+1}u_0 + b^z_{v,v-1}(\Omega)u + b^z_{v,v+1}(\Omega)u] + n_{v+1}(u,T,\Omega)[A_{v+1,v} + B_{v+1,v}u_0 + b^z_{v+1,v}(\Omega)u] + n_{v-1}(u,T,\Omega)[B_{v-1,v}u_0 + b^z_{v-1,v}(\Omega)u] = 0 \quad (28.1)$$

where, according to (8.75), (8.74) and (26.12)

$$b^z_{v,v-1}(\Omega) = 3B_{v,v-1}\cos^2\theta = 3vB_0\cos^2\theta$$

$$b^x_{v,v-1}(\Omega) = 3B_{v,v-1}\sin^2\theta\cos^2\varphi = 3vB_0\sin^2\theta\cos^2\varphi \tag{28.2}$$

$$a^z_{v,v-1}(\Omega) = \frac{3}{8\pi}A_{v,v-1}\cos^2\theta = \frac{3}{8\pi}vA_0\cos^2\theta$$

and

$$a^x_{v,v-1}(\Omega) = \frac{3}{8\pi}A_{v,v-1}\sin^2\theta\cos^2\varphi = \frac{3}{8\pi}vA_0\sin^2\theta\cos^2\varphi$$

On solving (28.1) subject to (28.2) and bearing in mind the conservation of the number of particles

$$\sum_{v=0}^{\infty} n_v(u, T, \Omega) = \frac{n}{4\pi} \tag{28.3}$$

we obtain

$$n_v(\theta) = \frac{n}{4\pi}\,\frac{A_0}{A_0 + B_0u_0 + 3B_0u\cos^2\theta} \times \left[\frac{B_0u_0 + 3B_0u\cos^2\theta}{A_0 + B_0u_0 + 3B_0u\cos^2\theta}\right]^v \tag{28.4}$$

Figure 6.3 shows plots of $n_v(\theta)$ for $v = 1, 3, 5$, and 10.

Polarisation of resultant luminescence

The intensity of luminescence polarised along the z and x axes and emitted as a result of transitions of the oscillator from level v to level $v-1$ is given by

$$W^z_{v,v-1} = h\nu_0\int_{\Omega}\{n_v a^z_{v,v-1}(\Omega) - (n_{v-1} - n_v)b^z_{v,v-1}(\Omega)u_0'\}\,d\Omega \tag{28.5}$$

$$W^x_{v,v-1} = h\nu_0\int_{\Omega}\{n_v a^x_{v,v-1}(\Omega) - (n_{v-1} - n_v)b^x_{v,v-1}(\Omega)u_0'\}\,d\Omega \tag{28.6}$$

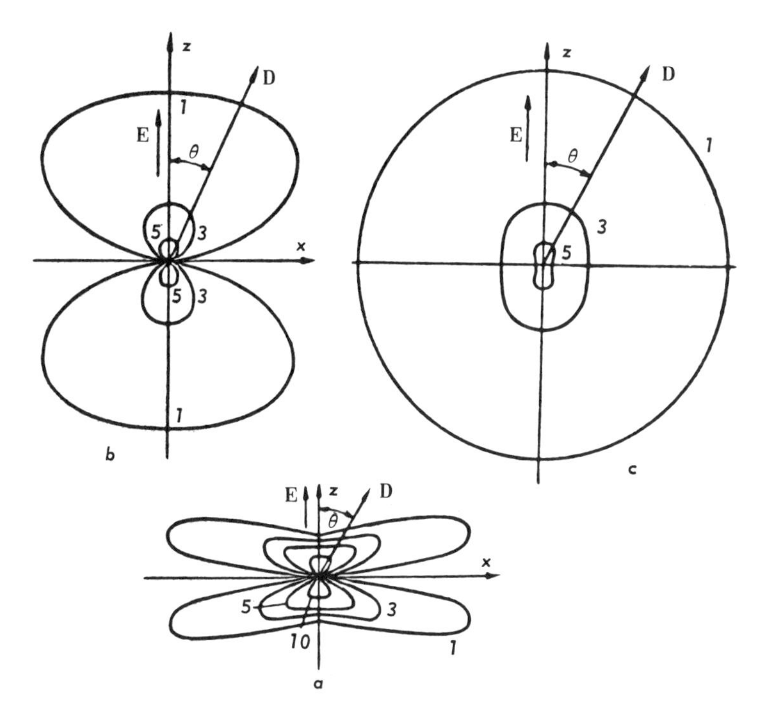

a – $B_0u = 5A_0$, $u_0 = 0$; b – $B_0u = 0.5\ A_0$, $u_0 = 0$; c – $B_0u = 0.1\ A_0$, $B_0u_0 = A$

Fig. 6.3 Dependence of the populations of the first, third, fifth and tenth levels of an oscillator on the angular position of its dipole moment

where $u_0' = \frac{1}{8\pi} u_0$ is the density of Planck radiation per unit solid angle and given polarisation.

Substituting (28.4) into (28.5) and (28.6), summing over v and integrating with respect to Ω, we obtain the simple expressions

$$W^z = \sum_{v=1}^{\infty} W^z_{v,v-1} = \frac{9}{8\pi} nB_0 uh\nu_0 \frac{1}{5} \tag{28.7}$$

$$W^x = \sum_{v=1}^{\infty} W^x_{v,v-1} = \frac{9}{8\pi} nB_0 uh\nu_0 \frac{1}{15} \tag{28.8}$$

which, together with (3.38) give the following expression for the polarisation of the luminescence emitted by a set of harmonic oscillators:

$$P = \frac{W^z - W^x}{W^z + W^x} = \frac{1}{2}$$

This result is valid for an arbitrary density of the exciting radiation and arbitrary temperature of the medium, and is identical with the polarisation obtained by the classical theory for a set of dipoles.

Polarisation of individual lines

The above value of the polarisation is valid for the resultant radiation emitted as a result of transitions from all energy levels. The polarisation of individual lines due to transitions from level v to level $v - 1$ depends on v, the intensity of illumination and the temperature, which is connected with the strong dependence of n_v (u, T, Ω) on θ (see Fig. 6.3).

The dependence of the polarisation of individual lines, $P_{v,v-1}$, on the intensity of exciting radiation can be found by substituting $u_0 = 0$, which corresponds to $T = 0$ in (28.4), (28.5) and (28.6). The expression for the intensity of luminescence then becomes

$$W^z_{v,v-1} = h\nu_0 \int_{\Omega} n_v a^z_{v,v-1}(\Omega)\, d\Omega$$

$$= -\frac{3}{16\pi} vnA_0 h\nu_0 \frac{1}{(\alpha u)^{1/2}} \int_{(\alpha u)^{1/2}}^{-(\alpha u)^{1/2}} \frac{x^{2(v+1)}\, dx}{(1 + x^2)^{v+1}} \quad (28.9)$$

$$W^x_{v,v-1} = h\nu_0 \int_{\Omega} n_v a^x_{v,v-1}(\Omega)\, d\Omega$$

$$= \frac{3}{32\pi} vnA_0 h\nu_0 \left\{ -\frac{1}{(\alpha u)^{1/2}} \int_{(\alpha u)^{1/2}}^{-(\alpha u)^{1/2}} \frac{x^{2v}\, dx}{(1 + x^2)^{v+1}} + \frac{1}{(\alpha u)^{3/2}} \int_{(\alpha u)^{1/2}}^{-(\alpha u)^{1/2}} \frac{x^{2(v+1)}\, dx}{(1 + x^2)^{v+1}} \right\} \quad (28.10)$$

where

$$\alpha = 3\frac{B_0}{A_0}, \qquad x = (\alpha u)^{1/2}\cos\theta$$

The integrals which enter into (28.9) and (28.10) can be evaluated exactly for any given v. However, for large v, integration leads to unwieldy results. We shall therefore find the polarisation of individual lines for a few special cases. On evaluating the integrals in (28.9) and (28.10) for $v = 1$ and 2 with the aid of (3.38), we find that the polarisations of the first two lines are

$$\begin{aligned} P_{10} = &\{7(\alpha u)^{-1/2} + 9(\alpha u)^{-3/2} - [(\alpha u)^{-1} + 1] \\ &\times [1 + 9(\alpha u)^{-1}]\tan^{-1}(\alpha u)^{1/2}\} \\ &\times\{(\alpha u)^{-1/2} + 3(\alpha u)^{-3/2} + [(\alpha u)^{-1} + 1] \\ &\times [1 - 3(\alpha u)^{-1}]\tan^{-1}(\alpha u)^{1/2}\}^{-1} \end{aligned} \tag{28.11}$$

$$\begin{aligned} P_{21} = &\{29(\alpha u)^{-1/2} + 78(\alpha u)^{-3/2} + 45(\alpha u)^{-5/2} \\ &- [(\alpha u)^{-1} + 1]^2[45(\alpha u)^{-1} + 3]\tan^{-1}(\alpha u)^{1/2}\} \\ &\times\{3(\alpha u)^{-1/2} + 22(\alpha u)^{-3/2} + 15(\alpha u)^{-5/2} \\ &+ [(\alpha u)^{-1} + 1]^2[3 + 15(\alpha u)^{-1}]\tan^{-1}(\alpha u)^{1/2}\}^{-1} \end{aligned} \tag{28.12}$$

These expressions give the exact values for the polarisations of the first and second lines for $u_0 = 0$ and arbitrary intensity u. When $\alpha u \ll 1$, we can expand $\tan^{-1}(\alpha u)^{1/2}$ into a series in powers of αu and obtain the following approximate expression for the polarisation

$$P_{10} = \frac{1}{2} - \frac{3}{14}(\alpha u) + \ldots \tag{28.11a}$$

$$P_{21} = \frac{2}{3} - \frac{5}{27}(\alpha u) + \ldots \tag{28.12a}$$

When $u \to \infty$ $(\alpha u \to \infty)$, all the terms in (28.11) and (28.12) with negative powers will vanish and the polarisation of both lines will tend to -1 (even for $\alpha u = 10^4$ we have $P_{10} = -0.95$, $P_{21} = -0.93$). This means that only the x component is represented in the first and second lines. It follows that as the

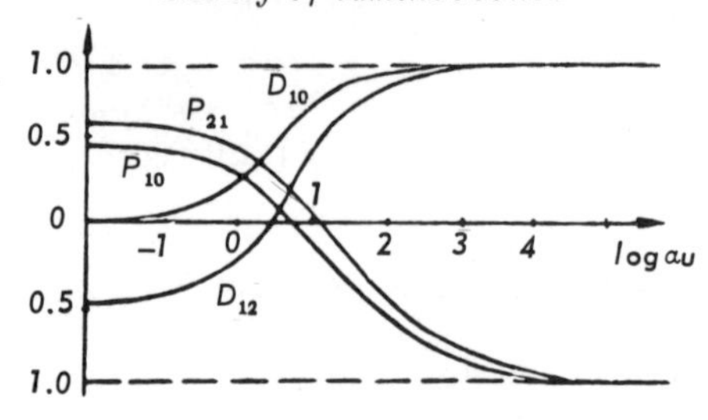

Fig. 6.4 Polarisation of luminescence and induced dichroism for the first and second lines as a function of the intensity of exciting radiation

intensity of the exciting radiation is increased continuously from zero to infinity, the polarisation varies from 1/2 and 2/3 to -1 (Fig. 6.4).

The limiting values of the polarisation of all the lines for $\alpha u \rightarrow 0$ can be obtained by expanding the integrand in (28.9) and (28.10) into a series in powers of x^2. If we confine our attention to the first terms in the expansion we obtain

$$W^z_{v,v-1} = \frac{3}{8\pi} vnA_0 h \nu_0 (\alpha u)^v \times \left[\frac{1}{2v+3} - \frac{v+1}{2v+5} (\alpha u) + \dots \right] \tag{28.13}$$

$$W^x_{v,v-1} = \frac{3}{8\pi} vnA_0 h \nu_0 (\alpha u)^v \times \left[\frac{1}{(2v+1)(2v+3)} - \frac{v+1}{(2v+3)(2v+5)} (\alpha u) + \dots \right] \tag{28.14}$$

which corresponds to

$$P_{v,v-1} = \frac{v}{v+1} - \frac{2v+1}{(v+1)(2v+5)} (\alpha u) + \dots \tag{28.15}$$

For negligible u, the limiting values of $P_{v,v-1}$ for successive lines are respectively equal to 1/2, 2/3, 3/4, . . . , 1.

The polarisations of individual lines obtained above are

consistent with the fact that the polarisation of the resultant luminescence is 1/2 for arbitrary densities of exciting radiation. This becomes clear when the polarisation of the resultant emission is expressed in terms of the polarisation of the individual lines

$$P = \frac{\sum_{v=1}^{\infty} W^z_{v,v-1} - \sum_{v=1}^{\infty} W^x_{v,v-1}}{\sum_{v=1}^{\infty} W^z_{v,v-1} + \sum_{v=1}^{\infty} W^x_{v,v-1}} = \frac{P_{10} + \sum_{v=2}^{\infty} \beta_v P_{v,v-1}}{1 + \sum_{v=2}^{\infty} \beta_v} \tag{28.16}$$

where

$$\beta_v = \frac{W^z_{v,v-1} + W^x_{v,v-1}}{W^z_{10} + W^x_{10}}$$

is the ratio of luminescence intensitites emitted in $v \to v-1$ transitions to the luminescence intensity in the first line.

It is evident from (28.13), (28.14) and (28.16) that all the β_v beginning with β_2 tend to zero as $u \to 0$, and the polarisation of the resultant luminescence is in the limit equal to 1/2. For large u, the negative polarisation of the first lines is compensated by the positive polarisation of lines due to transitions between higher energy levels, since the weighting of lines with large contributions to the resultant luminescence increases with increasing u.

The dependence of $P_{v,v-1}$ on the density of Planck radiation is different from the dependence of the polarisation on u. Suppose, for example, that the density of thermal radiation, u_0, is much greater than the density of the exciting radiation. On the first approximation, the distribution function (28.4) will then be

$$n_v(\Omega) = \frac{n}{4\pi} \frac{A_0}{A_0 + B_0 u} e^{-vh\nu_0/kT} \times \left\{ 1 + \frac{3u}{u_0} [v - (v+1) e^{-h\nu_0/kT}] \cos^2\theta \right\} \tag{28.17}$$

Substituting this expression into (28.5) and (28.6), and integrating with respect to angles, we obtain the following simple expressions for the rates of luminescence:

$$W^z_{v,v-1} = \frac{9}{8\pi} nvA_0 \left(\frac{u}{u_0}\right) e^{-\frac{vh\nu_0}{kT}} (1 - e^{-h\nu_0/kT}) \frac{1}{5} \quad (28.18)$$

$$W_{v,v-1} = \frac{9}{8\pi} nvA_0 \left(\frac{u}{u_0}\right) e^{-vh\nu_0/kT} (1 - e^{-h\nu_0/kT}) \frac{1}{15} \quad (28.19)$$

According to (28.18), (28.19) and (3.38), the polarisation $P_{v,v-1}$ of all the lines tends to 1/2 when the density of the exciting radiation is negligible in comparison with the density of the Planck radiation.

It should be noted that the polarisation of individual lines emitted by a harmonic oscillator cannot, in principle, be observed because all the lines are identical in frequency and overlap. However, for a slightly anharmonic oscillator, for which the above results are qualitatively valid, different transitions correspond to different frequencies and the polarisation of individual lines can be measured. If the frequencies are not too different and the lines overlap, different polarisations can be observed within the limits of a line profile.

Induced dichroism

We shall investigate dichroism by considering two beams of external radiation as in classical theory (Section 2). We shall assume that the first beam propagates along the x axis and produces an angular anisotropy in the distribution of the excited particles, while the second is parallel to the y axis and is polarised along the z or x axes. The energy density u' in the second beam will be taken to be much smaller than the energy density u in the first beam so that the distribution function is practically independent of u' and is given by (28.4).

The rates of absorption due to transitions of the oscillator between levels v and $v-1$ for radiation polarised along the z and x axes are respectively given by

$$W^z_{\text{abs}}(\nu_{v,v-1}) = h\nu_0 \times \int_\Omega [n_{v-1} b^z_{v-1,v}(\Omega) u' - n_v b^z_{v,v-1}(\Omega) u'] \, d\Omega \quad (28.20)$$

and

$$W^x_{\text{abs}}(\nu_{v,v-1}) = h\nu_0 \times \int_\Omega [n_{v-1} b^x_{v-1,v}(\Omega) u' - n_v b^x_{v,v-1}(\Omega) u'] \, d\Omega \quad (28.21)$$

If we sum these expressions over v between 1 and ∞, and integrate with respect to the angles, we obtain

$$W^z_{\text{abs}} = \sum_{v=1}^{\infty} W^z_{\text{abs}}(\nu_{v,v-1}) = nB_0 u' h\nu_0 \tag{28.22}$$

$$W^x_{\text{abs}} = \sum_{v=1}^{\infty} W^x_{\text{abs}}(\nu_{v,v-1}) = nB_0 u' h\nu_0 \tag{28.23}$$

The last formulae show that the absorption coefficient of a set of harmonic oscillators is independent of the polarisation of the absorbed radiation, and there is therefore no dichroism. This is valid only for the resultant absorption. The absorption coefficients of individual lines do depend on the distribution function. Consider the special case when there is no Planck radiation ($u_0=0$). Equations (28.20) and (28.21) can now be rewritten in the form

$$W^z_{\text{abs}}(\nu_{v,v-1}) = -\frac{3}{2} nvB_0 u' h\nu_0 \int_{(\alpha u)^{1/2}}^{-(\alpha u)^{1/2}} \frac{x^{2v}\,dx}{(1+x^2)^{v+1}} \tag{28.24}$$

$$W^x_{\text{abs}}(\nu_{v,v-1}) = \frac{3}{4} nvB_0 u' h\nu_0 \left\{ -\frac{1}{(\alpha u)^{1/2}} \int_{(\alpha u)^{1/2}}^{-(\alpha u)^{1/2}} \frac{x^{2(x+1)}\,dx}{(1+x^2)^{v+1}} + \frac{1}{(\alpha u)^{3/2}} \int_{(\alpha u)^{1/2}}^{-(\alpha u)^{1/2}} \frac{x^{2v}\,dx}{(1+x^2)^{v+1}} \right\} \tag{28.25}$$

Integrating these expressions for v equal to 1 and 2 with the aid of (2.42), we obtain the dichroism for the first two lines

$$D_{01} = \{[3(\alpha u)^{-3/2} + (\alpha u)^{-1/2}] + [(\alpha u)^{-1} + 1] \times [1 - 3(\alpha u)^{-1}] \tan^{-1}(\alpha u)^{1/2}\} \tag{28.26}$$

$$\times \{[(\alpha u)^{-1/2} - (\alpha u)^{-3/2}] + [1 - (\alpha u)^{-1}]^2 \tan^{-1}(\alpha u)^{1/2}\}^{-1}$$

$$D_{12} = \{(\alpha u)^{-1/2} + 14(\alpha u)^{-3/2} + 9(\alpha u)^{-5/2} + [(\alpha u)^{-1} + 1]^2 [1 - 9(\alpha u)^{-1}] \tan^{-1}(\alpha u)^{1/2}\} \times \{(\alpha u)^{-1/2} - 6(\alpha u)^{-3/2} - 3(\alpha u)^{-5/2} + [(\alpha u)^{-1} + 1]^2 [1 + 3(\alpha u)^{-1}] \tan^{-1}(\alpha u)^{1/2}\}^{-1} \quad (28.27)$$

If the energy density in the first beam which produces anisotropy tends to zero, the dichroism of the first line will also tend to zero, whilst $D_{12} \to -1/2$. As u increases, the dichroism of both lines increases and tends to +1 as $u \to \infty$. This form of the dichroism of individual lines is also connected with the strong dependence of the distribution function on the density of exciting radiation and the orientation of the oscillator dipole moments.

The limiting values of $D_{v,v+1}$ for $u \to 0$ can be found by expanding the integrand in (28.24) and (28.25) in powers of x^2 and retaining the first two terms. After integration with respect to angles this yields

$$W^z_{\text{abs}}(\nu_{v-1,v}) = 3nvB_0 u' h \nu_0 (\alpha u)^{v-1} \times \left[\frac{1}{2v+1} - \frac{v+1}{2v+3}(\alpha u) + \dots\right] \quad (28.28)$$

$$W^x_{\text{abs}}(\nu_{v-1,v}) = 3nvB_0 u' h \nu_0 (\alpha u)^{v-1} \times \left[\frac{1}{(2v-1)(2v+1)} - \frac{(v+1)(\alpha u)}{(2v+1)(2v+3)} + \dots\right] \quad (28.29)$$

and corresponds to

$$D_{v-1,v} = -\frac{v-1}{v} + \frac{(v+1)(2v-1)}{(2v+3)v^2}(\alpha u) + \dots \quad (28.30)$$

As $u \to 0$, the dichroism of the individual lines (D_{01}, D_{12}, D_{23}, ..) tends to the following limiting values: 0, -1/2, -2/3, -3/4, ..., -1. As u increases, $D_{v-1,v}$ also increases, and for small v may reach the maximum value of +1. The dichroism of the individual lines is not really in conflict with the fact that the resultant absorption does not exhibit dichroism.

The magnitude of $D_{v-1,v}$ will clearly depend on the density of Planck radiation. Since this radiation is completely isotropic, it tends to reduce the anisotropy in the distribution of excited oscillators due to plane-polarised exciting radiation. The result of this is that the dichroism of all lines tends to vanish for $(u_0 \gg u)$.

Just as in the case of the luminescence of individual lines, the absorption of a strictly harmonic oscillator cannot be observed experimentally. However, the above results may be used, at least qualitatively, in the interpretation of the optical properties of a slightly anharmonic oscillator.

29. NATURAL PROFILE OF A SPECTRAL LINE [68, 69]

Interaction equations

The quantum electrodynamic equations for the interaction of radiation with matter were given in Chapter 3. They are also valid for the harmonic oscillator but the mathematical calculations are then much simpler if the equations are written in a somewhat different form. Their derivation is similar to the derivation of (13.15) and will be given below in abbreviated form.

Consider a harmonic oscillator interacting with the electromagnetic field. Its motion can be described by the time-dependent Schroedinger equation for which the Hamiltonian may be written in the form

$$\mathbf{H} = \mathbf{H}_{osc} + \mathbf{H}_f + \mathbf{H}_{int} \tag{29.1}$$

where $\mathbf{H}_{osc}$ and $\mathbf{H}_f$ are the oscillator and field Hamiltonians and $\mathbf{H}_{int}$ is the interaction operator. In the non-relativistic approximation we can take

$$\begin{aligned} \mathbf{H}_{int} &= -\frac{e}{cm}(\mathbf{pA}) \\ &= -\frac{e}{cm}\sum_{\lambda}[\mathbf{q}_{\lambda}(\mathbf{pA}_{\lambda}) + \mathbf{q}_{\lambda}^{*}(\mathbf{pA}_{\lambda}^{*})] \end{aligned} \tag{29.2}$$

where e, m and $\mathbf{p}$ are the charge, mass and momentum of the oscillating charge and $\mathbf{A}$ is the vector potential which is taken in the form of a set of plane waves in volume V. If we expand the wave function for the oscillator + field system in terms of the eigenfunctions of $\mathbf{H}_{osc}$ and $\mathbf{H}_f$, we obtain

$$i\frac{h}{2\pi}C_{v\,(n_\lambda)} = (E_v + E_{(n_\lambda)})\,C_{v\,(n_\lambda)} - \left(\frac{he^2}{m^2V}\right)^{1/2}\sum_{v'\lambda}(\mathbf{p}\mathbf{e}_{\lambda'})_{v'v}$$

$$\times\left\{\sqrt{\frac{n_{\lambda'}}{\nu_{\lambda'}}}C_{v'\,(n_\lambda-\delta_{\lambda\lambda'})} + \sqrt{\frac{n_{\lambda'}+1}{\nu_{\lambda'}}}C_{v'\,(n_\lambda+\delta_{\lambda\lambda'})}\right\} \quad (29.3)$$

where $\mathbf{e}_\lambda$ is the unit polarisation vector of the photon with wave vector $\boldsymbol{\varkappa}_\lambda$, v is the number of the oscillator energy level and (n_λ) represents the set of quantum numbers for the field oscillators.

When the oscillator lies along the x axis

$$(\mathbf{p}\mathbf{e}_\lambda)_{vv'} = i\,\nu_{vv'}\,m\sqrt{\frac{h}{m\,\nu_0}}\left[\sqrt{\frac{v'}{2}}\delta_{v'-1,\,v} + \sqrt{\frac{v'+1}{2}}\delta_{v'+1,\,v}\right](\mathbf{x}_0\mathbf{e}) \quad (29.4)$$

where $\mathbf{x}_0$ is the unit vector of the x axis. Substituting (29.4) into (29.3) and summing over v, we obtain

$$i\frac{\partial}{\partial t}C_{v(n_\lambda)} = \frac{2\pi}{h}(E_v + E_{(n_\lambda)})C_{v(n_\lambda)} - a^*(v+1)^{1/2}\sum_{\lambda'}\sqrt{\frac{n_{\lambda'}}{\nu_{\lambda'}}}$$

$$\times(\mathbf{x}_0\mathbf{e}_{\lambda'})\,C_{v+1,\,(n_\lambda-\delta_{\lambda\lambda'})} - av^{1/2}\sum_{\lambda'}\sqrt{\frac{n_{\lambda'}+1}{\nu_{\lambda'}}}(\mathbf{x}_0\mathbf{e}_{\lambda'})\,C_{v-1,(n_\lambda+\delta_{\lambda\lambda'})} \quad (29.5)$$

where $a = i\left(\frac{\pi e^2\nu_0}{mV}\right)^{1/2}$. Virtual transitions are not taken into account in (29.5).

Probability amplitude for a given sequence of appearance of photons

Suppose that at time $t = 0$ the oscillator is in the energy level N and there are no photons in the surrounding space:

$$C_{v\,(n_\lambda^0)} = \delta_{vN} \quad (29.6)$$

The maximum value of N is determined by the limits of applicability of the dipole approximation in the visible region ($\lambda = 5{,}000$ Å and $m = m_e$). It may be of the order of 1,000.

In classical theory the line profile is evaluated under similar limitations. To simplify the calculations we shall suppose that the energy of level N is zero while the energy of the v-th level is

$$E_v = -(N-v)\,h\nu_0 \tag{29.7}$$

Suppose that after a sufficiently long interval of time the oscillator undergoes the successive transitions $N \to N-1$, $N-1 \to N-2$ and so on, so that it ends up in the zero level and N photons appear in the field. We shall denote their frequencies ν and wave vectors $\varkappa$ by ν_1, $\varkappa_1$, ν_2, $\varkappa_2, \ldots, \nu_N$, $\varkappa_N$. Since the frequencies $\nu_1, \nu_2, \ldots, \nu_N$ of all the photons are equal within the limits of the line width, it is not possible to establish experimentally which particular photon has the frequency ν_i and which has the frequency ν_j. In view of the indistinguishability of photons emitted by a harmonic oscillator, there is an uncertainty in the sequence in which photons with given parameters will appear. In fact, a photon of frequency ν_1 and wave vector $\varkappa_1$ may appear either as a result of a transition from level N to level $N-1$, or as a result of a transition from any other level j to the neighbouring lower-lying level. The solution of the problem, therefore, is correct only when it is independent of the interchange of subscripts of ν and $\varkappa$. This type of solution can be taken to be the sum of all the possible $C'_{0(n_\lambda)}$, each of which gives the probability that the oscillator has undergone a transition from level N to the zero level and N photons have appeared in the required sequence in the field. The solution is substantially simplified if, instead of trying to find the resultant quantity $C'_{0(n_\lambda)}$, one tries to determine $C'_{0(n_\lambda)}$. All the remaining terms are obtained simply by interchanging the subscripts of ν and $\varkappa$ in the expression for $C_{0(n_\lambda)}$.

Let us suppose that when the oscillator undergoes a transition from level N to level $N-1$, a photon of frequency ν_N is emitted, whilst the transition $N-1 \to N-2$ results in the emission of a photon of frequency ν_{N-1}, and so on. In accordance with (29.5), the amplitudes $C_{v(n_\lambda)}$ are given by the following set of equations:

$$i\frac{\partial}{\partial t}C_N = -aN^{1/2}\sum_{\lambda_N}\sqrt{\frac{n^0_{\lambda_N}+1}{\nu_N}}\,(\mathbf{x}_0\mathbf{e}_{\lambda_N})\,C_{N-1}$$

$$i\frac{\partial}{\partial t}C_{N-1}=2\pi(\nu_N-\nu_0)C_{N-1}-a^*N^{1/2}\sqrt{\frac{n^0_{\lambda_N}+1}{\nu_N}}(\mathbf{x}_0\mathbf{e}_{\lambda_N})C_N$$

$$-a(N-1)^{1/2}\sum_{\lambda_{N-1}}\sqrt{\frac{n^0_{\lambda_{N-1}}+1}{\nu_{N-1}}}(\mathbf{x}_0\mathbf{e}_{\lambda_{N-1}})C_{N-2}$$

. . . .

$$i\frac{\partial}{\partial t}C_{N-k}=\left[2\pi\sum_{j=N-k+1}^{N}(\nu_j-\nu_0)\right]C_{N-k}$$

$$-a^*(N-k+1)^{1/2}\sqrt{\frac{n^0_{\lambda_{N-k+1}}+1}{\nu_{N-k+1}}}(\mathbf{x}_0\mathbf{e}_{N-k+1})C_{N-k+1} \tag{29.8}$$

$$-(N-k)^{1/2}\sum_{\lambda_{N-k}}\sqrt{\frac{n^0_{\lambda_{N-k}}+1}{\nu_{N-k}}}(\mathbf{x}_0\mathbf{e}_{\lambda_{N-k}})C_{N-k-1}$$

. . . .

$$i\frac{\partial}{\partial t}C_0=\left[2\pi\sum_{j=1}^{N}(\nu_j-\nu_0)\right]C_0-a^*\sqrt{\frac{n^0_{\lambda_1}+1}{\nu_1}}(\mathbf{x}_0\mathbf{e}_{\lambda_1})C_1$$

In these expressions we have, for the sake of brevity, omitted the subscripts of C_v which characterise the state of the field.

The solution of (29.8) may be written in the form

$$C_N=e^{-N_0\gamma t}$$

$$C_{N-k}=(i)^k\prod_{j=0}^{k-1}g^*_{N-j}\sum_{l=0}^{k-l}Q_lP^{(k)}_l(-1)^l\times[e^{-\Gamma_{N-l}t}-e^{-\Gamma_{N-k}t}] \tag{29.9}$$

$$(k=1,2,3,\ldots,N)$$

where

$$\Gamma_N=N\gamma_0,\quad \Gamma_{N-l}=2\pi i\sum_{j}^{N}(\nu_j-\nu_0)+(N-l)\gamma_0$$

$$Q_0=1,\quad Q_l=\prod_{j=l}^{l}(\Gamma_{N-l}-\Gamma_{N-l+j})^{-1} \tag{29.10}$$

$$P^{(k)}_l=\prod_{j=1}^{k-l}(\Gamma_{N-l-j}-\Gamma_{N-l})^{-1}\quad(l=0,1,2,\ldots,k-1)$$

and

$$g^*_{N-j} = a^*(N-j)^{1/2} \sqrt{\frac{n^0_{\lambda_{N-j}} + 1}{\nu_{N-j}}} \, (\mathbf{x}_0 \mathbf{e}_{\lambda_{N-j}})$$

where $2\gamma_0$ is the classical line width. It can readily be seen that (29.9) satisfies the initial conditions (29.6). It can be shown that it is also the solution of equation (29.8).

Since all the Γ_j, except for Γ_0, contain a real part, it follows that as $t \to \infty$, all the amplitudes C_v, except C_0, will tend to zero. This means that after a sufficient interval of time, the oscillator is found in the zero state and N photons appear in the field. The amplitude C_0 is then given by

$$C_0 = -(i)^N \prod_{j=1}^{N} g^*_j \sum_{l=0}^{N-1} (-1)^l Q_l P_l^{(N)} e^{-\Gamma_0 t} \tag{29.11}$$

so that if we substitute $\beta_j = 2\pi i(\nu_j - \nu_0) - \gamma_0$, we obtain

$$C_0 = (i)^N \prod_{j=1}^{N} g^*_j Q_N e^{-\Gamma_0 t}$$

$$= \frac{(i)^N \prod_{j=1}^{N} g^*_j}{\beta_1(\beta_1 + \beta_2)(\beta_1 + \beta_2 + \beta_3) \ldots (\beta_1 + \beta_2 + \beta_3 + \cdots + \beta_N)} \times e^{-\Gamma_0 t} \tag{29.12}$$

Line shape

The expression (29.12) gives the probability amplitude for a definite sequence of appearance of the photons. When all the subscripts are interchanged, we obtain $N!$ amplitudes, so that the resultant and normalised amplitude is given by the sum

$$C'_0 = \frac{1}{\sqrt{N!}} \sum_{P_{\nu_j}} C_0 = \frac{1}{\sqrt{N!}} \prod_{j=1}^{N} g^*_j e^{-\Gamma_0 t} \sum_{P_{\nu_j}} Q_N \tag{29.13}$$

where P_{ν_j} represents summation over all the possible combinations of j and ν.

To evaluate the sum in (29.13), let us group all the terms at first in pairs and add, then group the result of this in threes and find the corresponding sums, and so on. Each such operation involves the grouping of terms which differ from each other only by the factor $^1/\beta_m$. For example,

$$\left(\frac{1}{\beta_1}+\frac{1}{\beta_2}\right)\frac{1}{(\beta_1+\beta_2)(\beta_1+\beta_2+\beta_3)\dots(\beta_1+\dots+\beta_N)}$$

$$=\frac{1}{\beta_1\beta_2(\beta_1+\beta_2+\beta_3)\dots(\beta_1+\dots+\beta_N)}$$

$$\left(\frac{1}{\beta_1\beta_2}+\frac{1}{\beta_1\beta_2}+\frac{1}{\beta_2\beta_3}\right)$$

$$\times\frac{1}{(\beta_1+\beta_2+\beta_3)(\beta_1+\beta_2+\beta_3+\beta_4)\dots(\beta_1+\dots+\beta_N)}$$

$$=\frac{1}{\beta_1\beta_2\beta_3(\beta_1+\beta_2+\beta_3+\beta_4)\dots(\beta_1+\dots+\beta_N)}$$

The sum in (29.13) consists of $N!$ terms. When the individual pairs are added, the number of terms becomes smaller by a factor of two. After the second operation is performed, the number of terms is reduced by a factor of 3 and after $N-1$ such operations, the result is

$$C_0=\frac{1}{\sqrt{N!}}\prod_{j=1}^{N}g_j^{*}e^{-\Gamma_0 t}\frac{1}{\beta_1\beta_2\beta_3\dots\beta_N} \tag{29.14}$$

To obtain the energy distribution of the emitted radiation, we must take the square of the modulus of C_0', sum the result over the polarisations and integrate over all directions of propagation. The final result is

$$W(\nu_1,\nu_2,\dots,\nu_N)=(2\gamma_0)^N\prod_{j=1}^{N}\frac{1}{4\pi^2(\nu_j-\nu_0)^2+\gamma_0^2} \tag{29.15}$$

This shows that the frequencies of all the photons emitted by the oscillators as a result of transitions from the N-th level to the zero level are independent of each other and are given by the same expression. This expression, in fact, defines the shape of the emission line as a whole. To find the energy distribution for a single photon, we must integrate (29.15) over $N-1$ frequencies. The result is the well-known classical formula for the natural line shape

$$W(\nu)=\frac{2\gamma_0}{4\pi^2(\nu-\nu_0)^2+\gamma_0^2}$$

$$=\frac{\gamma_0'}{\pi}\,\frac{1}{(\nu-\nu_0)^2+\gamma_0'^2} \tag{29.16}$$

or, in terms of the angular frequencies,

$$W(\omega)=\frac{\gamma_0}{\pi}\,\frac{1}{(\omega-\omega_0)^2+\gamma_0^2} \tag{29.17}$$

where $\gamma_0=\gamma_0/2\pi=2\pi e^2/3mc^3$; and $W(\nu)=2\pi W(\omega)$.

It follows that, according to the quantum theory of radiation, the natural line shape for a harmonic oscillator is independent of its initial energy and is identical with the line shape predicted by classical electrodynamics. The line shape for a set of non-interacting oscillators is therefore independent of the initial distribution function.

The above results were obtained on the assumption that the temperature of the surrounding medium was zero. Since the luminescence characteristics of the oscillator are independent of temperature (Section 28), it would appear that the luminescence line shape (in afterglow) is also independent of the background temperature emission and is given by (29.10). The results which we have obtained are consistent with the fact that the oscillator level width is proportional to the level number. This is also a consequence of the uncertainty relation if it is recalled that, whatever the initial conditions, the attenuation of a quantum-mechanical oscillator is always exponential (see (27.39)).

It is important that we should note that in the case of an even slightly anharmonic oscillator the above calculations lose their validity. The spectral line width is then equal to the sum of the upper and lower level widths.

30. SYSTEMS APPROACHING THE HARMONIC OSCILLATOR

Oscillator interacting with a medium

The general problem of the interaction of an oscillator with the surrounding medium is outside the scope of the present chapter since the oscillations are then anharmonic. In this section we shall confine our attention to one special case when non-radiative transitions proceed only between neighbouring levels and the transition probabilities satisfy the condition

$$d_{v,\,v+1} = (v+1)\,d_{01},\quad d_{v+1,\,v} = (v+1)\,d_{10} \qquad (30.1)$$

This is equivalent to the assumption that the operator representing the interaction of the oscillator with the medium is proportional to the position coordinate operator (or $d^n x/dt^n$). This occurs, for example, when the oscillator interacts weakly with the electric field due to the surrounding medium.

For a constant illumination, the balance equations are

$$\begin{aligned} &n_{v+1,\,v}\,[A_{v+1,\,v} + B_{v+1,\,v}(u_0+u) + d_{v+1,\,v}] \\ &+ n_{v-1}\,[B_{v-1,\,v}(u_0+u) + d_{v-1,\,v}] - n_v\,[B_{v,\,v+1}(u_0+u) \\ &+ d_{v,\,v+1} + A_{v,\,v-1} + B_{v,\,v-1}(u_0+u) + d_{v,\,v-1}] = 0 \end{aligned} \qquad (30.2)$$

If we solve these by analogy with (27.6), we obtain

$$n_v = n\frac{A_0 + d_{10} - d_{01}}{A_0 + B_0(u_0+u) + d_{10}}\left[\frac{B_0(u_0+u) + d_{01}}{A_0 + B_0(u_0+u) + d_{10}}\right]^v \qquad (30.3)$$

If we use the distribution function given by (30.3), we find that the rate of luminescence is

$$\begin{aligned} W_{\text{lum}} &= nB_0 uh\nu_0\frac{A_0}{A_0 + d_{10} - d_{01}} \\ &= W_{\text{abs}}\frac{A_0}{A_0 + d_{10} - d_{01}} \end{aligned} \qquad (30.4)$$

According to this expression W_{lum} is a function of the probabilities of non-radiative transitions (in contrast to the rate of absorption). When $(d_{10} - d_{01}) \gg A_0$, there is practically no luminescence. The energy yield can be found from (30.4):

$$\Gamma_{lum} = \frac{A_0}{A_0 + d_{10} - d_{01}} \tag{30.5}$$

The emission and absorption by an oscillator which, in addition to radiative friction, experiences a resistive force due to the medium, can also be treated classically. If the resistive force is proportional to the velocity of the charges, the integral rate of emission under stationary conditions is given by

$$W_{em}^{cl} = n \frac{\pi e^2}{3m} u(\nu) \frac{\gamma_0}{\gamma_0 + \gamma_{\tau p}} = W_{abs}^{cl} \frac{\gamma_0}{\gamma_0 + \gamma_{\tau p}} \tag{30.6}$$

This is predicted by (2.47). Since $A_0 = 2\gamma_0$, it follows from (30.5) and (30.6) that the classical and quantum-mechanical formulae will only be identical if we let $2\gamma_{fr} = d_{10} - d_{01}$.

Slightly anharmonic oscillator

It has already been noted in connection with the polarisation of luminescence, the dichroism and the natural shape of the spectral lines of an oscillator that even slight anharmonicity leads to essentially new effects. It is therefore of interest to compare the predictions of classical and quantum theories for an anharmonic oscillator. In general, the potential function of an anharmonic oscillator is given by a series in powers of distance. However, to elucidate some of the qualitative features, it is sufficient to retain only the first two terms, so that

$$V = \frac{1}{2} k_1 x^2 + \frac{1}{3!} k_2 x^3 \tag{30.7}$$

The classical motion of oscillators with this potential function has been investigated in detail by Viswanathan [70], who has shown that, to a high degree of accuracy, the solution is

$$x = x_3 + (x_2 - x_3)\, sn^2 (\mathrm{u}, k) \tag{30.8}$$

where $\mathrm{u} = \sqrt{-\frac{k_2}{m}(x_1 - x_3)}\, t$; $k^2 = \frac{x_2 - x_3}{x_1 - x_3}$ and x_1, x_2, x_3

are the roots of the equation

$$\frac{1}{2} m\dot{x}^2 + \frac{1}{2} k_1 x^2 - \frac{1}{6} k_2 x^2 = \mathrm{E} \qquad (30.9)$$

which represents the conservation of energy.

The function given by (30.8) can be expanded into a Fourier series

$$x = x_0 + x_{01} \cos \omega t$$
$$+ x_{02} \cos 2\omega t + \ldots + x_{0j} \cos j \omega t + \ldots,$$

where x_0 is the mean displacement of the oscillator from the equilibrium position, x_{01}, x_{0j} are the amplitudes of the fundamental oscillation and the j-th harmonic respectively and ω is the fundamental frequency. Calculations show that the intensities of the harmonics are much lower than the intensity of the fundamental oscillation. Moreover, the fundamental frequency ω is a decreasing function of the oscillator energy. As the oscillations become damped out, the frequency ω increases continuously together with the frequencies of the harmonics. It follows that the emission spectrum due to an anharmonic oscillator should, according to classical theory, consist of narrow equidistant bands whose width should increase with increasing initial energy.

In contrast to the harmonic oscillator, the distance between neighbouring energy levels of a quantum-mechanical anharmonic oscillator decreases with increasing level number and, at the same time, there are finite probabilities of transitions in which the vibrational quantum number v changes by 2, 3, 4, and so on. The emission spectrum is therefore found to consist of a series of bands whose separation decreases towards higher frequencies. We thus see that even here there are many discrepancies between classical and quantum theories.

A second shortcoming of the classical theory is that it does not predict the fine structure of bands but only indicates their widths. Classical theory allows all frequencies within the limits of a band. Quantum theory, on the other hand, predicts that each band should, in principle, have a fine structure. For example, the first band consists of lines due to the transitions $1 \rightarrow 0$, $2 \rightarrow 1$, $3 \rightarrow 2$, $v + 1 \rightarrow v$, and the second band is due to transitions $2 \rightarrow 0$, $3 \rightarrow 1$, $v + 2 \rightarrow v$ and so on.

Although in practice neighbouring lines tend to merge into a continuous band, this does not obviate the basic difference between the two theories. We thus see that even weak anharmonicity leads to a discrepancy between classical and quantum theories.

The place of the oscillator among other quantum-mechanical systems

It is well known that classical mechanics and electrodynamics cannot, in general, be used to describe atomic and molecular phenomena. The identity of classical and quantum-mechanical predictions with regard to the optical properties of the harmonic oscillator must therefore mean that the oscillator is an exception among other models of matter used in classical optics. The problem therefore arises as to whether there are systems in quantum theory whose optical properties are either partly or entirely the same as the properties of the oscillator.

The chief quantum-mechanical properties of the oscillator can be summarised as follows:

1. the number of energy states is unlimited;
2. the levels are equidistant;
3. transitions are possible only between neighbouring levels;
4. the Einstein coefficients are proportional to the level number;
5. all the dipole moment matrix elements are parallel to each other.

These statements can be regarded as a definition of the harmonic oscillator. If a quantum-mechanical system exhibits these properties, then it must be identified with a harmonic oscillator, and conversely if even one of these properties is not exhibited it must be of a different type.

Consider, to begin with, the properties which matter must have if its absorptive power is to be independent of the density of the exciting radiation. Suppose that a given system does have properties 2 to 5 but possesses a limited number of energy levels (cut-off oscillator). It is clear that the absorption coefficient of this system will be a function of the density u of the exciting radiation. In point of fact, as u increases, the population of the uppermost N-th level will increase. At the same time, particles occupying the N-th level will not be able to absorb the radiation and therefore the absorption coefficient will decrease with increasing density of the incident radiation.

Let us suppose now that the particles have all the properties of an oscillator with the exception of property 2, i.e. the energy levels are not equidistant. Here, one will observe a number of absorption lines, each characterised by its own absorption coefficient. The rate of absorption in any given line will depend only on the population of the corresponding pair of levels in accordance with the formula $W_{abs}(\nu_{ij}) = Bu(n_j - n_i)h\nu_{ij}$. Since, however, n_j and n_i are functions of only T and u, it follows that the absorption coefficient will also be a function of the density of the exciting radiation and of temperature.

If the system does not exhibit property 4, then the difference between the Einstein coefficients, $B_{v,v+1} - B_{v,v-1}$ will not be a constant, and this will again result in the dependence of the absorption coefficient on the distribution function and hence on u and T. Finally, if properties 3 and 5 are not exhibited the polarisation of the emitted luminescence and the natural line shape will depend on the mean energy of the dipole, although the absorptive power of the system at the fundamental frequency given by $h\nu_0 = E_{v+1} - E_v$ will not be affected.

Experiment shows us that the properties 1, 2 and 4 are not exhibited by real atoms. It follows that the independence of the absorption coefficient of the density of the exciting radiation is a property of an idealised model of matter, i.e. the harmonic oscillator. All real quantum-mechanical systems exhibit departures from Bouguer's law and induced dichroism.

Absorptive power is an important characteristic of matter and largely determines its other optical properties. If the polarisation of luminescence is to be constant, k must be independent of u and T since otherwise the relative contribution of the luminescence due to an individual particle will depend on u and this unavoidably means that the polarisation must depend on the entire luminescence. However, in contrast to the integral absorptive power, the polarisation is also sensitive to the directions of the dipole moments. If the matrix elements corresponding to different transitions between equidistant levels are not parallel, then the polarisation of the luminescence will depend on the distribution function and, through it, on u and T. The transition of a particle from one level to another will then be accompanied by rotation of the dipole moment and this will lead to depolarisation of the radiation.

It is important to note that the connection between the natural width of a spectral line emitted by a harmonic oscillator and the width of the energy levels is also quite exceptional. Whereas for other quantum-mechanical systems the line width is equal to the sum of the widths of the upper and lower levels between which the transition takes place, in the case of the harmonic oscillator the line width is equal to the difference between these widths. This result is valid only for the harmonic oscillator. The calculation performed in Section 29 is valid only if conditions 3 to 5 are satisfied. It has already been noted that for a weakly anharmonic oscillator the theory predicts completely different results. We may conclude that the anharmonic oscillator occupies a special place not only among classical but also among quantum-mechanical models of matter.

7

Absorption and Luminescence of a System of Particles with Two Energy Levels

31. DISTRIBUTION OF PARTICLES OVER ENERGY LEVELS FOR DIFFERENT MODES OF EXCITATION

Excitation by isotropic radiation

Consider a set of n particles with two non-degenerate energy levels. The particles are randomly distributed in space and the directions of their dipole moments are fixed. The energy level scheme is shown in Fig. 7.1. The Einstein coefficients for spontaneous and induced transitions will be denoted by A and B, and the probabilities of non-radiative transitions by d_{21} and d_{12}, respectively.

In the absence of external excitation, the particles are in complete thermodynamic equilibrium with the surrounding medium and the thermal radiation background. The populations of the excited and unexcited levels are independent of time, although the particles continuously undergo transitions from ground to excited states, and vice versa. The

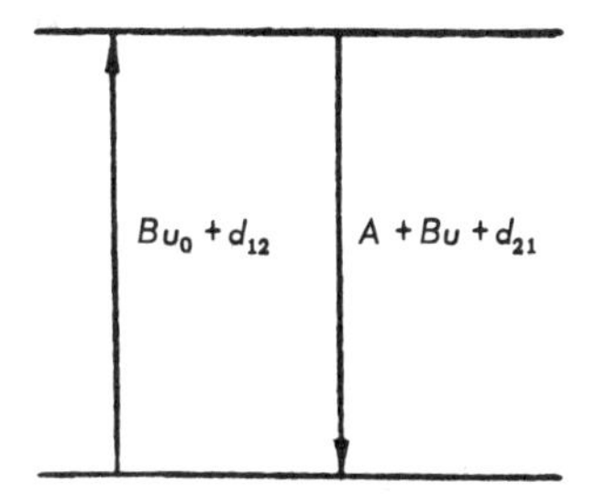

Fig. 7.1 Two-level scheme

principle of detailed balancing predicts that the transition probabilities are related by

$$\begin{gathered}(A + Bu_0)\, e^{-h\nu/kT} = Bu_0 \\ d_{12} = d_{21}\, e^{-h\nu/kT} = de^{-h\nu/kT}\end{gathered} \tag{31.1}$$

where u_0 is the density of equilibrium radiation and $d = d_{21}$. When the particles are excited by external radiation, their distribution over the energy levels is different from the Boltzmann distribution.

Consider, to begin with, the case when the external radiation does not give rise to anisotropy in the orientation distribution of the excited particles. If the particles do not rotate, this occurs only when the incident radiation is isotropic. However, calculations performed without taking into account anisotropy in the orientation distribution of the particles have a broader range of application. They are also valid if there is intensive Brownian motion and the system is illuminated by a directed beam of partly or fully polarised radiation (non-viscous solutions will be a random orientation distribution of excited and unexcited particles in this case). Moreover, we shall see below that the dependence of total absorption and luminescence on the density of the exciting radiation, the temperature of the medium and the non-radiative transitions for excitation by directed and isotropic radiation is very similar. It is therefore often useful to neglect the anisotropy in the orientation distribution of the particles because this leads to a simplification of the calculations and does not substantially affect the final general conclusions.

Let the density of the external exciting radiation be u, so that the balance equation can be written in the form

$$\frac{dn_2}{dt} = -n_2(A + Bu_0 + Bu + d) + n_1(Bu_0 + Bu + de^{-h\nu/kT}) \tag{31.2}$$

Under stationary conditions, the level populations are constant and therefore

$$n_2 = n_1 \frac{Bu_0 + Bu + de^{-h\nu/kT}}{A + Bu_0 + Bu + d} \tag{31.3}$$

Since the total number of particles per unit volume must be constant

$$n_1 + n_2 = n \tag{31.4}$$

we find that

$$n_2 = n \frac{Bu_0 + Bu + de^{-h\nu/kT}}{A + 2Bu_0 + 2Bu + d(1 + e^{-h\nu/kT})} \tag{31.5}$$

$$n_1 = n \frac{A + Bu_0 + Bu + d}{A + 2Bu_0 + 2Bu + d(1 + e^{-h\nu/kT})} \tag{31.6}$$

If we let u tend to zero in these expressions, we find that, in accordance with (31.1), (31.5) and (31.6), the distribution tends to the Boltzmann distribution

$$n_2^0 = n \frac{e^{-h\nu/kT}}{1 + e^{-h\nu/kT}} \qquad n_1^0 = n \frac{1}{1 + e^{-h\nu/kT}} \tag{31.7}$$

After the exciting radiation is switched on, the distribution function assumes the non-equilibrium form, and even for large excitation energies, when the probability Bu is greater than the remaining transition probabilities, $n_2 \to n_1 \to \frac{1}{2} n$. The level populations tend to the same limit for sufficiently high temperatures ($Bu_0 \gg A$). The distribution function is particularly sensitive to external excitation at low temperatures of the surrounding medium and low non-radiative

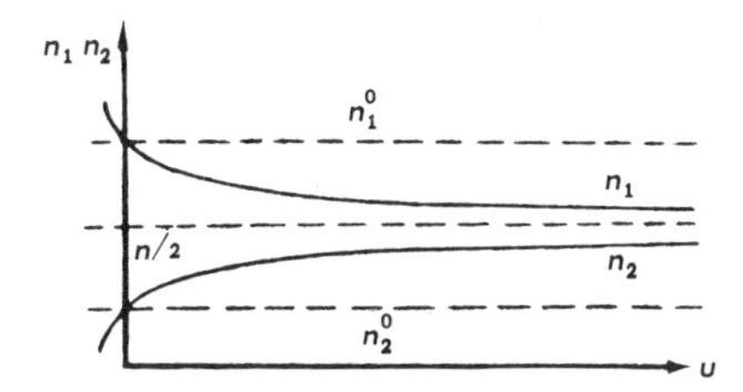

Fig. 7.2 Populations of the first and second levels as functions of the intensity of the exciting radiation

transition probabilities. The dependence of n_2 and n_1 on the density of the exciting radiation (positive and negative) is plotted in Fig. 7.2.

For purposes of comparison with subsequent calculations, let us now use (31.5) and (31.6) to establish the orientation distributions per unit solid angle. The results are

$$n_2(u, T, \Omega) = \frac{1}{4\pi} n_2 = \frac{\frac{1}{4\pi} n_2^0 + \frac{n}{4\pi} \frac{1}{6} \alpha u}{1 + \frac{1}{3} \alpha u} \tag{31.8}$$

$$n_1(u, T, \Omega) = \frac{1}{4\pi} n_1 = \frac{\frac{1}{4\pi} n_1^0 + \frac{n}{4\pi} \frac{1}{6} \alpha u}{1 + \frac{1}{3} \alpha u} \tag{31.9}$$

where n_j^0 is the total number of particles in the j-th level in the absence of external excitation and

$$\alpha = \frac{6B}{A + 2Bu_0 + d(1 + e^{-h\nu/kT})} \tag{31.10}$$

The parameter α, defined by the last expression, is an important characteristic of a system of particles with two energy levels. It determines unambiguously the minimum density of incident radiation for which various non-linear effects become appreciable, namely, departure from Bouguer's law, depolarisation of luminescence, induced dichroism, and so on. Since a similar parameter can be introduced for each pair of levels of any quantum-mechanical system

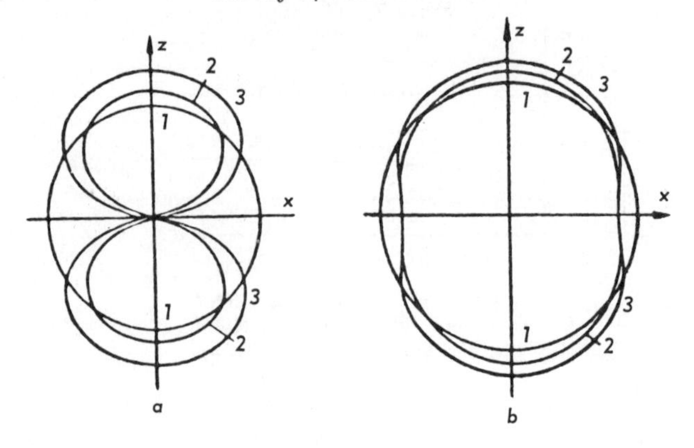

1 - isotropic radiation; 2 - natural radiation; 3 - plane-polarised radiation: a - $u_0 = d = 0$, $Bu = A$; b - $Bu_0 = Bu = A$, $d = 0$

Fig. 7.3 Angular distribution of excited particles during illumination

with a discrete spectrum, it follows that α can usefully be referred to as the non-linearity parameter.

Owing to (31.8) and (31.9), the level populations are linear functions of the density of the exciting radiation as long as $\alpha u \ll 1$, i.e. until one can assume that the denominators in (31.8) and (31.9) are practically independent of u.

Isotropic radiation does not produce an anisotropy in the orientation distribution of excited particles. The graphical representation of $n_2(u, T, \Omega)$ as a function of Ω therefore takes the form of a sphere of radius proportional to n_2. The cross-section of this sphere in the plane containing its centre is shown in Fig. 7.3.

Plane-polarised exciting radiation

Suppose that plane-polarised light is incident along the x axis and the electric vector is parallel to the z axis. The incident radiation will induce the transitions $1 \rightleftarrows 2$ whose probability will depend on the orientation of the particle dipole moment. According to (8.72)-(8.73), the probability of transitions with dipole absorption or emission of radiation is proportional to $\cos^2(\mathbf{DE})$, where $\mathbf{D}$ is the dipole moment of the particle and $\mathbf{E}$ is the electric vector in the absorbed or emitted radiation. In the present case, all the particles whose dipole moments are at an angle θ to the z axis are in identical conditions with respect to the incident radiation, so that the angular distribution of the particles has a symmetry axis parallel to the z axis. The differential Einstein coefficients can therefore be conveniently written down in terms of spherical coordinates with the polar axis lying along the z axis. By taking the components of **D** along the z and x axes, we obtain from (8.72) and (8.73) the differential Einstein coefficients for transitions with the absorption and emission of radiation polarised, respectively, along the z and x axes:

$$a^z(\Omega) = \frac{3}{8\pi} A\cos^2\theta, \quad b^z(\Omega) = 3B\cos^2\theta \tag{31.11}$$

$$a^x(\Omega) = \frac{3}{8\pi} A\sin^2\theta\cos^2\varphi, \quad b^x(\Omega) = 3B\sin^2\theta\cos^2\varphi \tag{31.12}$$

where A and B are the integral Einstein coefficients given by (8.76) and (8.77).

Under constant illumination, the distribution functions satisfy the balance equation

$$\begin{aligned} &n_2(u, T, \Omega)\,[A + Bu_0 + d + 3Bu\cos^2\theta] \\ &= n_1(u, T, \Omega)\,[Bu_0 + de^{-h\nu/kT} + 3Bu\cos^2\theta] \end{aligned} \tag{31.13}$$

Since for any direction

$$n_2(u, T, \Omega) + n_1(u, T, \Omega) = \frac{1}{4\pi} n \tag{31.14}$$

we have from (31.13)

$$n_2(u, T, \Omega) = \frac{\frac{1}{4\pi} n_2^0 + \frac{1}{8\pi} n\alpha u\cos^2\theta}{1 + \alpha u\cos^2\theta} \tag{31.15}$$

$$n_1(u, T, \Omega) = \frac{\frac{1}{4\pi} n_1^0 + \frac{1}{8\pi} n\alpha u\cos^2\theta}{1 + \alpha u\cos^2\theta} \tag{31.16}$$

According to these expressions, in spite of the initially random angular distribution, both the excited and unexcited particles have a definite dipole-moment orientation due to the anisotropy of the exciting radiation. This orientation depends on the intensity of this radiation, the temperature of the surrounding medium and the probabilities of radiative and non-radiative transitions. Figure 7.4 shows plots of n_2 as a function of θ for different values of Bu_0/A, Bu/A and d/A. It is evident from this figure that when $u_0 = d = 0$, the anisotropy in the distribution of the excited particles is particularly well defined for small u $(Bu/A \ll 1)$. n_2 is then approximately proportional to $\cos^2\theta$. Particles whose dipole moments are parallel to the electric vector in the incident radiation are preferentially excited. With increasing u, the anistropy decreases, and for very large u $(\alpha u \gg 1)$ there is practically no anisotropy since saturation sets in for practically all particles, whatever the orientation of the dipole

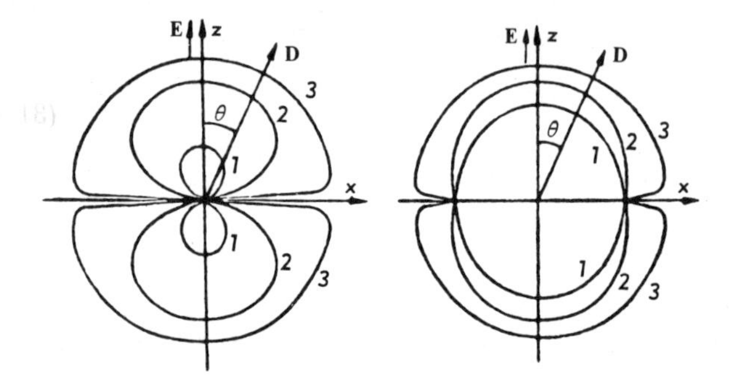

Fig. 7.4 Population of the second level as a function of the orientation of the particle dipole moment D_{21}. *Curves* 1-3 *correspond respectively to* $Bu/A = 0.1$, 1 *and* 100

moment $(n_2 = n_1)$. The small number of particles whose dipole moments lie in the xy plane form the only exception.

If u_0 and d are not zero, anisotropy of the angular distribution of excited particles is not well defined for small u. Planck radiation and interaction with the medium facilitate the attainment of a random distribution. A much higher density of exciting radiation is necessary to introduce appreciable anisotropy. Subsequent increase in u, however, will again be associated with a uniform orientation distribution of excited particles (see Fig. 7.4). It is important to note that if we substitute the mean value of $\cos^2\theta$, i.e. 1/3, into (31.15) and (31.16), we again obtain (31.8) and (31.9).

If we integrate (31.15) with respect to φ between 0 and 2π, and with respect to θ between 0 and π, we shall then obtain the total number of excited particles:

$$n_2 = \frac{n}{2}\left[1 + \frac{A + d(1 - e^{-h\nu/kT})}{A + 2Bu_0 + d(1 + e^{-h\nu/kT})} \times (\alpha u)^{1/2} \tan^{-1}(\alpha u)^{1/2}\right] \quad (31.17)$$

If $\alpha u < 1$, we can expand $\tan^{-1}(\alpha u)^{1/2}$ into a series, so that (31.17) takes the form

$$n_2 = n_2^0 + \frac{1}{2} n \alpha u \sum_{k=0}^{\infty} \frac{(-1)^k (\alpha u)^k}{(2k+3)} \quad (31.18)$$

where, as before, n_2^0 is the total number of excited particles under the conditions of thermodynamic equilibrium. As was to be expected, $n_2 \to n_2^0$ when $\alpha u \to 0$. If $u \to \infty$ $(\alpha u \to \infty)$, the

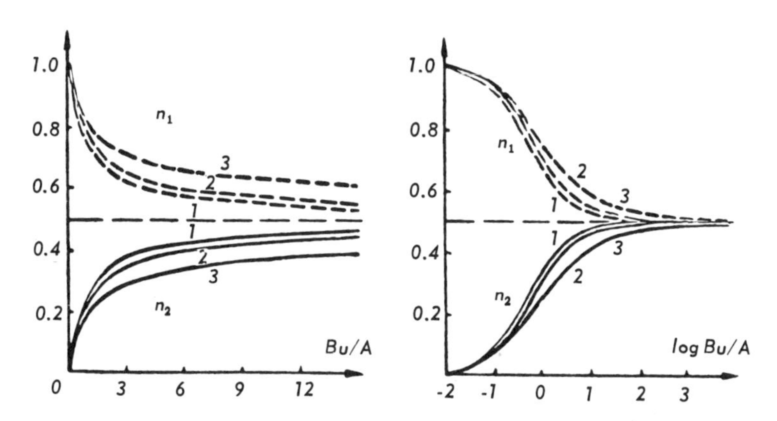

Fig. 7.5 Total populations of the first and second levels as functions of u *in the case of excitation by isotropic (1), natural (2) and plane-polarised (3) radiation*

second term in (31.17) vanishes and $n_2 \to \frac{1}{2}n$. A plot of n_2 and u for different values of u_0 and d is shown in Fig. 7.5.

Natural radiation

Let us suppose that the same system is excited by natural radiation propagating along the x axis. The distribution of the particles over the levels will then depend both on θ and

φ. This leads to complicated formulae for the rate of absorption and luminescence. To simplify the calculations let us transform to a new set of spherical coordinates with the polar axis parallel to the x axis. The direction of the vector **D** will then be determined by θ and φ' (Fig. 7.6).

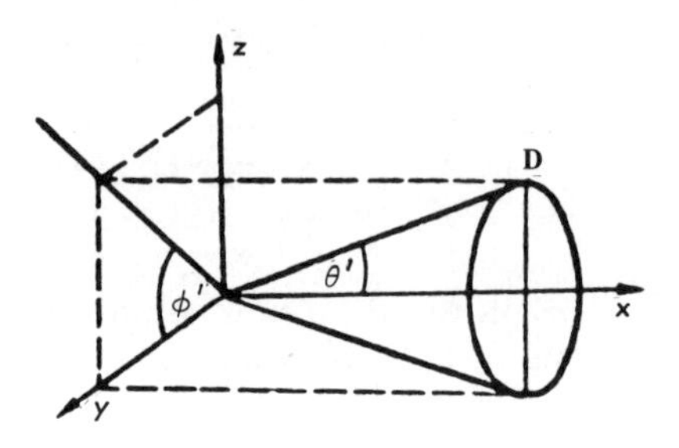

Fig. 7.6 Excitation by natural radiation

The choice of this set of coordinates is governed by the fact that in the case of excitation by natural radiation, the x axis is parallel to the symmetry axis of the orientation distribution of excited particles. The angular dependence of the distribution function in the new system of coordinates will therefore reduce to a dependence on the single angle θ'.

The flux of natural radiation acting on the particle whose orientation is specified by angles θ' and φ' can be resolved into two fluxes of equal intensities, polarised in mutually perpendicular planes. If the direction of vibrations in one of the component fluxes is parallel to the line of intersection of the planes Oxy and $O\mathbf{D}x$, the polarisation vector of the second flux will be perpendicular to the particle dipole moment **D**. Consequently, induced transitions will be stimulated by the first flux and the transition probabilities will be given by

$$B(\Omega')u = 3B\frac{1}{2}\cos^2\left(\frac{\pi}{2} - \theta'\right) = \frac{3}{2}Bu\sin^2\theta' \qquad (31.19)$$

In the case of natural incident radiation, (31.13) must be replaced by the following balance equation for the number of particles per unit solid angle:

$$n_2(u, T, \Omega')\left(A + Bu_0 + d + \frac{3}{2} Bu \sin^2\theta'\right) = n_1(u, T, \Omega)\left(Bu_0 + de^{-h\nu/kT} + \frac{3}{2} Bu \sin^2\theta'\right) \tag{31.20}$$

Solving this and bearing in mind (31.14), we obtain

$$n_2(u, T, \Omega) = \frac{\frac{1}{4\pi} n_2^0 + \frac{1}{16\pi} n \alpha u \sin^2\theta'}{1 + {}^1/_2\, \alpha u \sin^2\theta'} \tag{31.21}$$

$$n_1(u, T, \Omega) = \frac{\frac{1}{4\pi} n_1^0 + \frac{1}{16\pi} n \alpha u \sin^2\theta'}{1 + {}^1/_2\, \alpha u \sin^2\theta'} \tag{31.22}$$

If we substitute the mean value (2/3) of $\sin^2\theta'$ into these expressions, we again arrive at (31.8) and (31.9).

According to (31.21), natural radiation is similar to plane-polarised radiation in that it gives rise to anisotropy in the orientation distribution of the excited particles. Figure 7.3 shows the shape of $n_2(\Omega)$ in the xz plane for different modes of excitation. The circles (spheres in space) correspond to isotropic excitation. The function $n_2(\Omega)$ corresponding to plane-polarised and natural radiation is obtained by rotating curves 2 and 3 about the z and x axes, respectively.

To determine the total number of excited particles, we must integrate (31.21) with respect to the angles θ' and φ'

$$n_2 = \int_0^{2\pi} d\varphi' \int_0^{\pi} n_2(\Omega') \sin\theta'\, d\theta'$$

$$= \frac{n}{2}\left[1 - \frac{A + d(1 - e^{-h\nu/kT})}{A + B(2u_0 + 3u) + d(1 + e^{-h\nu/kT})} \times \frac{1}{a^{1/2}} \tanh^{-1} a^{1/2}\right] \tag{31.23}$$

where

$$a = \frac{\alpha u}{2 + \alpha u} = \frac{3Bu}{A + 2Bu_0 + d\,(1 + e^{-h\nu/kT}) + 3Bu} \tag{31.24}$$

The quantity a varies in the range between 0 and 1 as the density of the incident radiation is increased from 0 to ∞. If we expand $\tanh^{-1} a^{1/2}$ into a series we can readily verify that as $u \to 0$, $n_2 \to n_2^0$. If, on the other hand, $u \to \infty$, the second term in (31.23) vanishes and $n_2 \to n/2$.

It follows that for all the modes of excitation considered above, the maximum number of excited particles is $\frac{1}{2}n$. Figure 7.5 shows graphs of n_2 as a function of u for different modes of excitation. For a given density of excited radiation, n_2 is a maximum if the illumination is by natural radiation, or the excitation is anisotropic but the particles execute rapid rotations (this corresponds to the replacement of $\cos^2\theta$ and $\sin^2\theta'$ by their average values). When the system is illuminated by natural light, the particles whose dipole moments are at a small angle to the x axis are only weakly excited. The total population of the second level is therefore somewhat smaller than in the first case. Finally, plane-polarised radiation gives rise to weak excitation of particles whose dipole moments are at a small angle to the xy plane with the result that n_2 has its lowest value.

Change of level population due to introduction and removal of external excitation

According to (31.2) and (31.4), the populations of the second level after the illumination is switched on, and after it is switched off are respectively given by

$$\frac{d}{dt} n_2^{\text{gr}} + n_2^{\text{gr}} [A + 2B(u_0 + u) + d(1 + e^{-h\nu/kT})] = n[B(u_0 + u) + de^{-h\nu/kT}], \quad (31.25)$$

$$\frac{d}{dt} n_2^{\text{dec}} + n_2^{\text{dec}} [A + 2Bu_0 + d(1 + e^{-h\nu/kT})] = n[Bu_0 + de^{-h\nu/kT}]. \quad (31.26)$$

The solutions of these equations are

$$n_2^{\text{gr}} = Ce^{-t/\tau} + n_2^{\text{st}} \quad (31.27)$$

$$n_2^{\text{dec}} = C^0 e^{-t/\tau_0} + n_2^0 \quad (31.28)$$

where

$$\tau_0 = \frac{1}{A + 2Bu_0 + d(1 + e^{-h\nu/kT})} \tag{31.29}$$

$$\tau = \frac{1}{A + 2Bu_0 + d(1 + e^{-h\nu/kT}) + 2Bu} = \frac{3\tau_0}{3 + \alpha u} \tag{31.30}$$

n_2^0 and n_2^{st} are the populations of the second level in complete thermodynamic equilibrium and for constant illumination respectively (see equations (31.5) and (31.7)); C and C^0 are integration constants.

If the external source of radiation is switched on at time $t = 0$ and switched off at $t = T$, where $T_1 \gg \tau$, then if we determine C and C^0 from the initial conditions

$$\begin{aligned} n_2^{gr}(0) &= n_2^0 \\ n_2^{dec}(T_1) &= n_2^{st} \end{aligned} \tag{31.31}$$

and substitute into (31.27) and (31.28), we find that

$$n_2^{gr} = -(n_2^{st} - n_2^0)\, e^{-t/\tau} + n_2^{st} \tag{31.32}$$

$$n_2^{dec} = (n_2^{st} - n_2^0)\, e^{-(t-T_1)/\tau_0} + n_2^0 \tag{31.33}$$

It is evident from these formulae that the population of the second level increases or decreases strictly exponentially while the stationary conditions are being approached and after the excitation is switched off. The quantity τ_0 is the mean lifetime of the system in the excited state. In the visible region and for moderate temperatures, it is approximately given by

$$\tau_0 = \frac{1}{A + d}$$

and may vary within broad limits between a few seconds and 10^{-11} sec.

The greater the probability of non-radiative transitions, the faster will the system return to the state of thermodynamic equilibrium. As d increases there is a decrease not only in τ_0 but also in the pre-exponential factor $(n_2^{st} - n_2^0)$.

The probabilities of non-radiative transitions inhibit the departure of the system from the equilibrium state.

Temperature has a similar effect. However, this appears only when Bu_0 is comparable with $(A + d)$. The time necessary to reach the stationary conditions depends not only on the properties of the particles but also on their interaction with the medium. It is related through τ to the density of the exciting radiation. For $\alpha u \ll 1$, τ is practically identical with τ_0 and is independent of the sign of u. It follows that graphs of n_2^{gr} for positive and negative u are symmetrical with respect to n_2^0 (broken line in Fig. 7.7). With increasing

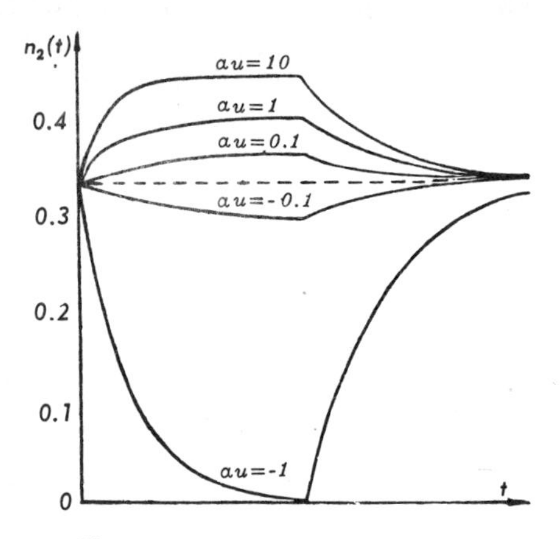

Fig. 7.7 Time-dependence of n_2^{gn} *and* n_2^{off} *in the case of excitation by positive and negative fluxes*

$|u|$ the magnitude of τ increases if $u < 0$ and decreases if $u > 0$. The symmetry of the n_2^{gr} curves for negative and positive u is then disturbed.

Maximum τ occurs when $u = -u_0$:

$$\tau_{max} = \frac{1}{A + d(1 + e^{-h\nu/kT})} \leqslant \frac{1}{A + 2d}$$

and tends to zero as $1/2Bu$ when $u \longrightarrow \infty$. The non-linearity parameter can be used to transform (31.32) so that it reads

$$n_2^{gr} = -\frac{1}{2}(n_1^0 - n_2^0)\frac{\alpha u}{3 + \alpha u} e^{-t/\tau} + n_2^{st} \qquad (31.32a)$$

When $Bu_0 \gg A + d$, the difference $n_1^0 - n_2^0$ will tend to zero. It follows that a positive flux of radiation will not be capable of appreciably changing $n_2{}^{gr}$ in comparison with $n_2^{st} \approx n_2^0$. Even for $\alpha u \to \infty$, the pre-exponential factor will not exceed 1/2 $(n_1^0 - n_2^0)$ which is very nearly zero. If under similar conditions the excitation is due to negative fluxes, $\alpha u \to -3$ and departures may be quite large. The population of the second level will change from $n/2$ to 0 as n changes from 0 to $-u_0$.

Pulsed excitation

In practice, pulsed excitation is often employed. Pulsed sources, e.g. high-power electrical discharges in gases, are capable of producing for short periods of time radiant energy fluxes of a magnitude which cannot be reached by constant intensity sources. Continuous illumination frequently leads to considerable heating and undesirable physical and mechanical processes in the specimen. Short pulses of radiation are therefore frequently very convenient.

Pulsed excitation with the aid of a phosphoroscope provides a means of cutting off rapidly decaying fluorescence and investigating phosphorescence separately, even though the spectra of both types of emission partly or completely overlap. Under certain conditions, pulsed excitation leads to an optico-acoustic phenomenon which is entirely absent under steady illumination.

Suppose that a system of particles with two energy levels is excited by pulsed illumination from time $t = 0$ onwards. The energy density is of the form shown in Fig. 7.8. The mark-to-space ratio is such that the radiation is on for time T_1 and off for time T_2. We shall assume, for the sake of simplicity, that the external radiation does not produce an anisotropy in the orientation distribution of excited particles.

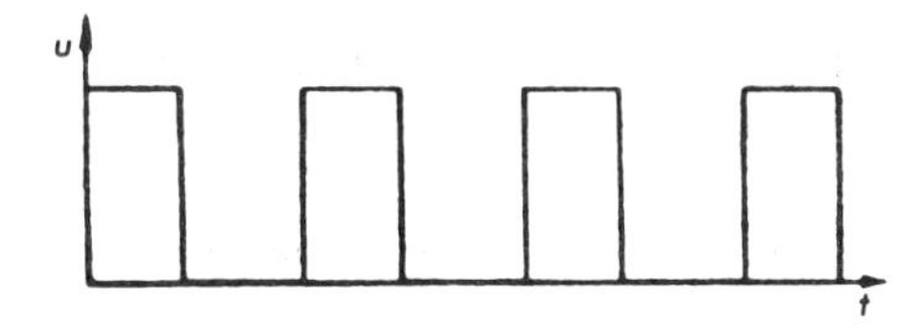

Fig. 7.8 Pulses of exciting radiation

The populations of the second level n_2^{on} and n_2^{off} will, as before, be given by (31.25) and (31.26). The solution of these equations for the j-th illumination and j-th dark period can be written in the form:

$$n_{2j}^{\text{on}} = C_j e^{-t/\tau} + n_2^{\text{st}} \tag{31.34}$$

$$n_{2j}^{\text{off}} = C_j^0 e^{-t/\tau_0} + n_2^0 \tag{31.35}$$

where the integration constants C_j and C_j^0 for each new period must be determined from the initial conditions.

In general, when T_1 and T_2 are of the same order as, or less than, τ and τ_0, the stationary conditions cannot be reached in one period and the system does not return to thermodynamic equilibrium with the surrounding medium after the source of radiation is switched off. To determine C_j and C_j^0 for any j, we can use the fact that at times $t = j(T_1 + T_2)$ and $t = j(T_1 + T_2) - T_2$ we have (from the continuity of $n_2(t)$):

$$\begin{gathered} n_{2j}^{\text{on}}\,[t = j(T_1 + T_2) - T_2] = n_{2j}^{\text{off}}[t = j(T_1 + T_2) - T_2] \\ n_{2j}^{\text{off}}\,[t = j(T_1 + T_2)] = n_{2(j+1)}^{\text{on}}[t = j(T_1 + T_2)] \end{gathered} \tag{31.36}$$

or, in view of (31.34) and (31.35),

$$\begin{gathered} C_j e^{[-j(T_1+T_2)+T_2]/\tau} + n_2^{\text{st}} = C_j^0\, e^{[-j(T_1+T_2)+T_2]/\tau_0} + n_2^0 \\ C_j e^{-j(T_1+T_2)/\tau_0} + n_2^0 = C_{j+1}\, e^{-j(T_1+T_2)/\tau} + n_2^{\text{st}} \end{gathered} \tag{31.36a}$$

Eliminating C_j^0, we obtain the recurrence relation

$$\begin{gathered} C_{j+1} = C_j \exp\left[T_2\left(\frac{1}{\tau} - \frac{1}{\tau_0}\right)\right] + \\ + (n_2^{\text{st}} + n_2^0)\left[\exp\left(-\frac{T_2}{\tau_0}\right) - 1\right] \exp\left(j\,\frac{T_1 + T_2}{\tau}\right) \end{gathered} \tag{31.37}$$

Using this j times and remembering that, according to (31.32), $C_1 = -(n_2^{\text{st}} - n_2^0)$, we have in the case of all j which begin with the value 1

$$C_{j+1} = (n_2^{\mathrm{st}} - n_2^0)\, e^{jT_2\left(\frac{1}{\tau} - \frac{1}{\tau_0}\right)}$$

$$\times \left[\left(e^{-\frac{T_2}{\tau_0}} - 1\right) \frac{1 - e^{-j\left(\frac{T_1}{\tau} + \frac{T_2}{\tau_0}\right)}}{1 - e^{-\left(\frac{T_1}{\tau} + \frac{T_2}{\tau_0}\right)}} - 1 \right] \tag{31.38}$$

If we find C_{j+1}^0 in a similar way, and substitute the result into (31.34) and (31.35), we obtain the following expressions

$$n_{2j}^{\mathrm{on}} = (n_2^{\mathrm{st}} - n_2^0) \left[\left(e^{-\frac{T_2}{\tau_0}} - 1\right) \frac{1 - e^{-\left(\frac{T_1}{\tau} + \frac{T_2}{\tau_0}\right)(j-1)}}{1 - e^{-\left(\frac{T_1}{\tau} + \frac{T_2}{\tau_0}\right)}} - e^{-(j-1)\left(\frac{T_1}{\tau} + \frac{T_2}{\tau_0}\right)} \right] e^{-t_j/\tau} + n_2^{\mathrm{st}} \tag{31.39}$$

$$n_{2j}^{\mathrm{off}} = (n_2^{\mathrm{st}} - n_2^0) \frac{(1 - e^{-T_1/\tau}) \left[1 - e^{-j\left(\frac{T_1}{\tau} + \frac{T_2}{\tau_0}\right)}\right]}{1 - e^{-\left(\frac{T_1}{\tau} + \frac{T_2}{\tau_0}\right)}} \times e^{-t_j^0/\tau_0} + n_2^0 \tag{31.40}$$

where $t_j = t - (j-1)(T_1 + T_2)$, $t_j^0 = t - (j-1)(T_1 + T_2) - T_1$ are measured from the beginning of the j-th period of illumination and the j-th dark space respectively [71]. The formula given by (31.36) cannot be extended to $j = 1$. The values of n_{21}^{on} must be taken from (31.32).

If we use the non-linearity parameter given by (31.10), we can rewrite the last two expressions in a form which is more convenient in practice:

$$n_{2j}^{\text{on}} = \frac{1}{2}(n_1^0 - n_2^0)\frac{\alpha u}{3+\alpha u}$$

$$\times\left[(e^{-T_2/\tau_0} - 1)\frac{1-e^{-(j-1)\xi}}{1-e^{-\xi}} - e^{-(j-1)\xi}\right]e^{-t_j/\tau}$$

$$+\frac{3n_2^0 + \frac{1}{2}n\alpha u}{3+\alpha u} \quad (31.41)$$

$$n_{2j}^{\text{off}} = \frac{1}{2}(n_1^0 - n_2^0)\frac{\alpha u}{3+\alpha u}\frac{(1-e^{-T_1/\tau})(1-e^{-j\xi})}{1-e^{-\xi}}$$

$$\times e^{-t_j^0/\tau_0} + n_2^0 \quad (31.42)$$

where

$$\xi = \frac{T_1}{\tau} + \frac{T_2}{\tau_0} \quad (31.43)$$

It is readily seen that the variable part of the populations of the second level is determined under pulsed excitation by the state of thermodynamic equilibrium preceding the excitation, the non-linearity parameter α and the density and sign of the exciting radiation. When $n_2^0 \ll n_1^0$ and $|\alpha u| \ll 1$, positive and negative radiation fluxes may give rise to considerable and symmetrical (with respect to n_2^0) changes in $n_2(t)$. At high temperatures, when $n_1^0 \rightarrow n_2^0$, only negative fluxes lead to appreciable changes in the level population.

By continuing the pulsed excitation for a sufficient time, it is possible to ensure that the system will reach a steady state, when an increase in n_2 during the illumination period is equal to the reduction in n_2 during a dark period. The steady-state condition sets in when n_{2j}^{on} and n_{2j}^{off} can be regarded as practically independent of j, i.e. when

$$j\xi = j\left(\frac{T_1}{\tau} + \frac{T_2}{\tau_0}\right) \gg 1 \quad (31.44)$$

According to (31.31), the condition for the steady state to be reached during illumination by continuous flux of radiation can be written in the form

$$t/\tau = j\left(\frac{T_1}{\tau} + \frac{T_2}{\tau_0}\right) \gg 1 \quad (31.45)$$

When τ is practically the same as τ_0, the time necessary for the attainment of steady-state conditions is the same both for continuous and pulsed excitation. When $\tau < \tau_0$, the steady-state conditions are reached later under pulsed conditions than under continuous conditions.

In steady state, (31.41) and (31.42) assume the simpler form

$$n_{2j}^{\text{on}} = \frac{1}{2}(n_1^0 - n_2^0)\frac{\alpha u}{3+\alpha u}\frac{e^{-\frac{T_2}{\tau_0}} - 1}{1 - e^{-\xi}} e^{-t_j/\tau} + \frac{3n_2^0 + \frac{1}{2}\alpha u n}{3 + \alpha u} \tag{31.46}$$

$$n_{2j}^{\text{off}} = \frac{1}{2}(n_1^0 - n_2^0)\frac{\alpha u}{3+\alpha u}\frac{1 - e^{-T_1/\tau}}{1 - e^{-\xi}} e^{-t_j^0/\tau_0} + n_2^0 \tag{31.47}$$

Graphs of the population of the second level as a function of time under pulsed excitation are shown in Fig. 7.9.

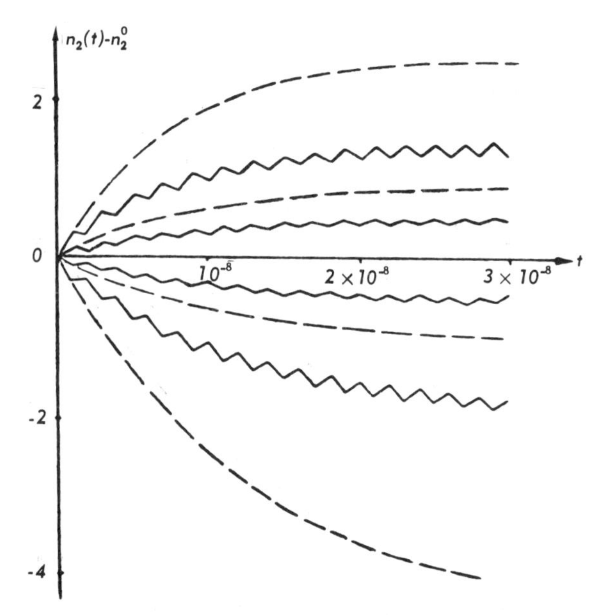

Fig. 7.9 Dependence of $n_2(t) - n_2^0$ on time in the case of illumination by pulsed positive and negative fluxes

32. ABSORPTION OF EXTERNAL RADIATION

Rate of absorption of isotropic plane-polarised and natural radiation

According to (17.21), the rate of absorption of incident radiation by all particles is given by

$$W_{\text{abs}} = h\nu \int_{\Omega} [n_1(u, T, \Omega)\, b(\Omega)\, u - n_1(u, T, \Omega)\, b(\Omega)\, u]\, d\Omega \quad (32.1)$$

If the exciting radiation does not give rise to anisotropy in the orientation distribution of the excited particles, then if we replace $b(\Omega)u$ by Bu in (32.1) and use (31.8)-(31.10), we find the rate of absorption of external radiation is given by

$$\begin{aligned} W^{\text{rand}}_{\text{abs}} &= nBuh\nu \frac{A + d(1 - e^{-h\nu/kT})}{A + 2B(u_0 + u) + d(1 + e^{-h\nu/kT})} \\ &= (n_1^0 - n_2^0)Buh\nu \frac{3}{3 + \alpha u} \end{aligned} \quad (32.2)$$

Next, if we substitute (31.15), (31.16) and (31.11) or (31.21), (31.22) and (31.18) into (32.1), and integrate with respect to the angles, we obtain the following expression for the rate of absorption of plane-polarised and natural positive radiation respectively:

$$W^{\text{p.pol}}_{\text{abs}} = 3(n_1^0 - n_2^0)Buh\nu \left(\frac{1}{\alpha u} - \frac{1}{(\alpha u)^{3/2}} \tan^{-1}(\alpha u)^{1/2}\right) \quad (32.3)$$

$$W^{\text{nat}}_{\text{abs}} = 3(n_1^0 - n_2^0)Buh\nu \frac{1}{\alpha u}[1 + (a^{1/2} - a^{1/2}) \tanh^{-1} a^{1/2}] \quad (32.4)$$

where a is given by (31.24).

It is important to note that whatever the mode of excitation, the rate of absorption of external radiation is proportional to the difference in the level populations prior to the exposure of the system to external radiation. It is evident from (31.7) that for a given frequency this difference is a function of only the temperature of the surrounding medium:

$$n_1^0 - n_2^0 = n\,\frac{1 - e^{-h\nu/kT}}{1 + e^{-h\nu/kT}} \tag{32.5}$$

At high temperatures, when $n_1^0 - n_2^0 \to 0$, external radiation is not absorbed and the medium becomes transparent.

When $\alpha u < 1$, the expressions for the rate of absorption can readily be expressed in the form of the power series:

$$W_{abs}^{rand} = (n_1^0 - n_2^0)\,Buh\nu\left[1 - \frac{1}{3}(\alpha u) + \frac{1}{9}(\alpha u)^2 \dots\right] \tag{32.6}$$

$$W_{abs}^{p.pol} = (n_1^0 - n_2^0)\,Buh\nu\left[1 - \frac{3}{5}(\alpha u) + \frac{3}{7}(\alpha u)^2 \dots\right] \tag{32.7}$$

$$W_{abs}^{nat} = (n_1^0 - n_2^0)\,Buh\nu\left[1 - \frac{2}{5}(\alpha u) + \dots\right] \tag{32.8}$$

It is evident from these expressions that when $\alpha u \ll 1$, the rate of absorption is proportional to the density of the incident radiation whatever its direction of propagation and polarisation. With increasing αu, the dependence of W_{abs} on u is no longer linear and the departure from linearity for given u is particularly well defined for plane-polarised light.

The maximum possible rate of absorption for $\alpha u \to \infty$ is independent of the mode of excitation and is given by

$$W_{abs}(\infty) = \frac{n}{2}\left[A + d\left(1 - e^{-h\nu/kT}\right)\right]h\nu \tag{32.9}$$

If we substitute this quantity into the expressions for the rate of absorption, we have

$$W_{abs}^{rand} = W_{abs}(\infty)\frac{\alpha u}{3 + \alpha u} \tag{32.10}$$

$$W_{abs}^{p.pol} = W_{abs}(\infty)\left[1 - \frac{1}{(\alpha u)^{1/2}}\tan^{-1}(\alpha u)^{1/2}\right] \tag{32.11}$$

$$W_{abs}^{nat} = W_{abs}(\infty)\left[1 - (a^{-1/2} - a^{1/2})\tanh^{-1} a^{1/2}\right] \tag{32.12}$$

Figure 7.10 shows plots of the rate of absorption as a function of u for different modes of excitation. It is evident from this figure that for a given value of u, the rate of absorption is a maximum when the particles are excited by isotropic radiation. W_{abs} increases somewhat more slowly when the

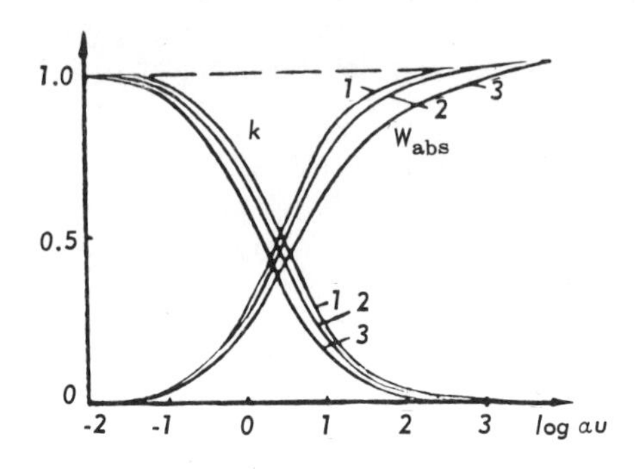

Fig. 7.10 Dependence of k *and* W_{abs} *on the intensity of incident radiation in the case of isotropic (1), natural (2), and plane-polarised (3) radiation*

excitation is due to natural radiation. The bottom curve refers to absorption of plane-polarised radiation.

The above result may seem unexpected because, in the case of isotropic excitation, the total number of excited particles is a maximum. However, the result can be understood if it is recalled that for isotropic excitation all particles participate equally in the absorption of the incident radiation. In the case of anisotropic illumination, a proportion of the particles is influenced to a greater extent by the electric field in the incident wave and rapidly reach excited states. Other particles absorb weakly even though they remain in the unexcited state because of the large angle between **D** and **E**.

We have assumed, so far, that the particles are excited by positive fluxes of radiation. The final formulae are, however, valid even for $u < 0$, and the rate of absorption can be investigated with the aid of (32.10). In (32.11) and (32.12), it is convenient to transform from functions of the imaginary argument $(\alpha u)^{1/2} = i(\alpha|u|)^{1/2}$ to functions of real variables:

$$W_{abs}^{p.pol} = W_{abs}(\infty)\left[1 - \frac{1}{(\alpha|u|)^{1/2}}\tanh^{-1}(\alpha|u|)^{1/2}\right] \quad (32.13)$$

$$W_{\text{abs}}^{\text{nat}} = W_{\text{abs}}(\infty)\,[1 - |a|^{-1/2}(1 - |a|\tan^{-1}|a|^{1/2}] \qquad (32.14)$$

The last expression is valid only if $a < 0$, which is not the case for all values of the negative radiation density.

According to (31.10), when $u < 0$, αu is negative and may vary within the limits

$$-3 < \alpha u < 0 \qquad (32.15)$$

whereas a varies between $-\infty$ and $+\infty$.

Let us now consider the absorption properties for negative radiation fluxes using (32.10). Analysis of (32.13) and (32.14) leads to similar results. We note that the quantity $\frac{3}{3+\alpha u}$ is positive for all values of u. Consequently, when $u < 0$, the rate of absorption is negative. In linear optics ($\alpha u \ll 1$), the rates of absorption of negative and positive radiation are practically equal in magnitude and opposite in sign. In the non-linear region, this is no longer the case. Since the absolute magnitude of the energy density in negative radiation cannot exceed u_0, it follows that to investigate W_{abs} for large $\alpha|u|$, one must at the same time take high temperatures of the surrounding medium. To find the limiting values of the rate of absorption, let us take the maximum value of $|u|$, i.e. u_0. We then have, from (32.2),

$$W_{\text{abs}}^{\text{rand}} = -nB|u|h\nu\,\frac{A + d(1 - e^{-h\nu/kT})}{A + d(1 + e^{-h\nu/kT})} \qquad (32.16)$$

It can readily be seen that with increasing temperature the absolute magnitude of the rate of absorption grows without limit. For large enough T (when $h\nu \ll kT$), we have the approximate formula

$$W_{\text{abs}}^{\text{rand}} = -n\,\frac{A}{A + 2d}\,kT \qquad (32.17)$$

It follows that, in contrast to excitation by positive radiation, excitation by negative radiation does not give rise to the saturation effect. It is important to note that this result is representative of the interaction between the medium and the source of radiation rather than of the interaction between the particles and the radiation. In point of fact, there are no radiative transitions when our assumption ($u_0 + u = 0$) is valid. The indefinite rise in W_{abs} means that negative fluxes from a cold source are fully compensated by the positive flux from a heated medium. Analysis of other special cases for which $u_0 + u > 0$ leads to similar results.

Absorption coefficients

Using (17.38), which relates the rate of absorption and the absorption coefficient, together with (32.2)-(32.4), we can readily find the following formulae for the absorption coefficients for positive radiation:

$$k^{\text{rand}} = \frac{1}{c}(n_1^0 - n_2^0)Bh\nu \frac{3}{3+\alpha u} = \frac{k^0}{1+\alpha u/3} \tag{32.18}$$

$$k^{\text{p.pol}} = \frac{3}{c}(n_1^0 - n_2^0)Bh\nu \left[\frac{1}{\alpha u} - \frac{1}{(\alpha u)^{3/2}} \tan^{-1}(\alpha u)^{1/2}\right] \tag{32.19}$$

$$k^{\text{nat}} = \frac{3}{c}(n_1^0 - n_2^0)Bh\nu \frac{1}{\alpha u}[1 + (a^{1/2} - a^{-1/2})\tanh^{-1} a^{1/2}] \tag{32.20}$$

The temperature dependence of the absorption coefficients and rates of absorption are the same and were investigated above. In contrast to this, the dependence of these quantities on the density of the exciting radiation is diametrically opposite. Whereas W_{abs} reaches its maximum value for $u \to \infty$, the absorption coefficient is a maximum for small u ($\alpha u \ll 1$). We then have the approximate result

$$k^{\text{rand}} \approx k^{\text{p.pol}} \approx k^{\text{nat}} \approx \frac{1}{c}(n_1^0 - n_2^0)Bh\nu = k^0 \tag{32.21}$$

With increasing $\bar{u}$, the absorption coefficient decreases and tends to zero as $\bar{u} \to \infty$ (in practice, for $\alpha u \gg 1$).

Figure 7.10 shows graphs of the absorption coefficients as functions of the density of exciting radiation. It is evident from this figure that departures from Bouguer's law are better defined (other things being equal) under the action of plane-polarised radiation. The formulae given by (31.10) and (32.19) enable us to estimate the minimum density of exciting radiation which leads to appreciable departures from Bouguer's law.

Suppose that it is possible to establish experimentally a change in the absorption coefficient by, say, 5%. According to (32.19), k decreases by 5.6% when $\alpha u_m = 0.1$, and therefore

$$u_{\min} = 0.1\left(\frac{1}{\alpha}\right) = \frac{1}{60}\,\frac{A + 2Bu_0 + d\,(1 + e^{-h\nu/kT})}{B}$$

$$= \frac{2\pi h \nu^3}{15c^3}\left[1 + \frac{d}{A}\,(1 + e^{-h\nu/kT})\right] + \frac{1}{30}u_0 \qquad (32.22)$$

In the absence of non-radiative transitions and of a background of thermal radiation, $u_{\min} = \dfrac{2\pi h\nu^3}{15c^3}$. In the visible part of the spectrum this corresponds to energy densities of roughly 2×10^{-14} erg cm^3 sec^{-1}. The density of Planck radiation for given polarisation assumes this value at a temperature of 8,400°K. When the incident radiation has a lower density, the departure from Bouguer's law is small and may be ignored. As one approaches the infrared region, it is found that $u_{\min}$ falls rapidly. For example, for $\nu = 2{,}000$ cm^{-1} (d and u_0 are assumed to be zero, as before), $u_{\min} = 2 \times 10^{-17}$ erg cm^3 sec^{-1}, which corresponds to a black-body temperature of 840°K.

In the radio-wave region ($\lambda = 5$ cm), a similar calculation yields the temperature of 0.084°K, i.e. Bouguer's law is never satisfied.

The above values of source temperature are valid for $Bu_0 = d = 0$. In the visible part of the spectrum Bu_0 is usually much less than A and this assumption is confirmed by experiment. With increasing λ, however, there is a rapid increase in Bu_0/A. At room temperature (300°K) it is equal to unity even for $\lambda = 70\ \mu$ and reaches 500 for $\lambda = 5$ cm. In the radio-frequency region, therefore, we must neglect the probability of spontaneous transitions rather than the quantity Bu_0. We then have $u_{\min} = \dfrac{1}{30}u_0$ and the departure from Bouguer's law depends not only on the density of the exciting radiation, but also on the temperature of the surrounding medium (the electromagnetic radiation background).

It is evident from (32.33) that departures from Bouguer's law are increasingly difficult to achieve as the temperature and the probabilities of non-radiative transitions increase. In the infrared region, the probabilities d may be higher than A by a few orders, and therefore the source temperature must be higher than the value given above (840°K) in order to achieve u_{min}.

It follows that as the density of the exciting radiation is varied from 0 to u_{min}, the absorption coefficient remains practically independent of u. At the same time, it follows from (32.16) that, depending on u_0, it may vary within very broad limits (from $k_{max} = \frac{1}{c} nBh\nu$ for $T = 0$ ($n_1^0 = n$) and zero when $u_0 \to \infty$ ($n_2^0 \to n_1^0$)). It is important to note that the non-linearity parameter, and hence the departure from Bouguer's law and other non-linear effects, are not directly connected with the lifetime of the excited state τ_0. When $u_0 = d = 0$, the non-linearity parameter is independent of τ_0, and in accordance with (31.10), is given by

$$\alpha = \frac{6B}{A} = \frac{3}{4\pi} \frac{c^3}{h\nu^3} \qquad (32.23)$$

Systems with long excited-state lifetimes are difficult to excite, and therefore the minimum incident densities for which non-linear effects become appreciable depend only on

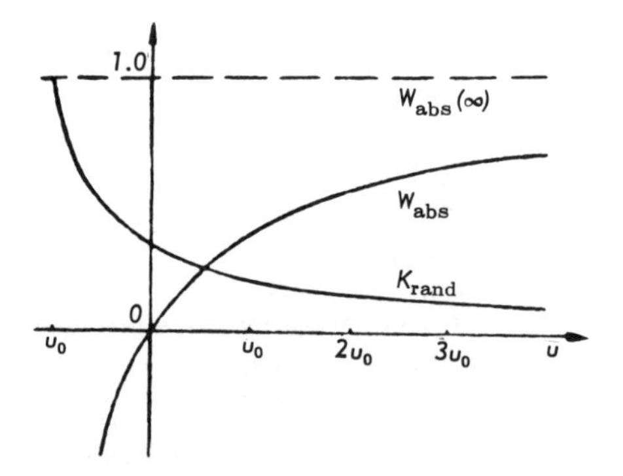

Fig. 7.11 Dependence of k *and* W_{abs} *on the intensity of exciting radiation for positive and negative values of* u

the level separation. This conclusion is valid when the medium is illuminated by a constant intensity source, but is not valid under pulsed excitation.

The formulae given by (32.18)-(32.20) are also valid for negative u. It follows from them that whatever the sign of u the absorption coefficient is always positive. If in the region of positive u, however, the absorption coefficient decreases with increasing u, then in the negative region it will increase with increasing $|u|$. In other words, as the density of the external radiation increases from $-u_0$ to ∞, the absorption coefficient decreases from some maximum value, which is a function of temperature, to zero (Fig. 7.11).

Specific values of k^{rand} can readily be found by writing down (32.18) in the explicit form

$$k^{\text{rand}} = \frac{1}{c} nBh\nu \frac{A + d(1 - e^{-h\nu/kT})}{A + 2B(u_0 + u) + d(1 + e^{-h\nu/kT})} \tag{32.24}$$

which leads to the approximate formula

$$k^{\text{rand}} = \frac{1}{c} nBh\nu \frac{A}{A + 2B(u_0 + u) + 2d} \tag{32.24a}$$

for $h\nu/kT \ll 1$.

Departures from Bouguer's law under pulsed excitation

The absorption coefficient of a system of particles for the j-th period of illumination is

$$k_j = \frac{1}{cu} W_{\text{abs}} = \frac{1}{c} (n_{1j}^{\text{on}} - n_{2j}^{\text{on}}) Bh\nu = \frac{1}{c} (n - 2n_{2j}^{\text{on}}) Bh\nu \tag{32.25}$$

Substituting from (31.46) for n_{2j}^{on}, we have in steady-state conditions:

$$k_j = \frac{3(n_1^0 - n_2^0)}{c(3 + \alpha u)} Bh\nu \times \left[1 + \frac{\alpha u}{3} \frac{1 - e^{-T_2/\tau_0}}{1 - e^{-\left(\frac{T_1}{\tau} + \frac{T_2}{\tau_0}\right)}} e^{-t_j/\tau} \right] \tag{32.26}$$

This shows that under pulsed excitation the absorption coefficient varies continuously with time and is a function of both τ and τ_0. If the dark period is large ($T_2 \gg \tau_0$), this formula then becomes

$$k = k^0 \frac{3}{3 + \alpha u}\left[1 + \frac{\alpha u}{3} e^{-t_j/\tau}\right] \tag{32.27}$$

where k^0 is the absorption coefficient for $\alpha u \ll 1$, and is given by (32.21). Under the above assumption, (32.27) is valid even for continuous illumination. It gives the absorption coefficient during the build-up process and under continuous illumination when $t_j \gg \tau$ (the second term in brackets can be neglected).

The absorption coefficient is a maximum for $t_j = 0$ ($k = k^0$). It decreases with increasing t_j, reaching its minimum value under stationary illumination

$$k_{st} = k^0 \frac{3}{3 + \alpha u}$$

If the illumination period T_1 is comparable with τ, then under pulsed excitation k will not reach k_{st}. The departure from Bouguer's law will then depend on τ. For given T_1, the departure from Bouguer's law will reach a maximum as τ decreases. Plots of k as a function of τ_0 for given t_j and a few values of the non-linearity parameter are given in Fig. 7.12. The values of τ are calculated from (31.30).

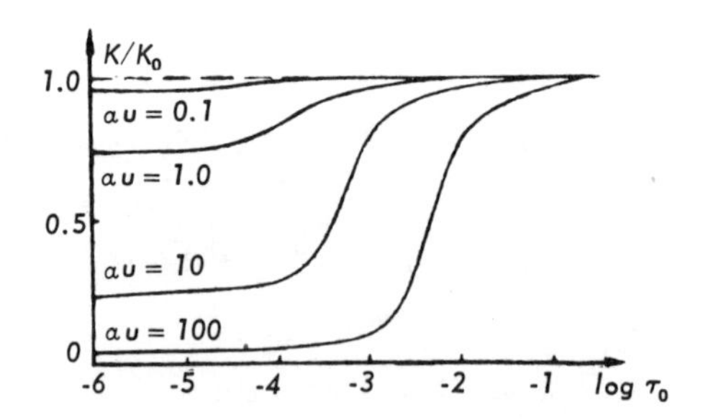

Fig. 7.12 Dependence of k *and* τ_0 *for* $t_j = 10^{-4}$ *sec*

Induced dichroism

Quantum-mechanical calculations of induced dichroism were outlined in Sections 17 and 28. We shall now show that, in contrast to a set of harmonic oscillators (Sections 2, 28), a system of particles with two energy levels exhibits induced dichroism. We shall elucidate the conditions under which its magnitude is small and can be neglected.

Consider, to begin with, the dichroism which appears under illumination by plane-polarised light propagating along the x axis. The distribution function is then determined by (31.15) and (31.16). The rates of absorption of the z and x components of the second flux propagating along the y axis are, respectively, given by

$$W^{z}_{\text{abs}} = h\nu \int_{\Omega} [n_1(u, T, \Omega) - n_2(u, T, \Omega)]\, b^{z}(\Omega)\, u'\, d\Omega \qquad (32.28)$$

$$W^{x}_{\text{abs}} = h\nu \int_{\Omega} [n_1(u, T, \Omega) - n_2(u, T, \Omega)]\, b^{x}(\Omega)\, u'\, d\Omega \qquad (32.29)$$

where, as before, u' is the energy density of the second flux and $|u'| \ll |u|$.

Using (31.11), (31.12), (31.15) and (31.16), and integrating with respect to the angles, we obtain

$$W^{z}_{\text{abs}} = 3\,(n_1^0 - n_2^0)\, Bu'h\nu \times \left[\frac{1}{\alpha u} - \frac{1}{(\alpha u)^{3/2}} \tan^{-1}(\alpha u)^{1/2}\right] \qquad (32.30)$$

$$W^{x}_{\text{abs}} = \frac{3}{2}\,(n_1^0 - n_2^0)\, Bu'h\nu \times \left[-\frac{1}{\alpha u} + \left(\frac{1}{(\alpha u)^{1/2}} + \frac{1}{(\alpha u)^{3/2}}\right) \tan^{-1}(\alpha u)^{1/2}\right] \qquad (32.31)$$

It is evident from these formulae that the rate of absorption, and consequently the absorption coefficient of a system of particles with two energy levels, which is excited by plane-polarised radiation, is a function of the polarisation of the absorbed radiation. When $\alpha u < 1$, (32.30) and (32.31) can be expanded into a series in powers of αu:

$$W^z_{\rm abs} = (n_1^0 - n_2^0)\, Bu'h\nu \times \left[1 - \frac{3}{5}(\alpha u) + \frac{3}{7}(\alpha u)^2 - \ldots\right] \tag{32.32}$$

$$W^x_{\rm abs} = (n_1^0 - n_2^0)\, Bu'h\nu \times \left[1 - \frac{3}{15}(\alpha u) + \frac{3}{35}(\alpha u)^2 - \ldots\right] \tag{32.33}$$

When $\alpha u \ll 1$, we have $W^z_{\rm abs} \to W^x_{\rm abs}$ as was to be expected, and this corresponds to the absence of dichroism. With increasing u, there is a decrease in the rate of absorption, and in the limit as $\alpha u \to \infty$ there is no absorption at all, in accordance with (32.30) and (32.31). $W^z_{\rm abs}$ is then found to tend to zero more rapidly than $W^x_{\rm abs}$ (Fig. 7.13). The rates of

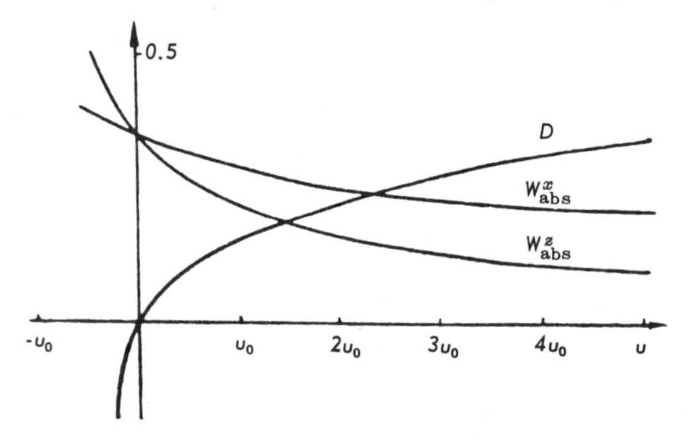

Fig. 7.13 Dependence of the rate of absorption of the second flux (polarised along the z *or* x *axis) and of induced dichroism on the magnitude and sign of* u

absorption also vanish at very high temperatures and small αu, when $n_2^0 \to n_1^0$. Substituting (32.30) and (32.31) into (2.42), we find that the induced dichroism is given by

$$D = \frac{W^x_{\rm abs} - W^z_{\rm abs}}{W^x_{\rm abs} + W^z_{\rm abs}} = \frac{(\alpha u + 3)\tan^{-1}(\alpha u)^{1/2} + 3(\alpha u)^{1/2}}{(\alpha u - 1)\tan^{-1}(\alpha u)^{1/2} + (\alpha u)^{1/2}} \tag{32.34}$$

from which it is evident that the dichroism is unambiguously determined by the product αu. As αu increases from 0 to ∞, the dichroism rises from zero to unity (Fig. 7.13).

It is important to note that dichroism is independent of the sign of the second flux of density u'. The second flux can be used as a 'probe' to determine the anisotropy of particles due to the effect of the first exciting flux.

If the anisotropy is produced by a negative flux, (32.34) can conveniently be written in the form

$$D=\frac{(3-\alpha|u|)\tanh^{-1}(\alpha|u|)^{1/2}-3(\alpha|u|)^{1/2}}{(\alpha|u|)^{1/2}-(1+\alpha|u|)\tanh^{-1}(\alpha|u|)^{1/2}} \qquad (32.35)$$

In the region of negative u, dichroism is negative since negative fluxes give rise to an increase in the absorption coefficient (Figs. 7.11, 7.13).

Dichroism is also produced when the particles are excited by natural radiation. The rates of absorption of the second flux are again given by (32.28) and (32.29). If we express the Einstein coefficient in the new set of coordinates (Fig. 7.6),

$$b^z(\Omega')=3B\sin^2\theta'\cos^2\varphi', \quad b^x(\Omega')=3B\cos^2\theta' \qquad (32.36)$$

and take (31.21) and (31.22) into account, we obtain the following expressions for the rates of absorption

$$W^z_{\text{abs}}=\frac{3}{2}(n_1^0-n_2^0)Bu'h\nu\frac{1}{2+\alpha u}\times\int_0^\pi\frac{\sin^3\theta'\,d\theta'}{1-a\cos^2\theta'}, \qquad (32.37)$$

$$W^x_{\text{abs}}=3(n_1^0-n_2^0)Bu'h\nu\frac{1}{2+\alpha u}\times\int_0^\pi\frac{\cos^2\theta'\sin\theta'\,d\theta'}{1-a\cos^2\theta'} \qquad (32.38)$$

where a is given by (31.24). These integrals can readily be reduced to standard forms by substituting $x=\sqrt{a}\cos\theta'$. The final result is

$$W^z_{\text{abs}}=3(n_1^0-n_2^0)Bu'h\nu\frac{1}{\alpha u}[1+(a^{1/2}-a^{-1/2})\tanh^{-1}a^{1/2}] \qquad (32.39)$$

$$W^x_{\text{abs}} = 6\,(n_1^0 - n_2^0)\,Bu'h\nu \times \frac{1}{a u}\,[a^{-1/2}\tanh^{-1} a^{1/2} - 1] \quad (32.40)$$

which yields

$$D = \frac{(3-a)\tanh^{-1}a^{1/2} - 3a^{1/2}}{(1+a)\tanh^{-1}a^{1/2} - a^{1/2}} \quad (a > 0)$$

$$D = \frac{(3+|a|)\tan^{-1}|a|^{1/2} - 3|a|^{1/2}}{(1-|a|)\tan^{-1}|a|^{1/2} - |a|^{1/2}} \quad (a < 0) \qquad (32.41)$$

Graphs of D as a function of u under excitation by plane-polarised and natural radiation are given in Fig. 7.14. It is

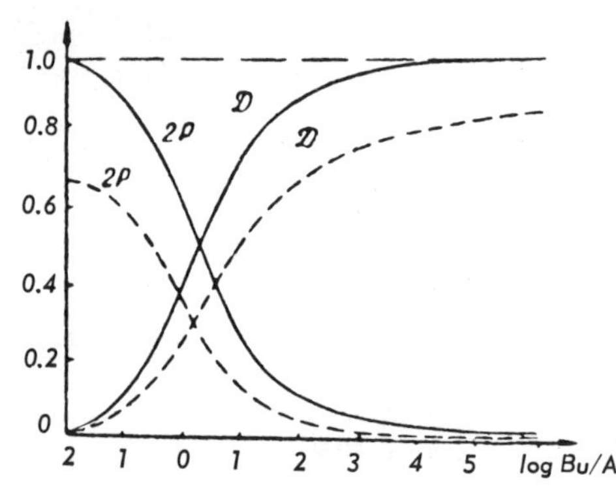

Fig. 7.14 Plots showing dichroism and polarisation of luminescence for illumination by natural and plane-polarised radiation

evident from this figure that dichroism increases more rapidly under plane-polarised illumination. The flux of natural radiation of equal density leads to a smaller dichroism. There is no dichroism under excitation by isotropic radiation.

33. RATE OF EMISSION AND POLARISATION OF LUMINESCENCE

Rate of emission and quantum yield of resultant luminescence

From (24.1), the rate of emission of the resultant luminescence propagating in all directions may be written in the form

$$W_{\text{lum}} = h\nu \int_{\Omega} \{n_2(\Omega)\,A - [n_1(\Omega)\,Bu_0 - n_2(\Omega)\,Bu_0]\}\,d\Omega \qquad (33.1)$$

If the distribution function is independent of the orientation of the particle dipole moments, the integration in (33.1) is reduced simply to multiplication by 4π. Substituting (31.8) and (31.9) into (33.1), we obtain the following expression for the rate of luminescence under isotropic excitation

$$W_{\mathrm{lum}}^{\mathrm{rand}} = nAh\nu \frac{Bu}{A + 2B(u_0 + u) + d(1 + e^{-h\nu/kT})} \tag{33.2}$$

$$= \frac{1}{2} nAh\nu \frac{\alpha u}{3 + \alpha u}$$

It is readily seen that the rate of luminescence is positive when $u > 0$, and negative for excitation by negative radiation. With increasing probability of non-radiative transitions, the rate of emission decreases and eventually vanishes. An analogous result is produced by increasing the temperature. As $u_0 \to \infty$, it is found that $W_{\mathrm{lum}}^{\mathrm{rand}} \to 0$.

According to (33.2), when $\alpha u \ll 1$, the rate of luminescence is practically directly proportional to the density of the exciting radiation. For very large αu, it reaches the maximum value

$$W_{\mathrm{lum}}(\infty) = \frac{1}{2} nAh\nu \tag{33.3}$$

It is characteristic that the limiting value of the rate of luminescence is independent of the temperature of the surrounding medium and of the probabilities of non-radiative transitions. This is an important difference as compared with the limiting value of the rate of absorption (32.9) which is very dependent on d and T. As $u \to \infty$, there is a saturation effect, when the rates of all the processes leading to the transformation of the incident radiation into luminescence and heat reach their limiting values. The rate of liberation of heat can be varied by changing d and T

$$Q = (n_2 d - n_1 d e^{-h\nu/kT}) h\nu = Q(\infty) \frac{\alpha u}{3 + \alpha u} \tag{33.4}$$

where

$$Q(\infty) = \frac{1}{2} nd(1 - e^{-h\nu/kT}) h\nu$$

is the limiting value of Q for $u \rightarrow \infty$. This leads to a change in the rate of absorption at constant rate of luminescence.

Using (33.3), we find that

$$W_{\text{lum}}^{\text{rand}} = W_{\text{lum}}(\infty)\frac{\alpha u}{3 + \alpha u} \tag{33.2a}$$

Comparison of this expression with (32.10) shows that the rates of absorption and luminescence are the same functions of αu.

Substituting (31.15) and (31.16), or (31.21) and (31.22), into (33.1), and integrating with respect to the angles, we find the rates of luminescence excited by plane-polarised and natural radiation respectively:

$$W_{\text{lum}}^{\text{p.pol}} = W_{\text{lum}}(\infty)\,[1 - (\alpha u)^{-1/2}\tanh^{-1}(\alpha u)^{1/2}] \tag{33.5}$$

$$W_{\text{lum}}^{\text{nat}} = W_{\text{lum}}(\infty)\,[1 - (a^{-1/2} - a^{1/2})\tanh^{-1}a^{1/2}] \tag{33.6}$$

Comparison of these expressions with (32.11) and (32.12) will readily show that whatever the mode of excitation, the rates of absorption and luminescence are the same functions of the non-linearity parameter (density of external radiation). The energy yield is therefore given by the same expression in all cases, namely

$$\gamma_{\text{en}} = \frac{W_{\text{lum}}(\infty)}{W_{\text{abs}}(\infty)} = \frac{A}{A + d(1 - e^{-h\nu/kT})} \tag{33.7}$$

Accordingly, γ_{en} is not directly dependent on the polarisation or the sign of the density of the exciting radiation, but is an internal characteristic of a particle and of its interaction with the medium. If there are no non-radiative transitions ($d = 0$), then $\gamma_{\text{en}} = 1$. The yield γ_{en} approaches unity for high temperatures (flames) and small ν (radio waves). This means that the presence of the external radiation will not, under these conditions, modify the distribution over the energy levels, and therefore the number of non-radiative excitations, i.e. the heat release, is zero.

By analogy with the luminescence energy yield, it is possible to define the energy yield through heat release:

$$\gamma_{\text{th}} = \frac{Q(\infty)}{W_{\text{abs}}(\infty)} = \frac{d(1 - e^{-h\nu/kT})}{A + d(1 - e^{-h\nu/kT})}$$

It is immediately evident that $\gamma_{en} + \gamma_{th} = 1$, i.e. under constant illumination the absorbed energy is transformed entirely into luminescence and heat. We note that the energy and the quantum yields of a system of particles with two levels are always the same, as in the case of the harmonic oscillator.

Dependence of the polarisation of luminescence on the intensity of exciting radiation

The resultant luminescence propagating in all directions was calculated in the preceding section. Our problem now is to find the polarisation of luminescence observed along the y axis. This may correspond to a real experiment. We shall assume, for simplicity, that the depolarisation factors, including, in particular, Brownian rotations, are absent. The effect of such rotations on the polarisation of luminescence was considered in Section 4. The polarisation of the luminescence of a system of particles with two energy levels corresponds in classical theory to the calculation of polarisation under the assumption that the same dipoles are responsible for both absorption and emission.

The rates of luminescence with electric vectors along the z and x axes per unit solid angle are, respectively, given by

$$W^{z}_{\text{lum}} = h\nu \int \{n_2(\Omega)\, a^z(\Omega) - [n_1(\Omega) - n_2(\Omega)]\, b^z(\Omega)\, u_0'\}\, d\Omega \tag{33.8}$$

$$W^{x}_{\text{lum}} = h\nu \int_{\Omega} \{n_2(\Omega)\, a^x(\Omega) - [n_1(\Omega) - n_2(\Omega)]\, b^x(\Omega)\, u_0'\}\, d\Omega \tag{33.9}$$

These expressions are written down by analogy with (33.1) and can be justified in a similar way. Since we are now interested only in the radiation which propagates along the y axis, the integral Einstein coefficients are replaced by the differential coefficients (31.11) and (31.12), and u_0 is replaced by $u_0' = \frac{1}{8\pi} u_0$, the density of equilibrium radiation per unit solid angle and given polarisation.

Integrating with respect to the angles, and using (31.5) and

(31.16), we find that the luminescence components for excitation by plane-polarised radiation are

$$W^z_{\text{lum}} = \frac{1}{8\pi} W_{\text{lum}}(\infty) \times \left[1 - \frac{3}{\alpha u} + \frac{3}{(\alpha u)^{3/2}} \tan^{-1}(\alpha u)^{1/2} \right] \quad (33.10)$$

$$W^x_{\text{lum}} = \frac{1}{8\pi} W_{\text{lum}}(\infty) \times \left[1 + \frac{3}{2\alpha u} - \frac{3}{2}\left(\frac{1}{(\alpha u)^{1/2}} + \frac{1}{(\alpha u)^{3/2}} \right) \tan^{-1}(\alpha u)^{1/2} \right] \quad (33.11)$$

It is readily seen that for high densities of the exciting radiation ($\alpha u \gg 1$), both components tend to the same limit. The luminescence becomes depolarised. Moreover, the part of the luminescence which propagates along the y axis (per unit solid angle) is then in the ratio of 1 to 4π as compared with the total luminescence. Consequently, when $\alpha u \gg 1$, the rate of luminescence is independent of the direction of propagation. The luminescence then becomes completely isotropic, just as in the case of equilibrium radiation.

When $\alpha u < 1$, (33.10) and (33.11) can be written in the form of the series

$$\begin{aligned} W^z_{\text{lum}} &= \frac{3}{8\pi} W_{\text{lum}}(\infty) \sum_{k=0}^{\infty} \frac{(-1)^k}{(2k+5)} (\alpha u)^k \\ &= \frac{3}{8\pi} W_{\text{lum}}(\infty) \left[\frac{1}{5}(\alpha u) - \frac{1}{7}(\alpha u)^2 + \ldots \right] \end{aligned} \quad (33.12)$$

$$\begin{aligned} W^x_{\text{lum}} &= \frac{3}{8\pi} W_{\text{lum}}(\infty) \sum_{k=0}^{\infty} \frac{(-1)^k}{(2k+3)(2k+5)} (\alpha u)^k \\ &= \frac{3}{8\pi} W_{\text{lum}}(\infty) \left[\frac{1}{15}(\alpha u) - \frac{1}{35}(\alpha u)^2 + \ldots \right] \end{aligned} \quad (33.13)$$

For very small αu, the two components are practically linear functions of u. With increasing αu, the dependence becomes non-linear, especially for W^z_{lum} (Fig. 7.15). Substituting (33.10) and (33.11) into (3.38), we obtain for the polarisation of the luminescence

$$P_{\text{p.pol}} = \frac{[3(\alpha u)^{-3/2} + (\alpha u)^{-1}]\tan^{-1}(\alpha u)^{1/2} - 3(\alpha u)^{-1}}{4/3 + [(\alpha u)^{-3/2} - (\alpha u)^{-1/2}]\tan^{-1}(\alpha u)^{1/2} - (\alpha u)^{-1}} \qquad (33.14)$$

which shows that the polarisation is uniquely determined by the non-linearity parameter (similarly to dichroism). When it is less than unity,

$$P_{\text{p.pol}} = \frac{1}{2} - \frac{3}{38}(\alpha u) + \dots \qquad (33.15)$$

When $\alpha u \ll 1$, which is practically always the case in the visible region, the polarisation reaches its maximum value of 0.5. With increasing u it decreases. For negative u we have $P > 0.5$, and for $\alpha u \gg 1$ $P \to 0$ (Fig. 7.15).

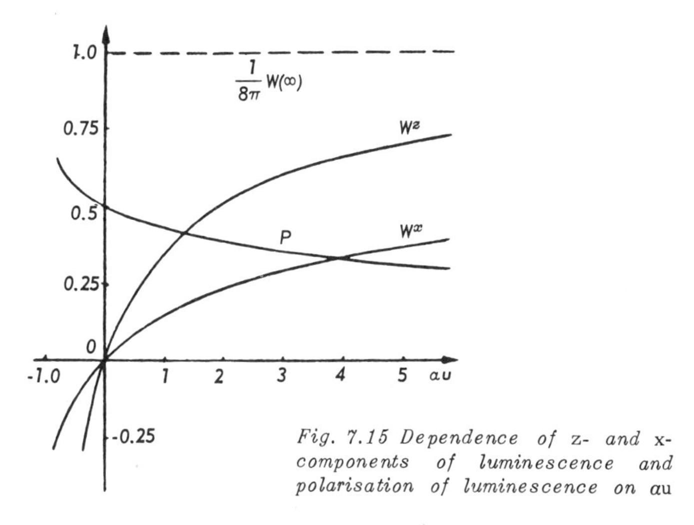

Fig. 7.15 Dependence of z- and x-components of luminescence and polarisation of luminescence on αu

It was noted in Section 31 that an increase in temperature would give rise to a reduction in anisotropy in the distribution of excited particles. According to (31.10) and (33.14), however, this not only does not give rise to depolarisation but, on the contrary, the polarisation increases with increasing u_0 (for $u =$ const) right up to 0.5. For $T \to \infty$ and finite u, the polarisation is always equal to 0.5. The dependence of the polarisation on temperature can readily be understood if it is recalled that luminescence is defined as the excess over and above the thermal background. At very high temperatures ($u/u_0 \ll 1$), the excited particles are almost isotropically distributed and the total emission is depolarised. This is connected not with the low degree of polarisation of the emitted luminescence, but with its small relative contribution to the resultant emission, which consists mainly of depolarised Planck radiation.

The effect of non-radiative transitions upon polarisation is similar to the effect of temperature. The parameter α decreases with increasing d and polarisation increases.

When the influence of the thermal radiation background is negligible, α can readily be expressed in terms of the quantum luminescence yield B_{lum}, the frequency and the density of the exciting radiation:

$$\alpha^0 = \alpha\,(T = 0) = \frac{3}{4\pi}\,\frac{c^3}{h\,\nu^3}\,B_{\mathrm{lum}} \tag{33.16}$$

It is evident from (33.14) and (33.16) that the polarisation of luminescence is not uniquely related to the lifetime of the excited state. The dependence of $P_{T=0}$ on τ during a transition from one system to another can only be due to the presence of non-radiative transition probabilities.

We now consider the polarisation of luminescence under excitation by natural radiation. As before, the general expression for the z and x components of luminescence propagating along the y axis are given by (33.8) and (33.9). However, the distribution function is given by (31.21) and (31.22). Substituting this into (33.8) and (33.9), and using the differential Einstein coefficients in the primed set of coordinates, we have, after integration with respect to the angles

$$W^z_{\mathrm{lum}} = \frac{1}{8\pi}\,W_{\mathrm{lum}}(\infty) \times \left[\frac{5}{2} - \frac{3}{2a} + \frac{3}{2a^{3/2}}(a-1)^2 \tanh^{-1} a^{1/2}\right] \tag{33.17}$$

$$W^x_{\mathrm{lum}} = \frac{1}{8\pi}\,W_{\mathrm{lum}}(\infty) \times \left[-2 + \frac{3}{a} + \frac{3}{a^{3/2}}(a-1)\tanh^{-1} a^{1/2}\right] \tag{33.18}$$

Substituting these equations in the form of series in powers of a, it is readily verified that for $u \to 0$ ($a \ll 1$), the z component is greater by a factor of 2 than the x component. This ratio decreases with increasing density of the exciting radiation, and tends to unity in the limit as $u \to \infty$. The absolute values are

$$W_{\text{lum}}^{z}(\infty) = W_{\text{lum}}^{x}(\infty) = \frac{1}{8\pi} W_{\text{lum}}(\infty)$$

Consequently, in this limiting case, the luminescence is again not only depolarised but completely isotropic.

Using (33.17) and (33.18), we find the following expressions for the polarisation:

$$P_{\text{nat}} = \frac{(a-1)\left[3a^{1/2} + (a-3)\tanh^{-1} a^{1/2}\right]}{a\left(\frac{1}{3}a + 1\right) + (a^2 - 1)\tanh^{-1} a^{1/2}} \tag{33.19}$$

$$P_{\text{nat}} = \frac{1}{3} - \frac{8}{63} a + \ldots \tag{33.19a}$$

When $a \ll 1$, (33.19a) yields 1/3 for the polarisation of the luminescence. This is the same as the result obtained, in the classical theory of the harmonic oscillator (see (3.45)). Figure 7.14 shows graphs of P as a function of the energy density of the exciting radiation for different modes of excitation. As was to be expected, $P_{\text{p.pol}}$ is always greater than P_{nat} for equal densities of external radiation.

8

Systems of Particles with an Arbitrary Number of Energy Levels

34. OPTICAL PHENOMENA IN A SYSTEM OF PARTICLES WITH THREE ENERGY LEVELS

The distribution function

Consider a system of particles with three energy levels (Fig. 8.1). We shall denote the probabilities of radiative and non-radiative transition by the symbols shown below

$$p_{ij} = A_{ij} + B_{ij}u^0_{ij} + d_{ij} + B_{ij}u_{ij} = p^0_{ij} + B_{ij}u_{ij} \tag{34.1}$$

$$p_{ji} = B_{ji}u^0_{ij} + d_{ji} + B_{ji}u_{ij} = p^0_{ji} + B_{ji}u_{ij} \ (i > j) \tag{34.2}$$

where p^0_{ij} and p^0_{ji} are the values of the transition probabilities in the absence of excitation at frequency ν_{ij}.

If there is no angular anisotropy in the distribution of excited and unexcited particles, the level populations will

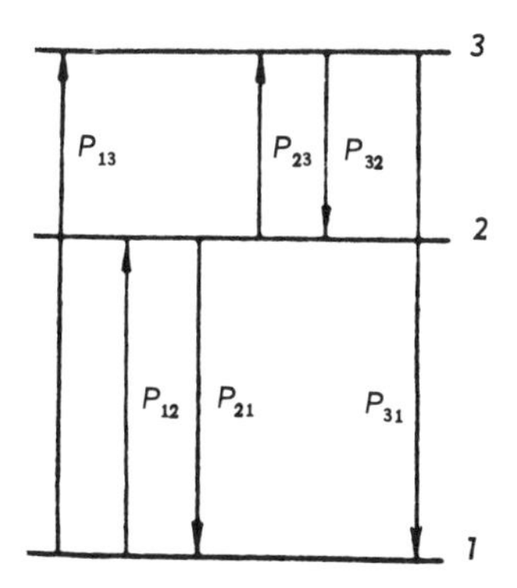

Fig. 8.1 A three-level system

satisfy the balance equations

$$\frac{dn_1}{dt} = -n_1(p_{12} + p_{13}) + n_2 p_{21} + n_3 p_{31} \tag{34.3}$$

$$\frac{dn_2}{dt} = -n_2(p_{21} + p_{23}) + n_1 p_{12} + n_3 p_{32} \tag{34.4}$$

$$n_1 + n_2 + n_3 = n \tag{34.5}$$

On solving these, we find that

$$n_1 = C_1 e^{-t/\tau_1} + C_2 e^{-t/\tau_2} + \frac{a_1}{\Delta} n \tag{34.6}$$

$$\begin{aligned} n_2 = C_1 \frac{p_{12} + p_{13} + p_{31} - \dfrac{1}{\tau_1}}{p_{21} - p_{31}} e^{-t/\tau_1} \\ + C_2 \frac{p_{12} + p_{13} + p_{31} - \dfrac{1}{\tau_2}}{p_{21} - p_{31}} e^{-t/\tau_2} + \frac{a_2}{\Delta} n \end{aligned} \tag{34.7}$$

$$\begin{aligned} n_3 = C_1 \frac{\dfrac{1}{\tau_1} - p_{21} - p_{12} - p_{13}}{p_{21} - p_{31}} e^{-t/\tau_1} \\ + C_2 \frac{\dfrac{1}{\tau_2} - p_{21} - p_{12} - p_{13}}{p_{21} - p_{31}} e^{-t/\tau_2} + \frac{a_3}{\Delta} n \end{aligned} \tag{34.8}$$

where we have used the notation

$$\begin{aligned} \Delta = p_{21}(p_{32}+p_{13}+p_{31}) + p_{12}(p_{32}+p_{23}+p_{31}) \\ + p_{13}(p_{23}+p_{32}) + p_{23}p_{31} \\ a_1 = p_{21}(p_{31}+p_{32}) + p_{31}p_{23} \\ a_2 = p_{12}(p_{31}+p_{32}) + p_{32}p_{13} \\ a_3 = p_{23}(p_{12}+p_{13}) + p_{21}p_{13} \end{aligned} \quad (34.9)$$

The quantities τ_1 and τ_2 are related to the transition probabilities by the formulae

$$\frac{1}{\tau_1} = b + \sqrt{b^2 - \Delta}, \quad \frac{1}{\tau_2} = b - \sqrt{b^2 - \Delta} \quad (34.10)$$

where

$$2b = p_{12} + p_{21} + p_{13} + p_{31} + p_{23} + p_{32}$$

The integration constants C_1 and C_2 may be found from the initial conditions.

Since (34.6)-(34.8) were obtained without introducing any assumptions about the transition probabilities, we can deduce from them the form of the variation in the level populations for different methods of excitation and during the return of the system to thermodynamic equilibrium. To obtain the equilibrium distribution over the levels, all the external exciting radiation fluxes must be equated to zero and the values of t must be such that all the time-dependent terms vanish. We then find from (34.6)-(34.8) that

$$n_1^0 = \frac{n}{1 + e^{-h\nu_{21}/\kappa T} + e^{-h\nu_{31}/\kappa T}} \quad (34.11)$$

$$n_2^0 = \frac{n e^{-h\nu_{21}/\kappa T}}{1 + e^{-h\nu_{21}/\kappa T} + e^{-h\nu_{31}/\kappa T}} \quad (34.12)$$

$$n_3^0 = \frac{n e^{-h\nu_{31}/\kappa T}}{1 + e^{-h\nu_{21}/\kappa T} + e^{-h\nu_{31}/\kappa T}} \quad (34.13)$$

since $p_{ji}^0 = p_{ij}\exp(h\nu_{ij}/kT)$. In these expressions and in our further analysis, we shall use the superscript 0 to represent the level populations in thermodynamic equilibrium.

In the presence of external excitation, the constant terms in (34.6)-(34.8) give the stationary distribution over the levels

$$n_1^{st} = \frac{n}{\Delta} [p_{21}(p_{31} + p_{32}) + p_{31}p_{23}] \qquad (34.14)$$

$$n_2^{st} = \frac{n}{\Delta} [p_{12}(p_{31} + p_{32}) + p_{32}p_{13}] \qquad (34.15)$$

$$n_3^{st} = \frac{n}{\Delta} [p_{23}(p_{12} + p_{13}) + p_{21}p_{13}] \qquad (34.16)$$

The level populations do not vary exponentially after the excitation has been switched on or off. Therefore, in contrast to a system with two energy levels, the mean lifetime of the excited state is not equal either to τ_1 or τ_2. If τ_1 and τ_2 are of the same order of magnitude, they have no individual physical meaning. Conversely, if $\tau_2 \gg \tau_1$, the luminescence afterglow of the system can be naturally divided into two components, namely, short-period luminescence which falls off practically as e^{-t/τ_1^0}, and long-period luminescence which falls off as e^{-t/τ_2^0}. In such cases, one speaks of fluorescence and phosphorescence, and τ_1^0 and τ_2^0 are the corresponding time constants.

Let us determine τ_1^0 and τ_2^0 for a system of particles with a metastable energy level, in which case

$$p_{21}^0 + p_{23}^0 \ll p_{31}^0 + p_{32}^0$$

In the absence of excitation, and when $T = 0$, we may assume that $p_{12}^0 + p_{13}^0 = 0$, so that from (34.9) and (34.10) we have

$$\frac{1}{\tau_1^0} \approx p_{31}^0 + p_{32}^0, \quad \frac{1}{\tau_2^0} \approx p_{21}^0 + p_{23}^0(1 - p_{32}^0\tau_1^0) \qquad (34.17)$$

It is evident from these formulae that the duration of fluorescence is determined by probabilities of transition from the third to the first and second levels. The duration of phosphorescence depends on all the transition probabilities and is of the order of

$$\frac{1}{p_{21}^0 + p_{23}^0}$$

It should be noted that while the steady state is being established, the quantities τ_1 and τ_2 may be complex. In particular, when

$$p_{12} = p_{23} = 0,$$

$$p_{21} = p_{31} + p_{32} + p_{13}$$

the square root is purely imaginary

$$\sqrt{b^2 - \Delta} = i\sqrt{p_{13}p_{32}}$$

and τ_1 and τ_2 are complex. This means that under certain conditions of excitation the level populations may undergo periodic pulsations.

Luminescence yield

By analysing the properties of a system of particles with three energy levels, we can obtain information about one of the most important problems in the theory of luminescence, i.e. the possible values of the energy yield [73]. Calculations show that the rate and the sign of the luminescence are very dependent on the mode of excitation. If the excitation is carried out at frequency ν_{31}, the luminescence in all three channels, $3 \rightarrow 1$, $3 \rightarrow 2$, $2 \rightarrow 1$, is positive. Excitation at frequency ν_{21} leads to a reduction in the number of particles in level 1 and an increase in the number of particles in level 2. The number of particles in level 3 can either increase or decrease. Positive luminescence is produced in channels $3 \rightarrow 1$ and $2 \rightarrow 1$ and negative luminescence in channel $3 \leftarrow 2$.

In this section we shall confine our attention to a detailed discussion of the special case when the particles are excited by an external flux of radiation of frequency ν_{32} (u_{32} = const, $u_{31} = u_{21} = 0$). Under the influence of external excitation, the system leaves the state of thermodynamic equilibrium and begins to luminesce at all frequencies. Once the stationary state has been reached, the level populations are given by

$$\left.\begin{aligned} n_1^{st} &= \frac{n_1^0 + nl_1u_{32}}{1 + {}^1/_3\,\alpha_{32}\,u_{32}} \\ n_2^{st} &= \frac{n_2^0 + nl_2u_{32}}{1 + {}^1/_3\,\alpha_{32}u_{32}} \\ n_3^{st} &= \frac{n_3^0 + nl_3u_{32}}{1 + {}^1/_3\,\alpha_{32}u_{32}} \end{aligned}\right\} \tag{34.18}$$

where α_{32} is a non-linearity parameter given by

$$\begin{aligned} \alpha_{32} &= [3(p_{21}^0 + 2p_{12}^0 + 2p_{13}^0 + p_{31}^0)\,B_{32}]\,[p_{21}^0(p_{32}^0 + p_{13}^0 + p_{31}^0) \\ &\quad + p_{12}^0(p_{32}^0 + p_{23}^0 + p_{31}^0) + p_{13}^0(p_{23}^0 + p_{32}^0) + p_{23}^0\,p_{31}^0]^{-1} \\ &= 3(l_1 + l_2 + l_3), \quad l_1 = \frac{1}{\Delta^0}\,(p_{21}^0 + p_{31}^0)\,B_{32} \end{aligned} \tag{34.19}$$

$$l_2 = l_3 = \frac{1}{\Delta^0}\,(p_{12}^0 + p_{13}^0)\,B_{32}$$

It can be readily shown with the aid of (34.18) that the rate of luminescence at frequency ν_{ij} may be written in the form

$$\begin{aligned} W_{\text{lum}}\,(\nu_{ij}) &= [n_i\,(A_{ij} + B_{ij}u_{ij}^0) - n_jB_{ij}u_{ij}^0]\,h\,\nu_{ij} \\ &= n\,\frac{(A_{ij} + B_{ij}u_{ij}^0)\,u_{32}h\,\nu_{ij}}{1 + {}^1/_3\,\alpha_{32}\,u_{32}}\,(l_i - l_je^{-h\nu_{ij}/\kappa T}) \end{aligned} \tag{34.20}$$

In view of (34.19) it follows that

$$\begin{aligned} W_{\text{lum}}\,(\nu_{21}) &= n\,\frac{(A_{21} + B_{21}u_{21}^0)\,p_{31}^0B_{32}u_{32}h\,\nu_{21}}{\Delta^0\,(1 + {}^1/_3\,\alpha_{32}\,u_{32})} \\ &\quad \times (e^{-h\nu_{31}/\kappa T} - e^{-h\nu_{21}/\kappa T}) \end{aligned} \tag{34.21}$$

$$\begin{aligned} W_{\text{lum}}\,(\nu_{31}) &= n\,\frac{(A_{31} + B_{31}u_{31}^0)p_{21}^0B_{32}u_{32}h\,\nu_{31}}{\Delta^0\,(1 + {}^1/_3\,\alpha_{32}u_{32})} \\ &\quad \times (e^{-h\nu_{21}/\kappa T} - e^{-h\nu_{31}/\kappa T}) \end{aligned} \tag{34.22}$$

$$W_{\text{lum}}(\nu_{32}) = n\,\frac{A_{32}(p^0_{12}+p^0_{13})\,B_{32}u_{32}h\,\nu_{32}}{\Delta^0(1+{}^1/_3\,a_{32}\,u_{32})} \qquad (34.23)$$

It is evident that there is no luminescence in the absence of external excitation ($u_{32}=0$). Luminescence vanishes if the temperature is zero, since then $p^0_{12} = p^0_{13} = 0, \exp(-h\,\nu_{ij}/kT)=0$. This means that all the particles occupy the lowest level and the presence of radiation of frequency ν_{32} cannot take them into the excited state.

In general, when all the transition probabilities are not zero, excitation at a particular frequency leads to luminescence at all frequencies.

The sign of luminescence whose frequency is equal to the frequency of the incident radiation is determined by the sign of u_{32}. When $u_{32}>0$, luminescence at this frequency is always positive. The sign of luminescence at the two other frequencies depends on the relationship between ν_{31} and ν_{21}. Since in our case $\nu_{31} > \nu_{21}$, it follows that for positive excitation we have

$$W_{\text{lum}}(\nu_{21}) < 0, \quad W_{\text{lum}}(\nu_{31}) > 0$$

Transition to illumination by negative radiation fluxes will lead to a change in the sign of the luminescence in all the channels.

In special cases, the luminescence can only appear at two frequencies. For example, if there are no direct $3 \to 1$ transitions, there is no luminescence at frequency ν_{21}. When $p^0_{21} = 0$ there is no luminescence at frequency ν_{31}. This is explained by the fact that in view of the above assumptions, and in accordance with (34.18),

$$\frac{n^{\text{st}}_2}{n^{\text{st}}_1} = e^{-h\nu_{21}/\kappa T} \;(\text{where } p^0_{31} = 0)$$

$$\frac{n^{\text{st}}_3}{n^{\text{st}}_1} = e^{-h\nu_{31}/\kappa T} \;(\text{where } p^0_{21} = 0)$$

and

$$n_i(A_{ij} + B_{ij}u_{ij}) - n_jB_{ji}u_{ij} = 0$$

By definition, the luminescence energy yield is given by

$$\gamma_{en} = \frac{W_{lum}(\nu_{21}) + W_{lum}(\nu_{31}) + W_{lum}(\nu_{32})}{(n_2^{st} - n_3^{st}) B_{32} u_{32} h\nu_{32}} \tag{34.24}$$

Substituting (34.18) and (34.21)-(34.24) into this expression we have, after some simple rearrangement,

$$\gamma_{en} = 1 - [d_{31} p_{21}^0 \nu_{31}/\nu_{32} + d_{32}(p_{21}^0 + p_{31}^0 e^{-h\nu_{32}/\kappa T}) - d_{21} p_{31}^0 \nu_{21}/\nu_{32}] [p_{21}^0 p_{32}^0 + p_{21}^0 p_{31}^0 + p_{23}^0 p_{31}^0]^{-1} \tag{34.25}$$

As in the case of a system of particles with two energy levels, the luminescence yield of the model under consideration is independent of the density of the incident radiation. If $d_{31} = d_{32} = d_{21} = 0$, the energy yield is equal to unity. This is an obvious result, since under the particular assumptions which we have introduced, the radiant energy cannot be transformed into heat. Characteristically, the energy of some of the emitted photons ($h\nu_{31}$) is greater than the energy of the absorbed photons ($h\nu_{32}$). This difference is, however, compensated by a reduction in the number of emitted photons with energy $h\nu_{21}$ (in comparison with thermal equilibrium).

Let us suppose that d_{31} and d_{32} are negligible in comparison with d_{21}, so that in accordance with (34.25), the energy yield may be greater than unity. As d_{31} and d_{32} increase (and d_{21} decreases), the ratio γ_{en} decreases to zero and becomes negative for certain ratios of transition probabilities. For this to occur it is necessary that positive luminescence at frequencies ν_{31} and ν_{32} should be quenched (large d_{31} and d_{32}), while the negative luminescence at frequency ν_{21} should reach a maximum ($d_{21} = 0$).

Thus, the energy yield of a system of particles with three levels may vary within the limits

$$-\varepsilon_1 \leqslant \gamma_{en} \leqslant 1 + \varepsilon_2$$

where ε_1 and ε_2 are positive numbers which depend on the temperature, the internal properties of the particles and the interaction with the surrounding medium. Analysis of other methods of excitation leads to similar conclusions.

Phosphorescence

The three-level model in which one of the levels is metastable is usually referred to as Jablonski's model. Sveshnikov

[74] has shown that this model is capable of explaining a considerable range of experimental data on the luminescence of organic phosphors.

Consider the emission by a system of particles with a metastable energy level, excited by radiation of frequency ν_{31} (first level normal, third level labile, second level metastable). We shall not be interested in the total intensity of the emission, and therefore for the sake of simplicity the anisotropy in the angular distribution of the particles will not be taken into account.

Since, in the case of organic phosphors, the frequencies ν_{31} and ν_{21} lie in the visible part of the spectrum, it follows that the influence of the thermal radiation background will be negligible at these frequencies.

If there is no excitation, all the particles occupy the first level. As soon as excitation is switched on, there are transitions from the lowest level to the labile level. The excited particle can then return to the normal state or reach the metastable state. Emission produced as a result of $1 \to 3$, $3 \to 1$ transitions occurring in immediate succession, is called fluorescence. Since it is rapidly damped out after the excitation is switched off, it can also be referred to as short-period emission.

A molecule occupying the metastable level can return to the normal state in two ways; if the probability of the $2 \to 3$ transition is not zero, it may return to the labile level and then undergo the $3 \to 1$ transition, or it can undergo direct transition from the metastable to the normal state. Two forms of phosphorescence correspond to these two possible ways in which the particle returns to the normal state. They are called α and β phosphorescence, respectively. The frequency of α phosphorescence is equal to the frequency of fluorescence, but the spectrum of β phosphorescence is shifted towards longer wavelengths.

Under constant illumination, we are unable to separate α phosphorescence from fluorescence either theoretically or experimentally. Consequently, α fluorescence is usually investigated experimentally with pulsed excitation (through a phosphoroscope). The total emission is recorded while the radiation is on, whereas during the dark pauses only phosphorescence is produced. A similar separation can be effected in the theoretical treatment.

After excitation has been switched off, the third and second level populations are given by (see (34.7) and (34.8))

$$n_3^{\mathrm{dec}}(t) = (1-\rho)\, n_3^{\mathrm{st}}\; e^{-t/\tau_1^0} + \rho\, n_3^{\mathrm{st}}\; e^{-t/\tau_2^0} \tag{34.26}$$

$$n_2^{\mathrm{dec}}(t) = -\frac{\tau_1^0}{\tau_2^0 - \tau_1^0}\, n_2^{\mathrm{st}}\; e^{-t/\tau_1^0} + \frac{\tau_2^0}{\tau_2^0 - \tau_1^0}\, n_2^{\mathrm{st}}\; e^{-t/\tau_2^0} \tag{34.27}$$

where

$$\rho = \tau_1^0 \tau_2^0 \frac{\left(\frac{1}{\tau_1^0} - p_{31}^0\right)(p_{23}^0 + p_{21}^0) - p_{32}^0 p_{21}^0}{(\tau_2^0 - \tau_1^0)(p_{23}^0 + p_{21}^0)} \approx \tau_1^0 \frac{p_{23}^0 p_{32}^0}{p_{23}^0 + p_{21}^0} \tag{34.28}$$

The number of particles in the second and third levels under stationary conditions can readily be found from (34.15) and (34.16) by setting $p_{12} = 0$ and recalling that excitation is carried out at frequency ν_{31} $(p_{13} = B_{13}u_{31})$:

$$n_3^{\mathrm{st}} = \frac{n}{\Delta}(p_{23}^0 + p_{21}^0) B_{13}u = n\,\tau_1\tau_2 (p_{23}^0 + p_{21}^0) B_{13}u_{31} \tag{34.29}$$

$$n_2^{\mathrm{st}} = \frac{n}{\Delta} p_{32}^0 B_{13}u = n\,\tau_1\tau_2\, p_{32}^0 B_{13}u_{31} \tag{34.30}$$

where

$$\Delta = \frac{1}{\tau_1\tau_2} = p_{31}^0(p_{23}^0 + p_{21}^0) + p_{21}^0 p_{32}^0 + (2p_{23}^0 + 2p_{21}^0 + p_{32}) B_{13}u_{31}$$

The quantities τ_1^0 and τ_2^0 determine the duration of fluorescence and phosphorescence and are related to the transition probabilities by (34.10), or approximately by (34.17). Usually, τ_1^0 is of the order of 10^{-8} sec while τ_2^0 is larger by four to eight orders of magnitude. It follows therefore that the first components in (34.26) and (34.27) vanish in a time of 10^{-6} sec, and thereafter the number of photons emitted as α

and β phosphorescence per unit time is given by

$$
\begin{aligned}
N^{\alpha}_{\text{phos}} &= \rho n_3^{\text{st}} A_{31} e^{-t/\tau_2^0} \\
&= np^0_{23} p^0_{32} \tau_1^0 \tau_1 \tau_2 A_{31} B_{13} u_{31} e^{-t/\tau_2^0}
\end{aligned}
\tag{34.31}
$$

$$
\begin{aligned}
N^{\beta}_{\text{phos}} &= \frac{\tau_2^0}{\tau_2^0 - \tau_1^0} n_2^{\text{st}} A_{21} e^{-t/\tau_2^0} \\
&= np^0_{32} \tau_1 \tau_2 A_{21} B_{13} u_{31} e^{-t/\tau_2^0}
\end{aligned}
\tag{34.32}
$$

It is easy to see that the time constants for α and β phosphorescence afterglow are equal and are determined by the mean lifetime of particles in the metastable level. As the temperature increases, the two phosphorescent bands vary in parallel. However, the relative proportions of α and β phosphorescence change very considerably with varying temperature.

Since

$$
p^0_{23} = p^0_{32} e^{-h\nu_{32}/\kappa T}
$$

the ratio of α and β phosphorescence is given by

$$
\frac{N^{\alpha}_{\text{phos}}}{N^{\beta}_{\text{phos}}} = \tau_1^0 p^0_{32} \frac{A_{31}}{A_{21}} e^{-h\nu_{32}/\kappa T}
\tag{34.33}
$$

or, in view of (34.1) and (34.17), by

$$
\frac{N^{\alpha}_{\text{phos}}}{N^{\beta}_{\text{phos}}} = \frac{p^0_{32} A_{31}}{A_{21}(p^0_{32} + A_{31} + d_{31})} e^{-h\nu_{32}/\kappa T}
\tag{34.33a}
$$

This shows that short-wave phosphorescence predominates at high temperatures, whereas at liquid-air temperatures, β phosphorescence is practically the only remaining component. This is completely confirmed by experiment.

The intensity of long-period afterglow is sensitive not only to temperature changes, but also to the appearance of various quenching factors leading to an increase in the probabilities of non-radiative transitions. The duration and intensity of fluorescence is then dependent on which particular level is involved in the deactivation of the particle. If, for example, there is an increase in d_{21}, the phosphorescence time constant τ_2^0 is reduced, and at the same time there is a reduction in N^{α}_{phos} and N^{β}_{phos} in accordance with (34.31) and (34.32).

The ratio of these two quantities will however remain constant. Conversely, an increase in d_{31} leads to a stronger quenching of short-wave phosphorescence. The ratio (34.33) decreases while the duration of afterglow remains unaltered.

Reduction in the duration of phosphorescence at constant temperature is therefore the main feature of deactivation of particles while they are in the metastable state. The experimental verification of these predictions has played an important role in the elucidation of the kinetics of fluorescence and phosphorescence. In particular, it has been shown that the particles cannot enter the metastable level directly but must pass through the labile state.

Equations (34.26) and (34.27) may be used in approximate calculations of fluorescence and phosphorescence under pulsed excitation. Usually, the period of illumination and the duration of the dark pause satisfy the inequality

$$\tau_1^0 \ll T \ll \tau_2^0$$

It follows that the fraction of second- and third-level populations which is responsible for phosphorescence is practically constant. The rapidly varying part of $n_2(t)$ (the first component in (34.27) is usually negligible and can be ignored. The first component in (34.24) can, in accordance with (34.28), form a considerable proportion of n_3^{st}. At the beginning of the illumination period, this part of the population reaches its maximum value and then rapidly vanishes at the beginning of the dark pause. It may therefore be considered that, on the average, the second- and third-level populations during the illumination period are equal to n_2^{st} and n_3^{st}, whereas during the dark pause, they are equal to n_2^{st} and n_3^{st}.

The total number of photons emitted by the particles per unit time during the illumination period is given by

$$N_{tot} = n_3 A_{31} + n_2 A_{21} = n \tau_1 \tau_2 B_{13} u_{31} \times [(p_{23}^0 + p_{21}^0) A_{31} + p_{32}^0 A_{21}] \quad (34.34)$$

In linear optics, where τ_1 and τ_2 are practically independent of the density of exciting radiation, N_{tot} is directly proportional to u_{31}. The resultant emission is more sensitive to the quenching of molecules in the labile level than in the metastable level, since with increasing d_{21} there is an increase in $p_{21}^0 = A_{21} + d_{21}$. Conversely, an increase in d_{31} leads only to a reduction in τ, and therefore in N_{tot}.

The relative proportions of α and β phosphorescence in the resultant emission may be found from (34.31), (34.32) and (34.34)

$$q_\alpha = \frac{N^\alpha_{\text{phos}}}{N_{\text{tot}}} = \tau_1^0 \frac{p^0_{23}\, p^0_{32}\, A_{31}}{(p^0_{23} + p^0_{21})\, A_{31} + p^0_{32} A_{21}} \qquad (34.35)$$

$$q_\beta = \frac{N^\beta_{\text{phos}}}{N_{\text{tot}}} = \frac{p^0_{32}\, A_{21}}{(p^0_{23} + p^0_{21}) A_{31} + p^0_{32}\, A_{21}} \qquad (34.36)$$

In these expressions we have allowed for the fact that during the period T the exponential e^{-t/τ_2^0} is very nearly equal to unity. The quantities q_α and q_β are not always convenient measures of the yield of α and β phosphorescence. For example, it is easy to see that the relative proportion of short-wave phosphorescence decreases with increasing d_{21} and d_{31}, whereas q_β is independent of the deactivation of the fluorescent state.

Non-linear effects

All the non-linear optical phenomena in the visible part of the spectrum were first discovered in molecules having a triplet level [75]. This is connected with the specific properties of the metastable state. We shall therefore discuss non-linear effects with the aid of Jablonski's model.

Suppose that the particles are excited by plane-polarised radiation of frequency ν and density u_{31}, propagating along the x-axis with the electric vector parallel to the z-axis. The transition probabilities between normal and labile levels depend on the density of the exciting radiation and the orientation of the particles. They are given by

$$\begin{aligned} p_{13} &= 3B_{13}u_{31}\cos^2\theta, \\ p_{31} &= A_{31} + d_{31} + 3B_{13}u_{31}\cos^2\theta \end{aligned} \qquad (34.37)$$

where θ is the angle between the z-axis and the particle dipole moment. The integral probabilities p^0_{32}, p^0_{23}, p^0_{21} $(p^0_{12} = 0)$ are connected with intramolecular processes but are independent of the orientation of the particles.

The distribution of the particles over the levels under

stationary conditions of illumination can readily be found from (34.14)-(34.16) if we substitute into them the appropriate transition probabilities and recall that $n_i(\Omega)$ are calculated per unit solid angle, so that instead of (34.5) we have

$$n_1(\Omega) + n_2(\Omega) + n_3(\Omega) = \frac{n}{4\pi} \tag{34.38}$$

If we introduce the non-linearity parameter

$$\alpha_{31} = \frac{3(2p_{23}^0 + 2p_{21}^0 + p_{32}^0)\, B_{31}}{(A_{31} + d_{31})\,(p_{21} + p_{23}) + p_{21}p_{32}} \tag{34.39}$$

we have from (34.14)-(34.16)

$$n_1(\Omega) = \frac{1}{4\pi}\, n\, \frac{1 + l_1 u_{31}\cos^2\theta}{1 + \alpha_{31} u_{31}\cos^2\theta} \tag{34.40}$$

$$n_2(\Omega) = \frac{1}{4\pi}\, n\, \frac{l_2 u_{31}\cos^2\theta}{1 + \alpha_{31} u_{31}\cos^2\theta} \tag{34.41}$$

$$n_3(\Omega) = \frac{1}{4\pi}\, n\, \frac{l_3 u_{31}\cos^2\theta}{1 + \alpha_{31} u_{31}\cos^2\theta} \tag{34.42}$$

where

$$l_2 = \frac{3}{\Delta^0}\, p_{32}^0 B_{31}$$

$$l_1 = l_3 = \frac{3}{\Delta^0}\, (p_{21}^0 + p_{23}^0)\, B_{31}$$

$$\Delta^0 = (A_{31} + d_{31})\,(p_{21}^0 + p_{23}^0) + p_{21}^0\, p_{32}^0$$

The quantities l_j are related to the non-linearity parameter by the simple expression

$$\alpha_{31} = l_1 + l_2 + l_3$$

According to (34.40)-(34.42), all the particles occupy the lowest level in the absence of external excitation. Comparison of these formulae with (31.15) and (31.16) shows that the angular distributions of excited particles in systems with two or three energy levels are determined by similar formulae. From the point of view of the dependence of optical characteristics on the density of exciting radiation, the difference between particles with two and three levels can be reduced to different values of the non-linearity parameter.

To elucidate the specific properties of molecules having a metastable energy level, α_{31} can conveniently be written in the form

$$\alpha_{31} = \alpha\beta \tag{34.43}$$

where

$$\alpha = \frac{6B_{31}}{A_{31} + d_{31}}$$

is the value which α_{31} would have in the absence of the second level (see (3.10) for $T = 0$) and

$$\beta = \frac{p_{21}^0 + p_{23}^0 + \frac{1}{2} p_{32}^0}{p_{21} + p_{23} + \frac{p_{21}p_{32}}{A_{31} + d_{31}}} \tag{34.44}$$

Usually $\alpha u_{31} \ll 1$ and therefore the polarisation of luminescence of a two-level system of particles approaches 0.5 and all the remaining non-linear effects are absent. For particles having metastable states, the non-linearity parameter for comparable densities of exciting radiation is higher by several orders of magnitude since $\beta \gg 1$. Suppose, for example, that in a given system, and for a given temperature, one observes only α-phosphorescence. In this typical case $p_{21} \ll p_{23} \ll p_{32}$ and therefore

$$\beta \approx \frac{\frac{1}{2} p_{32} + p_{23}}{p_{23}} \approx e^{(E_3 - E_2)/\kappa T}$$

If $E_3 - E_2 = 3{,}000\ \mathrm{cm}^{-1}$ (fluorescein), $\beta \approx 10^6$ at room temperature. The non-linearity effect will then set in at exciting radiation densities millions of times smaller than the densities which give rise to comparable departures from linearity in the two-level system.

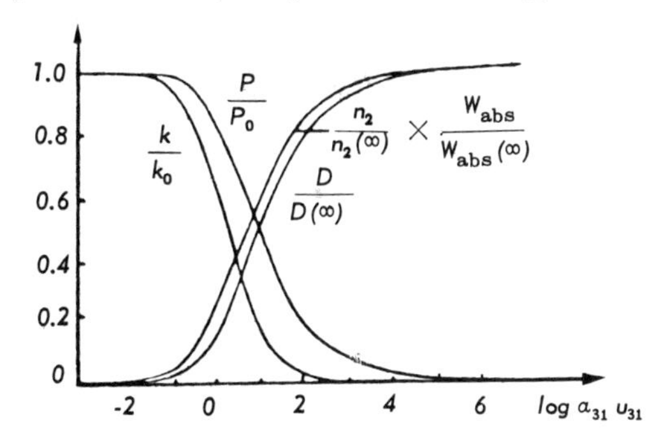

Fig. 8.2 Dependence of various optical characteristics (in relative units) on $\alpha_{31}u_{31}$

According to (34.41) and (34.42), we have for all directions (θ, ψ)

$$\frac{n_2(\Omega)}{n_3(\Omega)} = \frac{p_{32}^0}{p_{21}^0 + p_{23}^0} \tag{34.45}$$

Usually, p_{32}^0 is smaller than p_{21}^0 or p_{23}^0 by several orders of magnitude. It follows that excited particles occupy predominantly the metastable level (for large p_{21} and p_{23} the second level will no longer be metastable). The total number of molecules in this level is then

$$n_2 = \int_0^{2\pi} d\varphi \int_0^{\pi} n_2(\theta) \sin\theta \, d\theta = n \frac{p_{32}^0}{2p_{23}^0 + 2p_{21}^0 + p_{32}^0} \times \left[1 - \frac{1}{(\alpha_{31}u_{31})^{1/2}} \tan^{-1} (\alpha_{31}u_{31})^{1/2}\right] \tag{34.46}$$

It is easy to understand that under certain conditions practically all the particles may be found in the second level. Suppose that $p_{32}^0 \gg p_{23}^0 + p_{21}^0$, so that $n_2 = 0.85n$ for $\alpha_{31}u_{31} = 100$ and $n_2 = 0.95n$ for $\alpha_{31}u_{31} = 1{,}000$. It was shown in Section 31 that if the particles have two levels only, the maximum number of excited particles will not exceed $0.5n$. The presence of a triplet state introduces considerable modifications in the optical properties of the molecules. A plot of n_2 as a function of u_{31} is shown in Fig. 8.2. The rate of absorption of external radiation may be found from (34.40) and (34.42):

$$W_{abs} = h\nu_{31} \int\limits_{\Omega} [n_1(\Omega) - n_3(\Omega)]\, 3B_{31}u_{31} \cos^2\theta \, d\Omega$$
$$= W_{abs}(\infty) \left[1 - \frac{1}{(\alpha_{31} u_{31})^{1/2}} \tan^{-1} (\alpha_{31} u_{31})^{1/2}\right] \quad (34.47)$$

where

$$W_{abs}(\infty) = n \frac{(A_{31} + d_{31})(p^0_{21} + p^0_{23}) + p^0_{21} p^0_{32}}{2p_{23} + 2p_{21} + p_{32}} h\nu_{31} \quad (34.48)$$

is the rate of absorption for complete saturation by the incident radiation (when $\alpha_{31} u_{31} = 10{,}000$, $W_{abs} = 0.984\, W_{abs}(\infty)$).

In the absence of the metastable level ($p_{32} = p_{23} = p_{21} = 0$) the quantity $W_{abs}(\infty)$ is large and equal to $\frac{n}{2}(A_{31} + d_{31}) h\nu_{31}$ when $T = 0$ (see also (32.9)). As a result of accumulation of excited particles in the triplet level, the limiting rate of absorption is much smaller than this and tends to zero when $p_{21} = 0$, $p_{23} \to 0$. The substance becomes completely transparent at the frequency ν_{31} even for small u_{31}.

In view of (34.47), the absorption coefficient is given by

$$k = \frac{1}{Cu_{31}} W_{abs} = \frac{1}{Cu_{31}} W_{abs}(\infty) \left[1 - \frac{1}{(\alpha_{31}u_{31})^{1/2}} \times \tan^{-1}(\alpha_{31} u_{31})^{1/2}\right] \quad (34.49)$$

which resembles (32.19) and differs from it only in the values of $W_{abs}(\infty)$ and in the non-linearity parameter. Figure 8.2 plots k as a function of u_{31}.

To determine induced dichroism and the polarisation of luminescence it is necessary to carry out the same mathematical operations as in the preceding chapter. The final result is [76]

$$D = \frac{(\alpha_{31} u_{31} + 3) \tan^{-1} (\alpha_{31} u_{31})^{1/2} - 3(\alpha_{31} u_{31})^{1/2}}{(\alpha_{31} u_{31} - 1) \tan^{-1} (\alpha_{31} u_{31})^{1/2} - (\alpha_{31} u_{31})^{1/2}} \quad (34.50)$$

$$P = \{[3(\alpha_{31} u_{31})^{-3/2} - (\alpha_{31} u_{31})^{-1/2}] \tan^{-1} (\alpha_{31} u_{31})^{1/2} - 3 (\alpha_{31} u_{31})^{-1}\}$$
$$\times \{4/3 + [(\alpha_{31} u_{31})^{-3/2} - (\alpha_{31} u_{31})^{-1/2}] \tan^{-1}(\alpha_{31} u_{31})^{1/2} - (\alpha_{31} u_{31})^{-1}\}^{-1} \quad (34.51)$$

These formulae can also be obtained from (32.34) and (33.14) by replacing αu by $\alpha_{31}u_{31}$. It will be shown below that they are also valid for more complicated systems if the appropriate non-linearity parameter is substituted into them.

The dependence of dichroism and the polarisation of luminescence on the intensity of the exciting radiation is similar in two- and three-level systems of particles. Transitions from one kind of system to another require the appropriate replacement of the non-linearity parameter, and this shifts the D and P curves along the log u axis (see Fig. 8.2).

The non-linearity parameter varies not only from one kind of system to another, but also as a result of changes in the temperature and in the probabilities of non-radiative transitions. Graphs of the temperature dependence of D and P are

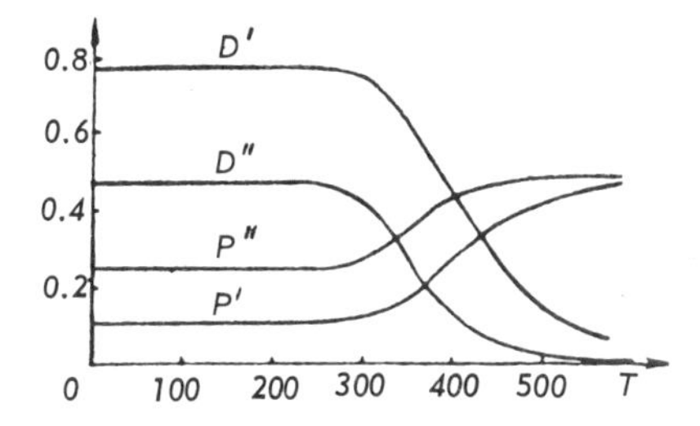

Fig. 8.3 Dependence of the polarisation of luminescence and dichroism on temperature. At t = 0 the product $\alpha_{31}u_{31}$ is equal to 100 (curves P' and D') and 10 (curves P'' and D'')

shown in Fig. 8.3. With increasing temperature there is an increase in the probability that the particles will leave the metastable level, $p^0_{23} = pe^{-h\nu/kT}$. The non-linearity parameter will therefore decrease and all non-linear effects will tend to zero. Consequently, the most favourable conditions for observing non-linear effects are achieved at low temperatures.

As can be seen from Fig. 8.2, the dependence of all the optical phenomena under consideration on the intensity of the incident radiation is similar and due to a common cause, i.e. the accumulation of particles in the metastable state. At exciting radiation densities where $\alpha_{31}u_{31} < 0.1$, non-linear effects are practically absent and need not be taken into account. Within these limits, radiation of frequency ν_{31} is transformed by a three-level system of particles just as by a set of classical harmonic oscillators. All non-linear effects begin at those values of u_{31} at which n_2 becomes an appreciable fraction of n.

The presence of observable phosphorescence indicates the possibility of a non-linear dependence of the rate of absorption, polarisation of luminescence and dichroism on u_{31}. The converse statement is not true since for small p_{23} and A_{21} phosphorescence may be absent even when non-linear effects are large. For large non-radiative transition probabilities, the accumulation of particles in the metastable state is impossible, and this leads to the quenching of phosphorescence and disappearance of non-linear phenomena.

It must be emphasised that investigation of non-linear effects yields new information about the properties of fluorescing molecules. For example, Cherdyshev and Vasserman [77] have used the temperature dependence of dichroism not only to find p^0_{32} but also to determine the activation energy of the metastable state.

It has already been pointed out that the non-linearity parameter is an important characteristic of quantum systems. It enters into the formulae for the absorption coefficient, the polarisation of luminescence, dichroism, the population of metastable levels and the phosphorescence yield. If α_{31} is determined experimentally for a particular non-linear effect, the result can be used to calculate the magnitude of all the remaining non-linear effects. Direct experimental determination of α_{31} is evidently impossible in the absence of a non-linear dependence of the optical characteristics of a medium on the density of the exciting light.

Some substances do not exhibit long-period afterglow. However, it does not follow that the metastable state is then necessarily absent, since phosphorescence may not occur because the probabilities of non-radiative deactivation of molecules from the metastable to the normal state are very high, and the probabilities of transition to the labile level are very small. The experimental demonstration of the presence of the metastable state without taking into account non-radiative effects may then turn out to be exceedingly difficult.

If the system does have a metastable level, non-linear optical effects are very dependent on the temperature of the surrounding medium. At low temperatures, they become appreciable even at relatively low exciting radiation densities. This fact should be used in the experimental verification of the presence of the metastable state.

Dependence of the polarisation of luminescence on the angle between the dipole moment of various transitions

In the last section we discussed the emission of radiation whose frequency was equal to the frequency of the exciting radiation. External radiation of frequency ν_{31} is, however, also accompanied by luminescence at frequency ν_{21}. Clearly, its polarisation will depend not only on the intensity of the exciting radiation but also on the mutual orientation of the matrix elements $\mathbf{D}_{31}$ and $\mathbf{D}_{21}$. For excitation by plane-polarised radiation, the rates of emission of the z- and x-components along the y axis are respectively given by

$$W^z_{\text{lum}}(\nu_{21}) = h\,\nu_{21}\int_0^{2\pi} d\,\varphi \int_0^{\pi} n_2(\Omega)\,a^z_{21}(\Omega)\sin\theta\,d\,\theta \tag{34.52}$$

$$W^x_{\text{lum}}(\nu_{21}) = h\,\nu_{21}\int_0^{2\pi} d\,\varphi \int_0^{\pi} n_2(\Omega)\,a^x_{21}(\Omega)\sin\theta\,d\,\theta \tag{34.53}$$

where $n_2(\Omega)$ is given by (34.41). The differential Einstein coefficients $a^z_{21}(\Omega)$ and $a^x_{21}(\Omega)$ are proportional to $\cos^2(\mathbf{D}_{21}\mathbf{z})$ and $\cos^2(\mathbf{D}_{21}\mathbf{x})$. In general, the directions of the matrix elements of the dipole moments $\mathbf{D}_{31}$ and $\mathbf{D}_{21}$ are not the same. Let the angle between them be denoted by ξ. For a given orientation of $\mathbf{D}_{31}$, the orientation of the vector $\mathbf{D}_{21}$ relative to $\mathbf{D}_{31}$ is defined by the two angles ξ and η. For different molecules with parallel directions of $\mathbf{D}_{31}$ and therefore the same value of $n_2(\Omega)$, the angle η may assume any value between 0 and 2π. Consequently, if we express $\cos^2(\mathbf{D}_{21}\mathbf{z})$ and $\cos^2(\mathbf{D}_{21}\,\mathbf{x})$ in terms of trigonometric functions of the angles θ, φ, ξ and η, we must average over the angle η. Thus

$$\begin{aligned} a^z_{21}(\Omega) &= \frac{3}{16\pi}\left[\cos^2\theta\,(2-3\sin^2\xi)+\sin^2\xi\right]A_{21} \\ a^x_{21}(\Omega) &= \frac{3}{16\pi}\left[\sin^2\theta\cos^2\varphi\,(2-3\sin^2\xi)+\sin^2\xi\right]A_{21} \end{aligned} \tag{34.54}$$

Substituting (34.41), (34.54) into (34.52) and (34.53) and integrating with respect to θ and φ, we find that

$$W^z_{\text{lum}}(\nu_{21}) = \frac{3}{32\pi} n \frac{A_{21} p_{32} h \nu_{21}}{2p^0_{21} + 2p^0_{23} + p^0_{32}} (\alpha_{31} u_{31})^{-3/2}$$

$$\times \left\{ (2 - 3\sin^2\xi) \left[\frac{2}{3} (\alpha_{31} u_{31})^{3/2} - 2(\alpha_{31} u_{31})^{1/2} + 2\tan^{-1}(\alpha_{31} u_{31})^{1/2} \right] \right.$$

$$\left. + 2\sin^2\xi \left[(\alpha_{31} u_{31})^{3/2} - (\alpha_{31} u_{31}) \tan^{-1}(\alpha_{31} u_{31})^{1/2} \right] \right\} \quad (34.55)$$

$$W^x_{\text{lum}}(\nu_{21}) = \frac{3}{32\pi} n \frac{A_{21} p^0_{32} h \nu_{21}}{2p^0_{21} + 2p^0_{23} + p^0_{32}} (\alpha_{31} u_{31})^{-3/2} \left\{ (2 - 3\sin^2\xi) \right.$$

$$\times \left[\frac{2}{3} (\alpha_{31} u_{31})^{3/2} + (\alpha_{31} u_{31})^{1/2} - (\alpha_{31} u_{31} + 1) \tan^{-1}(\alpha_{31} u_{31})^{1/2} \right]$$

$$\left. + 2\sin^2\xi \left[(\alpha_{31} u_{31})^{3/2} - \alpha_{31} u_{31} \tan^{-1}(\alpha_{31} u_{31})^{1/2} \right] \right\} \quad (34.56)$$

These expressions yield the following formula for the polarisation of luminescence [78]

$$P(\nu_{21}) = \frac{P^0_{21}(2 - 3\sin^2\xi)}{2 - P^0_{21}\sin^2\xi} \quad (34.57)$$

The quantity P^0_{21} is given by (34.51) and represents the value of $P(\nu_{21})$ when $\xi = 0$. Graphs of $P(\nu_{21})$ as a function of $\alpha_{31} u_{31}$ for a number of values of ξ are shown in Fig. 8.4.

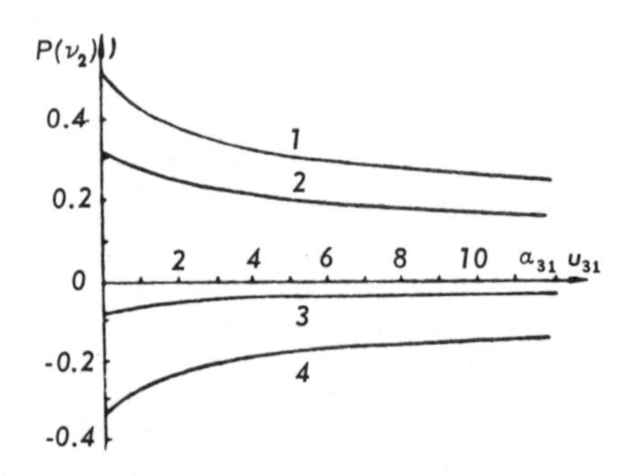

Fig. 8.4 Dependence of the polarisation of luminescence (β-phosphorescence) on $\alpha_{31}u_{31}$ for $\xi = 0°$ (curve 1), 30° (curve 2), 60° (curve 3) and 90° (curve 4)

If we substitute $P_{21}^0 = 1/2$ into (34.57), which is valid for $\alpha_{31} u_{31} \ll 1$, we obtain

$$P_{21} = \frac{2 - 3\sin^2\xi}{4 - \sin^2\xi} \tag{34.58}$$

which is identical with Levshin's formula (3.59) obtained within the framework of classical electrodynamics. We thus see that the classical formula (3.59) may be regarded as the zero-order approximation to the quantum-mechanical result (34.58). This approximation is valid whenever P_{21}^0 approaches 1/2. The conditions for the validity of (34.58) may also be formulated in a somewhat different way; Levshin's formula is valid within the same limits, as Bouguer's law. When the departure from Bouguer's law is small, we can use the first approximation

$$P^{(1)}(\nu_{21}) = \frac{2 - 3\sin^2\xi}{4\left[1 + \frac{3}{14}(\alpha_{31} u_{31})\right] - \sin^2\xi} \tag{34.59}$$

It is important to note that the above formulae may be used both in the study of β phosphorescence and in the interpretation of polarised luminescence spectra of molecules with two or more excited electronic states. In the latter case, depolarisation of luminescence will, as a result of saturation, occur at exciting radiation densities which are smaller by factors of the order of 10^6 than the densities necessary for the equal depolarisation of molecules with a metastable level. Levshin's formula (3.59) can therefore be used within wide limits in the interpretation of polarised spectra.

Interaction of particles with two beams of external radiation

In our discussion of non-linear effects exhibited by a system of particles having a metastable energy level, we neglected the thermal radiation background. It was also assumed that the excitation of external radiation occurred only at the single frequency ν_{31}.

These restrictions correspond to the usual experimental conditions under which non-linear effects are investigated, and simplify the mathematical analysis very considerably.

We shall consider in this section a more general case of the interaction of radiation with a system of particles. We shall use the above calculations to elucidate the effect of the thermal radiation background and the excitation by external radiation of other frequencies on the transformation of radiation by a given pair of levels.

1. Let us first consider the transformation of radiation of frequency ν_{21} and density u_{21} propagating along the x axis and polarised in the direction of the z axis. Suppose that external isotropic radiation of frequency ν_{31} and energy density u_{31} is also incident on the system. The transition probabilities p_{ij} are given by

$$\begin{gathered} p_{21} = p_{21}^0 + 3B_{21}u_{21}\cos^2\theta, \quad p_{12} = p_{12}^0 + 3B_{21}u_{21}\cos^2\theta \\ p_{31} = p_{31}^0 + B_{31}u_{31}, \quad p_{13} = p_{13}^0 + B_{13}u_{31} \\ p_{23} = p_{23}^0, \quad p_{32} = p_{32}^0 \end{gathered} \tag{34.60}$$

If we solve the corresponding balance equation, we find that under stationary conditions the populations of the first and second levels are given by

$$\begin{aligned} n_2 &= \frac{1}{4\pi}\,\frac{n_2(u_{21}=0) + l_2 u_{21}\cos^2\theta}{1 + \alpha_{21}u_{21}\cos^2\theta} \\ n_1 &= \frac{1}{4\pi}\,\frac{n_1(u_{21}=0) + l_1 u_{21}\cos^2\theta}{1 + \alpha_{21}u_{21}\cos^2\theta} \end{aligned} \tag{34.61}$$

where

$$n_2(u_{21}=0) = \frac{n}{\Delta}\left(p_{32}^0 p_{13}^0 + p_{12}^0 p_{31}^0 + p_{21}^0 p_{32}^0\right)$$

$$n_1(u_{21}=0) = \frac{n}{\Delta}\left(p_{31}^0 p_{23}^0 + p_{21}^0 p_{31}^0 + p_{21}^0 p_{32}^0\right)$$

$$l_2 = l_1 = \frac{3n}{\Delta}\left(p_{31} + p_{32}^0\right)B_{12}, \quad \alpha_{21} = \frac{3B_{21}}{\Delta}\left(2p_{32}^0 + 2p_{31} + p_{13} + p_{23}^0\right), \tag{34.62}$$

$$\Delta = p_{21}^0\left(p_{32}^0 + p_{13} + p_{31}\right) + p_{12}^0\left(p_{32}^0 + p_{23}^0 + p_{31}\right) + p_{13}\left(p_{23}^0 + p_{32}^0\right) + p_{23}^0 p_{31}.$$

The quantities n_1 $(u_{21}=0)$ and n_2 $(u_{21}=0)$ have a simple physical significance: they are respectively the total populations of the first and second levels in the absence of excitation at the frequency ν_{21} (i.e. when $u_{21}=0$). If we know the distribution function, we can calculate the absorption coefficient, the polarisation of the luminescence and the induced dichroism. Since similar calculations were described in adequate detail in the preceding sections, we shall not repeat them here and will proceed directly to a discussion of the final formulae.

The expressions for the absorption coefficient $k(\nu_{21})$, dichroism $D(\nu_{21})$ and polarisation $P(\nu_{21})$ observed along the y-axis are

$$k(\nu_{21}) = \frac{1}{c}\left[n_1(u_{21}=0) - n_2(u_{21}=0)\right] B_{21} h\nu_{21}\left[\frac{3}{\alpha_{21}u_{21}} - \frac{3}{(\alpha_{21}u_{21})^{3/2}}\tan^{-1}(\alpha_{21}u_{21})^{1/2}\right] \tag{34.63}$$

$$D(\nu_{21}) = \frac{(\alpha_{21}u_{21}+3)\tan^{-1}(\alpha_{21}u_{21})^{1/2} - 3(\alpha_{21}u_{21})^{1/2}}{(\alpha_{21}u_{21}-1)\tan^{-1}(\alpha_{21}u_{21})^{1/2} + (\alpha_{21}u_{21})} \tag{34.64}$$

$$P(\nu_{21}) = \left\{l_2A_{21}/\alpha_{21} - \left[n_2(u_{21}=0)(A_{21}+B_{21}u_{21}^0) - n_1(u_{21}=0)B_{21}u_{21}^0\right]\right\} \times\left[\frac{l_2}{\alpha_{21}}D(\nu_{21}) + \left[n_2(u_{21}=0)(A_{21}+B_{21}u_{21}^0) \right.\right.$$
$$\left.\left. n_1(u_{21}=0)B_{21}u_{21}^0\right]P_{21}^0\right]^{-1} D(\nu_{21})P_{21}^0 \tag{34.65}$$

where P_{21}^0 is a function of $\alpha_{21}u_{21}$ (see (34.51)).

Let us establish the conditions under which external radiation of frequency ν_{21} is transformed by a system of particles with three levels in the same way as in the case of a set of harmonic oscillators. To do this, consider equations (34.63)-(34.65) in the following special cases:

(a) $u_{31}=0$, $T=0$. Here the expressions given by (34.63)-(34.65) are respectively identical with (32.19), (32.34) and (33.14) for a system of particles with two energy levels $(\alpha_{21}=\alpha)$. The population of the third level is zero and it has no effect on the transformation of radiation of frequency ν_{21}. It is characteristic that under these assumptions we can, for an arbitrary ν_{21}, estimate the minimum exciting radiation density for which appreciable non-linear effects set in (see Section 33).

(b) $u_{31}=0$, $T\neq 0$. When $u_{31}=0$, the populations n_2 $(u_{21}=0)$ and n_1 $(u_{21}=0)$ are given by the Boltzmann formula, and consequently

$$n_2(u_{21}=0)(A_{21}+B_{21}u_{21}^0)-n_1(u_{21}=0)B_{21}u_{21}^0=0 \quad (34.66)$$

Substituting this into (34.65), we find that the polarisation is given by the function P_{21}^0. If in (34.63)-(34.65) we replace α_{21} by α, we again obtain the corresponding formulae for a system of particles with two energy levels.

Equation (34.62) shows that as the temperature increases there is an attendant increase in both the numerator and denominator in the expression for α_{21}. However, the denominator contains the terms $(u^0)^2$, while the numerator contains the density of Planck radiation in the first power only. At high temperatures, when $B_{ij}u_{ij}^0$ are comparable with or greater than $A_{ij}+d_{ij}$, the denominator increases more rapidly than the numerators and there is a reduction in α_{21}.

It follows that if for a given density of exciting radiation and $T=0$ there are no non-linear effects, then they will also be absent at higher temperatures. Conversely, if there are non-linear effects at low temperatures, they can always be reduced to zero as T increases, since as $T\to\infty$, $\alpha_{21}\to 0$.

(c) $u_{31}\neq 0$, $T=0$. The absorption coefficient and dichroism are, as before, given by (34.63) and (34.64), and (34.65) becomes

$$P(\nu_{21})=\frac{\dfrac{l_2}{\alpha_{21}}-n_2(u_{21}=0)}{\dfrac{l_2}{\alpha_{21}}D(\nu_{21})+n_2(u_{21}=0)P_{21}^0}D(\nu_{21})P_{21}^0 \quad (34.67)$$

where

$$\frac{l_2}{\alpha_{21}}(T=0)=n\frac{(A_{31}+d_{31}+A_{32}+d_{32})+B_{31}u_{31}}{2(A_{21}+d_{31}+A_{32}+d_{32})+3B_{31}u_{31}} \quad (34.68)$$

$$n_2(u_{21}=0)=n(A_{32}+d_{32})B_{31}u_{31}[(A_{21}+d_{21})(A_{32}+d_{32}+A_{31}+d_{31}+2B_{31}u_{31})+(A_{32}+d_{32})B_{31}u_{31}]^{-1} \quad (34.69)$$

As u_{31} increases from zero to infinity, the quantities given by (34.68) and (34.69) vary within the limits

$$\frac{n}{3} \leqslant \frac{l_2}{\alpha_{21}} \leqslant \frac{n}{2},$$

$$0 \leqslant n_2\,(u_{21} = 0) \leqslant \frac{n}{1 + 2\,\dfrac{A_{21} + d_{21}}{A_{32} + d_{32}}} \qquad (34.70)$$

By varying the values of the transition probabilities and of the quantity u_{31}, it is possible to obtain any combination of values of l_2/α_{21} and $n_2\,(u_{21} = 0)$ within the limits indicated by (34.20). Figure 8.5 shows graphs of the polarisation of luminescence as a function of the density u_{21} of the excited plane-polarised light. It is easy to see that the presence of the second flux of density u_{31} has an important effect on $P\,(\nu_{21})$. When $A_{21} + d_{21} < A_{31} + d_{31}$ the polarisation becomes negative for certain definite values of u_{31}. It is equal to 0.5 only when $\alpha_{21}u_{21} \ll 1$ and $n_2\,(u_{21} = 0) = 0$, that is, in the absence of excitation at frequency ν_{31}. Analysis of the special cases discussed above leads to the conclusion that a system of particles with three energy levels transforms radiation of frequency ν_{21} in the same way as a set of harmonic oscillators only in the absence of excitation at frequency ν_{31}, and provided $\alpha_{21}\,u_{21} \ll 1$, i.e. at exciting radiation densities not exceeding a certain value $u_{\min}$. As the temperature increases, non-linear effects become less important. In principle, for any u_{21} it is possible to select a temperature at which non-linear effects will be negligible.

2. Consider the excitation of particles by plane-polarised radiation of frequency ν_{31} and by anisotropic radiation of

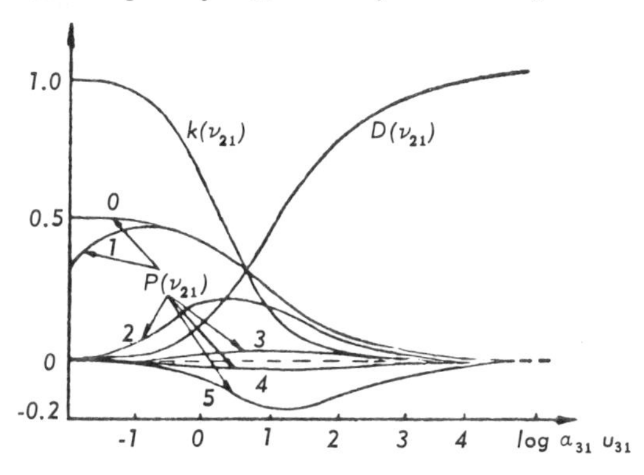

Fig. 8.5 Excitation at frequency ν_{31}

frequency ν_{21}. The formulae for the absorption coefficient, the induced dichroism and the polarisation of luminescence now become

$$k(\nu_{32}) = \frac{1}{c}\left[n_2(u_{32}=0) - n_3(u_{32}=0)\right] B_{32} h \nu_{32} \left[\frac{3}{\alpha_{32} u_{32}} - \frac{3}{(\alpha_{32} u_{32})^{1/2}} \tan^{-1} (\alpha_{32} u_{32})^{1/2}\right] \quad (34.71)$$

$$D(\nu_{32}) = \frac{(\alpha_{32} u_{32} + 3)\tan^{-1}(\alpha_{32} u_{32})^{1/2} - 3(\alpha_{32} u_{32})^{1/2}}{(\alpha_{32} u_{32} - 1)\tan^{-1}(\alpha_{32} u_{32})^{1/2} + (\alpha_{32} u_{32})^{1/2}} \quad (34.72)$$

$$P(\nu_{31}) = \left\{\frac{l_3}{\alpha_{32}} A_{32} - [n_3(u_{32}=0)(A_{32} + B_{32} u_{32}^0) - n_2(u_{32}=0) B_{32} u_{32}^0]\right\} \left\{\frac{l_3}{\alpha_{32}} A_{32} D(\nu_{32}) + [n_3(u_{32}=0)(A_{32} + B_{32} u_{32}^0) - n_2(u_{32}=0) B_{32} u_{32}^0] P_{32}^0\right\}^{-1} D(\nu_{32}) P_{32}^0 \quad (34.73)$$

where

$$n_3(u_{32}=0) = \frac{n}{\Delta_1^0}(p_{23}^0 p_{12} + p_{23}^0 p_{13}^0 + p_{23}^0 p_{13}^0)$$

$$l_3 = \frac{3n}{\Delta_1^0} B_{32}(p_{12} + p_{13}^0)$$

$$n_2(u_{32}=0) = \frac{n}{\Delta_1^0}(p_{32}^0 p_{13}^0 + p_{12} p_{31}^0 + p_{12} p_{32}^0)$$

$$\alpha_{32} = \frac{3}{\Delta_1^0} B_{32}[2(p_{12} + p_{13}^0) + p_{21} + p_{31}^0]$$

$$\Delta_1^0 = p_{21}(p_{32}^0 + p_{13}^0 + p_{31}^0) + p_{12}(p_{32}^0 + p_{23}^0 + p_{31}^0) + p_{13}^0(p_{23}^0 + p_{32}^0) + p_{23}^0 + p_{31}^0$$

$$p_{12} = p_{12}^0 + B_{21} u_{21}, \quad p_{21} = p_{21}^0 + B_{21} u_{21}$$

The expression for P_{32}^0 can be obtained from (31.51) by replacing $\alpha_{21} u_{21}$ by $\alpha_{32} u_{32}$.

Inspection of (34.71)-(34.73) shows that when $\alpha_{32} u_{32} \ll 1$ and $u_{21} = 0$, radiation of frequency ν_{32} is transformed by a system of particles with three levels in the same way as by a set of classical oscillators. Moreover, as before, the number of dipoles must be taken to be equal to the difference between the populations of the lower and upper levels of the pair under consideration, and the classical expression for the absorption coefficient must be multiplied by the oscillator strength.

It is important to note that the condition $u_{21}=0$ is not a necessary condition in this case. In fact, when $T=0$ we have $P(\nu_{32})$ equal to P_{32}^0 for all densities u_{21} of isotropic radiation. This is the specific feature of the effect of the lower level on the transformation of radiation by the upper pair of levels. However, if $T \neq 0$, we must again require the absence of excitation at frequency ν_{21}.

Additional excitation thus plays different roles depending on the properties of the system and the temperature of the medium. It is important only when isotropic radiation is capable of producing luminescence at the frequency under consideration. In point of fact, excitation at frequency ν_{31} gives rise to luminescence at frequency ν_{21}, while excitation at frequency ν_{21} can give rise to luminescence in the $3 \to 2$ channel only when $T \neq 0$. At $T=0$, isotropic emission of frequency ν_{21} has no effect on the nature of the transformation of radiation by the upper pair of levels.

3. A system of particles with three energy levels may be used to investigate other special cases of its interaction with external radiation, for example, the transformation of radiation of frequency ν_{31} in the presence of isotropic excitation of frequency ν_{21} or ν_{32}. We shall not consider these cases because they lead to the same conclusions as those already given in this section.

Essentially new results are obtained only for systems for which $\nu_{32}=\nu_{21}$, which is possible in the case of equal separation of the levels and when the second and third levels are the components of the degenerate level. In such systems, external radiations of frequency ν_{32} will simultaneously induce $1 \rightleftarrows 2$ and $1 \rightleftarrows 3$ transitions.

Optical method of alignment of the magnetic moments of particles

So far, when we have discussed transitions of particles from one stationary state to another, we were only interested in the change in energy. However, stationary states differ from each other not only in energy but also in the values of other physical parameters, including the magnetic moment. It is well known that, by applying magnetic and electric fields to an atom, it is possible to remove the degeneracy in the magnetic quantum number. This splits the energy levels and to each component produced as a result of this there corresponds a particular value of the magnetic moment. It follows that any change in the level population is connected with a change in the magnetic properties of the medium.

In thermodynamic equilibrium, sub-levels which appear as a result of Zeeman or Stark splitting are populated practically uniformly, since the splitting itself is usually small. The medium as a whole does not exhibit preferential orientation of the magnetic moment. This can only occur at temperatures approaching absolute zero. In 1949, Brossel and Kastler [79] put forward an optical method for the alignment of atoms. In principle, this method involves the exposure of atoms to polarised radiation in such a way that their final distribution over the levels is substantially different from the equilibrium distribution. A particle with three energy levels may serve as a simple model for the investigation of the phenomenon of optical alignment. The lower two levels may be looked upon as sub-levels of, say, the Zeeman splitting. The upper level corresponds to a set of sub-levels of the excited state.

When the system is excited at frequency ν_{31}, the particles undergo transitions to the uppermost level and then return to the first and second levels. As a result, the population of the second level is increased while that of the first is reduced. The atoms become aligned in the direction of the applied magnetic field, or in the opposite direction. The formulae for the level populations given in the present section can be used as they stand in the theoretical analysis of this phenomenon.

Optical alignment was first achieved experimentally in molecular beams of sodium and, later, mercury. Figure 8.6 shows the Zeeman splitting of the lower levels of sodium. Suppose that the sodium levels are excited by the resonance D_1 line with right-handed polarisation. Absorption of this radiation, results in transitions from the state ${}^2S_{1/2}\,(M=-{}^1/_2)$ to the magnetic sub-levels ${}^2P_{1/2}\,(M=+{}^1/_2,\ \Delta M=-1)$.

From these sub-levels the atoms can undergo the reverse transitions, in the downward direction, with the emission of

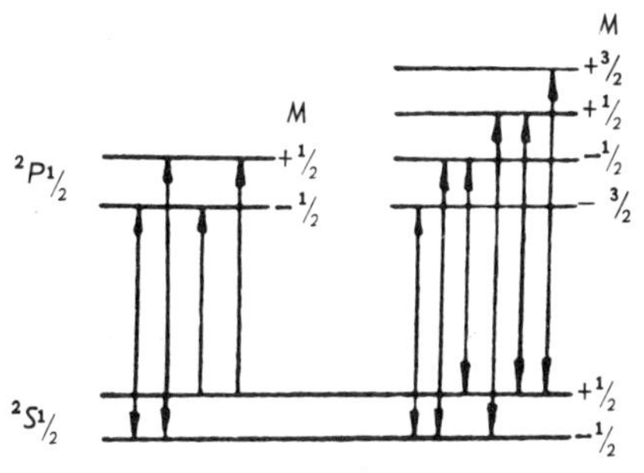

Fig. 8.6 Zeeman splitting of lower sodium levels

plane- or circularly-polarised radiation. This results in a gradual reduction in the occupation of the state $^2S_{1/2}$ with $M = -1/2$ and an increase in the occupation in the states with $M = +1/2$. The atoms become aligned along the magnetic field. In rarefied gases, where a redistribution of the atoms between the sub-levels is unlikely, the alignment may persist for considerable periods of time after the removal of excitation.

The degree of alignment

$$V = \frac{n_A - n_B}{n_A + n_B} \tag{34.74}$$

where n_A and n_B are the populations of the upper and lower sub-levels of the lower state, may reach 30-40%. Studies of phenomena associated with optical alignment in vapours and gases have led to the discovery of multi-photon transitions between Zeeman sub-levels, the elucidation of the effect of a buffer gas on the degree of alignment, the diffusion of radiation and other problems [80].

Rate of generation

The formulae (34.14)-(34.16) readily yield the rate of generation of radiation by a plane-parallel layer if the absorption coefficient at one of the frequencies is negative. Suppose that the particles are confined to the space between two plane-parallel plates whose reflection coefficients at frequency ν_{21} are r_{21} and r'_{21}. If the second level is metastable, excitation at frequency ν_{31} may be accompanied by the appearance of a negative absorption coefficient at frequency ν_{21}.

Generation of radiation will set in at a threshold value, u_{31}^{th}, of the incident radiation density.

In order to establish stationary generation it is necessary to satisfy the condition given by (20.9):

$$-k_{12}(\nu) = k_{12}^{loss} \tag{34.75}$$

where k_{12} is the negative absorption coefficient and k_{12}^{loss} is a coefficient characterising radiation losses. For a plane-parallel layer [81]

$$k_{12}^{loss} = \frac{1}{l} \ln \frac{1}{\sqrt{r_{12} r'_{12}}} + \rho_{12} \tag{34.76}$$

where l is the thickness of the layer and ρ_{12} is a parameter representing energy losses due to scattering and absorption by impurities. The first term in (34.76) is connected with the escape of radiation from the layer and was discussed in Section 20.

The condition for constant generation of radiation given by (34.75) can also be written in the form

$$n_2 g_1/g_2 - n_1 = \delta_{12} n \tag{34.77}$$

where

$$\delta_{12} = k_{12}^{loss} c\, \Delta\nu_{21}^{a} / \mathrm{n} n B_{12} h \nu_{21} \tag{34.78}$$

is a dimensionless parameter equal to the ratio of the loss coefficient to the absorption coefficient $k_{12}(\nu)$ when $n_1 = n$ $(\varkappa_{12} = k_{12}(n_1 = n))$.

If we neglect anisotropy in the orientation distribution of the excited particles, the absorption coefficient $k_{12}(\nu)$ can, with the aid of (34.15)-(34.16), be easily rewritten in the form

$$k_{12} = \frac{\mathrm{n}}{c} B_{12}(\nu_g)(n_1 - n_2 g_1/g_2) h \nu_{21} = \frac{k_{12}^{0}(\nu_g)}{1 + \alpha_{21}(\nu_g) u_{21}(\nu_g) \Delta\nu_{21}^{g}} \tag{34.79}$$

where $k_{12}^{0}(\nu_g)$ is the absorption coefficient for $u_{21} = 0$ at the generation frequency ν_g, $\Delta\nu_{21}^{g}$ is the width of the generated line and g_i are the statistical weights. Since $\Delta\nu_{21}^{g}$ may be much smaller than the width $\Delta\nu_{21}^{a}$ of the absorption line, the product $B_{12}u_{21}$ in (34.14)-(34.16) must be replaced by the integral

$$\int B_{12}(\nu) u_{21}(\nu)\, d\nu = B_{12}(\nu_g) \int u_{21}(\nu)\, d\nu = B_{12}(\nu_g) u_{21}(\nu_g) \Delta\nu_{21}^{g}$$

If the generation frequency coincides with the frequency corresponding to the centre of the absorption line, $B_{12}(\nu_g) = B_{12}/\Delta\nu^a_{21}$. We then have

$$k^0_{12}(\nu_g) = nB_{12}h\nu_{21}[p^0_{31}p^0_{21} + p^0_{31}p^0_{23} + p^0_{32}p^0_{21} - B_{13}u_{31}(p^0_{32}g_1/g_2 - p^0_{23}g_1/g_3 - p^0_{21}g_1/g_3)]n/c\,\Delta^0\Delta\nu^a_{21} \quad (34.80)$$

$$\alpha_{21}(\nu_g) = B_{12}[(p^0_{31} + p^0_{32})(1 + g_1/g_2) + p^0_{23} + B_{13}u_{31}(g_1/g_3 + g_1/g_2 + g_1^2/g_2g_3)][p^0_{31}(p^0_{21} + p^0_{23}) + p^0_{32}p^0_{21} + B_{13}u_{31}(p^0_{21} + p^0_{23})(1 + g_1/g_3) + B_{13}u_{31}p^0_{32}]^{-1}1/\Delta\nu^a_{21} \quad (34.81)$$

where Δ^0 is the value of (34.9) when $u_{21} = 0$. In these expressions we have also allowed for the fact that $p^0_{12} = p^0_{13} = 0$.

The non-linearity parameter α_{21} is always positive and decreases to a limiting value with increasing density of the incident pumping radiation. The value of $k^0_{12}(\nu_g)$ may be either positive or negative. Negative values of the absorption coefficient are produced at sufficiently large values of u_{31}. This is most easily achieved for molecules with a metastable energy level for which $p^0_{32} \gg p^0_{23} + p^0_{21}$. Substituting (34.78) into (34.75), we find that (see also Section 20)

$$u_{21}(\nu_g)\,\Delta\nu^g_{21} = -\frac{k^0_{12}(\nu_g) + k^{loss}_{12}}{\alpha_{21}(\nu_g)\,k^{loss}_{12}} = \frac{|k^0_{12}(\nu_g)| - k^{loss}_{12}}{\alpha_{21}(\nu_g)\,k^{loss}_{12}} \quad (34.82)$$

which represents the rate of generation, i.e. the amount of radiant energy liberated per unit volume per second,

$$W^{gen}_{21} = k^{loss}_{12}\,cu_{21}(\nu_g)\,\Delta\nu^g_{21}/n = c(|k^0_{12}(\nu_g)| - k^{loss}_{12})/\alpha_{21}(\nu_g)\,n \quad (34.83)$$

Equation (34.82) gives the density of generated radiation inside the resonator and is meaningful provided the numerator is greater than zero, i.e. when $|k^0_{12}(\nu_g)| > k^{loss}_{12}$. Consequently, $k^0_{12}(\nu_g) = -k^{loss}_{12}$ determines the minimum absolute magnitude of the negative absorption coefficient which is necessary for generation to occur. Equating the numerator of (34.82) to zero, and using (34.80), we are led to the following expression for the generation threshold:

$$u^{th}_{31} = [p^0_{31}(p^0_{21} + p^0_{23}) + p^0_{32}p^0_{21}](1 + \delta_{12})[p^0_{32}g_1/g_2 - (p^0_{21} + p^0_{23})g_1/g_3 - \delta_{12}(p^0_{21} + p^0_{23})(1 + g_1/g_3) - \delta_{12}p^0_{32}]^{-1}1/B_{13} \quad (34.84)$$

If we set the temperature equal to zero ($p_{23} = 0$) and recall that the second level is metastable, we can simplify this formula to read

$$u_{31}^{th} = p_{21}^{0}(1 + \delta_{12})/\eta B_{13}(g_1/g_2 - \delta_{12}) \qquad (34.85)$$

where $\eta = p_{32}^{0}/(p_{32}^{0} + p_{31}^{0})$. The value of u_{31}^{th} increases rapidly with increasing losses, i.e. with increasing δ_{12}. If $\delta_{12} \to g_1/g_2$, then $u_{31}^{th} \to \infty$, so that when $\delta_{12} > g_1/g_2$, generation is impossible whatever the rate of pumping. In the other limiting case, when $g_1 = g_2$, the probability of stimulated transitions induced by the threshold pumping rate is equal to $B_{13}u_{31}^{th} = p_{21}^{0}$.

Equations (34.78) and (34.81) may be used to find an explicit expression for the rate of generation. If we neglect the small terms ($p_{21}^{0} + p_{23}^{0} \ll p_{32}^{0}$, $B_{13}u_{31} \ll A_{31}$), we obtain

$$W_{21}^{g} = \eta n h \nu_{21} \frac{(g_1/g_2 - \delta_{12})}{1 + g_1/g_2} B_{13}(u_{31} - u_{31}^{th}) \qquad (34.86)$$

This shows that the rate of generation is proportional to the Einstein coefficient B_{13}, which determines the area under the absorption band for the pumping radiation and increases linearly with u_{31}. The maximum value of W_{21}^{gen} occurs at $\eta = 1$, i.e. in the absence of luminescence and non-radiative transitions in the $3 \to 1$ channel. The rate of generation increases with decreasing δ_{12} (decreasing k_{12}^{loss} and $\Delta\nu_{21}^{a}$, or increasing n and B_{12}).

In general, the rate of absorption of the pumping radiation is given by

$$W_{13}^{abs} = n B_{13}u_{31}h\nu_{31}[p_{31}^{0}p_{21}^{0} + p_{31}^{0}p_{23}^{0} + p_{32}^{0}p_{21}^{0} + B_{12}(\nu_g)u_{21}(\nu_g)\Delta\nu_{21}^{g}(p_{31}^{0}g_1/g_2 + p_{32}^{0}g_1/g_2 - p_{23}^{0}g_1/g_3)]/\Delta \qquad (34.87)$$

If we let $u_{21} = 0$, this formula will then determine the rate of absorption in the absence of generation. Substituting for u_{21} from (34.79), we obtain the expression for the rate of absorption under generation conditions as a function of the pumping intensity and all the parameters characterising the system of particles and the resonator. Since the second level is metastable, we have for $T = 0$, $B_{13}u_{31} \ll A_{21}$

$$W_{13}^{abs} = nB_{13}u_{31}h\nu_{31}(g_1/g_2 - \delta_{12})/(1 + g_1/g_2) \qquad (34.88)$$

which corresponds to a generation energy yield

$$\gamma_g = \eta(1 - 1/x)\,\nu_{21}/\nu_{31} \qquad (34.89)$$

where $x = u_{31}/u_{31}^{th}$ is the number of thresholds. The value of γ_g increases with increasing pumping intensity. In the limiting case, when $\eta = 0$, $x \gg 1$, practically all the energy, except for the Stokes losses in the $3 \to 2$ channel, is converted into the energy of the generated radiation.

It has already been shown (see Fig. 8.2) that in the absence of generation, intensive pumping leads to non-linear effects. The rate of absorption, and therefore the intensity of luminescence in all the channels, tend to saturate. The introduction of a medium into the resonator will substantially modify its optical properties. The appearance of appreciable $2 \rightleftarrows 1$ transition probabilities under the action of generated radiation is equivalent to the removal of the metastability of the second level. n_1 and n_2 will then no longer be functions of u_{31}, and hence the rate of absorption (34.88) will increase linearly with u_{31}.

Generation which occurs when the medium is introduced into the resonator is therefore connected not with the increase in the number of active molecules but, on the contrary, with a decrease in n_2 and an increase in n_1. It is precisely this increase in n_1 which leads to the sudden increase in the rate of absorption and the onset of generation.

It is important to note that at pumping densities for which $B_{13}u_{31}$ becomes comparable with A_{31}, there will be a nonlinearity in the $3 \to 1$ channel, and the rates of absorption and generation will tend to saturate. Such values of u_{31} are, however, practically inaccessible in the optical region and will not therefore be discussed here.

35. SPECTROSCOPIC CHARACTERISTICS OF PARTICLES WITH N ENERGY LEVELS [82]

General scheme based on the probabilistic method

Consider a set of n particles each of which has N energy levels. Suppose that the particles interact with external radiation, the Planck radiation and the surrounding medium. There are no restrictions on the exciting radiation other than that inherent in the probabilistic method (Section 16); it may be polarised, directed or isotropic.

The probabilities of transitions between levels i and j will, as before, be denoted by p_{ij}. In general, they depend on the density of the exciting radiation, the temperature of the medium and the direction of the matrix element of the dipole moment $\mathbf{D}_{ij}$. They can be expressed in the form

$$
\begin{aligned}
p_{ij} &= A_{ij} + B_{ij}u^0_{ij} + d_{ij} + p^u_{ij} = p^0_{ij} + p^u_{ij} \\
p_{ji} &= B_{ji}u^0_{ji} + d_{ji} + p^u_{ji} = p^0_{ji} + p^u_{ji} \quad (i > j)
\end{aligned}
\tag{35.1}
$$

where p^u_{ij} and p^u_{ji} represent the probabilities of transitions induced by the external radiation. If $u^{(\alpha)}_{ij}(\Omega')$ is the density of the exciting radiation per unit solid angle and given polarisation ($\alpha = 1, 2$), the quantities p^u_{ij} are given by

$$
p^u_{ij} = p^u_{ji} = \sum_{\alpha} \int_{\Omega'} b^{(\alpha)}_{ij}(\Omega, \Omega')\, u^{(\alpha)}_{ij}(\Omega')\, d\Omega' \tag{35.2}
$$

where $b^{(\alpha)}_{ij}(\Omega, \Omega')$ are the differential Einstein coefficients and Ω, Ω' are angles defining the direction of the dipole moment of the particle and of the electric vector in the incident radiation.

The angular dependence of the transition probability p_{ij} can conveniently be expressed with the aid of the four angles θ, φ, ξ_{ij} and η_{ij} (See Fig. 1.11). The first two angles determine the orientation of a fixed vector $\mathbf{D}_{kl}$. ξ_{ij} determines the angle between $\mathbf{D}_{ij}$ and $\mathbf{D}_{kl}$ and η_{ij} defines the position of $\mathbf{D}_{ij}$ on the surface of a cone whose axis is parallel to $\mathbf{D}_{kl}$.

In the case of an isotropic distribution, the number of particles whose vectors $\mathbf{D}_{kl}$ lie between θ and $\theta + d\theta$, and φ and $\varphi + d\varphi$, while the vectors $\mathbf{D}_{ij}$ lie in the interval between η_{ij} and $\eta_{ij} + d\eta_{ij}$, is

$$
\frac{d\Omega}{4\pi}\,\frac{d\eta_{ij}}{2\pi}\,n = \frac{1}{8\pi^2}\,n \sin\theta\, d\theta\, d\varphi\, d\eta_{ij}
$$

These particles are in identical conditions with respect to the exciting radiation. All the transition probabilities are equal and therefore the distribution functions under stationary illumination satisfy the following system of equations

$$n_i \sum_{\substack{j=1 \\ j \neq i}}^{N} p_{ij} - \sum_{\substack{j=1 \\ j \neq i}}^{N} n_j p_{ji} = 0 \quad (i = 1,2,3,\ldots, N) \tag{35.3}$$

The numbers n_i refer to unit solid angle $d\,\Omega$ and unit angle $d\,\eta_{ij}$. Since the total number of particles is constant and there are no rotations, we have

$$\sum_{j=1}^{N} n_j = \frac{1}{8\pi^2}\, n \tag{35.4}$$

Consequently, among the N equations in (35.3) there are $N-1$ linearly independent equations. Substituting an arbitrary but fixed n_k from (35.4) and (35.3) and rejecting the k-th equation, we have

$$\sum_{j \neq k} a_{ij} n_j = \frac{1}{8\pi^2}\, n c_i \quad (i \neq k) \tag{35.5}$$

where

$$a_{ij} = p_{ki} - p_{ji} \;\; (j \neq i), \;\; a_{ii} = p_{ki} + \sum_{j \neq i} p_{ij}, \;\; c_i = p_{ki} \tag{35.6}$$

The solution of (35.3) is

$$n_j = \frac{n}{8\pi^2} \frac{\mathrm{D}_j}{\mathrm{D}} \;\; (j \neq k), \;\; n_k = \frac{n}{8\pi^2}\left(1 - \frac{1}{\mathrm{D}} \sum_{j \neq k} \mathrm{D}_j\right) \tag{35.7}$$

where

$$\mathrm{D} = \begin{vmatrix} a_{11} \cdots & a_{1l} \cdots & a_{1,k-1} & a_{1,k+1} \cdots & a_{1m} \cdots & a_{1N} \\ \cdots & \cdots & \cdots & \cdots & \cdots & \cdots \\ a_{l1} \cdots & a_{ll} \cdots & a_{l,k-1} & a_{l,k+1} \cdots & a_{lm} \cdots & a_{lN} \\ \cdots & \cdots & \cdots & \cdots & \cdots & \cdots \\ a_{k-1,1} \cdots & a_{k-1,l} \cdots & a_{k-1,k-1} & a_{k-1,k+1} \cdots & a_{k+1,m} \cdots & a_{k-1,N} \\ a_{k+1,1} \cdots & a_{k+1,l} \cdots & a_{k+1,k-1} & a_{k+1,k+1} \cdots & a_{k+1,m} \cdots & a_{k+1,N} \\ \cdots & \cdots & \cdots & \cdots & \cdots & \cdots \\ a_{m1} \cdots & a_{m,l} \cdots & a_{m,k-1} & a_{m,k+1} \cdots & a_{mm} \cdots & a_{mN} \\ \cdots & \cdots & \cdots & \cdots & \cdots & \cdots \\ a_{N1} \cdots & a_{Nl} \cdots & a_{N,k-1} & a_{N,k+1} \cdots & a_{Nm} \cdots & a_{NN} \end{vmatrix} \tag{35.8}$$

is the determinant of Equation (35.5). The determinants D_j

are obtained by replacing the corresponding column by one consisting of the coefficients c_i. The order of the determinants is $(N-1)$ (the k-th row and k-th column are absent).

The distribution of particles over the energy levels is subject to certain general laws which may be established with the aid of (35.7). We shall show that the populations of any pair of levels (m, l) may be written in the form

$$\bar{n}_m = \frac{n}{8\pi^2}\frac{\mathrm{D}_m}{\mathrm{D}} = \frac{n}{8\pi^2}\frac{\mathrm{D}_m(u_{ml}=0)+\Delta_m p^u_{ml}}{\mathrm{D}(u_{ml}=0)+\Delta_{ml}p^u_{ml}} = \frac{n_m(u_{ml}=0)+l_m p^u_{ml}}{1+\alpha_{ml}p^u_{ml}} \tag{35.9}$$

$$n_l = \frac{n}{8\pi^2}\frac{\mathrm{D}_l}{\mathrm{D}} = \frac{n}{8\pi^2}\frac{\mathrm{D}_l(u_{ml}=0)+\Delta_l p^u_{ml}}{\mathrm{D}(u_{ml}=0)+\Delta_{ml}p^u_{ml}} = \frac{n_l(\bar{u}_{ml}=0)+l_l p^u_{ml}}{1+\alpha_{ml}p^u_{ml}} \tag{35.10}$$

where $l_l = l_m$ and $\bar{n}_m(\bar{u}_{ml}=0)$ and $n_l(\bar{u}_{ml}=0)$ represent the populations of the m-th and l-th levels in the absence of excitation at frequency ν_{ml}. In the special case when all the $\nu_{ij}=0$, the quantities $n_m(\bar{u}_{ml}=0)$ and $n_l(u_{ml}=0)$ are given by Boltzmann's formula. The parameters

$$l_m = l_l = \frac{\bar{\Delta}_m}{\mathrm{D}(u_{ml}=0)}, \quad \alpha_{ml} = \frac{\Delta_{ml}}{\mathrm{D}(u_{ml}=0)} \tag{35.11}$$

depend on all the transition probabilities and all the ν_{ij}, but are independent of u_{ml} (p^u_{ml}).

To prove (35.9) and (35.10), let us substitute the explicit expressions for D_m and D_l as functions of p^u_{lm} and p^u_{ml}. Since (35.5) has a unique solution, and k has been chosen arbitrarily, it follows that without loss of generality we may assume that $m \neq k$ and $l \neq k$. According to (35.1) and (35.6), p^u_{ml} enters only into the four coefficients a_{ml}, a_{lm}, a_{ll} and a_{mm}. We can also write

$$\mathrm{D}_m = (-1)^{2l'} a_{ll} M^l_l + (-1)^{m'+l'} a_{ml} M^m_l + R_m \tag{35.12}$$

$$D_l = (-1)^{m'+l'} a_{lm} L^l_m + (-1)^{2m'} a_{mm} L^m_m + R_l \tag{35.13}$$

where M and L are the minors which are obtained from D_m and D_l by crossing out one row (upper index) and one column (lower index). The quantities R_m, R_l and the minors are independent of p^u_{ml} or p^u_{lm}. Since the k-th row and k-th column are absent from the determinant, it follows that $l' = l$ when $l < k$ and $l' = l - 1$ when $l > k$. A similar connection exists between m' and m.

The minors M^l_l and M^m_l consist of the same rows and columns as L^l_m and L^m_m and differ from each other only in the different position of the column consisting of the coefficients c_i. Therefore, by completing $(m' - l' - 1)$ interchanges of columns in M^l_l and M^m_l, we obtain

$$\begin{aligned} M^l_l &= (-1)^{m'-l'-1} L^l_m \\ M^m_l &= (-1)^{m'-l'-1} L^m_m \end{aligned} \tag{35.14}$$

Substituting this into (35.12), we have, in view of (35.8)

$$\begin{aligned} D_m &= (-1)^{m'+l'-1} a_{ll} L^l_m + (-1)^{2m'-1} a_{ml} L^m_m + R_m \\ &= D_m (u_{ml} = 0) + [(-1)^{m'+l'-1} L^l_m + L^m_m] p^u_{ml} \end{aligned} \tag{35.12a}$$

$$D_l = D_l (u_{ml} = 0) + [L^m_m + (-1)^{m'+l'-1} L^l_m] p^u_{ml} \tag{35.13a}$$

It is evident from the latter formulae that the coefficients of p^u_{ml} are equal, i.e. $\Delta_m = \Delta_l$.

In order to establish the validity of (35.9) and (35.10), it will be sufficient to show in addition that the determinant D does not contain terms proportional to $(p^u_{ml})^2$ and can therefore be written in the form

$$D = D (u_{ml} = 0) + \Delta_{ml} p^u_{ml}$$

Let us isolate in D those terms which include products of the coefficients a_{ml}, a_{lm}, a_{ll} and a_{mm} (the remaining coefficients are independent of p^m_{ml}):

$$\begin{aligned} D &= (-1)^{2l+2(m'-1)} a_{ll} a_{mm} M^{lm}_{lm} + (-1)^{2(m'+l')-1} a_{ml} a_{lm} M^{ml}_{lm} + R \\ &= (p^u_{ml} p^u_{lm} M^{lm}_{lm} - p^u_{ml} p^u_{lm} M^{ml}_{lm}) + R' \end{aligned} \tag{35.15}$$

Since the minors M_{lm}^{lm} and M_{lm}^{ml} of the determinant D are equal, it follows that terms containing $(p_{ml}^{u})^2$ cancel out.

The populations of any pair of levels can therefore be written in the form given by equations (35.9) and (35.10). It also follows from (35.15) that none of the quantities p_{ij} or $B_{ij}u_{ij}$ enters D in a power greater than the first. According to (35.9) and (35.10), at high enough densities of the exciting radiation at frequency ν_{ml}, the populations of the m-th and l-th levels are equal and tend to the common limit l_m/α_{ml} as $u_{ml} \to \infty$. When $u_{ml} = 0$ and $n_m = n_l$, this equality remains valid for any u_{ml}(or p_{ml}^{u}).

The rates of absorption of external radiation of frequency ν_{ml} and of luminescence propagating in all directions and having an arbitrary polarisation can, by analogy with our preceding discussion (see Section 24), be written in the form

$$W_{\text{abs}}(\nu_{ml}) = h\nu_{ml} \int_0^{2\pi} d\eta_{ml} \int_0^{2\pi} d\varphi \int_0^{\pi} p_{ml}^{u} (n_l - n_m) \sin\theta \, d\theta \quad (35.16)$$

$$W_{\text{lum}}(\nu_{ml}) = h\nu_{ml} \int_0^{2\pi} d\eta_{ml} \times \int_0^{2\pi} d\varphi \int_0^{\pi} \{A_{ml}n_m - (n_l - n_m) B_{ml}u_{ml}^{0}\} \sin\theta \, d\theta \quad (35.17)$$

Substituting (35.9) and (35.10) into these formulae, we find that

$$W_{\text{abs}}(\nu_{ml}) = h\nu_{ml} \int_0^{2\pi} d\eta_{ml} \int_0^{2\pi} d\varphi \int_0^{\pi} p_{ml}^{u} \frac{n_l(u_{ml}=0) - n_m(u_{ml}=0)}{1 + \alpha_{ml}p_{ml}^{u}} \times \sin\theta \, d\theta \quad (35.18)$$

$$\begin{aligned} W_{\text{lum}}(\nu_{ml}) = {} & h\nu_{ml} \int_0^{2\pi} d\eta_{ml} \int_0^{2\pi} d\varphi \\ & \times \int_0^{\pi} \frac{n_m(u_{ml}=0)(A_{ml} + B_{ml}u_{ml}^{0}) - n_l(u_{ml}=0) B_{ml}u_{ml}^{0}}{1 + \alpha_{ml}p_{ml}^{u}} \sin\theta \, d\theta \\ & + h\nu_{ml} \int_0^{2\pi} d\eta_{ml} \int_0^{2\pi} d\varphi \int_0^{\pi} \frac{l_m p_{ml}^{u} A_{ml}}{1 + \alpha_{ml} p_{ml}^{u}} \sin\theta \, d\theta \end{aligned} \quad (35.19)$$

As can be seen from (35.18), the rate of absorption is largely determined by the difference $n_l\,(u_{ml}=0)-n_m\,(u_{ml}=0)$.

If in the absence of excitation at the frequency under consideration the populations of the upper and lower levels are equal, there will be no absorption, whatever the density of the incident radiation u_{ml}. When $p^u_{ml}\to\infty$ and $n_l\,(u_{ml}=0)$ $\neq n_m\,(u_{ml}=0)$, the rate of absorption tends to a finite limit.

The expression for the rate of luminescence (35.19) can naturally be divided into two parts. The first part may be different from zero even in the absence of excitation at the frequency ν_{ml}. This part represents luminescence due to excitation of the particles at other frequencies. If it is excited only at the single frequency ν_{ml}, the first term in (35.19) is zero. In this case, the populations of the level are connected by the Boltzmann formula $n_m\,(u_{ml}=0)=$ $n_l\,(u_{ml}=0)\exp(-h\nu_{ml}/kT)$. The numerator in the integrand is zero, since $A_{ml}+B_{ml}u^0_{ml}=B_{ml}u^0_{ml}\exp(h\nu_{ml}/kT)$. The first term in (35.19) will also tend to zero for very large p^u_{ml}, while the second represents luminescence due to excitation at frequency ν_{ml}. It tends to zero when $u_{ml}=0$, and to a finite limit when $p^u_{ml}\to\infty$. In contrast to the rate of absorption, $W_{\text{lum}}(\nu_{ml})$ is not equal to zero when $n_l\,(u_{ml}=0)=n_m\,(u_{ml}$ $=0)$. Similarly, it can be shown that the rate of luminescence propagating in a given direction and polarised along mutually perpendicular vectors $\mathbf{e}_\alpha$ $(\alpha=1,2)$ is

$$W^\alpha_{\text{lum}}(\nu_{ml})=h\nu_{ml}\int\limits_0^{2\pi}d\eta_{ml}\int\limits_0^{2\pi}d\varphi\int\limits_0^{\pi}[a^{(\alpha)}_{ml}n_m$$

$$-(n_l-n_m)\,b^{(\alpha)}_{ml}u'^{0}_{ml}]\sin\theta\,d\theta=h\nu_{ml}\int\limits_0^{2\pi}d\eta_{ml}\int\limits_0^{\pi}d\varphi$$

$$\times\int\limits_0^{\pi}\frac{n_m\,(u_{ml}=0)\,(a^{(\alpha)}_{ml}+b^{(\alpha)}_{ml}u'^{0}_{ml})-n_l\,(u_{ml}=0)\;b^{(\alpha)}_{ml}u'^{0}_{ml}}{1+\alpha_{ml}p^u_{ml}}$$

$$\times\sin\theta\,d\theta+h\nu_{ml}\int\limits_0^{2\pi}d\eta_{ml}\int\limits_0^{\pi}d\varphi\int\limits_0^{\pi}\frac{\alpha_m\,p^u_{ml}\,a^{(\alpha)}_{ml}}{1+\alpha_{ml}\,p^n_{ml}}\sin\theta\,d\theta \tag{35.20}$$

In these expressions

$$a^\alpha_{ml}=\frac{3}{8\pi}A_{ml}\cos^2(\mathbf{D}_{ml},\;\mathbf{e}_\alpha)$$

and

$$b^{\alpha}_{ml} = 3B_{ml}\cos^2(\mathbf{D}_{ml}\,\mathbf{e}_{\alpha})$$

are the differential Einstein coefficients which are functions of six angles, three of which (η_{ml}, φ, θ) define the orientation of the particle, two determine the direction of observation and the sixth is determined by the choice of $\mathbf{e}_1$ and $\mathbf{e}_2$. Since in (35.20) the integration is carried out only with respect to the first three angles, the rate of luminescence, and hence its polarisation, are functions of three angles, i.e. they depend on the direction of observation and the orientation of the vector $\mathbf{e}_{\alpha}$ in the plane perpendicular to this direction. By calculating $W^{(1)}_{\text{lum}}(\nu_{ml})$ and $W^{(2)}_{\text{lum}}(\nu_{ml})$ from (15.20), it is easy to determine the polarisation of the luminescence. The expressions for the rates of absorption of an additional beam of low density u'_{ml} ($u'_{ml} \ll u_{ml}$, $u'_{ml} \ll u_{ij}$) (see Section 32), polarised along $\mathbf{e}_1$ and $\mathbf{e}_2$, can by analogy with (35.18) be written in the form

$$W^{(\alpha)}_{\text{abs}}(\nu_{ml}) = h\nu_{ml}\int_0^{2\pi} d\eta_{ml}\int_0^{2\pi} d\varphi\int_0^{\pi}(n_l - n_m)\,B_{ml}\,u'_{ml}$$

$$\times\cos^2(\mathbf{D},\mathbf{e}_{\alpha})\sin\theta\,d\theta = h\nu_{ml}\int_0^{2\pi} d\eta_{ml}\int_0^{2\pi} d\varphi\int_0^{\pi}[n_l\,(u_{ml}=0)$$

$$-n_m\,(u_{ml}=0)]\,\frac{B_{ml}\,u'_{ml}\cos^2(\mathbf{D}_{ml}\,\mathbf{e}_{\alpha})}{1+\alpha_{ml}\,p^{u}_{ml}}\sin\theta\,d\theta \qquad (35.21)$$

Since the density u_{ml} of the exciting radiation enters only into the numerator of equation (35.21), it follows that when $p^{u}_{ml}\to\infty$ the rate of absorption $W^{\alpha}_{\text{abs}}(\nu_{ml})$ tends to zero (see (35.18)). Knowing $W^{(1)}_{\text{abs}}$ and $W^{(2)}_{\text{abs}}$, it is possible to determine the induced dichroism of a system of particles in a given direction.

The formulae given above for the rate of absorption and luminescence can be simplified quite considerably in certain special cases. For example, in gases and non-viscous solutions, external radiation does not give rise to anisotropy in the angular distribution of the excited particles, and therefore integration with respect to the angles is unnecessary. In other cases, the angular dependence of n_j may be simple enough for the integrals to become expressible in terms of known functions. One such example is discussed overleaf.

Limits of applicability of the theory of harmonic oscillators

It was shown in Chapter 7 that non-linear optical effects arise during the excitation of particles by plane-polarised and natural or isotropic radiation. The maximum departure from linearity takes place when the incident radiation is plane-polarised, so that, to establish the limits of applicability of the classical theory of the harmonic oscillator, it is sufficient to consider the transformation of plane-polarised radiation by a pair of levels m, l of a quantum-mechanical system with an arbitrary number of energy levels.

Suppose that the incident radiation has a frequency ν_{ml} and propagates along the x axis, and is polarised along the z axis. Excitation at all other frequencies will, for the sake of simplicity, be assumed to be isotropic. The transition probabilities p^u_{ij} are, in this case, given by

$$p^u_{ij} = B_{ij}u_{ij}, \quad p^u_{ml} = 3\,B_{ml}\,u_{ml}\cos^2\theta \tag{35.22}$$

Observation of the luminescence and dichroism will be carried out along the y axis, as before, while the unit vectors $\mathbf{e}_1$ and $\mathbf{e}_2$ will be taken to be parallel to the z and x axes. We then have

$$\begin{aligned} \cos^2(\mathbf{D}_{ml}\,\mathbf{e}_1) &= \cos^2(\mathbf{D}_{ml}\,,\,\mathbf{z}) = \cos^2\theta, \\ \cos^2(\mathbf{D}_{ml}\,\mathbf{e}_2) &= \cos^2(\mathbf{D}_{ml}\,\mathbf{x}) = \sin^2\nu\cos^2\varphi \end{aligned} \tag{35.23}$$

Substituting (35.22) and (35.23) into (35.18)-(35.21), it is easy to complete the integration with respect to the angles. We can now use the above results together with (2.42) and (3.38) to obtain the following formulae for the absorption coefficient, the dichroism and the polarisation:

$$\begin{aligned} k(\nu_{ml}) = \frac{1}{cu_{ml}}\,W_{\text{abs}}(\nu_{ml}) &= (n^0_l - n^0_m)\,f_{ml}\,\frac{\pi e^2}{3mc}\left[\frac{3}{(\alpha'_{ml}u_{ml})}\right. \\ &\left. - \frac{3}{(\alpha'_m u_m)^{3/2}}\tan^{-1}(\alpha_{ml}u_{ml})^{1/2}\right] \end{aligned} \tag{35.24}$$

$$D(\nu_{ml}) = \frac{(3+\alpha'_{ml}u_{ml})\tan^{-1}(\alpha'_{ml}u_{ml})^{1/2} - 3(\alpha'_{ml}u_{ml})^{1/2}}{(\alpha'_{ml}u_{ml}-1)\tan^{-1}(\alpha'_{ml}u_{ml})^{1/2} + (\alpha'_{ml}u_{ml})^{1/2}} \tag{35.25}$$

$$P(\nu_{ml})=\frac{\frac{l'_m}{\alpha'_m}A_{ml}-[n^0_m(A_{ml}+B_{ml}u^0_{ml})-n^0_l B_{ml}u^0_{ml}]}{\frac{l'}{\alpha'_{ml}}A_{ml}D(\nu_{ml})+[n^0_m(A_{ml}+B_{ml}u^0_{ml})-n^0_l B_{ml}u^0_{ml}]P^0_{ml}} \times D(\nu_{ml})P^0_{ml}, \quad (35.26)$$

where

$$P^0_{ml}=\frac{(3+\alpha'_{ml}u_{ml})\tan^{-1}(\alpha'_{ml}u_{ml})^{1/2}-3(\alpha'_{ml}u_{ml})^{1/2}}{\frac{4}{3}(\alpha'_{ml}u_{ml})^{3/2}-(\alpha'_{ml}u_{ml})^{1/2}+(1-\alpha'_{ml}u_{ml})\tan^{-1}(\alpha'_{ml}u_{ml})^{1/2}} \quad (35.27)$$

f_{ml} is the oscillator strength, $l'_m = l_m B_{ml}$, $\alpha'_{ml} = \alpha_{ml}B_{ml}$, and n^0_m and n^0_l are total populations of the m-th and l-th levels in the absence of the excitation frequency ν_{ml}.

Equations (35.24)-(35.27) are valid for any value of N (beginning with 2) and become identical under the appropriate assumptions to the corresponding formulae obtained earlier for systems with two or three energy levels. The dependence of the various optical characteristics on the density of the exciting radiation given by these formulae is illustrated graphically in Fig. 8.5. These curves are based on (34.63)-(34.65) which are analogous to (35.24)-(35.27) and differ from them only by the presence of l_2, and α_{21} in place of l'_m and α'_{ml}.

It is evident from the above formulae and from Fig. 8.5 that the transformation of external radiation by a pair of levels of a system of particles with an arbitrary number of energy levels is, in general, different from the transformation of radiation by a set of harmonic oscillators. The absorptive power and the polarisation of the emitted luminescence tend to zero for large $\alpha'_{ml}u_{ml}$ and at the same time dichroism appears. The polarisation of the emitted luminescence may be negative at certain values of the parameters. It varies with u_{ml}, and with excitation at other frequencies, in accordance with a complicated law. If we replace the quantum-mechanical system of particles by a set of harmonic oscillators of frequency ν_{ml}, then, under the conditions of excitation and observation which we have discussed here we would obtain the following results: 1. the absorption coefficient would be independent of u_{ml}, 2. there would be no dichroism and 3. the polarisation of the emitted radiation would be constant and equal to 0.5.

The limits of applicability of the classical theory of the harmonic oscillator can thus be established by finding the conditions under which (35.24)-(35.26) yield $k(\nu_{ml}) = \text{const}$ (as u_{ml} varies), $D(\nu_{ml}) = 0$ and $P(\nu_{ml}) = 0.5$. This occurs when

$$\alpha'_{ml} u_{ml} \ll 1 \tag{35.28}$$

$$n^0_m (A_{ml} + B_{ml} u^0_{ml}) - n^0_l B_{ml} u^0_{ml} = 0 \tag{35.29}$$

The inequality given by (35.28) imposes limitations on the energy density of the exciting radiation. It follows from this condition that in all systems (other than the harmonic oscillator, which we shall not consider), non-linear effects will unavoidably appear at certain definite densities of the exciting radiation. The minimum densities $u_{\min}$ at which there are appreciable departures from linearity are unambiguously determined by the non-linearity parameters α'_{ml} which depend on the transition probabilities of the system. It is only in the one simple case when the system has only two energy levels that α'_{ml} is determined by the level separation, since $\alpha'_{ml}\,(u^0 = d = 0)$ is then independent of the transition probabilities and is unambiguously determined by the frequency (see Section 32). In a system of particles with three or more energy levels, neither the level separation nor the lifetime of the excited state provides any criterion for the unambiguous definition of α'_{ml} and, consequently, $u_{\min}$. The only criterion that can be used for this is the parameter α'_{ml} itself. A change in the transition probability in any channel affects the magnitude of α'_{ml} and of $u_{\min}$ at which non-linear effects set in.

It was shown in Section 34 for a system of particles with three energy levels that α'_{ml} becomes large when the system has metastable levels in which excited particles accumulate.

The temperature dependence of α'_{ml} is very complicated. However, beginning with a certain temperature, the magnitude of α'_{ml} decreases with increasing temperature and tends to zero as $T \to \infty$. This is connected with the fact that the numerator in the expression for α^0_{ml} includes products of $N-1$ transition probabilities, which contain the thermal radiation density u^0_{ij} while the numerator u_{ij} contains only $N-2$ probabilities. In fact, it follows from (35.15) that all the probabilities $B_{ij}u_{ij}$ enter into the expression for D as first powers only. It follows that D contains terms with the products of $N-1$ quantities $B_{ij}u^0_{ij}$, while Δ_{ml} does not contain $B_{ml}u^0_{ml}$ (see (35.9)). It follows that the thermal radiation background tends to reduce all non-linear effects. However, at high temperatures, the difference $n^0_l - n^0_m$ is also found to decrease, and therefore there is a decrease in the absorptive power and the rate of luminescence. The medium becomes more transparent at the frequency under consideration.

The expression given by (35.29) imposes limitations on the additional illumination at other frequencies, which can appreciably affect the nature of the transformation of external radiation by a particular pair of levels m, l. If all the u_{ij}, apart from u_{ml}, are zero, the numbers of n^0_m and n^0_l satisfy Boltzmann's formula, and (35.29) is an identity at all temperatures. In the presence of additional illumination, the effect is unimportant in the one special case when $n^0_m = u^0_{ml} = 0$. Under these conditions, (35.29) is also satisfied. The absence of upper level populations for $u^0_{ml} = 0$ and $u_{ij} \neq 0$ can easily be established experimentally. When n^0_m is zero, there is no luminescence in the channel under consideration before excitation at frequency ν_{ml}.

We conclude that the limits of applicability of the classical theory of the harmonic oscillator as applied to the transformation of incident radiation by a pair of levels of the quantum-mechanical system is determined by the following two conditions: 1. the absence of additional illumination of the system at other frequencies which gives rise to luminescence in the channel under consideration ν_{ml}, and 2. the density of exciting radiation u_{ml} must be such that $\alpha'_{ml}u_{ml} < 1$.

To find the parameter α'_{ml}, it is sufficient to write out the determinant of the system of balance equations. In general, this may be written in the form

$$D = D(u_{ml} = 0) + \Delta_{ml}p^u_{ml} \tag{35.30}$$

The quotient $\Delta_{ml}/D(u_{ml} = 0)$ multiplied by the Einstein coefficient B_{ml} is the required non-linearity parameter.

If the above conditions are satisfied, the pair of levels can be replaced by a set of $n^0_l - n^0_m$ classical dipoles of frequency λ_{ml} and oscillator strength f_{ml}. The transformation of radiation by a particular pair of levels can be carried out within the framework of classical electrodynamics. The values of f_{ml}, ν_{ml} and $(n^0_l - n^0_m)$ must be introduced artificially into classical theory, i.e. either from quantum mechanics or from experiment.

References

1. HEITLER, W., The quantum theory of radiation, Oxford University Press (1954).
2. BORN, M. and WOLF, E., Optics, Pergamon Press (1959).
3. KRAVETS, T. P., Collected papers, Izd. Akad, Nauk SSSR (1959).
4. VOL'KENSHTEIN, M. V., Molecular optics, GITTL (1951).
5. LEVSHIN, V. L., Photoluminescence of liquids and solids, GITTL (1951).
6. JABLONSKI, A., Z. Phys., V96, p236 (1935).
7. SOLOV'EV, K. N., Optika Spektrosk., V10, p737 (1961).
8. VAVILOV, S. I., Collected papers, VII, p58.
9. FEOFILOV, P. P., Physical basis of polarised emission, Consultants Bureau (1961).
10. GURINOVICH, G. P. and SEVCHENKO, A. N., Trudy Inst. Fiz. Mat., Minsk, No. 11 (1957).
11. PERRIN, F., Annln. Phys., V12, p222 (1959).
12. BORN, M. and OPPENHEIMER, R. Annln. Phys., V84, p457 (1927).
13. EL'YASHEVICH, M. A., Atomic and molecular stectroscopy, Fizmatgiz (1962).
14. EL'YASHEVICH, M. A., Usp. fiz. Nauk, V71, p156 (1960).
15. BLOKHINTSEV, D. I., Quantum mechanics, Allyn and Bacon (1965).
16. SOMMERFELD, A., Atomic structure and spectral lines, Methuen (1934).
17. REBANE, K. K., Trudy Inst. Fiz. Astr. Est. SSR, No. 16 (1961).
18. STEPANOV, B. I., Zh. eksp. teor. Fiz., V15, p435 (1945).
    RICE, O., Phys. Rev., V33, p748 (1929).
19. CRAMER, H., Elements of probability theory and some of its applications, Wiley (1954).
20. STEPANOV, B. I. and KAZACHENKO, L. P., Optika Spektrosk., V12, p131 (1962).
21. REBANE, K. K., Optika Spektrosk., V9, p557 (1960); Trudy Inst. Fiz. Mat., Akad. Nauk Est. SSR, No. 14 (1961).

REBANE, K.K., SIL'D, O.I. and KHIZHNYAKOV, V.V., Trudy Inst. Fiz. Mat., Akad. Nauk Est. SSR, No. 13 (1960).
REBANE, K.K., PURGA, A.P., SIL'D, O.I. and KHIZHNYAKOV, V.V., Trudy Inst. Fiz. Mat., Akad. Nauk Est. SSR., V114, pp31, 48 (1961).
22. LAX, M., J. Chem. Phys., V20, p1752 (1952).
23. KUBAREV, S.I., Dokl. Akad. Nauk SSSR, V130, p1067 (1960); Izv. Akad. Nauk SSSR, Ser. Fiz., V24, p775 (1960); Optika Spektrosk., V9, p3 (1960); ibid., V10, p535 (1961).
MEYER, J.J.G., Physica, V21, p253 (1955).
MARHAM, J.J., Rev. mod. Phys., V31, p956 (1959).
ABARENKOV, I.V., Dissertation, Leningrad State University (1959).
KUBO, R. and TOYOZAWA, Y., Prog. theor. Phys., V13, p160 (1955).
24. APANASEVICH, P.A., Trudy Inst. Fiz. Mat., Minsk, No. 2, p55 (1957); No. 3, p72 (1958); Optika Spektrosk., V5, p97 (1958).
25. STEPANOV, B.I. and APANASEVICH, P.A., Dokl. Akad. Nauk SSSR, V115, p488 (1957).
26. WEISSKOPF, V. and WIGNER, E., Z. Phys., V63, p54; V65, p18 (1930).
27. APANASEVICH, P.A., Trudy Inst. Fiz. Mat., Minsk (1957); Dissertation, Inst. Fiz. Mat., Minsk (1958); Optika Spektrosk., V5, p97 (1958).
28. AKHIEZER, A.I. and BERESTETSKII, V.B., Elements of quantum electrodynamics, Consultants Bureau (1962).
29. APANASEVICH, Optika Spektrosk., V5, p97 (1958).
30. EINSTEIN, A., Phys. Z, V18, p121 (1917).
31. LANDAU, L., Z. Phys., V45, p430 (1927).
32. BLOCH, F., Phys. Z., V29, p58 (1928).
33. APANASEVICH, P.A., Izv. Akad. Nauk SSSR. Ser. Fiz., V24, p509 (1960); Dissertation, Akad. Nauk Byel. SSR (1958).
34. NEPORENT, B.S. and BAKHSHIEV, N.G., Optika Spektrosk., V5, p634 (1958).
35. LEVSHIN, L.V. and GORSHKOV, V.K., Optika Spektrosk., V10, p759 (1961).
36. KHAPALYUK, A.P. and STEPANOV, B.I., Izv. Akad. Nauk Byel. SSR., Ser. Fiz., No. 4 (1961); Optika Spektrosk., V13, p714 (1962).
37. STEPANOV, B.I. and RUBANOV, A.S., Dokl. Akad. Nauk SSSR, V128, p517 (1959); Dokl. Akad. Nauk Byel. SSR, V4, p373 (1960); in symposium Molecular spectroscopy, Izd. leningr. gos. Univ., p20 (1960); also see reference 84.
38. FABRIKANT, V.A., VUDYNSKII, M.M. and BUTAEVA, F.A., Byull. izobr., No. 20 (1959).
FABRIKANT, V.A., VUDYNSKII, M.M. and BUTAEVA, F.A., Byull. izobr., No. 20 (1959).
39. BASOV, N.G. and PROKHOROV, A.M., Usp. Fiz. Nauk, V57, p485 (1955); Zh. eksp. teor. Fiz., V28, p249 (1955); Dokl. Akad. Nauk SSSR, V101, p47 (1955).
40. BASOV, N.G. and PROKHOROV, A.M., Usp. fiz. Nauk, V57, p485 (1955).
ZEIGER, H.J. and TOWNES, C.H., Phys. Rev., V95, p282 (1954).
41. In symposium Studies in experimental and theoretical physics, G.S., Landsberg memorial volume, p62, Izd. Akad. Nauk SSSR (1959).
42. ABREKOV, V.I., PESIN, M.S. and FABELINSKII, I.L., Zh. eksp. teor. Fiz., V39, p892 (1960).
43. MAIMAN, T., Phys. Rev. Letters, V4, p564 (1960); Nature, V187, p493 (1960).
44. SOROKIN, P.P. and STEVENSON, M.J., Phys. Rev. Letters, V5, p557 (1960); Aviation Week, V73, p73 (1960); Science News Lett., V78, p434 (1960).
45. BASOV, N.G. and OSIPOV, B.D., Optika Spektrosk., V4, p795 (1958).
46. STEPANOV, B.I. and RUBANOV, A.S., Dokl. Akad. Nauk Byel. SSR, V4, p373 (1960); see also reference 84.
47. SINGER. J.R., Masers, Wiley (1959).
TROUP, G., Masers and lasers, Methuen (1962).
LENGYEL, B.A., Lasers, generation of light by stimulated emission, Wiley (1962).

HEAVENS, O.S., Optical masers, Methuen (1964).
BIRNBAUM, G., Optical masers, Academic Press (1964).
ZVEREV, G.M., KARLOV, N.V., KORNIENKO, L.S., MANENKOV, A.M. and PROKHOROV, A.M., Usp. fiz. Nauk, V77, p61 (1962); Dokl. Akad. Nauk Byel. SSR, V142, p1282; ibid., V145, p650 (1962); Optika Spektrosk., V12, pp443, 553; ibid., V13, pp282, 714 (1962); ibid., V15, p57 (1963); Dokl. Akad. Nauk Byel. SSR, V6, pp147, 151, 223, 288, 297, 335, 418, 633, 629, 768 (1962); Trudy Inst. fiz. Khim., V5, p59 (1962); Izv. Akad. Nauk Byel. SSR. Ser. Fiz., No. 1, pp45, 132 (1961); No. 1, p42; No. 2, p18; No. 3, p23 (1962).
Quantum electronics issue, Proc. IEEE, V51, No. 1 (1963).
'Chemical lasers', Appl. Optics, Suppl. 2 (1965).
ALLEN, L. and JONES, D.G.C., Adv. Phys., V56, p479 (1965).
HEAVENS, O.S., Brit. J. Appl. Phys., V17, p287 (1966).
ANDERSON, O.T. and BEDESEM, M.P., Am. J. Phys., V4, p296 (1966).

48. ALENTSEV, M.N., ANTONOV-ROMANOVSKII, V.V., STEPANOV, B.I. and FOK, N.V., Zh. eksp. teor. Fiz., V28, p253 (1955); Optika Spetrosk., V1, p125 (1956).
49. STEPANOV, B.I., FOK, M.V. and KHAPALYUK, A.P., Dokl. Akad. Nauk SSSR, V105, p50 (1955).
    STEPANOV, B.I., Trudy Inst. Fiz. Mat., Minsk, No.1 (1956); Usp. fiz. Nauk, V58, p8 (1956).
50. STEPANOV, B.I., Dokl. Akad. Nauk SSSR, V112, p839 (1957); Izv. Akad. Nauk SSSR, Ser. Fiz., V22, p1034 (1958); ibid., V22, p1367 (1958).
    KAZACHENKO, L.P. and STEPANOV, B.I., Optika Spektrosk., V2, p339 (1957).
    NEPORENT, B.S., Dokl. Akad. Nauk SSSR, V119, p682 (1958).
51. STEPANOV, B.I., KRAVTSOV, L.A. and RUBINOV, A.N., Dokl. Akad. Nauk Byel. SSR, V6, p14 (1962).
52. BORISEVICH, N.A. and GRUZINSKII, V.V., Izv. Akad. Nauk Byel. SSR. Ser. Fiz.-Tekhn., No.3, p46 (1961); Dokl. Akad. Nauk Byel. SSR, V4, p380 (1960); Izv. Akad. Nauk SSSR. Ser. Fiz., V24, p545 (1960).
53. LEVSHIN, V.L., Optika Spektrosk., V11, p362 (1961).
54. KRATSOV, L.A., Dokl. Akad. Nauk Byel. SSR, V5, p331 (1961).
55. VOLOD'KO, L.V., SEVCHENKO, A.N. and UMREIKO, D.S., Vestsi Akad. Navuk Byel. SSR. Ser. Fiz.-Tekhn., No. 1 (1961).
    DOMBI, J., HEVESI, J. and HORVAI, R., Acta Phys. Chem. Szeqed, V5, p20 (1959).
56. STEPANOV, B.I. and SAMSON, A.M., Izv. Akad. Nauk SSSR. Ser. Fiz., V24, p502 (1960).
    SAMSON, A.M., Optika Spektrosk., V5, p500 (1959); ibid., V8, p89 (1960); Trudy Inst. fiz. Khim., V1, No.1 (1958); ibid., V2, Nos. 6 and 8 (1959); Dokl. Akad. Nauk Byel. SSR, V1, No.2 (1958); ibid., V3, No.5 (1959); Izv. Akad. Nauk SSSR. Ser. Fiz., V22, p1399 (1958); Vestsi Akad. Navuk Byel. SSR. Ser. Fiz.-Tekhn., No.3, p16 (1959); see also reference 84.
57. GALANIN, M.D., Trudy FIAN, V12, p3, Moscow, Izd. Akad. Nauk SSSR (1960).
    FORSTER, F., Naturwiss, V33, p116 (1946); Annln. Phys., V2, p55 (1947).
58. KUBAREV, S.I., Optika Spektrosk., V10, p535 (1961).
59. STEPANOV, B.I. and APANASEVICH, P.A., Izv. Akad. Nauk SSSR. Ser. Fiz., V22, p1380 (1958); Optika Spektrosk., V7, p437 (1959); Dokl. Akad. Nauk SSSR, V116, p722 (1957).
60. STEPANOV, B.I. and APANASEVICH, P.A., Optika Spektrosk., V7, p437 (1959).
61. COURANT, R. and HILBERT, D., Methods of mathematical physics, Vol. 1., Interscience (1962).
62. FOK. V.A., Uchen. Zap. leningr. gos. Univ., No. 14; ibid., ser. fiz., No. 3, p5 (1937).
63. SCHIFF, L.I., Quantum mechanics, McGraw-Hill (1955).

64. STEPANOV, B.I. and GRIBKOVSKII, V.P., Dokl. Akad. Nauk SSSR, V121, p 446 (1958).
65. GRIBKOVSKII, V.P., APANASEVICH, P.A. and STEPANOV, B.I., Trudy Inst. Fiz. Mat., Minsk, No. 3, p 131 (1959).
66. REBANE, K.K. and SIL'D, O.I., Optika Spektrosk., V13, p 465 (1962).
67. GRIBKOVSKII, V.P. and STEPANOV, B.I., Optika Spektrosk., V8, p 176 (1960).
68. GRIBKOVSKII, V.P., Vestsi Akad. Navuk Byel. SSR. Ser. Fiz.-Tekhn., No. 1, p 43 (1960).
69. FEDOROV, F.I., Uchen. Zap. leningr. gos. Univ., Ser. Fiz., No. 8, No. 146, p 33 (1952).
70. VISWANATHAN, K.S., Proc. Ind. Acad. Sci., V46, p A203 (1957).
71. STEPANOV, B.I. and RUBANOV, A.S., Trudy Inst. fiz. Khim., V2, p 52 (1959).
72. SLOBODSKAYA, P.V. and STEPANOV, B.I., Trudy GOI, V23, No. 133 (1951).
73. FOK, M.V. and KHAPALYUK, A.P., Dokl. Akad. Nauk SSSR, V105, p 50 (1955); see also references 83 and 84.
74. SVESHNIKOV, B.Ya., Zh. eksp. teor. Fiz., V18 (1948); ibid., V105, p 1208 (1955).
75. SVESHNIKOV, B.Ya., Dokl. Akad. Nauk SSSR, V51, p 675 (1946).
CHERDYNTSEV, S.V., Zh. eksp. teor. Fiz., V18, p 352 (1948).
CHERDYNTSEV, S.V. and VASSERMAN, I.I., Zh. eksp. teor. Fiz., V18, p 360 (1948).
76. FEOFILOV, P.P., Dissertation, GOI (1943);
STEPANOV, B.I. and GRIBKOVSKII, V.P., Optika Spektrosk., V8, p 224 (1960).
77. CHERDYSHEV, S.V. and VASSERMAN, I.I., Zh. eksp. teor. Fiz., V18, p 360 (1948).
78. GRIBKOVSKII, V.P., Dokl. Akad. Nauk Byel. SSR, V4, p 199 (1960).
79. BROSSEL, J. and KASTLER, A., Compt. Rend., V229, p 1213 (1949).
80. SKROTSKII, G.V. and IZYMOVA, T.G., Usp. fiz. Nauk, V73, p 423 (1961).
81. IVANOV, A.P., STEPANOV, B.I., BERKOVSKII, B.I., and KATSEV, I.L., Dokl. Akad. Nauk Byel. SSR, V6, p 147 (1962).
82. GRIBKOVSKII, V.P., Dokl. Akad. Nauk Byel. SSR, V4, p 284 (1960).
83. STEPANOV, B.I., Luminescence of complex molecules, GITTL (1956).
84. STEPANOV, B.I., Spectroscopy of negative light fluxes, GITTL (1956).
85. PRINGSHEIM, P., Fluorescence and phosphorescence, Wiley (1949).

INDEX